Dynamic Spectrum Access and Management in Cognitive Radio Networks

Are you involved in designing the next generation of wireless networks? With spectrum becoming an ever scarcer resource, it is critical that new systems utilize all available frequency bands as efficiently as possible. The revolutionary technology presented in this book will be at the cutting edge of future wireless communications.

Dynamic Spectrum Access and Management in Cognitive Radio Networks provides you with an all-inclusive introduction to this emerging technology, outlining the fundamentals of cognitive radio-based wireless communication and networking, spectrum sharing models, and the requirements for dynamic spectrum access. In addition to the different techniques and their applications in designing dynamic spectrum access methods, you'll also find state-of-the-art dynamic spectrum access schemes, including classifications of the different schemes and the technical details of each scheme. This is a perfect introduction for graduate students and researchers, as well as a useful self-study guide for practitioners.

Ekram Hossain is an Associate Professor in the Department of Electrical and Computer Engineering at the University of Manitoba, Winnipeg, Canada. He received his Ph.D. in Electrical Engineering from the University of Victoria, Canada, in 2000. His current research interests include design, analysis, and optimization of wireless/mobile communication networks, cognitive radio systems, distributed systems, and mobile computing. Dr. Hossain serves as an Editor for the *IEEE Transactions on Mobile Computing*, the *IEEE Transactions on Wireless Communications*, the *IEEE Transactions on Vehicular Technology*, *IEEE Wireless Communications*, *IEEE Communications Surveys and Tutorials*, and several other international journals. He is a registered professional engineer in the province of Manitoba, Canada.

Dusit Niyato is an Assistant Professor in the Division of Computer Communications, School of Computer Engineering at Nanyang Technological University (NTU), Singapore. He received his Ph.D. from the University of Manitoba, Canada, in 2008. His current research interests include wireless communications and networking.

Zhu Han is an Assistant Professor of Electrical and Computer Engineering at the University of Houston. He was awarded his Ph.D. in Electrical Engineering from the University of Maryland, College Park, and worked for a period in industry as an R & D Engineer for JDSD. His research interests include wireless resource allocation and management, wireless communications and networking, game theory, network security, and wireless multimedia.

Dynamic Spectrum Access and Management in Cognitive Radio Networks

EKRAM HOSSAIN
University of Manitoba

DUSIT NIYATO
Nanyang Technological University (NTU)

ZHU HAN
University of Houston

CAMBRIDGE
UNIVERSITY PRESS

University Printing House, Cambridge CB2 8BS, United Kingdom

One Liberty Plaza, 20th Floor, New York, NY 10006, USA

477 Williamstown Road, Port Melbourne, VIC 3207, Australia

314-321, 3rd Floor, Plot 3, Splendor Forum, Jasola District Centre, New Delhi - 110025, India

103 Penang Road, #05-06/07, Visioncrest Commercial, Singapore 238467

Cambridge University Press is part of the University of Cambridge.

It furthers the University's mission by disseminating knowledge in the pursuit of education, learning and research at the highest international levels of excellence.

www.cambridge.org
Information on this title: www.cambridge.org/9780521898478

First published 2009

A catalogue record for this publication is available from the British Library

Library of Congress Cataloging in Publication data
Hossain, Ekram, 1971–
Dynamic spectrum access and management in cognitive radio networks / Ekram Hossain, Dusit Niyato, and Zhu Han.
p. cm.
Includes Bibliographical References And Index.
ISBN 978-0-521-89847-8 (hardback)
1. Cognitive radio networks. 2. Radio resource management (Wireless communications)
I. Niyato, Dusit. II. Han, Zhu, 1974– III. Title.
TK5103.4815.H67 2009
621.384 – dc22 2009002060

ISBN 978-0-521-89847-8 Hardback

Contents

Preface

Frequency spectrum is a limited resource for wireless communications and may become congested owing to a need to accommodate the diverse types of air interface used in next generation wireless networks. To meet these growing demands, the Federal Communications Commission (FCC) has expanded the use of the unlicensed spectral band. However, since traditional wireless communications systems also utilize the frequency bands allocated by the FCC in a static manner, they lack adaptability. Also, many studies show that while some frequency bands in the spectrum are heavily used, other bands are largely unoccupied most of time. These potential spectrum holes result in the under-utilization of available frequency bands.

The concepts of software-defined radio and cognitive radio have been recently introduced to enhance the efficiency of frequency spectrum usage in next generation wireless and mobile computing systems. Software radio improves the capability of a wireless transceiver by using embedded software to enable it to operate in multiple frequency bands using multiple transmission protocols. Cognitive ratio, which can be implemented through software-defined radio, is able to observe, learn, optimize, and intelligently adapt to achieve optimal frequency band usage. Through dynamic spectrum access, a cognitive wireless node is able to adaptively and dynamically transmit and receive data in a changing radio environment. Therefore, techniques for channel measurement, learning, and optimization are crucial in designing dynamic spectrum access schemes for cognitive radio under different communication requirements.

In fact, cognitive radio based on dynamic spectrum access has emerged as a new design paradigm for next generation wireless networks. Cognitive radio aims at maximizing the utilization of the limited radio bandwidth while accommodating the increasing number of services and applications in wireless networks. The driving force behind this cognitive radio technology is the new spectrum licensing paradigm initiated by the FCC, which will be more flexible to allow unlicensed (or secondary) users to access the spectrum as long as the licensed (or primary) users are not interfered with. This new spectrum licensing paradigm will improve the utilization of the frequency spectrum and enhance the performance of wireless systems. Dynamic spectrum access (DSA) or opportunistic spectrum access (OSA) is the key approach in a cognitive radio network which is adopted by a cognitive radio user (i.e. an unlicensed user) to access the radio spectrum opportunistically. Development of dynamic spectrum access-based cognitive radio technology has to deal with technical and practical considerations as well as regulatory requirements. Therefore, there is increasing interest in this technology from researchers in both academia and industry, and engineers in the wireless industry, as well as from spectrum policy makers.

Design, analysis, and optimization of dynamic spectrum access require multidisciplinary knowledge, namely, knowledge of wireless communication and networking, signal processing, artificial intelligence (e.g. for learning), decision theory, optimization, and economic theory. A comprehensive introduction to the dynamic spectrum access and spectrum management problem in a cognitive radio network therefore needs to cover the basic concepts/theories for designing dynamic spectrum access methods as well as state-of-the-art of dynamic spectrum access and management methods.

This book provides a comprehensive treatment of the dynamic spectrum access and spectrum management problem in cognitive radio networks. The topics covered include the following: introduction to cognitive radio and the basic concepts of dynamic spectrum access; the analysis, design, and optimization of dynamic spectrum access techniques for cognitive radio; and state-of-the-art of dynamic spectrum access techniques. The key features of this book are as follows:

- a unified view of dynamic spectrum access for cognitive radio networks,
- coverage of a wide range of techniques for design, analysis, and optimization of dynamic spectrum access for cognitive radio networks,
- comprehensive treatment of state-of-the-art dynamic spectrum access techniques, and
- outlining the key research issues related to dynamic spectrum access.

The book is divided into three parts: Part I (Introduction), Part II (Techniques for design, analysis, and optimization of dynamic spectrum access and management), and Part III (Dynamic spectrum access and management). Part I comprises Chapters 1 and 2, which provide an introduction to the different wireless technologies and cognitive radio networks. The topics covered include the basics of cellular wireless, wireless local area network (WLAN), wireless metropolitan area network (WMAN), wireless personal area network (WPAN), wireless regional area network (WRAN) technology and related standards; the basic components, features, and potential applications of cognitive radio; an introduction to dynamic spectrum access; the motivations for cognitive radio-based wireless access technology; and the research issues in the different layers of the protocol stack for a cognitive radio network.

In Part II, different techniques which can be applied to the problem of the design, analysis, and optimization of dynamic spectrum access mechanisms in cognitive radio networks are introduced. In Chapter 3, the signal processing techniques (e.g. techniques for parameter estimation, filtering, and prediction) which are required for a wireless node to observe and gain knowledge of the ambient radio environment in order to access the radio spectrum dynamically are described. Optimization techniques, which are useful to obtain the optimal dynamic spectrum management scheme, are discussed in Chapter 4. Major variations of optimization techniques (e.g. unconstrained and constrained optimization, non-linear optimization, combinatorial optimization) are presented. Also, stochastic optimization based on dynamic programming, the Markov decision process (MDP), and stochastic programming for dynamic spectrum access in a random radio environment are discussed. In Chapter 5, game theory techniques are discussed in the context of designing dynamic spectrum access methods. Game theory is an attractive tool to model the dynamic spectrum sharing problem in a cognitive wireless network. The basics of different game theory models, namely, non-cooperative, repeated, cooperative (i.e. bargaining and coalition), and evolutionary game models are presented. Intelligent algorithms (i.e. learning techniques) are fundamental to cognitive radio design, and provide the ability to observe, learn, plan, and optimize the decision of dynamic spectrum access in a cognitive radio network. An introduction to different machine learning techniques including supervised learning, unsupervised learning, and reinforcement learning is provided in Chapter 6. Example applications of these techniques are discussed in the context of cognitive radio. Intelligent techniques based on genetic algorithms and fuzzy control are also discussed.

Part III deals with the modeling, design, and analysis of dynamic spectrum access and management schemes in cognitive radio networks. In Chapter 7, different models of spectrum access/sharing are discussed. Different approaches for spectrum sensing, which is a key component of dynamic spectrum access, are reviewed. A comprehensive review of the different medium access control (MAC) protocols developed for dynamic spectrum access is presented. Dynamic spectrum access and management architectures can be either centralized or distributed. In the centralized approach, a central controller collects information about the radio environment and controls the spectrum access by cognitive radio users. In Chapter 8, different schemes for centralized dynamic spectrum access are reviewed. Chapter 8 also introduces the concept of spectrum auction, which requires a centralized auctioneer for dynamic spectrum access and management. If there is no central controller available in a cognitive radio network, the cognitive radio users have to make their spectrum access decisions independently in a distributed manner. Chapter 9 introduces the concept of distributed dynamic spectrum access. Two major approaches to distributed dynamic spectrum access, namely, the cooperative and the non-cooperative approach are discussed. Chapter 10 deals with the application of learning algorithms in distributed dynamic spectrum access. The signaling protocols required to support distributed spectrum access are also reviewed. While dynamic spectrum access and management can be designed by considering only the technical issues, the economic issues are also important, especially under the new spectrum licensing paradigm. Chapters 11 and 12 address the economic aspects of spectrum sharing and management in a cognitive radio network. Chapter 11 introduces the concept of spectrum trading and reviews the different spectrum trading schemes. Through spectrum trading, the licensed users (or primary service provider) are able to sell a portion of the unused spectrum to the unlicensed users (or secondary service provider). The basic economic theories which can be used in dynamic spectrum access are reviewed. Chapter 12 provides an extensive review of economics-inspired dynamic spectrum access and management models. The first two models are based on competitive dynamic spectrum access and pricing. The concept of *collusion* among cognitive radio users in the context of spectrum trading is discussed. A spectrum trading model for IEEE 802.22-based wireless regional area networks (WRANs) is presented.

To use this book, if the reader is familiar with wireless technologies and the concept of cognitive radio, the first chapter can be skipped. Also, if the reader is an experienced researcher, some of the chapters in Part II can also be skipped. Since each chapter is quite independent, skipping any chapter will not affect following the rest of the book.

The authors would like to acknowledge the research support from Telecommunications Research Laboratory (TR*Labs*), Winnipeg, Canada, the Natural Sciences and Engineering Research Council (NSERC) of Canada, and Boise State University, Idaho, USA, during the writing of this book.

Part I

Introduction

1 Wireless communications systems

Wireless communications technology has become a key element in modern society. In our daily life, devices such as garage door openers, TV remote controllers, cellular phones, personal digital assistants (PDAs), and satellite TV receivers are based on wireless communications technology. Today the total number of users subscribing to cellular wireless services have surpassed the number of users subscribing to the wired telephone services. Besides cellular wireless technology, cordless phones, wireless local area networks (WLANs), and satellites are being extensively used for voice- as well as data-oriented communications applications and entertainment services.

In 1895, Guglielmo Marconi demonstrated the feasibility of wireless communications by using electromagnetic waves. In 1906, the first radio broadcast was done by Reginald Fessenden to transmit music and voice over the air. In 1907, the commercial trans-Atlantic wireless transmission was launched. In 1946, the first public mobile telephone systems were introduced in several American cities. The first analog cellular system, the Nordic Mobile Telephone System (NMT), was introduced in Europe in 1981. In 1983, the first cellular wireless technology, the advanced mobile phone system (AMPS), was deployed for commercial use. During the last two decades there has been significant research and development in wireless communications technology. In fact, today it has emerged as the most flourishing branch of development in the area of telecommunications.

The various wireless communications systems available today differ in terms of data rate of transmission, geographical coverage area, transmission power, and mobility support for users. Wireless communications systems can be broadly categorized as follows:

- *high-power wide area systems (or cellular systems)*, which support mobile users roaming over a wide geographic area;
- *low-power local area systems*, for example cordless telephone systems, which are implemented with relatively simpler technology;
- *low-speed wide area systems*, which are designed for mobile data services with relatively low data rates (e.g. paging systems);
- *high-speed local area systems*, which are designed for high speed and local communications (e.g. wireless LANs).

The first two categories are voice-oriented systems while the remaining two are data-oriented systems. Although different types of wireless systems have different

Table 1.1 Radio frequency bands.

	Designation	Frequency
ELF	Extremely low frequency	3 Hz–30 Hz
SLF	Superlow frequency	30 Hz–300 Hz
ULF	Ultralow frequency	300 Hz–3000 Hz
VLF	Very low frequency	3 kHz–30 kHz
LF	Low frequency	30 kHz–300 kHz
MF	Medium frequency	300 kHz–3000 kHz
HF	High frequency	3 MHz–30 MHz
VHF	Very high frequency	30 MHz–300 MHz
UHF	Ultrahigh frequency	300 MHz–3000 MHz
SHF	Superhigh frequency	3 GHz–30 GHz
EHF	Extremely high frequency	30 GHz–300 GHz

transmission rate, power, coverage, and mobility requirements, similar challenges exist for the design and implementation of these systems. These challenges include radio resource allocation/management and medium access control, rate control, handoff and mobility management, quality of service (QoS) provisioning, and security. This chapter intends to provide an introduction to the different wireless communications and networking technologies, to cover the basics of spectrum management, and the radio propagation characteristics and channel models for wireless communications, and finally to summarize the common research challenges in wireless communications systems.

1.1 Radio frequency bands and spectrum management

1.1.1 Radio frequency bands

Wireless communications systems are built based on the transmission of electromagnetic waves (i.e. radio waves) with frequencies in the range 3 Hz–300 GHz. These radio waves are transmitted and received through antennas which transform the radio frequency electrical energy to electromagnetic energy and vice versa. Radio waves with different frequencies have different propagation characteristics, each of which is suitable for a specific wireless application. For example, low-frequency radio waves are suitable for long-range communications and high-frequency radio waves are more suitable for short-range but high-speed wireless communications. The frequency of radio waves can be divided into different groups/bands as shown in Table 1.1.

Various wireless applications and services use different radio frequencies (Table 1.2). For example, 535 kHz–1.7 MHz is used for AM radio transmission, 54 MHz–88 MHz and 470 MHz–800 MHz are used for television (TV) signal transmission, and 88 MHz–108 MHz is used for FM broadcast.

1.1.2 Spectrum management

Interference can occur when radio waves are transmitted simultaneously from multiple sources over the same frequency. Frequency/spectrum management is required to

Table 1.2 Licensed spectrum allocations in the USA.

Service/system	Frequency
AM radio	535–1605 KHz
FM radio	88–108 MHz
Broadcast TV (channels 2–6)	54–88 MHz
Broadcast TV (channels 7–13)	174–216 MHz
Broadcast TV (UHF)	470–806 MHz
Broadband wireless	746–764 MHz, 776–794 MHz
3G wireless	1.7–1.85 GHz, 2.5–2.69 GHz
1G and 2G cellular	806–902 MHz
Personal communications systems	1.85–1.99 GHz
Wireless communications service	2.305–2.32 GHz, 2.345–2.36 GHz
Satellite digital radio	2.32–2.325 GHz
MMDS	2.15–2.68 GHz
Satellite TV	12.2–12.7 GHz
LMDS	27.5–29.5 GHz, 31–31.3 GHz
Fixed wireless services	38.6–40 GHz

control the transmission of radio waves to avoid interference among wireless users. Traditional spectrum management techniques, as defined by the Federal Communications Commission (FCC), are based on the command-and-control model. In this model, radio frequency bands are licensed to the authorized users by the government. The common method for allocation is referred to as a “spectrum auction.” In a spectrum auction, the government opens a radio frequency band for bidding and could specify a certain type of wireless technology/application for this particular radio frequency band (e.g. TV or cellular service). Any user/company interested in using this radio frequency band submits the bid (e.g. the amount of money it is willing to pay to the government to obtain the license). The government (i.e. the auctioneer) determines the winning user/company, which is generally the user/company offering the highest bid. The licensed user is authorized to use the radio frequency band under certain rules and regulations (e.g. etiquette for spectrum usage) specified by the government. The duration of the license issued to an authorized user is also determined by the government. While most of the spectrum is managed under this command-and-control scheme, there are some spectrum bands that are reserved for industrial, scientific, and medical purposes, referred to collectively as the industrial, scientific, and medical (ISM) radio band. This ISM band can also be used for data communication. However, since there is no control on this ISM band, the data communication could be interfered with by any ISM equipment. The allocation of this ISM band is shown in Table 1.3.

This command-and-control-based spectrum management framework can guarantee that the radio frequency spectrum will be exclusively licensed to an authorized user (i.e. licensee) who wins the radio frequency bidding and can use the spectrum without any interference. However, this command-and-control model can cause inefficient spectrum usage, as shown in the report from the spectrum Policy Task Force (SPTF) of the FCC in 2002 [4]. The spectrum management inefficiency arises due to the fact that an authorized user may not fully utilize the spectrum at all times in all locations. Also,

Table 1.3 Unlicensed spectrum allocations in the USA.

Band	Usage	Frequency
ISM band I	Cordless phones, 1G WLANs	902–928 MHz
ISM band II	Bluetooth®, 802.11b, 802.11g WLANs	2.4–2.4835 GHz
ISM band III	Wireless PBX	5.725–5.85 GHz
U-NII1 band I	Indoor systems, 802.11a WLANs	5.15–5.25 GHz
U-NII band II	Short-range outdoor systems, 802.11a WLANs	5.25–5.35 GHz
U-NII band III	Long-range outdoor systems, 802.11a WLANs	5.725–5.825 GHz

regulatory requirements put limitations on the wireless technology that can use the licensed spectrum, and this may prevent an authorized user from changing their wireless transmission techniques and services according to market demand.

To meet the spectrum demand of emerging wireless applications/services, there was a request to FCC to change the spectrum management policy to make it more flexible. The recommendations to change the spectrum management policy are as follows [5]: (1) improve flexibility of spectrum usage; (2) take all dimensions and related issues of spectrum usage into the policy; and (3) support and encourage efficient use of the spectrum. The objectives behind these recommendations were to improve both the technical and economic efficiency of spectrum management. From a technical perspective, spectrum management needs to ensure the lowest interference and the highest utilization of the radio frequency band. The economic aspects of spectrum management relate to the revenue and satisfaction of the spectrum licensee. To achieve economic efficiency, an economic model needs to be integrated into the spectrum management framework. Pricing is an important issue to achieve efficient spectrum management. The spectrum owners or service providers can compete or cooperate with each other in offering prices to the wireless service users to achieve the highest revenue. In this regard, service providers may be required to provide a quality-of-service (QoS) guarantee to users.

Different spectrum management models have been introduced in the literature for different radio frequency bands and different wireless applications [5]. These spectrum management models have improved the flexibility of spectrum usage and have also opened up new opportunities for the different wireless technologies to utilize the radio spectrum more efficiently. To exploit these opportunities, the wireless transceivers need to be more intelligent to access the radio spectrum. Such an intelligent wireless transceiver is referred to as a *cognitive radio*. The analysis and design of smart spectrum access techniques by cognitive radio is the main focus of this book.

1.2 Wireless protocols

In wireless communications, one of the most important components is the protocol. In this case, the communication system is divided into subsystems, i.e. layers. Different protocols are used in different layers for different tasks. Data is transferred among

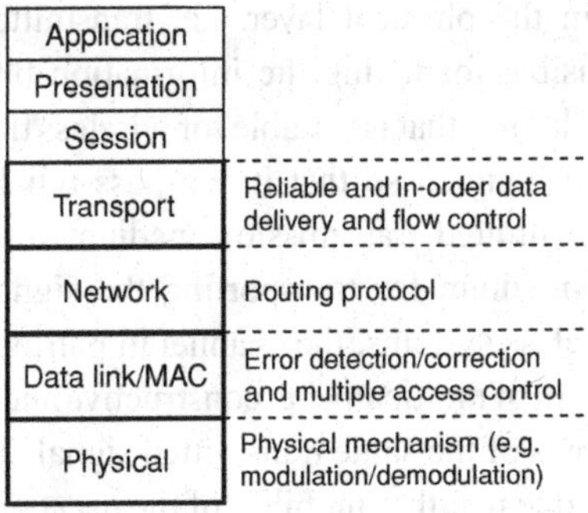

Layer	Function
Application	
Presentation	
Session	
Transport	Reliable and in-order data delivery and flow control
Network	Routing protocol
Data link/MAC	Error detection/correction and multiple access control
Physical	Physical mechanism (e.g. modulation/demodulation)

Figure 1.1 Protocol stack.

the layers and processed by the corresponding protocols, which specify the rules or conventions for such transfer (i.e. data transmission and reception). The protocol in a particular layer relies on the protocol in the lower layer to perform more primitive functions. Also, this protocol provides services to protocols in the higher layer. According to the open systems interconnection (OSI) model, the five important layers of wireless communications are as follows (Figure 1.1):

- *Physical layer*: Protocols in this layer provides a physical mechanism for transmitting signal bits between the transmitter and receiver. In wireless systems, protocols in this layer perform the modulation and demodulation of electromagnetic waves used for transmission.
- *Data-link layer*: Since the wireless links can often be unreliable, one function of the data-link layer is to perform error detection and/or correction. A part of the data-link layer, called the medium access control (MAC) sublayer, is responsible for allowing data packets to be sent over the shared media without undue interference with other transmissions. This aspect is referred to as multiple-access communications.
- *Networking layer*: Protocols in this layer (i.e. routing protocols) are responsible for determining the routing of data from the source to its destination. The Internet protocol (IP) is the default protocol at this layer to provide the routing function across multiple networks. Since the end systems can be mobile, and therefore the associations among the nodes continually changing, wireless systems place greater demand on the network layer.
- *Transport layer/host-to-host layer*: This layer is responsible for reliable and in-order data delivery on an end-to-end basis and flow control in the network to ensure that the network does not become congested. The most commonly used protocol in this layer is the transmission control protocol (TCP). TCP was primarily designed for wired networks for end-to-end transmission rate control. This protocol needs to be modified to achieve satisfactory performance in a wireless communication environment.
- *Application layer*: Protocols in this layer contain the logic needed to support the various user applications. For each different type of application, such as file transfer or Internet browsing, a separate module is required which is peculiar to that application.

There are three basic components in the physical layer, i.e. transmitter, channel, and receiver. The transmitter is responsible for taking the information bits from the information source and converting it into a form that is suitable for wireless transmission. The transmitter shapes and modulates the signal so that it can pass reliably through the channel, while efficiently using the limited transmission medium resources (i.e. the radio spectrum). The channel is a medium for transporting the signal produced by the transmitter to the receiver. In wireless systems, the channel impairments include channel distortion, which may take the form of multipath, i.e. constructive and destructive interference between many received copies of the same transmitted signal. The channel characteristics are usually time-varying, due to either mobility of the users or changes in the propagation environment. Interference and noise impact the quality of transmission in a wireless medium. This interference is produced (accidentally or intentionally) by other sources whose output signals occupy the same frequency band as or an adjacent band to the transmitted signal. On the other hand, receiver noise is produced by the electronic components in the receiver. After the signal is transmitted over the channel, the receiver receives it and produces an estimate of the transmitted signal (or information bit). In wireless systems, the receiver frequently estimates the time-varying nature of the channel in order to implement compensation techniques. The system also implements error detection and/or error correction techniques to improve the reliability of the wireless channel.

In the data-link layer, an error detection mechanism is used to detect the error in the received data which is caused by noise and interference. In addition, an error correction mechanism will provide an ability to correct the erroneous data. For error correction, an automatic repeat-request (ARQ) protocol and/or a forward error correction (FEC) coding mechanism can be used. With ARQ, the transmitter sends the data using an error detection code, for example the cyclic redundancy check (CRC) code. The receiver checks the received data for error, and requests retransmission if an error is detected. However, if there is no error detected, the receiver will send an acknowledgement (ACK) to inform the transmitter of successful data reception. With FEC, the transmitter encodes the data with an error correcting code (ECC). The coded data is transmitted to the receiver. The receiver checks the received data. If an error is detected, the receiver will attempt to correct the error in the original data by using the ECC.

The MAC protocol describes the method for multiple wireless users to share the channel. Four different methods for multiple-access communications are as follows:

- *Frequency-division multiple access (FDMA)*: The spectrum is shared by assigning specific frequency channels to specific users on either a permanent or a temporary basis.
- *Time-division multiple access (TDMA)*: All users are allowed to access all of the available spectrum, but users are assigned specific time intervals during which they can access it, on either a temporary or a permanent basis.
- *Code-division multiple access (CDMA)*: Users are allowed to use the available spectrum, but their signal must be spread (encrypted) with a specific code to distinguish it from other signals.

- *Space-division multiple access (SDMA)*: The spectrum is shared among the users by exploiting the spatial distribution of users' mobiles through the use of smart directional antennas that minimize the interference between users' mobiles.

1.3 Radio propagation characteristics and channel models

1.3.1 Radio propagation

For transmission of a wireless signal, radio frequency electrical energy from the transmitter is converted into electromagnetic energy by the antenna and radiated into the surrounding environment. For reception of a signal, electromagnetic energy impinging on the antenna is converted into radio frequency electrical energy and fed into the receiver. The quality of the received signal depends heavily on the radio propagation characteristics. This radio propagation is generally site-specific and can vary significantly depending on the terrain, frequency of operation, velocity of the mobile terminal, interference sources and other dynamic factors. Also, the radio propagation depends on the frequency of operation. At lower frequencies (<500 MHz) in the radio spectrum, the signal strength loss is less than that of higher frequency. However, due to the larger wavelengths, the antenna size for communications equipment operating in the lower frequency bands should be larger. On the other hand, at higher frequencies, it is possible to use low power transmitters (of about 1 W) to provide adequate signal coverage over, for example, a few floors of a multistorey building or a few kilometers outdoors in a line-of-sight (LOS) situation. The antenna sizes are also of the order of an inch, which means that the design of transmitters and receivers can be compact and requires lower power.

Radio propagation in open areas is very different from radio propagation in indoor and urban areas. In open areas across small distances or free space, the signal strength falls as the square of the distance. In other terrain, the signal strength often falls at a much higher rate as a function of distance depending on the environment and radio frequency. In urban areas, the shortest direct path (LOS path) between transmitter and receiver is usually blocked by buildings and other terrain features. Similarly, walls, floors, and other objects within buildings obstruct LOS communications. Such scenarios are called non-LOS (NLOS) or obstructed LOS (OLOS) scenarios, where the signal is usually carried by a multiplicity of indirect paths with various signal strengths. The signal strengths of NLOS and OLOS paths depend on the distance they have traveled, the obstacles they have reflected from or passed through, and the location of objects around the transmitter and the receiver. Because signals from the transmitter arrive at the receiver via a multiplicity of paths, with each taking a different time to reach the receiver, the resulting channel has an associated multipath delay spread that affects the reception of data. The maximum data rate that can be supported by a channel is affected by the multipath structure of the channel and the fading characteristics of the multipath components. The rate of fluctuations in the channel is referred to as the Doppler spread of the channel. This fluctuation is caused by the movement of the transmitter, receiver, or objects in between, and influences the signaling scheme and the receiver design.

The basic channel propagation mechanisms can be summarized as follows [6]:

- *Free-space or line-of-sight (LOS) propagation:* This mechanism corresponds to a clear transmission path between the transmitter and the receiver. Satellite communications generally rely on line-of-sight paths between the transmitter and satellite, and between the satellite and receiver.
- *Reflection:* This situation arises when electromagnetic waves are incident upon a surface with dimensions that are very large compared to the wavelength. Upon reflection or transmission, a ray attenuates by factors that depend on the frequency, the angle of incidence, and the nature of the medium (e.g. material properties, thickness, homogeneity, etc.). This phenomenon often dominates radio propagation in indoor applications.
- *Diffraction:* When electromagnetic waves are forced to travel through a small slit, they tend to spread out on the far side of the slit. This is referred to as diffraction. Due to diffraction, electromagnetic waves can bend over hills and around buildings to provide communications. Electromagnetic waves that are incident upon the edges of buildings, walls, and other large objects can be viewed as secondary sources. This can result in propagation into shadowed regions, since the diffracted field can reach the receiver, even though it is not in the line of sight of the transmitter. However, this secondary source suffers a much greater loss than that experienced via reflection or direct transmission. Consequently, diffraction is an important phenomenon outdoors, where signal transmission through buildings is virtually impossible. It is less important in indoor scenarios, where the diffracted signal is extremely weak compared to a reflected signal.
- *Scattering:* Irregular objects, such as walls with rough surfaces and furniture (indoors), and vehicles, foliage, and so on (outdoors), scatter rays in all directions in the form of spherical waves. Propagation in many directions results in reduced power levels, especially far from the scatterer. As a result, this phenomenon is not that significant unless the receiver or transmitter is located in a highly cluttered environment.

The importance of these three propagation mechanisms depends on the particular propagation scenario. As a result of reflection, diffraction, and scattering mechanisms, radio propagation can be roughly characterized by three nearly independent phenomenon: path loss variations with distance, slow log-normal shadowing, and fast multipath fading.

The channel can be characterized in the large and small scales. A large-scale propagation model characterizes signal strength over large transmitter–receiver separation distances and predicts the mean signal strength. As the mobile moves away from the transmitter, the local average received signal will gradually decrease and can be predicted by large-scale propagation models. A small-scale propagation model or fading model characterizes the rapid fluctuations of the received signal strength over very short travel distances (a few wavelengths) or short time durations (of the order of seconds). As a mobile moves over very small distances, the instantaneous received signal strength may fluctuate rapidly giving rise to small-scale fading since the received signal is a sum of many contributions coming from different directions. In small-scale fading, the received signal power may vary by as much as three or four orders of magnitude (30 or 40 dB) when the receiver is moved by only a fraction of a wavelength.

1.3.2 Channel models

Accurate characterization of the radio channel through key parameters and mathematical models is important for predicting signal coverage, achievable data rates, and the specific performance attributes of alternative signaling and reception schemes. This channel model can be used to analyze the interference in different systems, and determine the optimum locations for installing base station antennas.

Large-scale channel model

For large-scale radio propagation, channel models for wireless communications take one of two forms: physical models and statistical models. In physical models, the exact physics of the propagation environment is captured in which the site geometry has to be taken into consideration. Although this physical model provides the most reliable estimates of propagation behavior, it is computationally intensive. On the other hand, a statistical model is based on measuring the propagation characteristics in a variety of environments. The channel model is built by considering the measured statistics for a particular class of environments. These statistical models are easier to describe and to be used than the physical models, but do not provide the same accuracy.

In the physical approach, a model of the environment is built up which includes the terrain, buildings, and other features that may affect radio propagation. With this physical model, the possible propagation paths are determined. This process is referred to as ray tracing. For the different paths, the path losses between the transmitter and the receiver are estimated.

When the transmitter and the receiver have an LOS path between them, the free-space propagation model is used. With this model, the relationship between the transmit power and received power is given by

$$P_r(d) = \frac{P_t G_t G_r \lambda^2}{(4\pi d)^2 L},$$

where d is the separation between transmitter and receiver antennas (in meters), G_t is the transmitter antenna gain, G_r is the receiver antenna gain, L is the system loss factor not related to propagation (e.g. filter losses, antenna losses; $L \geq 1$), λ is the wavelength of the transmitted signal in meters, $G = \frac{4\pi A_e}{\lambda^2}$, where A_e is the effective aperture of the antenna ($= \lambda^2/(4\pi)$, for isotropic antenna), $\lambda = \frac{c}{f}$ (c = light speed, m/s; f = carrier frequency, Hz). This model is valid for values of d which are in the far-field of the transmitting antenna. The far-field or Fraunhofer distance of a transmitting antenna is given by: $d_f = \frac{2D^2}{\lambda}$, D = Largest physical linear dimension of the antenna.

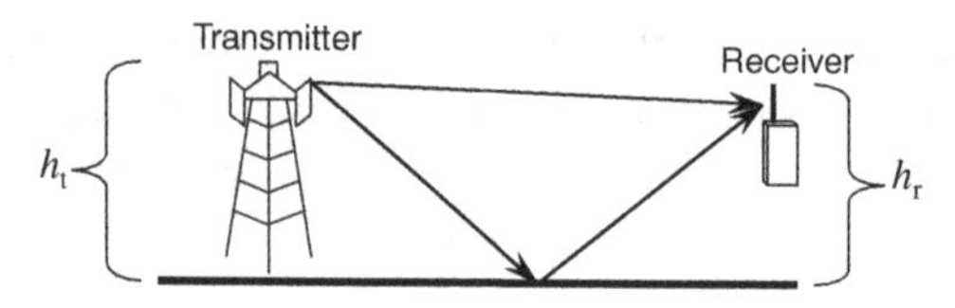

Figure 1.2 Two-ray channel model.

For the free-space propagation model, with $L = 1$, since $\frac{P_t}{P_r} = \frac{(4\pi d)^2}{G_r G_t \lambda^2}$, the signal attenuation (or path loss) is given by

$$\overline{P}_L(dB) = 10\log(A) + 20\log(d) - 20\log(\lambda),$$

where $A = \frac{(4\pi)^2}{G_t G_r}$. That is, for the same antenna dimensions and separation, the lower the carrier wavelenth λ (i.e. the longer the carrier frequency f), the higher is the free space path loss. Note that the received power in free space in terms of received power at a reference point d_0 is given by

$$P_r(d) = P_r(d_0)\left(\frac{d_0}{d}\right)^2, \quad d \geq d_0 \geq d_f.$$

A simple two-path model for an outdoor environment can be used to illustrate the effect of transmitting and receiving antenna heights. Here, the base station and the mobile terminal are both assumed to be at an elevation above the earth, which is modeled as a flat surface in between the base station and the mobile terminal (Figure 1.2). Usually there is an LOS component between the base station and the mobile terminal which carries the signal as in free space. There will be also another path over which the signal travels that consists of a reflection off the flat surface of the earth. The two paths travel different distances based on the height of the base station antenna, h_t, and the height of the mobile terminal antenna, h_r, and result in the addition of signals either constructively or destructively at the receiver.

The relationship between the transmit power and the received power for the two-ray model can be approximated by:

$$P_r(d) = P_t G_t G_r \frac{h_t^2 h_r^2}{d^4}, \tag{1.1}$$

where P_r and P_t are respectively the receive and transmit power, G_r and G_t denote respectively the antenna gains of the receiver and transmitter, and d is the distance between the transmitter and receiver [6].

In the statistical approach, the propagation characteristics are empirically approximated on the basis of measurements in certain general types of environments, such as urban, suburban, and rural environments. The statistical approach is broken down into two components: a component based on the estimate of the average path loss and a component representing local variations. The average large-scale path loss for an arbitrary transmitter–receiver separation d is given by [6]:

$$\overline{P}_L(d) = \overline{P}_L(d_0) + 10n\log\frac{d}{d_0}, \tag{1.2}$$

where n is the path loss exponent, d_0 is the reference distance, and $\overline{P}_L(d)$ is the path loss at distance d. Therefore, the average received power will be given by

$$\overline{P}_r(d) = P_t - \overline{P}_L(d). \tag{1.3}$$

The actual loss, which will vary around the mean value $\overline{P}_L(d)$, is referred to as *shadow fading*. It is primarily caused by the signal being blocked from the receiver by buildings (in outdoor areas), walls (inside buildings), as well as other objects in the environment. At any value of d, the path loss $P_L(d)$ at a particular location is random and log-normally distributed (normal given in dB) about the mean distance-dependent value. That is, $P_L(d)$ dB $= \overline{P}_L(d) + X_\sigma = P_L(d_0) + 10\, n \log(d/d_0) + X_\sigma$, where X_σ is a zero-mean Gaussian distributed random variable (in dB) with standard deviation σ (also in dB). Consequently, $P_r(d)$ dBm $= P_t$ dBm $- P_L(d)$ dB. Log-normal shadowing implies that measured signal levels at a specific transmitter–receiver separation have a Gaussian (normal) distribution about the distance-dependent mean (i.e. $\overline{P}_r(d)$), where the measured signal levels have values in dB units. Typical values of σ lie in the range 8–12 dB.

Small-scale channel model and fading

Small-scale radio propagation is generally modeled by considering the multipath spread due to the channel scattering. In this case, scattering by randomly located scatterers (e.g. traffic signs, lamp posts) gives rise to different paths with different path lengths, resulting in a multipath delay-spread. Non-overlapping scatterers give rise to distinct multipaths, which are characterized by their locations in the scattering medium. Signals reflected by scatterers located in the same ellipse will experience the same propagation delay. Signals that are reflected by scatterers located on different ellipses will arrive at the receiver with different delays. The three major effects due to scattering are as follows:

- Rapid changes in signal strength over a small travel distance or time interval. Multipath waves combine at the receiver antenna to give a resultant signal which can vary widely in amplitude and phase depending on the distribution of the intensity and the relative propagation time of the waves and the bandwidth of the transmitted signal.
- Random frequency modulations due to varying Doppler shifts on different multipath signals.
- Time dispersion (echoes) caused by multipath propagation delays. If the differential delay spread is large compared to the symbol interval, the received signals corresponding to the successive transmitted symbols will overlap giving rise to intersymbol interference (ISI).

Also, as the mobile terminal moves, the position of each scatterer with respect to the transmitter and the receiver may change. This phenomenon is referred to as fading. The factors which influence multipath fading are as follows [6]:

- *Multipath propagation:* Due to reflecting objects and scatterers in a wireless propagation environment, multiple versions of the transmitted signal arrive at the receiving antenna which are displaced with respect to one another in time and spatial orientation. In this way, the mobile channel introduces a delay spread into the received signal. That is, the received signal has a longer duration than that of the transmitted signal due to the different delays of the signal paths (time dispersion).

- *Speed of the mobile:* Due to the relative motion between the transmitter and the receiver, each multipath wave experiences an apparent shift in frequency (Doppler shift).
- *Speed of surrounding objects:* If objects in the radio channel are in motion, they induce a time-varying Doppler shift on the multipath signal components.
- *Transmission bandwidth of the signal and bandwidth of the multipath channel:* The bandwidth of the multipath channel can be quantified by the coherence bandwidth, which is related to the specific multipath structure of the channel. If the radio signal bandwidth is greater than the channel bandwidth, the received signal will be distorted but the small-scale signal fading will not be significant. If the radio signal bandwidth is less than the channel bandwidth, the amplitude of the signal will vary rapidly but the signal will not be distorted in time.

Small-scale channel fading can be categorized according to the time dispersion and frequency dispersion. In this case, the multipath delay spread leads to time dispersion and frequency-selective fading, while the Doppler spread leads to frequency dispersion and time-selective fading. The categorization based on multipath time delay spread results in flat fading and frequency-selective fading. If the mobile radio channel has a constant gain and linear phase response over a bandwidth which is greater than the bandwidth of the transmitted signal then the received signal will undergo flat fading. In the case of frequency-selective fading, the channel possesses a constant-gain and linear phase response over a bandwidth that is smaller than the bandwidth of the transmitted signal. Therefore, certain frequency components in the received signal spectrum have greater gains than others. Depending on the value of the Doppler spread, the radio channel may experience either fast fading or slow fading. In the case of fast fading, the channel impulse response changes within the symbol duration, while in the case of slow fading the channel impulse response changes at a much slower rate than the transmitted baseband signal. With slow fading, the channel may be assumed to be static over one or several symbol intervals.

Three common statistical channel models for small-scale fading are described by Rayleigh, Rician, and Nakagami-*m* distributions. In the Rayleigh channel model, there is no line-of-sight path between the transmitter and the receiver, and the probability density function, which characterizes the received signal amplitude, is given by

$$f_{\mathrm{R}}(r) = \begin{cases} \dfrac{r}{\sigma^2} \exp\left(-\dfrac{r^2}{2\sigma^2}\right), & 0 \le r \le \infty, \\ 0, & r < 0, \end{cases} \tag{1.4}$$

where $2\sigma^2$ $(= \frac{1}{2}E(r^2))$ is the average power of the received signal. For a Rayleigh fading channel, the probability density function of the received signal power is given by an exponential distribution as follows:

$$f_{\mathrm{P}}(p) = \frac{1}{\overline{P}} \mathrm{e}^{-\frac{p}{\overline{P}}}, \tag{1.5}$$

where $\overline{P} = \sigma^2$.

If there is a line-of-sight path between the transmitter and the receiver, the probability density function of the received signal amplitude is given by a Rician distribution as follows:

$$f_{\mathrm{R}}(r) = \frac{r}{\sigma^2}\exp\left(-\frac{r^2}{2\sigma^2}\right)\cdot\exp\left(-\frac{a_0^2}{2\sigma^2}\right)\cdot I_0\left(\frac{a_0 r}{\sigma^2}\right)$$
$$= \frac{r}{\sigma^2}\exp\left(-\frac{r^2+a_0^2}{2\sigma^2}\right) I_0\left(\frac{a_0 r}{\sigma^2}\right), \qquad (1.6)$$

where a_0^2 is the power of the LOS component and $I_0(\cdot)$ is the zero-order modified Bessel function of the first kind, given by

$$I_0(x) = \frac{1}{2\pi}\int_0^{2\pi}\exp(x\cos\theta)\mathrm{d}\theta. \qquad (1.7)$$

The Rician distribution is often described in terms of the parameter K (*Rician factor*), which is defined as the ratio between the deterministic signal power (i.e. the power of the LOS component) and the total power of all other scattered components:

$$K = \frac{a_o^2}{2\sigma^2}. \qquad (1.8)$$

As the dominant signal becomes weaker (K approaches zero), the composite signal has an envelope that is Rayleigh distributed. As $K \to \infty$, there is no fading (the wireless channel approaches an AWGN channel).

The Rician model is often applicable in an indoor environment, whereas the Rayleigh model characterizes outdoor settings. The Rician model also becomes more suitable in smaller cells or in more open outdoor environments.

The probability distribution function (pdf) for the instantaneous power varying about the local mean power $\bar{P}$ is given by

$$f_{\mathrm{P}}(p) = \frac{1+K}{\bar{P}}\mathrm{e}^{-K}\mathrm{e}^{-\frac{1+K}{\bar{P}}p} I_0\left(\sqrt{\frac{4K(1+K)}{\bar{P}}p}\right). \qquad (1.9)$$

where $\overline{P}$ is the local mean power $\overline{P} = (1/2)E(r^2)$.

The Nakagami-m pdf of the envelope r is given by

$$f_{\mathrm{R}}(r) = \frac{2m^m r^{2m-1}}{\Gamma(m)\Omega^m}\exp\left(-\frac{mr^2}{\Omega}\right), \quad m \geq 1/2, \qquad (1.10)$$

where $m = E^2[r^2]/\mathrm{Var}[r^2]$, $\Omega = E[r^2]$, $\Gamma(m) = \int_0^\infty x^{m-1}\mathrm{e}^{-x}\mathrm{d}x$, $R_{\mathrm{rms}} = \sqrt{\Omega}$.

For Nakagami-m fading, pdf of the power is given by

$$f_{\mathrm{P}}(p) = \left(\frac{m}{\overline{P}}\right)^m\left(\frac{p^{m-1}}{\Gamma(m)}\right)\exp\left(-\frac{mp}{\overline{P}}\right), \qquad (1.11)$$

where $\bar{P} = \frac{1}{2}E[r^2] = \frac{1}{2}\Omega$.

Since the Rician distribution contains a Bessel function, while the Nakagami distribution does not, the Nakagami distribution often leads to closed-form analytical expressions and insights that could not be obtained otherwise.

Nakagami-m distribution can model fading conditions that are either more or less severe than Rayleigh fading, where m is the parameter of the distribution. In particular, $m = 1$ implies Rayleigh fading, and $m = \infty$ implies no fading. The Nakagami-m probability density function of the power can be expressed as follows:

$$f_{\mathrm{R}}(r) = \left(\frac{m}{\overline{\overline{P}}}\right)^{m} \left(\frac{p^{m-1}}{\Gamma(m)}\right) \left(-\frac{mp}{\overline{P}}\right). \tag{1.12}$$

The relationship of parameter m of the Nakagami-m distribution to the Rician distribution is as follows: $m = (1+K)^2/(2K+1)$, where K is the Rician factor. In this case, $m = 1$ corresponds to $K = 0$, i.e. a Rayleigh distribution. However, the case $1/2 \leq m < 1$ is not covered by Rician distribution.

Integrating large-scale fading with small-scale fading

For a transmitter–receiver separation of d, the statistically varying received power can be expressed as

$$P_{\mathrm{r}}(d) = \alpha^2 10^{x/10} g(d) P_{\mathrm{t}} G_{\mathrm{t}} G_{\mathrm{r}}, \tag{1.13}$$

where α and x are random variables used to represent, respectively, the shadow fading and multipath effects, $g(d) = kd^{-n}$, and $f_\alpha(\alpha) = (\alpha/\sigma_\alpha^2)\mathrm{e}^{-\alpha^2/2\sigma^2}, \quad \alpha \geq 0$ with $E[\alpha^2] = 2\sigma_\alpha^2$. Therefore,

$$\begin{aligned} P_{\mathrm{r}}(d)\,\mathrm{dBm} &= 10\log_{10}\alpha^2 + x + 10\log_{10} g(d) + P_t\,\mathrm{dBm} + 10\log_{10} G_t G_r \\ &= \tilde{P}_r(d)\,\mathrm{dBm} + 10\log_{10}\alpha^2, \end{aligned} \tag{1.14}$$

where $\tilde{P}_{\mathrm{r}}(d)$ dBm is a random variable (r.v.) representing the statistically varying long-term received power in dBm due to shadow fading which is a Gaussian random variable with average value $\overline{P_{\mathrm{r}}}(d)$ dBm. Here $\overline{P_{\mathrm{r}}}(d)\,\mathrm{dBm} = 10\log_{10} g(d) + P_{\mathrm{t}}\,\mathrm{dBm} + 10\log_{10} G_{\mathrm{t}} G_{\mathrm{r}}$ with $E[\alpha^2] = 1$ implying $\sigma_\alpha^2 = 1/2$ and $E[x] = 0$ ($\overline{P_{\mathrm{r}}}(d)$ dBm is called the *area mean power*). That is,

$$f_{\tilde{P}_{\mathrm{r}}(d)\,\mathrm{dBm}}(p) = \frac{1}{\sqrt{2\pi}\sigma} \exp\left(-\frac{p - \overline{P_{\mathrm{r}}}(d)\,\mathrm{dBm}}{2\sigma^2}\right) \tag{1.15}$$

and

$$P_{\mathrm{r}}(d) = \alpha^2 \tilde{P}_{\mathrm{r}}. \tag{1.16}$$

It can be shown [7] that, for a Rayleigh fading channel, the instantaneous received power follows an exponential distribution with average value given by the local mean power, that is,

$$f_{P_{\mathrm{r}}}(p) = \frac{1}{\tilde{P}_{\mathrm{r}}} \mathrm{e}^{-\frac{p}{\tilde{P}_{\mathrm{r}}}}. \tag{1.17}$$

Packet-level channel modeling

A simple model for the packet error process in a wireless channel is a two-state (*good* state and *bad* state) Markov chain model (i.e. the Gilbert–Elliott model). For a Rayleigh fading channel, the packet error process can be modeled by using a two-state Markov chain with transition probability matrix $\mathbf{M}_c$ given by [8]

$$\mathbf{M}_c = \begin{bmatrix} \epsilon & 1-\epsilon \\ 1-\kappa & \kappa \end{bmatrix} \tag{1.18}$$

where ϵ and $1-\epsilon$ are the probabilities that the transmission of packet k is successful, given that the transmission of packet $(k-1)$ was successful or unsuccessful, respectively. The duration of the good and bad periods are geometrically distributed. Therefore, the distribution of residual life (time to end of duration) of both bad and good periods are memoryless and have the same distribution as their respective typical distributions. These two geometric distributions can be parameterized by arbitrary parameters, whereas in an independent and identically distributed (iid) channel model the parameters are complements of each other. This two-state Markov channel model has been used for wireless protocol (e.g. ARQ, TCP) performance evaluation in fading channels [9].

Given the matrix $\mathbf{M}_c$, the channel properties are completely characterized. The steady-state probability that a packet error occurs is given by $P_e = (1-\kappa)/(2-\epsilon-\kappa)$, and the average length of an error burst is $(1-\kappa)^{-1}$. For a Rayleigh fading channel with a fading margin F (i.e. the maximum fading attenuation with respect to the mean signal power which still allows correct reception), $P_e = 1 - e^{-1/F}$ and the Markov parameter is $\kappa = 1 - (Q(\theta, \rho\theta) - Q(\rho\theta, \theta))/(e^{1/F} - 1)$, where $\theta = \sqrt{(2/F)/(1-\rho^2)}$. Here $Q(\cdot,\cdot)$ is Marcum's Q function (given in (1.19)), and $\rho = J_0(2\pi f_m T)$ is the Gaussian correlation coefficient, where J_0 is the zero-order Bessel function (given in (1.20)) of two successive samples of the complex amplitude of a fading channel with maximum Doppler shift f_m. Here $f_m = v/\lambda$ is the ratio of mobile speed and carrier wavelength taken T seconds apart (T is the packet duration). Therefore, for a given packet duration T, maximum Doppler shift f_m, and average packet error rate P_e, the Markov parameters ϵ and κ can be obtained from

$$Q(x,y) = \int_y^\infty e^{-\frac{(x^2+w^2)}{2}} I_0(xw) w \, dw. \tag{1.19}$$

$$J_0(x) = \sum_{k=0}^{\infty} \frac{(-\frac{1}{4}x^2)^k}{k!\Gamma(k+1)}. \tag{1.20}$$

Note that

$$\Gamma(z) = \int_0^\infty t^{z-1} e^{-t} dt. \tag{1.21}$$

The gamma function $\Gamma(z)$ satisfies the recurrence relation

$$\Gamma(z+1) = z\Gamma(z). \tag{1.22}$$

When n is an integer, $\Gamma(n+1) = n!$.

In the above two-state channel model, different values of P_e and $f_m T$ result in different degrees of correlation in the fading process. When $f_m T$ is small, the fading process is very correlated (i.e. there will be long bursts of packet errors). On the other hand, for large values of $f_m T$, the channel error events are almost independent (i.e. there will be short bursts of packet errors).

To model the packet error process in a fading channel, a more general finite-state Markov channel (FSMC) model can be used. This model is particularly suitable to model wireless packet transmissions using adaptive modulation and coding (AMC). This FSMC model can be used to analyze radio channels with non-independent fading (and hence bursty channel errors). For a slowly varying Nakagami-m fading channel modeled as an FSMC, each state of the Markov chain corresponds to one transmission mode for AMC. With AMC, the SNR at the receiver is divided into $J+1$ non-overlapping intervals by thresholds Ψ_j ($j \in \{0, 1, \ldots, J\}$), where $\Psi_0 = -\infty < \Psi_1 < \cdots < \Psi_{J+1} = \infty$. The channel is said to be in state j if $\Psi_j \leq \gamma < \Psi_{j+1}$, for which B_j bits can be transmitted per symbol. To avoid any possible transmission error, no packet is transmitted when $\gamma < \Psi_1$. The probability of using transmission mode j (i.e. $\Pr(j)$) is given by

$$\Pr(j) = \frac{\Gamma(m, m\Psi_j/\overline{\gamma}) - \Gamma(m, m\Psi_{j+1}/\overline{\gamma})}{\Gamma(m)}, \tag{1.23}$$

where $\overline{\gamma}$ is the average SNR, m is the Nakagami fading parameter ($m \geq 0.5$), $\Gamma(m)$ is the Gamma function, and $\Gamma(m, \gamma)$ is the complementary incomplete Gamma function, given by

$$\Gamma(a, x) = \frac{1}{\Gamma(a)} \int_x^{\infty} \mathrm{e}^{-t} t^{a-1} \mathrm{d}t, \quad (a > 0). \tag{1.24}$$

Assuming that the channel is slowly fading (i.e. transitions occur only between adjacent states), the state transition matrix for the FSMC can be expressed as follows [10]:

$$\mathbf{M}_c = \begin{bmatrix} M_{0,0} & M_{0,1} & & & \\ M_{1,0} & M_{1,1} & M_{1,2} & & \\ & \ddots & \ddots & \ddots & \\ & & M_{J-1,J-2} & M_{J-1,J-1} & M_{J-1,J} \\ & & & M_{J,J-1} & M_{J,J} \end{bmatrix}, \tag{1.25}$$

where each row of matrix $\mathbf{M}_c$ corresponds to a transmission mode. The transition probability from state j to j', $M_{j,j'}$, can be obtained as follows:

$$M_{j,j+1} = \frac{L_{j+1} \times T}{\Pr(j)}, \quad j = 0, \ldots, J-1, \tag{1.26}$$

$$M_{j,j-1} = \frac{L_j \times T}{\Pr(j)}, \quad j = 0, \ldots, J, \tag{1.27}$$

$$M_{j,j} = \begin{cases} 1 - M_{j,j+1} - M_{j,j-1}, & 1 < j < J, \\ 1 - M_{0,1}, & j = 0, \\ 1 - M_{J,J-1}, & j = J, \end{cases} \tag{1.28}$$

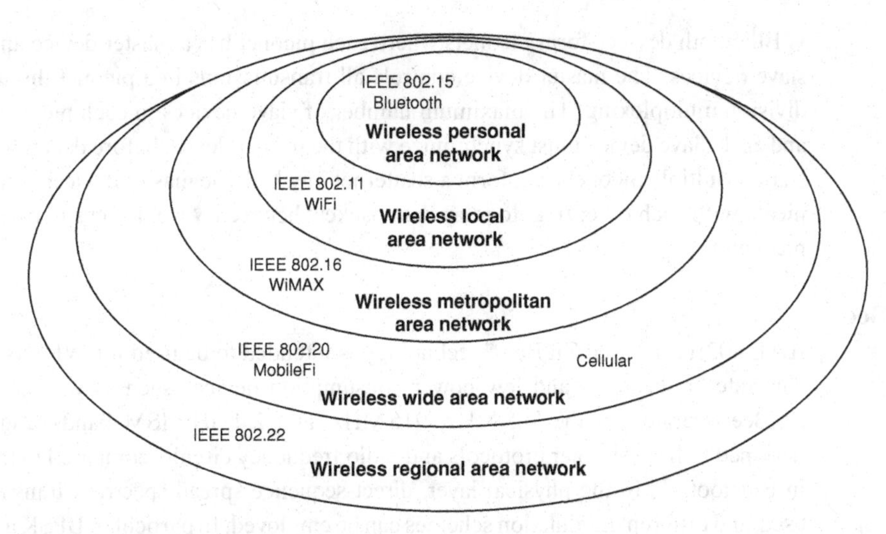

Figure 1.3 Different wireless technologies.

where T is the packet length. L_j is the level-crossing rate at Ψ_j corresponding to state j and can be estimated from

$$L_j = \sqrt{2\pi \frac{m\Psi_j}{\overline{\gamma}} \frac{f_{\mathrm{m}}}{\Psi(m)}} \left(\frac{m\Psi_j}{\overline{\gamma}}\right)^{m-1} \exp\left(-\frac{m\Psi_j}{\overline{\gamma}}\right), \tag{1.29}$$

where f_{m} is the maximum Doppler shift.

1.4 Wireless communications technologies

Current wireless communications technologies can be classified by the corresponding transmission range or coverage area as shown in Figure 1.3.

1.4.1 Wireless PAN technologies

Wireless personal area network (WPAN) technology is a short-range wireless networking technology. Two major WPAN technologies are Bluetooth® and ZigBee®, which are both based on the IEEE 802.15 standards.

Bluetooth

Bluetooth® is a standard for low-power and short-range wireless communications. The standard was designed to operate in the 2.4–2.4835 GHz ISM band. Depending on the transmission power (and hence the transmission range), Bluetooth devices are divided into three classes – class 1, class 2, and class 3. For class 1, 2, and 3 devices, the transmit power is 1, 2.5, and 100 mW, and the corresponding transmission range is 1, 10, and 100 meters, respectively [11]. The standard supports data rates up to 1 Mbps (in version 1.2) and 3 Mbps (in version 2.0). In the physical layer, adaptive frequency-hopping spread spectrum (AFH) is used.

Bluetooth devices form piconets where each piconet has a master device and several slave devices. The master device controls all transmissions in a piconet through time-division multiplexing. The maximum number of slave devices in each piconet is seven and each slave device must synchronize with the master device before data transmission starts. Multiple piconets can form a scatternet in which the master devices can communicate with each other (e.g. to relay data packets between slave devices in two different piconets).

ZigBee

IEEE 802.15.4-based ZigBee® technology is intended to be used for WPANs with low data rate applications and low power consumption devices such as wireless sensors. ZigBee operates on the 868 MHz, 915 MHz, and 2.4 GHz ISM bands. ZigBee was designed to have simpler protocols and radio frequency circuits compared to those used in Bluetooth®. In the physical layer, direct-sequence spread spectrum transmission is used, and different modulation schemes can be employed. In particular, BPSK and QPSK modulation can be used to support data rates of 40 kbps and 250 kbps per channel in the 868 and 915 MHz bands, and 2.4 GHz band, respectively [12]. The transmission range is between 10 and 75 meters and the maximum transmit power is 1 mW. ZigBee supports both beacon-enabled and non-beacon networks. In a beacon-enabled network, a ZigBee router broadcasts the beacon signals periodically, and ZigBee nodes are allowed to transmit based on the beacon signal. In a non-beacon network, an unslotted carrier-sense multiple access with collision avoidance (CSMA/CA) protocol is used. ZigBee has the capability of being power efficient. A ZigBee node can switch to sleep mode (e.g. turn off the radio transceiver) between two consecutive beacon signals to reduce power consumption.

1.4.2 Wireless LAN technologies

Wireless LAN (WLAN) technology was designed to provide high-speed wireless connectivity in a local area. The most common architecture of WLANs is based on a central controller called an access point (i.e. infrastructure-based architecture) in which the traffic of all WLAN users goes through the access point. The access point can be connected to a wired network for receiving incoming and sending outgoing Internet traffic. In another architecture, WLANs can form an ad hoc network in which each WLAN client transmits data to another client directly. Since the coverage area of a WLAN is small (e.g. around 100 meters), the mobility of the WLAN users is rather limited. IEEE 802.11 defines the most common WLAN standards.

IEEE 802.11 standards (e.g. 802.11, 802.11b, 802.11g, 802.11a) define a set of protocols for WLANs operating in the industrial, scientific, and medical (ISM) bands of 5.15–5.35 GHz and 5.725–5.825 GHz. The physical layer transmissions can be based on direct sequence spread spectrum (DSSS), frequency-hopping spread spectrum (FH-SS), or orthogonal frequency-division multiplexing (OFDM) techniques. In the medium access control (MAC) layer, both contention-free and contention-based protocols are supported through the point coordination function (PCF) mode and the distributed coordination function (DCF) mode, respectively. Specifically, a polling protocol is used in the PCF mode and a carrier sense multiple access with collision avoidance (CSMA/CA) protocol is used in the DCF mode.

IEEE 802.11 operates at the 2.4 GHz band with data rates up to 2 Mbps. IEEE 802.11a operates on the 5 GHz band and can achieve data rates up to 54 Mbps. IEEE 802.11b and 802.11g operate at the 2.4 GHz band with data rates of 11 and 54 Mbps, respectively. The emerging IEEE 802.11n technology is expected to provide data rates up to 248 Mbps by employing the multiple-input multiple-output (MIMO) antenna technique in the physical layer. Also, IEEE 802.11n uses advanced techniques in medium access control (MAC) layer techniques such as frame aggregation [13].

1.4.3 Cellular wireless networks

In a cellular network, a service area is divided into cells and a base station is used in each cell to provide wireless communication services to the mobile users in that cell. Cellular users can move between cells, and, therefore, handoff and mobility management are important functionalities in any cellular network. Several cellular wireless technologies have been developed, the most common of which are Global System for Mobile Communications (GSM), general packet radio service (GPRS) technologies, and code-division multiple access (CDMA)-based technologies.

Evolution of cellular wireless technology

The first generation (1G) cellular and cordless phones were based on analog technology using frequency modulation (FM). NMT (Nordic Mobile Telephone), AMPS (Advanced Mobile Phone Service) and TACS (Total Access Communications Systems) were the three primary analog cellular radio systems standards. The second generation (2G) systems refer to the digital cellular and cordless systems which employ digital modulation and advanced call processing capabilities. GSM (Global System for Mobile Communication), which was the first European digital cellular radio system, IS-54/IS-136 TDMA (time-division multiple access)-based North American digital cellular systems, the CDMA (code-division multiple access)-based IS-95 system, and the Japanese PDC (Personal Digital Cellular Radio) are examples of 2G cellular wireless systems. As an evolution of 2G systems, a wide range of 2.5G standards have been developed to allow each of the major 2G technologies (GSM, CDMA, and IS-136) to be upgraded incrementally to provide more efficient Internet data services. Examples of 2.5G technologies include HSCSD (high speed circuit switched data), GPRS (general packet radio service), EDGE (enhanced data rates for GSM evolution), and IS-95B.

Third generation (3G) systems have already been launched in the market. The main attribute of these systems is the support for high-speed packet data services up to 2 Mbps for wireless Internet access. Studies and standardization efforts on 3G systems are being carried out under the names IMT-2000 (International Mobile Telecommunications in the year 2000), UMTS (Universal Mobile Telecommunication System) and MBS (Mobile Broadband System). The majority of the RTT (radio transmission technology) proposals for IMT-2000 chose DS/CDMA (direct sequence code-division multiple access) as the leading multiple access technique. Activities on the harmonization of the technical specifications for the different RTT candidates are being carried out with an aim to achieve a single converged 3G standard. Two major 3G technology camps working in this direction are referred to as W-CDMA (wideband CDMA)/UMTS and cdma2000.

GSM technology

GSM is the first digital cellular wireless technology and operates in the 850–900 MHz and 1800–1900 MHz bands. GSM systems use TDMA as the multiple access technique to coordinate transmissions from multiple users. The transmission time in each of the 200 kHz channels is divided into frames each of size 4.615 ms, and each frame consists of eight time slots. The maximum transmission range is 35 kilometers and four different cell sizes, i.e. macro, micro, pico, and umbrella cells, are supported in the standard. With Gaussian minimum-shift keying (GMSK) as the modulation technique, GSM supports total data rate 270.833 kbps per channel [6]. GSM technology, which is often referred to as a 2G technology, primarily offers circuit-switched voice communications services.

GPRS is a separate packet-swiched data network within GSM for wireless packet data services. GPRS uses multiple channel coding schemes in the physical layer. This offers a flexible radio channel allocation in which from one to eight time slots can be allocated to one connection in one TDMA frame. Depending on the coding rate, GPRS can support data rates up to 171.2 kbps [6] for an individual user. GPRS defines a special multiframe structure, which is divided into blocks comprised of the same time slot in four consecutive TDMA frames for transmission of a single radio link level frame.

The EDGE standard was designed to provide a 3G evolution option for GSM systems. Two components of EDGE, namely, enhanced circuit switched data (ECSD) and EGPRS (enhanced GPRS), define the enhancements for circuit-mode and packet-mode data, respectively, and increase data rates to over 384 kbps (for a single dedicated user on a single GSM channel) through the use of adaptive modulation and coding schemes. In EGPRS, nine modulation and coding schemes (MCSs) are defined – four code rates with GMSK and five code rates with 8-PSK modulation. Each of the MCSs is designed to operate efficiently on part of the dynamically varying signal-to-interference plus noise ratio (SINR) range. At the transmitter, the MCS is selected depending on the channel conditions measured at the receiver. For radio link level error control, the *incremental redundancy* transmission concept [14] is used. These technologies are often referred to as 2.5G cellular wireless technologies.

CDMA-based cellular wireless technologies

CDMA-based cellular wireless systems use spread-spectrum transmission technique to support voice and data communications. The Interim Standard 95 (IS-95), also referred to as cdmaOne, is the first digital cellular CDMA technology designed for circuit-switched voice and data communications over 1.25 MHz channels. Each IS-95 channel supports 64 users with a per user throughput of 14.4 kbps. IS-95B is the 2.5G evolution of IS-95, which increases the per user throughput up to 115.2 kbps. This is achieved through code aggregation where up to eight codes can be assigned to a data connection for the duration of a data burst.

The 3G evolution for both 2G cellular CDMA systems and GSM systems have resulted in two CDMA standard streams – wideband CDMA (WCDMA) and several variants of cdma2000. The frequency bands used by 3G networks include the following: 2500–2690 MHz, 1710–1885 MHz, and 806–960 MHz. While the variants of cdma2000 (e.g. cdma2000 1xRTT, cdma2000 1xEV, cdma2000 3xRTT) are based on fundamentals of IS-95 and IS-95B technologies, WCDMA, which is also referred to as Universal Mobile Telecommunication System (UMTS), is based on GSM technology and its evolutions (i.e. EDGE). The Third Generation Partnership Project (3GPP) and Third Generation Partnership Project 2 (3GPP2) are the standardization bodies responsible for WCDMA and cdma2000 standards, respectively.

The WCDMA air interface uses a radio channel with a bandwidth of 5 MHz to support data rates up to 2.048 Mbps per user using variable direct sequence spread spectrum chip rates. The cdma2000 1xRTT radio interface uses a channel of bandwidth 1.25 MHz to support packet data rates of up to 144 kbps per user. The cdma2000 1xEV technology also uses a radio channel with a bandwidth of 1.25 MHz and it provides options for installing radio channels with data only (cdma2000 1xEV-DO) or with data and voice (cdma2000 1xEV-DV). While the cdma2000 1xEV-DO air interface can provide throughput of the order of several hundred kbps per user, the cdma2000 1xEV-DV air interface can offer data rates of up to 144 kbps. The cdma2000 3xRTT technology combines three adjacent 1.25 MHz radio channels into a channel of bandwidth 3.75 MHz to support data rates over 2 Mbps [15]. Both WCDMA and cdma2000 air interfaces support frequency-division duplexing (FDD) and time-division duplexing (TDD) modes.

High-speed downlink packet access (HSDPA) and high-speed uplink packet access (HSUPA) technologies are an evolution of the WCDMA standard that enhances the capabilities of 3G by enabling higher data transfer rates to support packet-based multimedia services. HSDPA has been included in release 5 of the 3GPP standards and is currently being deployed in 3G networks worldwide. HSDPA provides downlink data rates of up to 14.4 Mbps, which can significantly increase the capacity of mobile networks. The HSUPA technology, which is also known as FDD enhanced uplink (EUL), has been introduced in release 6 of the 3GPP standards and can provide uplink transfer rates of up to 5.76 Mbps [15].

The UMTS Long Term Evolution (LTE) standard is currently being developed by 3GPP. This standard intends to support an instantaneous downlink peak data rate of 100 Mb/s within a 20 MHz downlink spectrum allocation and an instantaneous uplink peak data rate of 50 Mb/s within a 20 MHz uplink spectrum allocation through the introduction of new transmission schemes and advanced multi-antenna technologies [16].

The EVDO Rev C (also called DORC) standard is being developed by 3GPP2 with an aim to increasing the peak data rates of up to 200 Mbps in the downlink. It will support flexible and dynamic channel bandwidth scalability from 1.25 MHz up to 20 MHz [16].

1.4.4 Wireless MAN technologies

Wireless metropolitan area network (WMAN) technology based on broadband wireless access [17] provides an alternative to last-mile wired broadband access (e.g. through

DSL and cable modem), especially in areas where wired infrastructure is not available. There are two major standards for WMAN technology – IEEE 802.16 (WiMAX) and IEEE 802.20 (MobileFi).

IEEE 802.16/WiMAX technology

The IEEE 802.16 standard [18] incorporates several advanced radio transmission techniques such as orthogonal frequency-division multiple access (OFDMA), adaptive modulation and coding (AMC), adaptive forward error correction (FEC), and hybrid automatic repeat request (H-ARQ). With these techniques, IEEE 802.16 can support data rates up to 100 Mbps with a maximum transmission range of 30 km. Also, in the standard, a QoS framework is defined to support different types of user traffic.

The IEEE 802.16 standard supports two architectures, namely, last-mile access and mesh network architectures. In last-mile access architecture, which is similar to the cellular system architecture, the base station (BS) controls the transmissions to/from the subscriber stations (SSs). In the mesh network architecture (i.e. IEEE 802.16j), SSs can relay traffic from other SSs to the base station in a multihop fashion. The IEEE 802.16 QoS framework supports four service classes, namely, unsolicited grant service (UGS), real-time polling service (rtPS), non-real-time polling service (nrtPS), and best-effort (BE). While the UGS class is for constant-bit-rate traffic, the rtPS class is for real-time traffic such as voice over IP (VoIP) and video, and the nrtPS class is for traffic which requires throughput guarantee. The BE service class is for traffic which does not require any QoS guarantee.

IEEE 802.16 with a single carrier air interface operates in the 10–66 GHz band with line-of-sight communication. IEEE 802.16a with OFDM/TDMA and OFDMA air interfaces operates in the 2–11 GHz band with non-line-of-sight communication. The mobile WiMAX technology based on the IEEE 802.16e standard supports broadband wireless access for high mobility users.

IEEE 802.20/MobileFi technology

IEEE 802.20-based MobileFi technology [19] is designed for mobile broadband wireless access. It intends to support IP-based mobility functions (e.g. IP roaming and handoff) for broadband wireless users with mobility up to 250 km/hr. The standard supports radio channels with bandwidths of 1.25 and 5 MHz in the frequency band lower than 3.5 GHz. The air interface is based on OFDM with adaptive modulation (i.e. QPSK, 8PSK, 16QAM, and 64QAM modulation modes) and the minimum downlink and uplink data rates are 4 Mbps and 800 kbps per cell, respectively. The scope of the standard covers the physical, MAC, and logical link control layers.

1.4.5 Wireless RAN technology

IEEE 802.22-based wireless regional area networks (WRANs) [20] are intended to support a wider transmission range (e.g. 50–100 km) using the frequency spectrum of the TV band. In North America, the range of operating frequencies for IEEE 802.22 networks will be 54–862 MHz, while frequencies in the range of 41–910 MHz will be supported in the standard for international use. The IEEE 802.22 network supports point-to-multipoint stationary connections. The base station in a WRAN cell controls all transmissions to and from the consumer premise equipments (CPEs). OFDMA will be used along with adaptive modulation and coding in the physical layer to achieve a spectral efficiency in the range of 0.5 to 5 bits/symbol/Hz depending on channel quality. To enhance system throughput, the standard will support channel bonding techniques for which multiple 6 MHz channels can be used simultaneously for transmission. Since IEEE 802.22 networks will operate in the same frequency band as that of the TV service, and multiple 802.22 networks may also exist in the same or overlapping area, interference management will be crucial in these networks. Therefore, dynamic spectrum sensing and spectrum access based on the cognitive radio concept will be employed in IEEE 802.22-based WRANs.

1.5 Advanced wireless technologies

1.5.1 OFDM technology

Orthogonal frequency-division multiplexing (OFDM) is a technique of transmitting multiple digital signals simultaneously over a large number of orthogonal subcarriers. Based on the fast Fourier transform algorithm to generate and detect the signal, data transmission can be performed over a large number of carriers that are spaced apart at precise frequencies. The frequencies (or tones) are orthogonal to each other. Therefore, the spacing between the subcarriers can be reduced and hence high spectral efficiency can be achieved. OFDM transmission is also resilient to interference and multipath distortion which causes inter-symbol interference (ISI) [21].

Orthogonal frequency-division multiple access (OFDMA) is a multiple access scheme implemented based on OFDM. In OFDMA, different users are allocated with different subcarriers, and hence multiple users can transmit their data simultaneously. QoS can be achieved in OFDMA by allocating different number of subcarriers to the users with different QoS requirements. OFDM can be combined with the CDMA scheme (i.e. multi-carrier code-division multiple access (MC-CDMA) or OFDM-CDMA). In this case, different codes are assigned to different users for concurrent transmissions.

OFDM and OFDMA are used in the emerging wireless standards including IEEE 802.11a/WiFi, IEEE 802.16/WiMAX, IEEE 802.20/MobileFi, and IEEE 802.22.

1.5.2 MIMO technology

To enhance the performance of wireless transmission, multiple-input multiple-output (MIMO) or multiple antennas can be used to transmit and receive the radio signals. Data transmitted from multiple antennas will experience different multipath fading, and at the receiver these different multipath signals are received by multiple antennas. By using advanced signal processing techniques, multipath signals at the receiver can be combined to reconstruct original data. MIMO systems take advantage of this spatial diversity to achieve higher data rates or lower bit error rates. There are two basic types of MIMO systems, namely, the space-time coding MIMO system (for diversity maximization) and spatial multiplexing MIMO system (for data rate maximization).

In a space-time coding MIMO system, a single data stream is redundantly transmitted over multiple antennas by using suitably designed transmit signals. At the receiver, multiple copies of the signal are received which are used to construct the original data. Space-time coding can be categorized into space-time trellis coding (STTC), in which trellis code is transmitted over multiple antennas and multiple time-slots, and space-time block-coding (STBC) in which a block of data is transmitted over multiple antennas. Both STTC and STBC can achieve diversity gain which improves the error performance. While STTC can also achieve coding gain (i.e. results in lower error rate), STBC can be implemented with less complexity. Instead of transmitting the same data over multiple antennas, a spatial multiplexing MIMO system transmits different data streams over multiple antennas. In this case, the number of transmit antennas is equal or larger than that of receive antennas and the data rate can increase by a factor of the number of transmit antennas [22].

Multiuser MIMO (MU-MIMO) was proposed to support data transmission from multiple users simultaneously. In this case, data from different users can be transmitted over different antennas. In this MU-MIMO, MIMO broadcast channels and MIMO multiple access channels are used for downlink and uplink transmissions, respectively. Alternatively, space-division multiple access (SDMA) can exploit the information of users' locations to adjust the transmission and reception parameters to achieve the best path gain in the direction of each user. Phased array antenna techniques are generally used for SDMA.

MIMO is an optional feature in the IEEE 802.16/WiMAX standard while it will be used in the IEEE 802.11n standard.

1.5.3 Beamforming technique

Beamforming is a technique in signal processing used for directional data transmission and reception. Using beamforming, the signal is received/transmitted in a particular direction to improve the receive/transmit gain. When transmitting, a beamforming transmitter controls the phase and amplitude of the signal to create a pattern of constructive and destructive interference in the waveform. When receiving, these signals are combined in such a way that the expected pattern of radiation is preferentially observed. Different weights can be assigned to the different signals so that the decoded information has the smallest error probability. In a conventional beamforming system, these weights are fixed and can be obtained according to the location of the receive antenna and the direction of the signal. Alternatively, the weights can be adaptively adjusted by considering the characteristics of the received signal to mitigate the interference from unwanted sources.

1.5.4 Ultra-wideband technology

Ultra-wideband (UWB) technology was developed for short-range and high-bandwidth wireless communications. A UWB transmitter transmits data over a large bandwidth

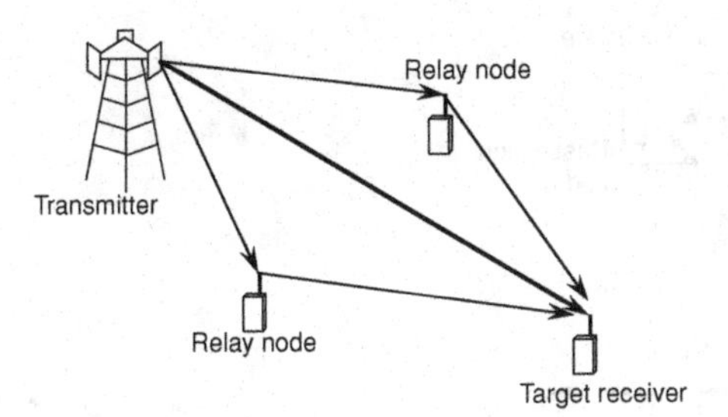

Figure 1.4 Cooperative diversity.

(more than 500 MHz). Since UWB transmission uses very low transmit power, it does not interfere with other narrow band transmissions. Two implementation approaches for UWB are based on OFDM and pulse transmission. In the OFDM approach, the transmission bandwidth is divided into a number of subcarriers, and the OFDM technique is used for data transmission on these subcarriers [23]. On the other hand, in the pulse-based approach, data is encoded into a series of pulses which are transmitted at very high frequency over the large bandwidth. UWB is the transmission technique adopted for IEEE 802.15.3a standard-based WPAN technology.

1.5.5 Cooperative diversity

Cooperative diversity [24, 25] is a communication technique which utilizes both the direct signal from the transmitter to the receiver and the relayed signal from a relay node to the receiver in order to achieve a diversity gain (Figure 1.4). The receiver in a cooperative diversity system is able to decode the information from the direct and relayed signals. The concept of cooperative diversity is similar to MIMO in which the transmission performance can be improved by using spatial diversity. Unlike MIMO, cooperative diversity uses distributed antennas deployed at the relay nodes (which are referred to as virtual antennas). Since these distributed antannas are physically separated, it is more likely that multiple antennas will experience independent and different fading conditions. This results in diversity in transmission which can enhance the performance of information decoding. Two types of cooperative diversity are amplify-and-forward (AF) (i.e. non-regenerative) cooperative diversity and decode-and-forward (DF) (i.e. regenerative) cooperative diversity. With AF cooperative diversity, each cooperative (relay) node simply amplifies and forwards the received signal towards the target receiver. In contrast, with DF cooperative diversity, a relay node decodes and re-encodes the received signal before forwarding the data to the target receiver. While the AF cooperative diversity is simpler to implement and could also lead to maximum diversity order, the DF cooperative diversity can prevent error propagation.

1.6 Multihop wireless networking

In traditional wireless systems, communication is mostly performed in a single hop, e.g. from mobile to base station or vice versa. However, due to the limitation in transmit

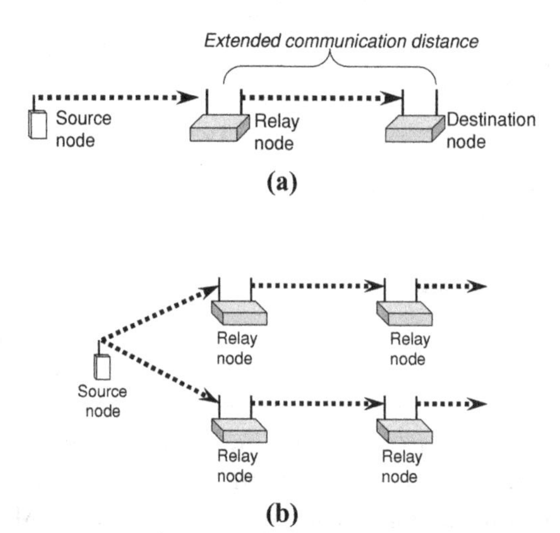

Figure 1.5 Advantages of multihop networks: (a) extension of coverage area and (b) enhancement of throughput.

power, single-hop communication limits the distance between the transmitter and the receiver. This limitation can be removed through multihop communications. In a multihop communication network, the nodes can cooperatively receive and forward (i.e. relay) traffic from upstream nodes to downstream nodes toward the destination. Wireless ad hoc, mesh, and sensor networks exploit multihop communications. Multihop communications can also achieve higher network reliability since traffic from one source can be relayed through multiple routes simultaneously. Some of the advantages of multihop networking are shown in Figure 1.5. In Figure 1.5(a), the range of wireless transmission is extended using relay node. The throughput and reliability of the transmission can be enhanced since the data can be transmitted over multiple routes (Figure 1.5(b)).

1.6.1 Wireless sensor networks

A wireless sensor network is composed of wireless sensors distributed in a geographical area. These sensors are designed to measure and aggregate target signal from the ambient environment. The sensors are equipped with wireless transceivers to transmit measured data and receive sensory data from other sensors. In such a scenario, sensory data from the points of measurements can be delivered to the sink node, which is connected to the external network, through multihop communications. Wireless sensor networks can be used for both military applications (e.g. military surveillance) and civil applications (e.g. environmental monitoring, inventory tracking [26], and healthcare service [27, 28, 29]).

One of the most important issues in sensor network design is energy conservation. Due to the requirements of small size and low cost, the sensor nodes are usually hardware-constrained and the capacity of energy storage (e.g. battery) and amount of energy supply are limited. For these reasons, protocols for wireless sensor networks must be carefully designed so that both the radio resources and the energy resources are efficiently managed. A typical sensor node is designed so that it can switch among different operating modes – active, standby, and sleep mode (e.g. from active to standby or sleep), each of which has different levels of energy consumption to conserve energy when the activity in the network is low. Different aspects of wireless sensor network design have been extensively investigated in the recent literature. These include sensor network modeling [30], security issues [31], data aggregation techniques [32], clustering algorithms [33], routing protocols [34], node placement strategies [35], clock synchronization [36], key management schemes [37], multimedia transmissions [38], and implementation [39] of sensor networks.

1.6.2 Wireless ad hoc networks

A wireless mobile ad hoc network (MANET) [40, 41] is an autonomous collection of mobile nodes which communicate with each other over relatively bandwidth constrained wireless links. Since the nodes are mobile, the network topology may change rapidly and unpredictably over time. Therefore, a source node may need to communicate to a destination node in a multihop fashion. Also, since the topology of the network changes frequently, the nodes need to self-organize to establish network connectivity to support various mobile applications. Ad hoc networks can support diverse military applications where moving battle units need to communicate with each other to exchange data. In addition, a special type of ad hoc network, which is referred to as a vehicular ad hoc network (VANET) [42, 43, 44, 45] can be used to support intelligent transportation system (ITS) applications. A VANET can be designed to exchange traffic information among vehicles to enhance safety and efficiency of transportation. VANETs can also be used for many military and public-safety applications [46].

Due to the autonomous and self-organizing nature of an ad hoc network, design of MAC and routing protocols poses significant research challenges. In the MAC layer, due to the broadcast nature of transmission, hidden and exposed terminal problems can arise, which may result in transmission collisions and performance degradation. In the routing layer, due to rapid and unpredictable topology changes, routing algorithms may need to be invoked frequently to find a suitable route from the source node to the destination node. Different aspects of design, analysis, and optimization of protocols for wireless mobile ad hoc networks have been studied, namely, MAC protocol design [47], QoS provisioning [48], clustering [49], QoS routing [50], mobile management [51], Transmission Control Protocol (TCP) [52], multicasting [53], and security [54]. A survey on the application of game theory in ad hoc network protocol design can be found in [55].

1.6.3 Wireless mesh networks

Similar to a wireless mobile ad hoc network, a wireless mesh network uses multihop communication to relay user traffic from a source node to a destination node or an Internet gateway. However, in a wireless mesh network, there are several stationary nodes, which are referred to as mesh routers, forming a mesh backbone. Each mesh

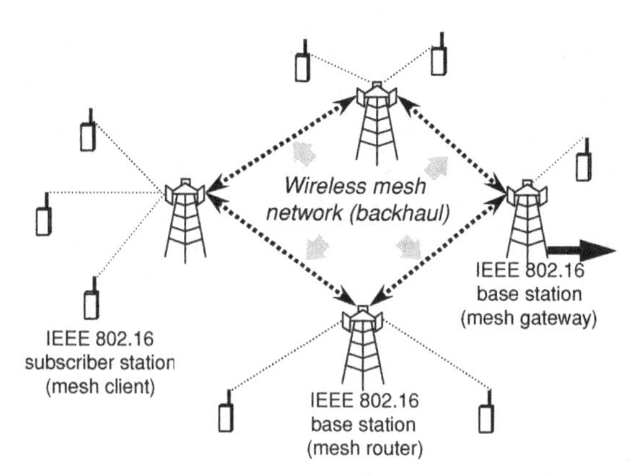

Figure 1.6 An infrastructure wireless mesh network based on IEEE 802.16 technology.

router can support multiple mesh clients, where the mesh clients can be either stationary or mobile. For example, in an IEEE 802.11-based wireless mesh network, the access points (APs) form a mesh backbone for the user nodes (i.e. mesh clients). Potential applications of wireless mesh networks include backhaul support for cellular wireless networks [56] or mobile hot spots, community networks, and different monitoring and surveillance applications. The problems of efficient MAC and routing protocol design and radio resource management for wireless mesh networks have been addressed in the literature [57, 58]. A survey on wireless mesh networks can be found in [59].

An example of an IEEE 802.16-based infrastructure wireless mesh network is shown in Figure 1.6 [60]. In this figure, IEEE 802.16 base stations (i.e. mesh routers) form a backhaul network to relay traffic from IEEE 802.16 subscriber stations (i.e. mesh clients) and from other base stations to the external network (e.g. Internet) through a mesh gateway. When used as a backhaul for a cellular network, the cellular base stations are connected to the IEEE 802.16 base stations.

1.7 Radio resource management in wireless networks

A radio resource management (RRM) framework is required for a wireless network to achieve the desired network objective (or wireless applications/services requirement) under the constraint on available radio resources (e.g. the radio spectrum in terms of frequency band or time slot, or transmission power). The major components of an RRM framework are shown in Figure 1.7. In the system model shown in Figure 1.7, users are divided into multiple classes depending on their QoS requirements (e.g. delay, throughput, loss). The major components of this RRM framework are described next.

1.7.1 Admission control

In a wireless network, admission control is required to limit the number of ongoing users in the system so that the limited available radio resource is not overwhelmed due

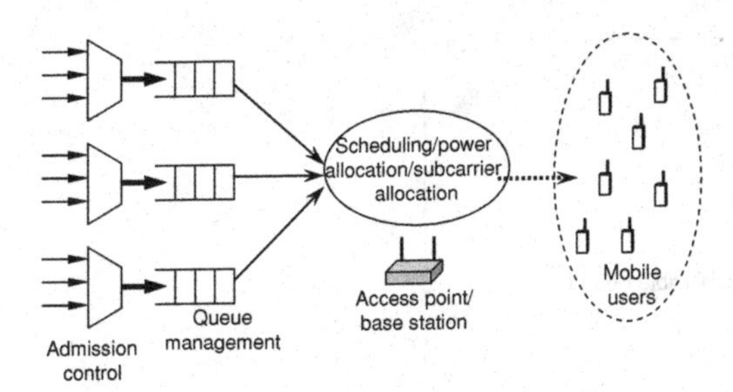

Figure 1.7 Radio resource management in wireless networks.

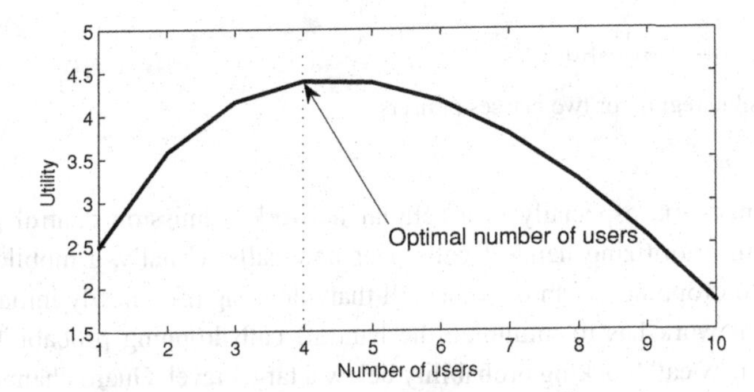

Figure 1.8 Illustration of optimal number of admitted users.

to the acceptance of too many users in the system. There are two major objectives of admission control: to ensure the QoS guarantee for both ongoing and incoming users, and to prioritize different users in different service classes. Again, acceptance of too few users may result in an under utilization of the radio resources. This tradeoff is illustrated in Figure 1.8 in terms of the variation in total network utility with the number of users in the network. With a small number of admitted users in the network, the utility first increases, since all admitted users receive satisfactory QoS performance. However, as the number of admitted users increases to the point where the available radio resources are not enough to accommodate all the user traffic, the QoS performance for the users starts degrading.

Through admission control, prioritization in QoS performances can be achieved among different classes of users. For example, video and voice traffic generally have tighter QoS requirements (e.g. in terms of delay) than data traffic. Due to limited available radio resources, admission control must limit the number of admitted users in each class so that the users' QoS performance can be maintained at the target level. For two different classes of user, the admissible region can be shown as in Figure 1.9. This admission region is defined as the combination of the maximum number of admittable users in each service class. From Figure 1.9, it is observed that the maximum number of admittable users of class 1 is smaller than that of class 2. Admission control must operate in this admissible region to maximize the utilization/revenue of the system. A survey of admission control schemes for wireless networks can be found in [61] and [62].

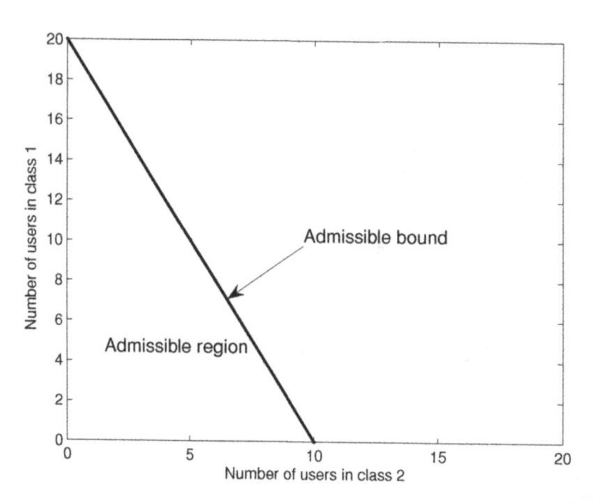

Figure 1.9 Admissible region for two classes of users.

In a mobile network, especially in a cellular network, admission control plays an important role in prioritizing handoff calls over new calls. Usually, a mobile user is more sensitive to dropping off an ongoing call than blocking off a newly initiated call. Therefore, the network has to minimize the handoff call dropping probability while maintaining the new call blocking probability below a target level. Guard channel-based admission control [63] is one popular scheme for cellular networks. In the guard channel scheme, both new call and handoff calls are admitted as long as the number of occupied channels is less than a predefined admission control threshold. However, if the number of occupied channels is equal to or higher than this threshold, only handoff calls are accepted and incoming new calls are blocked. The generalized version of guard channel scheme, namely, the fractional guard channel scheme, was proposed in [64], in which rather than completely blocking a new call, a new call is accepted with a certain probability. A number of other call admission control schemes for cellular mobile networks were also proposed in the literature [65, 66, 67, 68, 69].

1.7.2 Queue management

In a wireless transmitter, incoming packets are buffered in a queue before they are transmitted in the air. Queue management refers to the actions taken for the incoming packets. For example, if the queue becomes congested (i.e. the buffer is almost full), incoming packet(s) could be dropped. This packet dropping through queue management may signal the higher layer protocol (e.g. TCP) to reduce the transmission rate so that the congestion in the transmission queue can be avoided. This mechanism is referred to as active queue management [70, 71, 72]. Alternatively, upon detection of the possibility of congestion, a congestion signal can be sent to the higher layer protocol directly to reduce the transmission rate [73]. For example, in the case of video traffic, the video encoding rate can be reduced when the number of packets in the queue becomes larger than a predefined threshold.

1.7.3 Traffic scheduling

The traffic scheduler in a wireless transmitter is responsible for selecting packets from the transmission queue based on a scheduling policy to transmit in the air. The scheduling policy should consider the QoS requirements of user traffic as well as the wireless channel condition [74]. The traffic scheduling mechanism can be either work-conserving or non-work-conserving. In case of work-conserving scheduling, the transmission channel is never idle as long as the transmission queues are non-empty. On the other hand, in case of non-work-conserving scheduling, the transmission channel can be idle even though there are packets buffered in the transmission queues.

Some of the popular traffic scheduling policies are round-robin, weighted round-robin, generalized processor sharing (GPS), weighted-fair queuing, and virtual clock scheduling schemes [75]. These packet scheduling policies were developed primarily for wired networks where transmission error is negligible compared to that in a wireless network. Similar scheduling schemes can be designed for wireless networks considering the time-varying and location-dependent nature of the wireless channel errors. Due to the time and location-dependent nature of channel variation, fairness is an important issue in wireless packet scheduling [76, 77]. A survey of wireless packet scheduling can be found in [75].

1.7.4 Medium access control (MAC)

While traffic scheduling is used to choose among the queues (e.g. in a mobile terminal or in a base station/access point) or users for wireless transmission, a MAC protocol controls access of the users to the shared radio channel. Also, the MAC protocol must support service differentiation and QoS guarantee for different types of traffic/user in the network. A MAC protocol can be categorized as either a centralized protocol or a distributed protocol. In the case of centralized MAC, a central controller (e.g. base station) controls channel access of the users (e.g. which channel, which user, and when to transmit). In this case, all users in the network must be able to communicate their QoS requirements to the central controller.

In case of a distributed MAC protocol, all users can access the channel independently. Therefore, collisions can occur if more than one user transmits data on the same channel at the same time. Distributed MAC protocols can be designed based on carrier sensing and collision avoidance mechanisms [78]. Distributed MAC protocols based on the IEEE 802.11 standard use this principle [79]. In particular, before the user transmits data, it has to sense the channel. Since many users may sense the channel to be idle at the same time, each user must perform collision avoidance (e.g. through random backoff) before data transmission. In case of an unsuccessful transmission, which could be due to collision and/or channel error, a collision or error recovery must be performed by the MAC protocol, for example, through retransmission.

The problem of design and analysis of MAC protocols for different types of wireless networks has been extensively studied in the literature. A survey of MAC protocols for cellular, wireless ad hoc, and sensor networks can be found in [80], and [81], respectively.

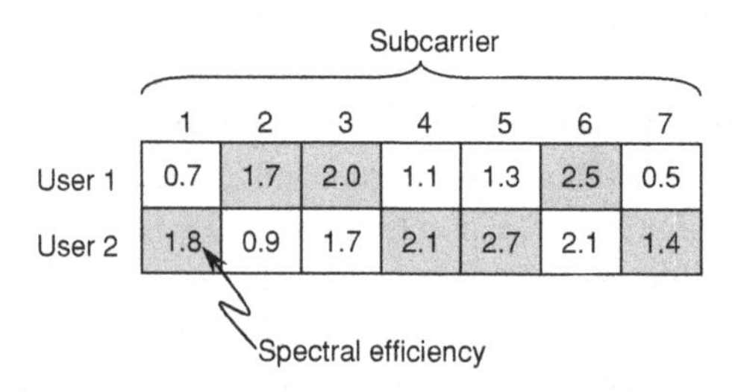

Figure 1.10 Subchannel allocation in OFDMA system.

MAC design issues for ultra-wideband wireless systems were addressed in [82]. MAC issues for multimedia traffic in wireless networks were surveyed in [83].

1.7.5 Subcarrier allocation

In OFDMA, a subcarrier or subchannel (i.e. a group of subcarriers) is allocated to a user for transmission. To achieve the best system performance (e.g. the highest system throughput), an optimization problem can be formulated to obtain an optimal assignment of subcarriers among the users. This optimization exploits multiuser diversity. In particular, in a multiuser OFDMA system, different subcarriers can have different channel quality for different users (e.g. due to the frequency-selective fading). An example of multiuser diversity and subcarrier allocation is shown in Figure 1.10. In this case, given the channel quality of each user on each subcarrier, the spectral efficiency (in bits/Hz) can be estimated. Due to the frequency-selective fading, this spectral efficiency is different for different users and different subcarriers. To achieve the highest system throughput, subcarrier 1 should be assigned for the transmission of user 2, since the spectral efficiency of this subcarrier 1 for user 2 is higher than that of user 1 (1.8 and 0.7). To obtain the optimal assignment by solving this assignment problem of subcarrier allocation, standard techniques (e.g. the Hungarian method [84]) can be used.

When the wireless channel experiences fast fading (e.g. due to high mobility), the channel quality and spectral efficiency (e.g. obtained from channel estimation) can vary in each transmission frame. In such a case, the subcarrier allocation algorithm can be invoked in every frame to optimize the system throughput. Two major issues in subcarrier allocation are QoS support and fairness [85, 86, 87]. While subcarrier allocation must ensure that the QoS requirements of all users can be satisfied, fairness in subcarrier assignment has to be also achieved although some users may experience deep channel fading (i.e. poor channel quality and subsequently low spectral efficiency).

1.7.6 Power control

In a CDMA system, power control is very important not only to achieve the highest transmission rate but also to limit the interference to other users. In a multiuser CDMA network, all users share the same transmission bandwidth. In an uplink transmission scenario, the SINR for user i at the base station receiver is given as follows:

$$\gamma_i = \frac{P_i h_i}{\sum_{j \neq i} h_j P_j + I_o + \sigma^2}, \tag{1.30}$$

where P_i is the transmit power of user i, h_i is the channel gain, I_o is the interference from users in other cells, and σ^2 is the noise power. The objective of power control is to maximize the system throughput (i.e. the sum of the throughput of all users) [88]. An optimization problem can be formulated to obtain optimal transmit powers for all users. However, this solution is viable only when the base station or the central controller can control the transmit power of all users and the users obey to do so. In contrast, if the users are independent and rational to maximize their individual throughputs, a non-cooperative game can be formulated to obtain an equilibrium solution for the transmit power allocation problem [89, 90]. This equilibrium solution ensures that none of the users will unilaterally deviate from the solution.

1.8 Next generation heterogeneous wireless access networks and cognitive radio

The evolving future generation wireless networks will have the following attributes:

- *High transmission rate*: New wireless applications and services, e.g. video and file transfer, require higher data rate to reduce the data transmission time and support a number of users. Many advanced techniques in the physical layer have been developed to increase the data rate without increasing spectrum bandwidth and transmit power requirement.
- *QoS support*: Various types of traffic, e.g. voice, video, and data, will be supported by the next generation wireless system. Service differentiation and QoS support are required to prioritize different types of traffic according to the performance requirement. Radio resource management framework has to be designed to efficiently access the available spectrum.
- *Cross-layer design*: In order to reduce the overhead in the protocol stack, the cross-layer design concept was proposed to create a link among different protocols in different layers [91, 92]. Through cross-layer design, the wireless system can be optimized throughout the protocol stack, so that the overall system performance in terms of data rate, error, and radio resource utilization is improved.
- *Integration of different wireless access technologies*: Next generation wireless networks will use the IP technology to glue the different wireless access technologies to a converged wireless system. In this converged network, multi-interface mobile units will be common. With multiple radio interfaces, a mobile should be able to connect to different wireless networks using different access technologies simultaneously. For example, a mobile can connect to a WLAN through the IEEE 802.11-based radio interface. However, when this mobile moves out of range of the WLAN, it can connect

to a cellular network (e.g. using a 3G air interface) or a WiMAX network to resume the communication session. Such a heterogeneous wireless access network provides two major advantages: it enhances the data transmission rate since multiple data streams can be transmitted concurrently, and it enables seamless mobility through providing wireless connectivity anytime and anywhere.

Many different research issues arise in such a heterogeneous wireless access environment. These include network selection [93], bandwidth allocation [94], admission control [95], QoS support [96, 97], vertical handoff [98, 99, 100], and routing [101]. When multiple wireless networks are available, the mobile unit must choose a set of networks to connect in order to maximize its utility. From the network service providers' point of view, a method is required to efficiently allocate the available bandwidth to users in different service areas so that radio resource utilization, and hence the revenue, is maximized. Also, admission control and mobility management are essential to ensure QoS and seamless connectivity.

- *Software-defined radio and cognitive radio*: Frequency spectrum is the scarcest radio resource and its efficient usage will be crucial in next generation wireless communications systems. With the traditional approach of allocating frequency spectrum to wireless transceivers, it has been observed that some frequency bands are heavily used in some locations, while others are lightly utilized elsewhere. Spectrum opportunities are thus generated, which can be exploited by intelligent and adaptive cognitive wireless transceivers implemented through software-defined radio. With software-defined radio, a wireless transceiver can change its transmission parameters such as transmit power and operating frequency band. Dynamic spectrum access techniques, which will be the key for implementing cognitive radio, need to be developed for different wireless systems with different capabilities and different operating environments. This adaptability of wireless transmission in a cognitive radio can be achieved by employing intelligent algorithms to observe, learn, optimize, and make decisions [102]. In essence, the mobile units or wireless nodes will need to use the "cognitive" capabilities in all layers of the protocol stack. Cognitive radio is envisioned to be a significant component in next generation wireless systems.

  Designing dynamic spectrum access methods for cognitive radio requires knowledge from multiple scientific and engineering disciplines to achieve the desired design objectives. These disciplines include traditional wireless communications and networking, optimization and game theory, machine learning, and economics.
- *Integration of cognitive radio concepts in traditional wireless systems*: Cognitive radio and dynamic spectrum access techniques can be integrated into traditional wireless communications systems to achieve better flexibility of radio resource usage so that the system performance can be improved. For example, load balancing/dynamic channel selection in traditional cellular wireless systems and WLANs, distributed subcarrier allocation in OFDM systems, and transmit power control in UWB systems can be achieved by using dynamic spectrum access-based cognitive radio techniques.

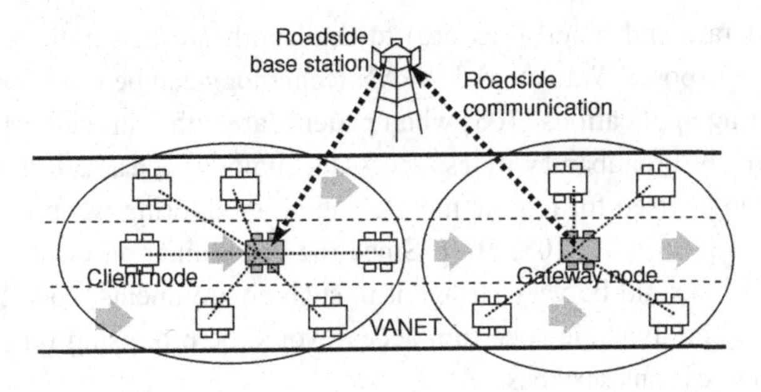

Figure 1.11 Wireless communications in intelligent transportation systems.

- *Emergence of cognitive radio-based wireless applications and services*: Emerging wireless services and applications, a few of which are described below, can take advantage of cognitive radio:
  1. *Future generation wireless Internet services*: Next generation wireless Internet is expected to provide seamless QoS guarantee to mobile users for a variety of multimedia applications. Cognitive radio technology based on dynamic spectrum access will facilitate provisioning of these future generation wireless Internet services.
  2. *Wireless intelligent transportation systems (ITS)*: A wireless intelligent transportation system refers to an integrated wireless communication and software system to facilitate information exchange and processing to improve safety and efficiency of transportation by vehicles. Wireless communication can be used to provide traffic information to the vehicles to avoid road congestion. An example of application of wireless technologies in ITS is shown in Figure 1.11. In this system, a vehicle can communicate with the roadside base station to download/upload traffic-related information which would be useful for route selection. Also, the vehicles can communicate among each other to exchange information locally by forming a VANET. While IEEE 802.11-based radio can be used for vehicle-to-vehicle communications [103], WiMAX radio can be used for vehicle-to-roadside communications [104]. High mobility of the vehicular nodes and application requirements (e.g. stringent delay requirements for safety and traffic warning applications) make wireless communication and mobility management very challenging in an ITS environment. Since mobility is a key feature in such a communication environment, fast and efficient dynamic spectrum access would improve system performance.
  3. *Wireless ehealth services*: These services are designed to provide data networking among patient, physician, and medical staff in different hospital environments. Wireless communications and networking technology is being applied to improve the efficiency and usability of healthcare services. Wireless technology can provide the mobility, availability, and capability advantages to different healthcare services (also referred to as wireless telemedicine services). In a remote patient monitoring system, biosignal sensors attached to patients can transmit monitored

data (e.g. heart rate and blood pressure) to the healthcare center for diagnostic and monitoring purposes. WLAN and WPAN technology can be used for wireless patient monitoring applications [105] when patients are either in the hospital or at home. Alternatively, broadband wireless access technology such as cellular network and WiMAX can be used for remote mobile patient monitoring when patients are in an ambulance [106, 107, 108, 109]. Since the constraints on electromagnetic interference (EMI) could be very stringent in such environments, cognitive radio technology based on dynamic spectrum access would be promising for providing wireless communications services.

4. *Public safety services*: Communications services for public safety can take advantage of the cognitive radio technology based on dynamic spectrum access to achieve the desired service objectives (e.g. prioritizing emergency calls over other commercial service calls).

Research challenges in developing these cognitive radio-based applications will be discussed in the next chapter.

1.9 Summary

In this chapter, an overview of the different wireless technologies has been presented. The principle of current spectrum management based on command-and-control has been discussed, and its limitations have been pointed out. New spectrum management schemes are expected to be standardized, which would allow more flexible and more efficient spectrum usage. Such a spectrum management scheme would enable dynamic or adaptive spectrum access. To be able to adaptively access the radio spectrum and optimize system performance, a wireless transceiver should have the ability to observe, learn, plan, and schedule the spectrum usage. These capabilities are the basic requirements for a cognitive radio. Dynamic spectrum access-based cognitive radio is emerging as a new paradigm for designing wireless communications and networking systems.

2 Introduction to cognitive radio

Cognitive radio is a new paradigm of designing wireless communications systems which aims to enhance the utilization of the radio frequency (RF) spectrum.[1] The motivation behind cognitive radio is the scarcity of the available frequency spectrum, increasing demand, caused by the emerging wireless applications for mobile users. Most of the available radio spectrum has already been allocated to existing wireless systems, however, and only small parts of it can be licensed to new wireless applications. Nonetheless, a study by the Spectrum Policy Task Force (SPTF) of the Federal Communications Commission (FCC) has showed that some frequency bands are heavily used by licensed systems in particular locations and at particular times, but that there are also many frequency bands which are only partly occupied or largely unoccupied [110]. For example, spectrum bands allocated to cellular networks in the USA [111] reach the highest utilization during working hours, but remain largely unoccupied from midnight until early morning.

The major factor that leads to inefficient use of the radio spectrum is the spectrum licensing scheme itself. In traditional spectrum allocation based on the command-and-control model, where the radio spectrum allocated to licensed users is not used, it cannot be utilized by unlicensed users and applications [5]. Due to this static and inflexible allocation, legacy wireless systems have to operate only on a dedicated spectrum band, and cannot adapt the transmission band according to the changing environment. For example, if one spectrum band is heavily used, the wireless system cannot change to operate on another more lightly used band.

The right to access the spectrum (or license) is generally defined by frequency, space, transmit power, spectrum owner (i.e. licensee), type of use, and the duration of license. Normally, a license is assigned to one licensee, and the use of spectrum by this licensee must conform to the specification in the license (e.g. maximum transmit power, location of base station). In the current spectrum licensing scheme, the license cannot change the type of use or transfer the right to other licensee. This limits the use of the frequency spectrum and results in low utilization of the frequency spectrum. Essentially, due to the current static spectrum licensing scheme, *spectrum holes* or spectrum opportunities (Figure 2.1) arise. Spectrum holes are defined as frequency bands which are allocated

[1] Note that the terms radio frequency (RF) spectrum and radio spectrum are used interchangeably. We will generally use radio spectrum in this book.

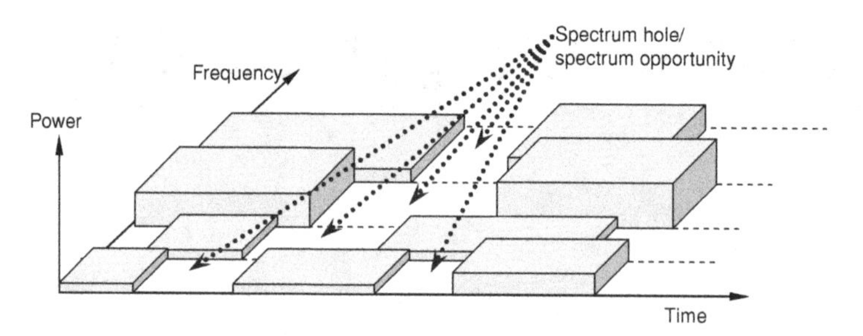

Figure 2.1 Spectrum hole (or spectrum opportunity).

to, but in some locations and at some times not utilized by, licensed users, and, therefore, could be accessed by unlicensed users [102].

The limitations in spectrum access due to the static spectrum licensing scheme can be summarized as follows:

- *Fixed type of spectrum usage:* In the current spectrum licensing scheme, the type of spectrum use cannot be changed. For example, a TV band which is allocated to National Television System Committee (NTSC)-based analog TV cannot be used by digital TV broadcast or broadband wireless access technologies. However, this TV band could remain largely unused in many locations due to cable TV systems.
- *Licensed for a large region:* When a spectrum is licensed, it is usually allocated to a particular user or wireless service provider in a large region (e.g. an entire city or state). However, the wireless service provider may use the spectrum only in areas with a good number of subscribers, to gain the highest return on investment. Consequently, the allocated frequency spectrum remains unused in other areas, and other users or service providers are prohibited from accessing this spectrum.
- *Large chunk of licensed spectrum:* A wireless service provider is generally licensed with a large chunk of radio spectrum (e.g. 50 MHz). For a service provider, it may not be possible to obtain license for a small spectrum band to use in a certain area for a short period of time to meet a temporary peak traffic load. For example, a cdma2000 cellular service provider may require a spectrum with bandwidth of 1.25 MHz or 3.75 MHz to provide temporary wireless access service in a hotspot area.
- *Prohibit spectrum access by unlicensed users:* In the current spectrum licensing scheme, only a licensed user can access the corresponding radio spectrum and unlicensed users are prohibited from accessing the spectrum even though it is unoccupied by the licensed users. For example, in a cellular system, there could be areas in a cell without any users. In such a case, unlicensed users with short-range wireless communications would not be able to access the spectrum, even though their transmission would not interfere with cellular users.

In order to improve the efficiency and utilization of the available spectrum, these limitations are being remedied by modifying the spectrum licensing scheme. The idea is to make spectrum access more flexible by allowing unlicensed users to access the radio spectrum under certain restrictions. Since the legacy wireless systems were designed to

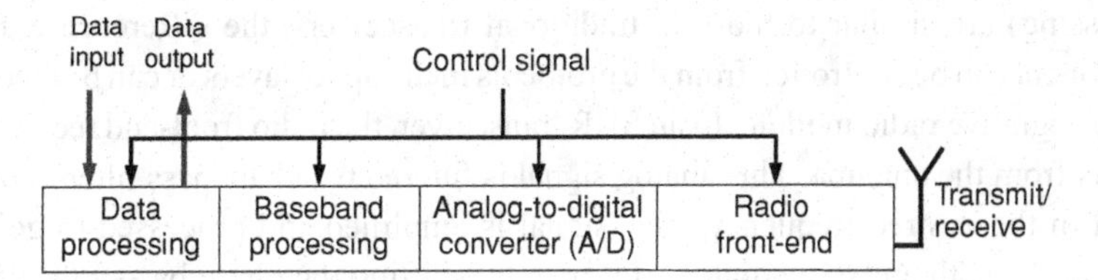

Figure 2.2 SDR transceiver.

operate on a dedicated frequency band, they are not able to utilize the improved flexibility provided by this spectrum licensing scheme. Therefore, the concept of *cognitive radio* has been introduced. The main goal of cognitive radio is to provide adaptability to wireless transmission through dynamic spectrum access so that the performance of wireless transmission can be optimized, as well as enhancing the utilization of the frequency spectrum. The major functionalities of a cognitive radio system include spectrum sensing, spectrum management, and spectrum mobility. Through spectrum sensing, the information of the target radio spectrum (e.g. the type and current activity of the licensed user) has to be obtained so that it can be utilized by the cognitive radio user. The spectrum sensing information is exploited by the spectrum management function to analyze the spectrum opportunities and make decisions on spectrum access. If the status of the target spectrum changes, the spectrum mobility function will control the change of operational frequency bands for the cognitive radio users.

2.1 Software-defined radio

A software-defined radio (SDR) is a reconfigurable wireless communication system in which the transmission parameters (e.g. operating frequency band, modulation mode, and protocol) can be controlled dynamically. This adjustability function is achieved by software-controlled signal-processing algorithms. SDR is a key component to implementing cognitive radios. The main functions of SDR are as follows [112]:

- *Multiband operation:* SDR will support wireless data transmissions over a different frequency spectrum used by different wireless access systems (e.g. cellular band, ISM band, TV band).
- *Multistandard support:* SDR will support different standards (e.g. GSM, WCDMA, cdma2000, WiMAX, WiFi). Also, different air interfaces within the same standard (e.g. IEEE 802.11a, 802.11b, 802.11g, or 802.11n in WiFi standard) can be supported by SDR.
- *Multiservice support:* SDR will support multiple types of services, e.g. cellular telephony or broadband wireless Internet access.
- *Multichannel support:* SDR will be able to operate (i.e. transmit and receive) on multiple frequency bands simultaneously.

The general structure of an SDR transceiver is shown in Figure 2.2. While most of the components in SDR (i.e. data processing, analog-to-digital converter, and baseband processing) are similar to those in traditional transceivers, the difference is that each component can be controlled from the protocols in the upper layers or can be reconfigured by the cognitive radio module. In an SDR transceiver, the radio front-end receives analog signals from the antenna. This analog signal is filtered by a bandpass filter to obtain the signal in the desired frequency. This signal is amplified and processed to generate an in-phase (I) path and a quadrature (Q) path by shifting the phase by $-\pi/2$. Both I and Q path signals are then converted to digital data. The sampling rate of A/D has to be chosen to satisfy the conditions of Nyquist's sampling theorem. However, the sampling rate should be minimized to reduce the signal processing overhead. The sampling rate and the parameters of the analog and digital filters as well as the signal-processing algorithms can be reconfigured according to the operating frequency band and the wireless air interface technology.

With SDR, the transmission parameters in a wireless transceiver can be reconfigured according to the communication specifications and requirements [112]:

- The radio transceiver parameters (e.g. the operational standard and frequency band) can be set before the system is delivered to the customer. However, the parameters cannot be modified after the system is configured. Although dynamic reconfiguration of the system is not supported in this case, one SDR transceiver model can be sold to many customers with different requirements.
- The radio transceiver parameters can be occasionally reconfigured (e.g. a few times during the system's life time), for example when the network structure changes or when a new base station is added.
- The radio transceiver parameters can be changed on a connection basis. For example, when a user wants to initiate a wireless Internet connection, the transceiver can choose from the different wireless access networks available (e.g. GSM, WiFi or WiMAX), based on network availability, performance, and price.
- The radio transceiver parameters can be dynamically changed on a time-slot basis. For example, the transmission power can be altered when the level of interference changes. The unlicensed user(s) can change the operating frequency band when the activity of the licensed user(s) is detected.

Next generation mobile phones will be designed to support multiple wireless access technologies so that a mobile user can have the flexibility to switch between different networks. In [113], a design approach was discussed for an SDR transceiver for multistandard mobile phones. The space and power consumption as well as the scalability of the equipment were identified as the major design constraints. An SDR platform, namely, Kansas University Agile Radio (KUAR), was developed in [114]. The platform consists of a power supply, control processor, and a digital board of programmable signal processor, analog-to-digital and digital-to-analog converters, RF transceiver, and antennas. The radio front-end supports an operating frequency of 5–6 GHz. The digital board is an embedded PC running on Linux. A field-programmable gate array (FPGA) is used in this digital board to provide flexibility of implementing the signal processing algorithm. For the software part, KUAR has radio control and management programs, namely *Boot*, *Policy*, *Ops*, and *QoS*. The *Boot* program is used to load related radio modules used by other components. The *Ops* and *QoS* programs are used to measure network parameters, e.g. the traffic load in the protocol stacks and the RF environment of the network. The *Ops* program can also change the frequency band and the modulation mode (e.g. QPSK, QAM-16, and QAM-64). The *Policy* program is used to control the transmission parameters within the regulatory rules. This KUAR platform was used to implement a WiMAX 802.16a experimental transmitter and receiver.

Recently, SDR platforms to support a broader operational spectrum have been proposed. For example, in [115] the receiver was designed to act as a signal conditioner for A/D converters so that a particular spectrum range can be sampled efficiently. This SDR platform supports a frequency spectrum of 800 MHz to 5 GHz. Other implementations and experimental platforms of software-defined radio were developed in [116, 117, 118, 119, 120, 121, 122].

2.2 Cognitive radio features and capabilities

Cognitive radio, which is implemented based on the software-defined radio, provides mechanisms for intelligent spectrum sensing, spectrum management, and spectrum access for cognitive radio users (e.g. unlicensed users). The term "cognitive radio" was defined in [102] as follows: "Cognitive radio is an intelligent wireless communication system that is aware of its ambient environment. This cognitive radio will learn from the environment and adapt its internal states to statistical variations in the existing RF stimuli by adjusting the transmission parameters (e.g. frequency band, modulation mode, and transmit power) in real-time and [in an] on-line manner." A cognitive radio network enables us to establish communications among cognitive radio nodes/users. The communication parameters can be adjusted according to change in the environment, topology, operating conditions, or user requirements [1]. The two main objectives of cognitive radio are: (1) to achieve highly reliable and highly efficient wireless communications, and (2) to improve the utilization of the frequency spectrum.

2.2.1 Cognitive radio architecture

The protocol architecture of cognitive radio is shown in Figure 2.3. In the physical layer, the RF front-end is implemented based on software-defined radio (i.e. the SDR transceiver). The adaptive protocols in the MAC, network, transport, and application layers should be aware of the variations in the cognitive radio environment. In particular, the adaptive protocols should consider the traffic activity of primary users, the transmission requirements of secondary users, and variations in channel quality, etc. To link all modules, a cognitive radio control is used to establish interfaces among the SDR transceiver, adaptive protocols, and wireless applications and services. This cognitive radio module uses intelligent algorithms to process the measured signal from the physical layer, and receive information on transmission requirements from the applications to control the protocol parameters in the different layers.

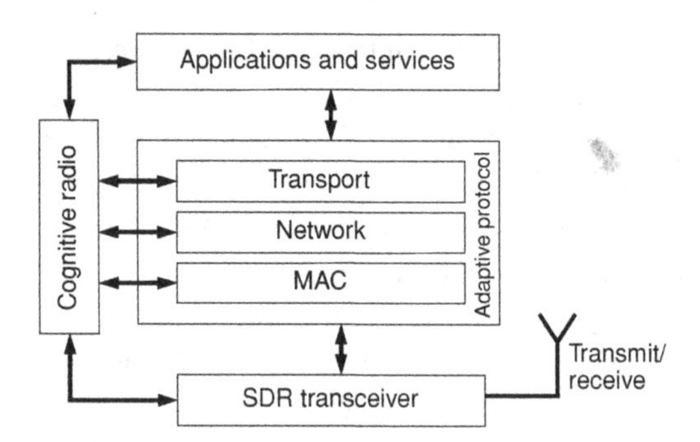

Figure 2.3 Cognitive radio protocol stack.

2.2.2 Functions of cognitive radio

The main functions of cognitive radio to support intelligent and efficient dynamic spectrum access are as follows:

- *Spectrum sensing:* The goal of spectrum sensing is to determine the status of the spectrum and the activity of the licensed users by periodically sensing the target frequency band. In particular, a cognitive radio transceiver detects an unused spectrum or spectrum hole (i.e. band, location, and time) and also determines the method of accessing it (i.e. transmit power and access duration) without interfering with the transmission of a licensed user.

  Spectrum sensing can be either centralized or distributed. In centralized spectrum sensing, a sensing controller (e.g. access point or base station) senses the target frequency band, and the information thus obtained is shared with other nodes in the system. Centralized spectrum sensing can reduce the complexity of user terminals, since all the sensing functions are performed at the sensing controller. However, centralized spectrum sensing suffers from location diversity. For example, the sensing controller may not be able to detect an unlicensed user at the edge of the cell. In distributed spectrum sharing, unlicensed users perform spectrum sensing independently, and the spectrum sensing results can be either used by individual cognitive radios (i.e. non-cooperative sensing) or shared with other users (i.e. cooperative sensing). Although cooperative sensing incurs a communication and processing overhead, the accuracy of spectrum sensing is higher than that of non-cooperative sensing.
- *Spectrum analysis:* The information obtained from spectrum sensing is used to schedule and plan spectrum access by the unlicensed users. In this case, the communication requirements of unlicensed users are also used to optimize the transmission parameters. Major components of spectrum management are spectrum analysis and spectrum access optimization. In spectrum analysis, information from spectrum sensing is analyzed to gain knowledge about the spectrum holes (e.g. interference estimation, duration of availability, and probability of collision with a licensed user due to sensing error). Then, a decision to access the spectrum (e.g. frequency, bandwidth, modulation mode, transmit power, location, and time duration) is made by optimizing the system

performance given the desired objective (e.g. maximize the throughput of the unlicensed users) and constraints (e.g. maintain the interference caused to licensed users below the target threshold).

- *Spectrum access:* After a decision is made on spectrum access based on spectrum analysis, the spectrum holes are accessed by the unlicensed users. Spectrum access is performed based on a cognitive medium access control (MAC) protocol, which intends to avoid collision with licensed users and also with other unlicensed users. The cognitive radio transmitter is also required to perform negotiation with the cognitive radio receiver to synchronize the transmission so that the transmitted data can be received successfully. A cognitive MAC protocol could be based on a fixed allocation MAC (e.g. FDMA, TDMA, CDMA) or a random access MAC (e.g. ALOHA, CSMA/CA) [5].
- *Spectrum mobility:* Spectrum mobility is a function related to the change of operating frequency band of cognitive radio users. When a licensed user starts accessing a radio channel which is currently being used by an unlicensed user, the unlicensed user can change to a spectrum band which is idle. This change in operating frequency band is referred to as spectrum handoff. During spectrum handoff, the protocol parameters at the different layers in the protocol stacks have to be adjusted to match the new operating frequency band. Spectrum handoff must try to ensure that the data transmission by the unlicensed user can continue in the new spectrum band.

2.2.3 Dynamic spectrum access

Implementation of cognitive radio will be based on dynamic spectrum access by the unlicensed users. Dynamic spectrum access can be defined [1] as a mechanism to adjust the spectrum resource usage in a near-real-time manner in response to the changing environment and objective (e.g. available channel and type of applications), changes of radio state (e.g. transmission mode, battery status, and location), and changes in environment and external constraints (e.g. radio propagation, operational policy).

There are three major models of dynamic spectrum access [123], namely, commons-use, shared-use, and exclusive-use models. In the commons-use model, the spectrum is open for access to all users. This model is already in use in the ISM band [124]. In the shared-use model, licensed users (i.e. primary users) are allocated the frequency bands which are opportunistically accessed by the unlicensed users (i.e. secondary users) when they are not occupied by the primary users. In the exclusive-use model, a licensed user can grant access of a particular frequency band to an unlicensed user for a certain period of time [125]. This model is more flexible than the traditional command-and-control spectrum licensing model, since the type of use and the licensee of the spectrum can be dynamically changed.

With opportunistic spectrum access, a secondary user can exploit unused in-band segments without causing interference to the active primary users. There are two approaches for opportunistic spectrum access: *spectrum underlay* and *spectrum overlay*. The spectrum underlay approach constrains the transmission power of secondary users so that they

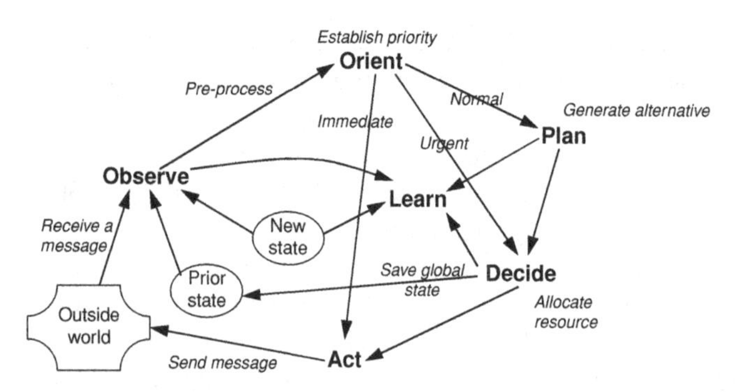

Figure 2.4 Cognitive cycle.

operate below the interference temperature limit of primary users. One possible approach is to transmit the signals in a very wide frequency band (e.g. UWB transmission) so that a high data rate is achieved with extremely low transmission power. It is based on the worst-case assumption that the primary users transmit all the time. Therefore, it does not exploit spectrum white space (i.e. spectrum holes). The spectrum overlay approach (or opportunistic spectrum access) does not necessarily impose any severe restriction on the transmission power by secondary users. It allows secondary users to identify and exploit the spectrum holes defined in space, time, and frequency. This approach is compatible with the existing spectrum allocation, and, therefore, the legacy systems can continue to operate without being affected by the cognitive radio users. However, the basic etiquettes for secondary users need to be defined by the regulatory bodies to ensure compatibility with legacy systems.

In the commons-use model, dynamic sharing can be between homogeneous networks (e.g. IEEE 802.11a operating in the 5 GHz UNII band) or between heterogeneous networks (e.g. coexistence between IEEE 802.11b and 802.15.1 (Bluetooth) networks [126, 127]). When all the networks in a heterogeneous environment have cognitive/adaptive capabilities (i.e. all coexisting networks have equal incentives to adapt), it is referred to as symmetric sharing. On the other hand, when there is one or more network without cognitive/adaptive capabilities (e.g. coexistence of legacy technology with cognitive radio technology, coexistence of powerful 802.11 networks with low-power 802.15.4 networks [128, 129, 130]), this is referred to as asymmetric spectrum sharing.

Dynamic spectrum access is divided into two major phases, namely, spectrum exploration (sensing and analysis) and spectrum exploitation (decide and handoff). Different design techniques can be used in these phases, some examples of which will be given in the following chapters.

2.2.4 Components of cognitive radio

The major functions of cognitive radio, which are required to adapt the transmission parameters according to the changing environment, can be represented through a "cognitive cycle" (Figure 2.4) [131]. The different components in a cognitive radio transceiver which implement these functionalities are shown in Figure 2.5.

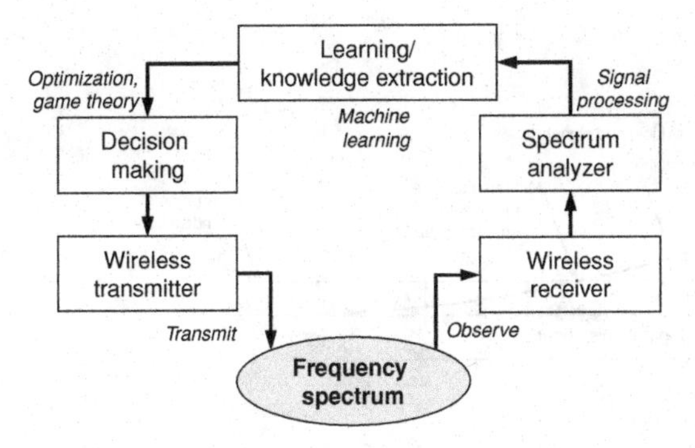

Figure 2.5 Components in a cognitive radio node.

- *Transmitter/receiver:* A software-defined radio-based wireless transceiver is the major component with the functions of data signal transmission and reception. In addition, a wireless receiver is also used to observe the activity on the frequency spectrum (i.e. spectrum sensing). The transceiver parameters in the cognitive radio node can be dynamically changed as dictated by higher layer protocols.
- *Spectrum analyzer:* The spectrum analyzer uses measured signals to analyze the spectrum usage (e.g. to detect the signature of a signal from a licensed user and to find spectrum holes for unlicensed users to access). The spectrum analyzer must ensure that the transmission of a licensed user is not interfered with if an unlicensed user decides to access the spectrum. In this case, various signal-processing techniques can be used to obtain spectrum usage information.
- *Knowledge extraction/learning:* Learning and knowledge extraction use the information on spectrum usage to understand the ambient RF environment (e.g. the behavior of licensed users). A knowledge base of the spectrum access environment is built and maintained, which is subsequently used to optimize and adapt the transmission parameters to achieve the desired objective under various constraints. Machine learning algorithms from the field of artificial intelligence can be applied for learning and knowledge extraction.
- *Decision making:* After the knowledge of the spectrum usage is available, the decision on accessing the spectrum has to be made. The optimal decision depends on the ambient environment – that is, it depends on the cooperative or competitive behavior of the unlicensed users. Different techniques can be used to obtain an optimal solution. For example, optimization theory can be applied when the system can be modeled as a single entity with a single objective. In contrast, game theory models can be used when the system is composed of multiple entities each with its own objective. Stochastic optimization may be applied when the states of the system are random.

2.2.5 Interference temperature

The FCC Spectrum Policy Task Force has recommended a new method in measuring interference [4]. The FCC has also introduced a new interference measure, namely,

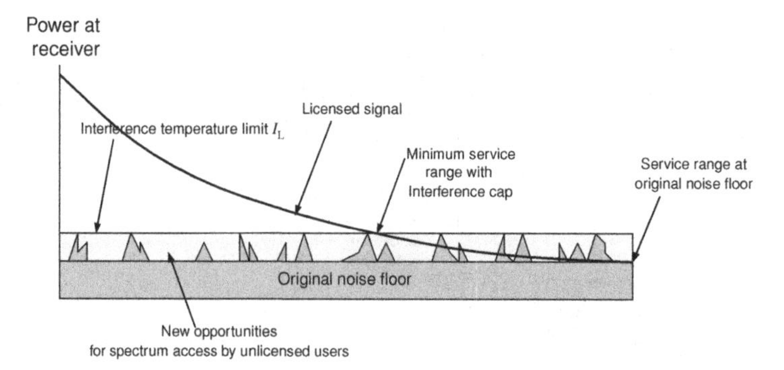

Figure 2.6 Interference temperature model.

interference temperature I_T, to quantify and manage the sources of interference (e.g. transmissions from unlicensed users) (Figure 2.6). Based on this measure, the interference temperature limit I_L is used as a bound for the interference caused to the licensed users in a particular frequency band at a particular location. The concept of interference temperature is similar to that of noise temperature, but is extended to include both (random) noise and (deterministic) interference from other sources of signals. The interference temperature is a metric measured at the receiver which reflects the quality of the signal reception, and can provide accurate and effective information on the interference in the target frequency band so that an unlicensed user can evaluate the feasibility of accessing that frequency band.

The interference temperature, which has the unit of degrees Kelvin, can be expressed as follows:

$$I_T(f_c, W) = \frac{P_I(f_c, W)}{KW}, \tag{2.1}$$

where $P_I(f_c, W)$ is the average interference power in Watts for a bandwidth W Hz centered at frequency f_c and K is Boltzmann's constant (i.e. $K = 1.38 \times 10^{-23}$) Joules per Kelvin degree. Two models of interference temperature, i.e. ideal and generalized, were presented in [132].

Ideal interference temperature model

Let the transmit power of an unlicensed user over a particular frequency band be denoted by P. This frequency band contains signals from n licensed users, where the signal from user i has bandwidth W_i centered at frequency f_i. Then, the transmission of an unlicensed user must ensure the following interference temperature limit for the licensed receiver [133]:

$$I_T(f_i, W_i) + \frac{M_i P}{KW_i} \leq I_L(f_i) \tag{2.2}$$

for $i \in \{1, \ldots, n\}$, where $0 \leq M_i \leq 1$ is a multiplicative attenuation factor due to path loss and fading in the link between the unlicensed transmitter and the licensed receiver. An example with $n = 3$ is shown in Figure 2.7.

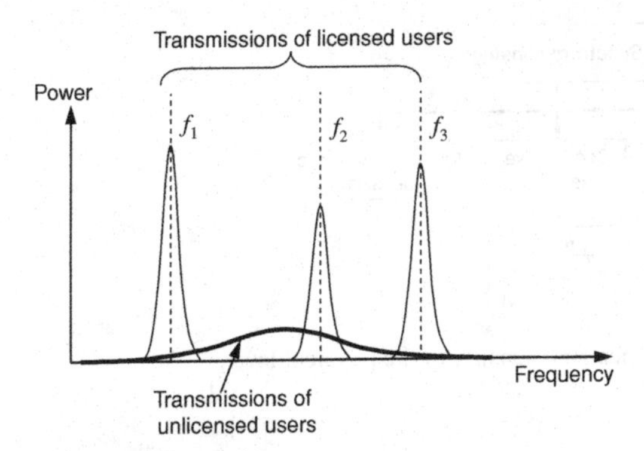

Figure 2.7 Interference temperature for licensed users.

In this ideal interference temperature model, the signal from an unlicensed user has to be distinguished from that of a licensed user. Also, measurement of interference must be based on the presence of a licensed signal. For this, the signal waveform from licensed users has to be known in order to measure the interference at the correct point in time. In particular, the interference must be measured when a licensed user is transmitting. If the interference is measured when there is no signal from any licensed users, the measured interference temperature will be incorrect.

Generalized interference temperature model

In the generalized interference temperature model, information about the signal from a licensed user is not available and hence the signals from licensed and unlicensed users cannot be differentiated. Therefore, the generalized interference temperature model defines interference for the entire frequency range, and the constraint for the interference temperature limit can be defined as follows [133]:

$$I_{\mathrm{T}}(f_{\mathrm{c}}, W) + \frac{MP}{KW} \leq I_{\mathrm{L}}(f_{\mathrm{c}}). \tag{2.3}$$

In this case, the constraint is expressed in terms of f_{c} and W, where f_{c} is the center frequency and W is the bandwidth of the radio spectrum used by the unlicensed user.

Capacity under interference temperature

The interference temperature limit dictates the upper bound for the transmit power of an unlicensed user in a particular frequency band. Given this upper bound of transmit power, the spectrum capacity r (i.e. rate) of an unlicensed user can be estimated from [134]

$$r = W \log\left(1 + \frac{S}{\sigma^2 + P_{\mathrm{I}}}\right), \tag{2.4}$$

where W is the bandwidth of the frequency band, S is the received signal power of the unlicensed user, σ^2 is the noise power, and P_{I} is the interference power at the receiver of the unlicensed user due to the transmission of the licensed user.

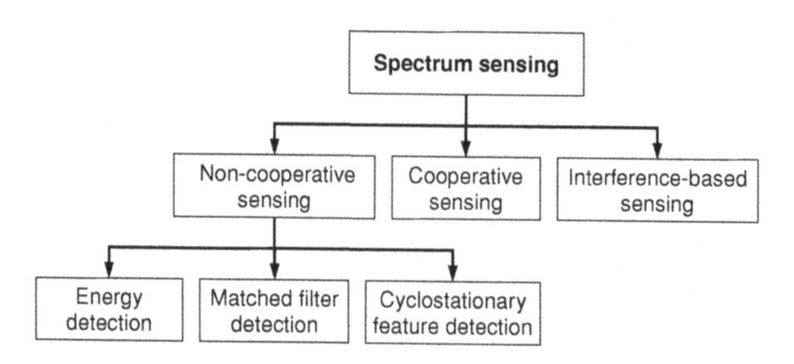

Figure 2.8 Different types of spectrum sensing in the physical layer.

2.2.6 Spectrum sensing

The objective of spectrum sensing is to detect the presence of transmissions from licensed users. There are three major types of spectrum sensing, namely, non-cooperative sensing, cooperative sensing, and interference-based sensing (Figure 2.8). These will be described below.

Non-cooperative transmitter sensing

Non-cooperative spectrum sensing is used by an unlicensed user to detect the transmitted signal from a licensed user by using local measurements and local observations. The model for signal detection at time t can be described as [135]

$$x(t) = \begin{cases} n(t), & H_0, \\ h \times s(t) + n(t), & H_1, \end{cases} \tag{2.5}$$

where $x(t)$ is the received signal of an unlicensed user, $s(t)$ is the transmitted signal of the licensed user, $n(t)$ is the additive white Gaussian noise (AWGN), and h is the channel gain. Here, H_0 and H_1 are defined as the hypotheses of not having and having a signal from a licensed user in the target frequency band, respectively.

The performance of a spectrum sensing technique is generally measured in terms of the probability of correct detection (P_d), the probability of false alarm (P_f), and the probability of miss (P_m). Mathematically, $P_d = \text{Prob}\{\text{decision} = H_1|H_1\}$, $P_f = \text{Prob}$ $\{\text{decision} = H_1|H_0\}$, and $P_m = \text{Prob}\{\text{decision} = H_0|H_1\}$.

The three different methods in non-cooperative sensing are as follows:

- *Matched filter detection or coherent detection:* Matched filter detection [14] is generally used to detect a signal by comparing a known signal (i.e. a template) with the input signal. A matched filter will maximize the received SNR for the measured signal. Therefore, if the information of the signal from a licensed user is known (e.g. modulation and packet format), a matched filter is an optimal detector in stationary Gaussian noise [136]. Since a template is used for signal detection, a matched filter requires only a small amount of time to operate. However, if this template is not available or is incorrect, the performance of spectrum sensing degrades significantly. Matched filter detection is suitable when the transmission of a licensed user has pilot, preambles, synchronization word or spreading codes, which can be used to construct the template for spectrum sensing.

- *Transmitter energy detection:* Energy detection is the optimal method for spectrum sensing when the information from a licensed user is unavailable [136]. In the case of energy detection, the output signal from a bandpass filter is squared and integrated over the observation interval. A decision algorithm compares the integrator output with a threshold [137] to decide whether a licensed user exists or not. In general, the energy detection performance deteriorates (e.g. P_{m} increases) when the SNR decreases.

  An energy detection algorithm was proposed in [138] for a non-fading environment and the expressions for probability of detection P_{d} and probability of false alarm P_{f} were obtained as follows: $P_{\mathrm{d}} = Q\left(\sqrt{2\gamma}, \sqrt{\lambda}\right)$ and $P_{\mathrm{f}} = \Gamma(m, \lambda/2)/\Gamma(m)$, where γ is the SNR of the received signal, λ is the energy detection threshold, $\Gamma(\cdot)$ and $\Gamma(\cdot, \cdot)$ are the complete and incomplete gamma functions, respectively, and $Q(\cdot)$ is the generalized Marcum Q-function (given by (1.19) in Chapter 1). In the presence of shadowing and multipath fading, the probability of detection can be obtained from [135]: $P_{\mathrm{d}} = \int_x Q\left(\sqrt{2\gamma}, \sqrt{\lambda}\right) f_\gamma(x)\mathrm{d}x$, where $f_\gamma(x)$ is the probability distribution function of SNR under fading.

  The two shortcomings of energy detection are: (1) it is susceptible to the uncertainty of noise power, and (2) it can only detect the presence of the signal but cannot differentiate the type of signal (e.g. signals from secondary users sharing the same channel with the primary user). Therefore, the detection error would be high in presence of signal sources other than the licensed user.
- *Cyclostationary feature detection:* The transmitted signal from a licensed user generally has a periodic pattern. This periodic pattern is referred to as cyclostationarity, and can be used to detect the presence of a licensed user [139]. A signal is cyclostationary (in the wide sense) if the autocorrelation is a periodic function. With this periodic pattern, the transmitted signal from a licensed user can be distinguished from noise, which is a wide-sense stationary signal without correlation. In general, cyclostationary detection can provide a more accurate sensing result and it is robust to variations in noise power. However, the detection is complex and requires long observation periods to obtain the sensing result. A pattern recognition scheme based on a neural network can be used to implement cyclostationary feature detection for spectrum sensing [140].

In order to improve the overall performance of spectrum sensing, multiple detection methods can be integrated in a single unlicensed system (e.g. fast and fine sensing [141]). For example, energy detection can be used to perform a quick scan of a wide range of spectrum bands. The results from energy detection can be used to eliminate the spectrum bands with high energy densities (e.g. due to the transmission of licensed users). Then, feature detection can be applied to a few candidate bands with low energy densities to search for the unique features of a signal from a licensed user [142]. A comparison of the different spectrum sensing techniques will be provided in Chapter 3.

Cooperative sensing

An unlicensed transmitter may not always be able to detect the signal from a licensed transmitter due to its geographic separation and channel fading. For example, in Figure 2.9, the transmitter and receiver of the unlicensed user cannot detect the signal

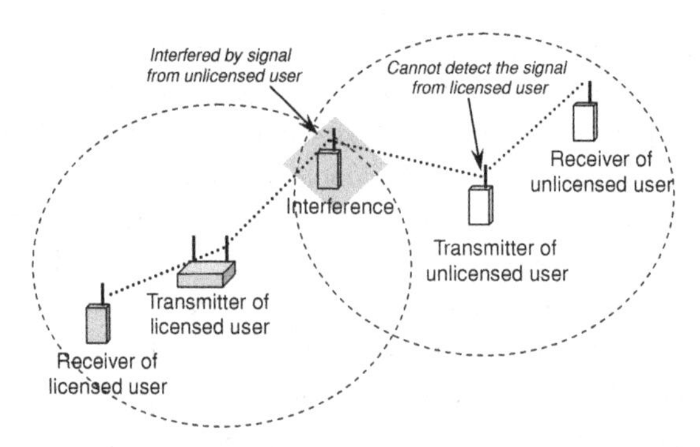

Figure 2.9 Hidden node problem.

from the transmitter of the licensed user since they are out-of-range. This is referred to as the hidden node problem. In this case, when the transmitter of the unlicensed user transmits, it will interfere with the receiver of the licensed user.

To solve the hidden node problem in non-cooperative transmitter sensing, cooperative spectrum sensing can be used. In cooperative sensing, spectrum sensing information from multiple unlicensed users are exchanged among each other to detect the presence of licensed users. The cooperative spectrum sensing architecture can be either centralized or distributed [143, 144]. Using cooperative exchange of spectrum sensing information, the hidden node problem can be solved and the detection probability can be significantly improved in a heavily shadowed environment. However, this incurs a greater communication and computation overhead compared with non-cooperative sensing. For cooperative sensing, two different networks (i.e. a sensor network and an operational network) can be deployed to perform spectrum sensing and access [145], respectively. In this case, the sensor network collects spectrum usage information of licensed users which can be processed by a central controller. Then a spectrum usage map is created and distributed to the operational network of unlicensed users for optimizing the spectrum access. Further discussion on collaborative sensing will be provided in Chapter 3.

Interference-based sensing

Interference-based sensing was proposed by the FCC [146]. In this case, the sensing algorithm will measure the noise/interference level (from all sources of signals) at the receiver of the licensed user. This information is used by an unlicensed user to control the spectrum access (e.g. by computing expected interference level) without violating the interference temperature limit. Alternatively, an unlicensed transmitter may observe the feedback signal from a licensed receiver to gain knowledge on the interference level.

2.2.7 Spectrum analysis and spectrum decision

Spectrum analysis is required for the characterization of different spectrum bands in terms of operating frequency, bandwidth, interference, primary user activity, and channel capacity. For example, in the spectrum underlay approach, based on the interference temperature limit at the primary receiver and operating frequency, the permissible transmission power at the cognitive radio can be determined. Subsequently, the channel capacity can be estimated. Spectrum analysis models can be based on either current spectrum sensing results or spectrum usage history.

The spectrum analysis architecture can be either local or cooperative. A cooperative architecture, which can be either centralized or distributed, can improve the accuracy of the spectrum usage model. A cooperative architecture requires exchanging information among cognitive radios (and hence causes additional overhead) and may suffer from the scalability problem.

The spectrum decision deals with whether or not to transmit taking into account the fact that spectrum sensing/detection could be erroneous, and in case of transmission how to exploit the spectrum holes (i.e. what modulation and power level to use, how to share the spectrum holes among cognitive radios). This is primarily a medium access control problem for a cognitive radio and it also requires the price offered/charged by the primary user/service provider to be considered. Also, spectrum access decisions may need to be communicated between the cognitive nodes and the intended receivers.

Spectrum decisions can be made based on either a local or a global optimization criterion. In case of local optimization, the spectrum access decision is made in a non-cooperative (i.e. distributed) way and, in the case of global optimization, a cooperative spectrum access decision is made either in a centralized or a distributed way.

In a non-cooperative/local spectrum access strategy each cognitive node is responsible for its own decision. If the miss-detection probability is large, the access policy should be conservative. If the false alarm probability is large, the access policy should be aggressive. Therefore, the access strategy can be jointly optimized with the sensing strategy.

Game theory is a powerful tool to analyze the spectrum sharing problem in a non-cooperative (i.e. competitive) spectrum access scenario. Different game theory models will be described in Chapter 5. The survey of game theory to telecommunications can be found in [147]. A non-cooperative spectrum access strategy has minimal communication requirements (and hence a lower overhead), but it may result in poor spectrum utilization.

In a cooperative centralized strategy, a centralized server maintains a database of spectrum availability and access information (based on the information received from a group of secondary users, say, through a dedicated control channel). Therefore, spectrum management is simpler and coordinated and enables efficient spectrum sharing. Classical optimization theory can be used to solve the cooperative spectrum access problem in a centralized setting. This will be discussed in detail in Chapter 4.

A cooperative distributed strategy relies on cooperative local actions throughout the network in order to achieve a performance close to the global optimal performance. However, this approach may suffer due to the hidden node problem and large control overheads. Cooperative game theory techniques can be used to model and solve the spectrum access problem in such scenarios. In both the centralized and the distributed strategies, the primary user may or may not cooperate.

Once a decision is made to access the spectrum opportunities, several issues related to radio link control and resource management (i.e. how to access) need to be resolved. These include pulse shaping, transmission power control, selection of the number of spectrum bands to access and the set of appropriate bands (e.g. subcarriers in an OFDM system), adaptive modulation and coding (e.g. bit loading in the OFDM subcarriers), etc.

2.2.8 Potential applications of cognitive radio

Cognitive radio concepts can be applied to a variety of wireless communications scenarios, a few of which are described below:

- *Next generation wireless networks:* Cognitive radio is expected to be a key technology for next generation heterogeneous wireless networks. Cognitive radio will provide intelligence to both the user-side and provider-side equipments to manage the air interface and network efficiently. At the user-side, a mobile device with multiple air interfaces (e.g. WiFi, WiMAX, cellular) can observe the status of the wireless access networks (e.g. transmission quality, throughput, delay, and congestion) and make a decision on selecting the access network to connect with. At the provider-side, radio resource from multiple networks can be optimized for the given set of mobile users and their QoS requirements. Based on the mobility and traffic pattern of the users, efficient load balancing mechanisms can be implemented at the service provider's infrastructure to distribute the traffic load among multiple available networks to reduce network congestion.
- *Coexistence of different wireless technologies:* New wireless technologies (e.g. IEEE 802.22-based WRANs [148]) are being developed to reuse the radio spectrum allocated to other wireless services (e.g. TV service). Cognitive radio is a solution to provide coexistence between these different technologies and wireless services. For example, IEEE 802.22-based WRAN users can opportunistically use the TV band when there is no TV user nearby or when a TV station is not broadcasting [149]. Spectrum sensing and spectrum management will be crucial components for IEEE 802.22 standard-based WRAN technology to avoid interference to TV users and to maximize throughput for the WRAN users.
- *eHealth services:* Various types of wireless technologies are adopted in healthcare services to improve efficiency of the patient care and healthcare management. However, using wireless communication devices in healthcare application is constrained by EMI (electromagnetic interference) and EMC (electromagnetic compatibility) requirements. Since the medical equipments and biosignal sensors are sensitive to EMI, the transmit power of the wireless devices has to be carefully controlled. Also, different biomedical devices (e.g. surgical equipment, diagnostic and monitoring devices) use RF transmission. The spectrum usage of these devices has to be carefully chosen to avoid interference with each other. In this case, cognitive radio concepts can be applied. For example, many wireless medical sensors are designed to operate in the

ISM (industrial, scientific, and medical) band, which can use cognitive radio concepts to choose suitable transmission bands to avoid interference.

- *Intelligent transportation system:* Intelligent transportation systems (ITS) will increasingly use different wireless access technologies to enhance the efficiency and safety of transportation by vehicles. Two different types of communications scenarios arise in an ITS system – vehicle-to-roadside (V2R) communication and vehicle-to-vehicle (V2V) communication. In vehicle-to-roadside communications, information is exchanged between the roadside unit (RSU) and the onboard unit (OBU) in a vehicle. In vehicle-to-vehicle communications, a special form of ad hoc network, namely, a vehicular ad hoc network (VANET), is formed among vehicles to exchange safety-related information. High mobility of the vehicles and rapid variations in network topologies pose significant challenges to efficient V2R and V2V communications. Cognitive radio concepts can be used in both OBUs and RSUs so that they can adapt their transmissions to cope with the rapid variations in the ambient radio frequency environment [150]. With multi-radio capabilities at the OBUs, they should be able to adaptively choose the radio to communicate with the RSUs.
- *Emergency networks:* Public safety and emergency networks can take advantage of the cognitive radio concepts to provide reliable and flexible wireless communication. For example, in a disaster scenario, the standard communication infrastructure may not be available, and therefore, an adaptive wireless communication system (i.e. an emergency network) may need to be established to support disaster recovery. Such a network may use the cognitive radio concept to enable wireless transmission and reception over a broad range of the radio spectrum.
- *Military networks:* With cognitive radio, the wireless communication parameters can be dynamically adapted based on the time and location as well as the mission of the the soldiers. For example, if some frequencies are jammed or noisy, the cognitive radio transceiver can search for and access alternative frequency bands for communication. Also, location-aware cognitive radio can control the transmitted waveform in a particular region to avoid interference to the high priority military communication systems [151].

Understanding the key concepts and design techniques for dynamic spectrum access would be fundamental for the researchers, communications engineers, and application developers to implement the above cognitive radio-based applications.

2.3 Research challenges in cognitive radio

2.3.1 Issues in spectrum sensing

Spectrum sensing gives rise to several physical and MAC layer research issues. While the physical layer issues are mostly related to signal processing, the MAC layer issues are related to optimization of spectrum sensing.

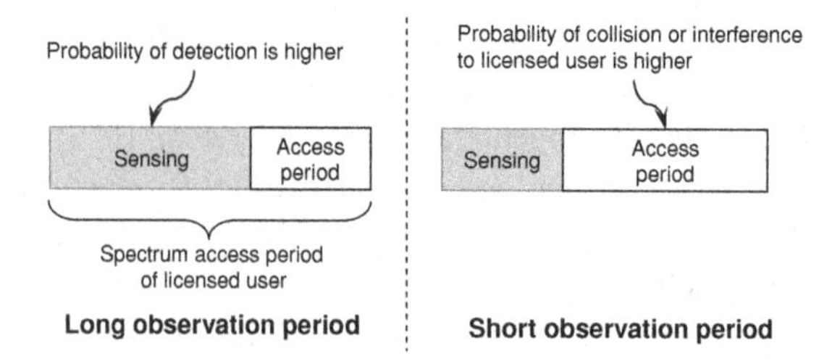

Figure 2.10 Tradeoff between spectrum sensing time and user throughput.

Sensing interference limit

One objective of spectrum sensing, especially for interference-based sensing, is to obtain the status of the spectrum (i.e. idle/occupied), so that the spectrum can be accessed by an unlicensed user under the interference constraint. The challenge lies in the interference measurement at the licensed receiver caused by transmissions from unlicensed users. First, an unlicensed user may not know exactly the location of the licensed receiver which is required to compute interference caused due to its transmission. Second, if a licensed receiver is a passive device, the transmitter may not be aware of the receiver.

Spectrum sensing in multiuser networks

Multiple users, both licensed and unlicensed, may share the radio spectrum in a network. Also, multiple networks can coexist for which transmissions in one network may interfere with transmissions in other networks. In such a case, coordinated and cooperative spectrum sensing would be preferred since it can detect the spectrum access status by the licensed users in different locations in the network. The spectrum sensing information can be used to obtain a spectrum map which can be utilized by the unlicensed users to make spectrum access decisions.

Optimizing the period of spectrum sensing

In spectrum sensing, the longer the observation period, the more accurate will be the spectrum sensing result. However, during sensing, a single-radio wireless transceiver cannot transmit in the same frequency band. Consequently, a longer observation period will result in lower system throughput (Figure 2.10). This performance tradeoff can be optimized to achieve an optimal spectrum sensing solution. If the accuracy of spectrum sensing is low, collision and interference to the transmissions by licensed users could occur to degrade the performances of both licensed and unlicensed users. Classical optimization techniques (e.g. convex optimization) can be applied to obtain the optimal solution.

Spectrum sensing in multichannel networks

Multichannel transmission (e.g. OFDM-based transmission) would be typical in a cognitive radio network. However, the number of available channels would be larger than the number of available interfaces at the radio transceiver. Therefore, only a fraction of the available channels can be sensed simultaneously. Selection of the channels (among all available channels) to be sensed will affect the performance of the system. For spectrum sensing, a channel which is mostly occupied by the licensed user(s) should be less preferred to a channel which is occasionally occupied. In a multichannel environment, selection of the channels should be optimized for spectrum sensing to achieve the optimal system performance under hardware constraints at the cognitive radio transceiver.

2.3.2 Spectrum management issues

The main objective of spectrum management is to observe and control access to the spectrum holes by the unlicensed users. The research issues in spectrum management are related to spectrum analysis and spectrum access decisions.

Issues in spectrum analysis

Based on the spectrum sensing result, spectrum analysis is performed to estimate spectrum quality. One of the issues here is how to quantify the quality of the spectrum opportunity which can be accessed by an unlicensed user. This quality can be characterized by the signal-to-noise ratio (SNR), the average duration and correlation of the availability of spectrum holes. The information about this spectrum quality available to a cognitive radio user can be imprecise and noisy. Learning algorithms from artificial intelligence is one of the candidate techniques that can be used by the cognitive radio users for spectrum analysis.

Issues in the spectrum access decision

Some of the research issues related to the spectrum access decision of an unlicensed user are described below:

- *Decision model:* A decision model is required for spectrum access. The complexity of this decision model depends on the parameters considered during spectrum analysis (e.g. the average duration of spectrum holes and the SNR of the target frequency band) as well as the utility of the cognitive user obtained through accessing the spectrum holes. The decision model becomes more complex when an unlicensed user has multiple objectives. For example, an unlicensed user may intend to maximize its throughput while minimizing the interference caused to the licensed user. Stochastic optimization methods (e.g. the Markov decision process) will be an attractive tool to model and solve the spectrum access decision problem in a cognitive radio environment.
- *Competition/cooperation in a multiuser environment:* When multiple users (both licensed and unlicensed users) are in the system, their preference will affect the spectrum access decision. These users can be cooperative or non-cooperative in accessing the spectrum. In a non-cooperative environment, each user has its own objective, while in a cooperative environment, all users can cooperate to achieve a single objective (Figure 2.11). For example, multiple unlicensed users may compete with each other to access the radio spectrum (e.g. O1, O2, O3, and O4 in Figure 2.11) so that their individual throughput is maximized. During this competition between unlicensed users, all of them have to ensure that the interference caused to the licensed user(s) is

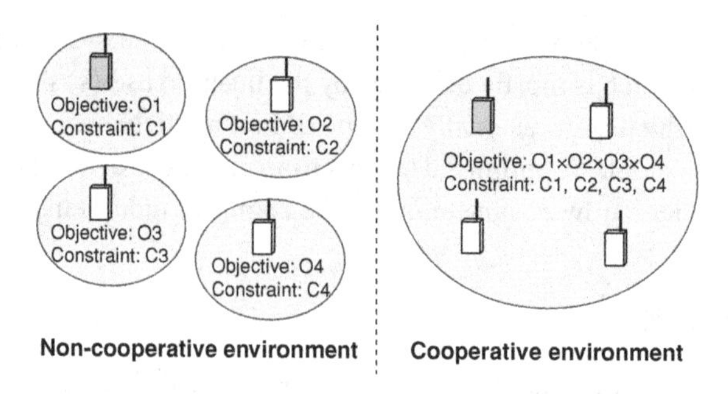

Figure 2.11 Cooperative and non-cooperative spectrum access.

maintained below the corresponding interference temperature limit. Non-cooperative game theory is the most suitable tool to obtain the equilibrium solution for the spectrum access problem in such a scenario. This equilibrium solution would guarantee that none of the users will unilaterally deviate to achieve higher payoff (which could degrade the payoff of other users).

In a cooperative environment, the cognitive radios cooperate with each other to make the spectrum access decision to maximize a common objective function under given constraints. In such a scenario, a central controller may coordinate the spectrum management. Alternatively, the cognitive radios can communicate among themselves in a distributed manner and a mechanism for information exchange, negotiation, and synchronization would therefore be required.

- *Distributed implementation of spectrum access control:* In a distributed multiuser environment, for non-cooperative spectrum access, each user can reach an optimal decision independently by observing the historical behavior/action of other users in the system. Therefore, a distributed algorithm is required for an unlicensed user to make the decision on spectrum access autonomously. The stability and convergence property of the distributed algorithm needs to be evaluated to ensure that it achieves the desired solution for an unlicensed user.

2.3.3 Spectrum mobility issues

The spectrum mobility functions in a cognitive radio network allow an unlicensed user to change its operating spectrum dynamically based on the spectrum condition. The research issues in spectrum mobility relate to switching between available frequency bands to provide smooth spectrum access without interruption to both application and service due to the change in the radio environment.

- *Search for the best frequency bands:* A cognitive radio must keep track of available frequency bands so that if necessary (e.g. a licensed user is detected), it can switch immediately to other frequency bands. During transmission by an unlicensed user, the condition of the frequency band has to be observed. In a similar way to spectrum sensing, this would of course incur some overhead. The observation can be performed in a proactive manner or in an on-demand basis [152]. In the proactive approach, the condition of the available channels is periodically observed and the knowledge about these channels is continuously updated. When a spectrum handoff is required, this knowledge can be used by an unlicensed user to switch to a new frequency band. In contrast, with an on-demand approach, channel observation can be performed only when an unlicensed user needs to switch the channel. While the proactive approach incurs a larger overhead due to periodic observation, the latency of spectrum handoff would be smaller in this case.

- *Protocol stack adaptation:* Since the latency due to spectrum handoff could be high, the modification and adaptation of other components in the protocol stack are required. For example, when an unlicensed user switches channel, the TCP timer at the transport layer can be frozen to avoid any mis-interpretation of the delay incurred for the acknowledgement message. A cross-layer optimized framework for protocol adaptation has to be developed to cope with spectrum mobility.
- *Self-coexistence and synchronization:* When an unlicensed (or secondary) user performs spectrum handoff, two issues have to be taken into account. First, the target channel must not currently be used by any other secondary user (i.e. the self-coexistence requirement), and the receiver of the corresponding secondary link must be notified of the spectrum handoff (i.e. the synchronization requirement). For the self-coexistence issue, a spectrum broker can be used to manage spectrum allocation. However, this issue becomes challenging when such a broker is not available in the system. For synchronization, the MAC protocol must be designed with provision for spectrum handoff information exchange.

2.3.4 Network layer and transport layer issues

The major issue in the network layer is the design of the routing protocol for multihop cognitive radio networks. The routing algorithm needs to be integrated with spectrum management in order to improve the overall system performance. It was observed that the performance of the cross-layer approach is much better than the performance of the decoupled approach for which routing algorithm and spectrum management perform separately [153, 154]. In the cross-layer approach, the routing algorithm (e.g. dynamic source routing (DSR)) considers the spectrum allocation in each hop. In the decoupled approach [144], the shortest-path and spectrum sharing algorithms are performed independently. Although integration of routing algorithms and spectrum management can achieve better performance, many issues related to routing in the network layer remain to be investigated.

- *Message broadcasting mechanism*: Given the routing metric, a routing algorithm requires information exchange among the network nodes to choose the best route. However, since in a cognitive radio network a common channel for this information exchange may not be available, an efficient message broadcasting mechanism based on dynamic spectrum access would be required.
- *Routing metrics and route selection*: Due to the activity of licensed users, connectivity between the nodes of unlicensed users can be dynamically changed. Also, the routing

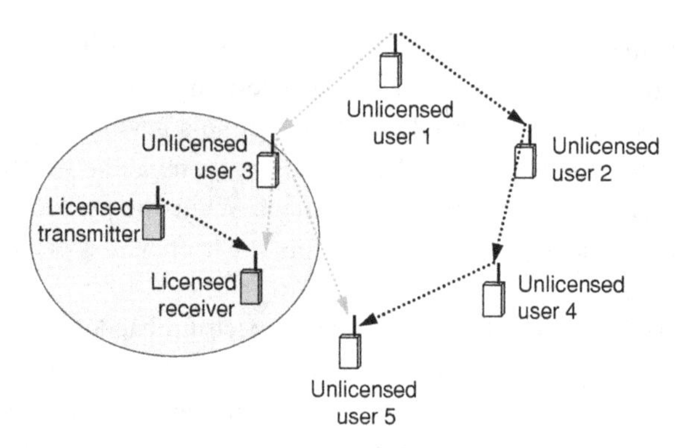

Figure 2.12 Routing in the presence of licensed users.

metric has to be jointly defined based on network parameters (e.g. hop-counts and link quality) and spectrum management parameters (e.g. interference to licensed users and the average number of available channels). For example, in Figure 2.12, unlicensed user 1 may require to use the route through users 2 and 4 to the destination user 5 for which the number of hops is more than that of the route through user 3 (i.e. three hops versus two hops).This is because using the shorter route may cause interference to the licensed user. However, during periods when the licensed transmitter is inactive, the route through unlicensed user 3 can be selected.

In the transport layer, modifications have to be made to TCP and UDP working in a cognitive radio environment to avoid performance degradation [155]. Congestion control in TCP depends on the packet loss rate and the round-trip-time (RTT). With a wireless link, a TCP sender could mis-interpret packet loss due to wireless error as a sign of congestion. As a result, congestion avoidance and slow start mechanisms would be invoked. Some variants of TCP were proposed to cope with this problem in a wireless environment. For example, an indirect-TCP (I-TCP) splits a connection of wireless link from the wired link, so that each TCP connection can be optimized for wireless and wired networks, respectively. A survey of TCP variants for wireless networks can be found in [156].

In cognitive radio networks, many factors such as the transmit power, the bandwidth of the spectrum hole, and the interference level can affect the packet loss rate. Similarly, RTT can be affected by the delay due to spectrum sensing and spectrum handoff, for which when the current channel becomes unusable for an unlicensed user, the unlicensed user has to search for a new channel. Therefore, the transport layer protocol has to consider these effects to optimize end-to-end rate control.

2.3.5 Cross-layer design for cognitive radio networks

Communication systems are usually designed based on a layered architecture. The layered architecture promotes the modularity, standardization, and expandability of the system. Since the communication functionalities are separated into different layers, with

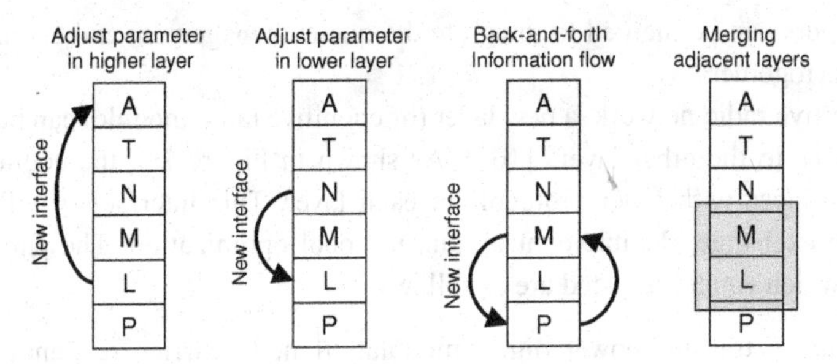

Figure 2.13 Cross-layer interactions.

those that are closely related grouped in one layer, each of the layers can be designed independently. In addition, the components in each layer can be updated, changed, and expanded independently without any side effect on other components. The layered structure provides explicit abstraction of the functions in each layer and supports well-defined interfaces between the layers. Therefore, components developed or implemented by different manufacturers can interoperate with each other.

However, since the different components are ordered in the layered structure, each component has to operate sequentially (e.g. a lower layer has to wait until a higher layer has finished processing). This sequential operation results in computational overhead and latency. With the layers being isolated, components in one layer may not be able to access information in other layers. As a result, the interaction among the components in the different layers is limited. The performance of the entire system may therefore not be optimized due to the lack of global information exchange. Also, some components can have redundant functions (e.g. error control and recovery in both MAC and physical layers). Wireless systems with layered protocol structures may lack adaptability with channel variations. Since the interaction between the layers has to be performed only through a standard interface, the communication system may not be able to respond immediately when the radio environment changes. As a result, the overall system performance could be degraded.

Some of the limitations of layered protocol structure in a communication system can be mitigated through cross-layer design and optimization methods [157, 158]. One objective of cross-layer design is to increase the information flow between components in the different layers. Cross-layer optimization aims to improve the overall performance of a communication system through optimization of decision variables and parameters in different layers simultaneously.

The inter-layer interactions which can be exploited for cross-layer design are shown in Figure 2.13. The interaction between non-adjacent layers can be defined in either downward/upward or both directions [159]. For example, the wireless transmission rate in the data link level can be reported to the video encoding software in the application layer so that the video encoding rate can be adjusted. When the transmission quality degrades (e.g. due to channel fading), the video encoding rate may need to be reduced, which would result in a degraded video quality [160]. Layers can be merged to tightly integrate the communication functions in these layers [91]. For example, the MAC and data link layers can be merged to integrate the queue management and channel access mechanisms together.

In a cognitive radio network, a new layer (or cognitive radio module) can be designed that interfaces to the other layers [161]. As shown in Figure 2.3, the cognitive radio module interfaces with other protocols in each layer. This interfacing will facilitate information exchange, flexible control, and protocol optimization. The parameters in each layer which can be adjusted are as follows:

- *Physical layer:* transmit power, digital modulation mode, carrier frequency, spectrum bandwidth, processing gain, channel coding type and rate, duty cycle, and waveform of the transmitted signal.
- *MAC layer:* packet type, packet size, data rate, channel/time slot allocation, scheduling scheme, retransmission probability, and MAC protocol.
- *Network and transport layers:* routing metric and network scheduling algorithm in the routing layer, and congestion control parameters (e.g. TCP window size) and rate control parameter (e.g. bucket size of token bucket rate control) in the transport layer.
- *Application layer:* source coding (e.g. video compression) and encryption algorithm.

The major challenges in applying cross-layer design methods to cognitive radio are as follows:

- *Multi-objective system:* The components in different protocol layers have different objectives. For example, in the application layer, high-quality video is required to satisfy the users. However, transmitting high-quality video could incur congestion in the network due to the large amount of data to be transmitted and the capacity limitations of the radio spectrum. Such tradeoffs in different protocol layers have to be compromised to achieve the optimal performance. For this, the techniques of multi-objective optimization may be applied [162].
- *Information access and parameter control:* To support intelligent algorithms in a cognitive radio module, access to the information and control of the protocol parameters in the different layers would be required. However, the level of information access and parameter control should be limited in order to maintain the layered structure of the protocol stack. Cognitive radio design must compromise between the modularity and adaptability of the protocol stack.
- *Complexity:* When the concept of cross-layer optimization is applied to the design of cognitive radio, the complexity issue becomes important. As the number of control parameters increases, the complexity of optimization increases. In contrast, in a fast-changing radio environment, the solution of the optimization formulation must be obtained within a limited amount of time. The design of cognitive radio must compromise between the optimality of system performance and the complexity of the cognitive radio system.

An example of cross-layer optimization in cognitive radio networks was presented in [163]. In a cognitive radio environment, the information gathered (e.g. signal quality) by a cognitive radio could often be imprecise. Therefore, the radio interface and protocols in a cognitive radio have to be designed to process noisy and inaccurate information. In [163], fuzzy logic was used as a solution to process imprecise information in a

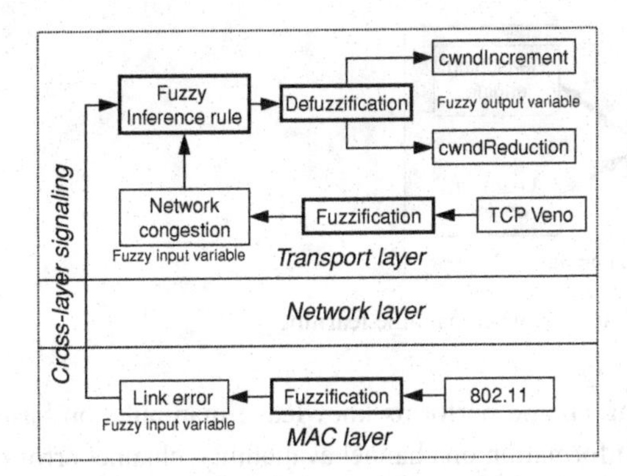

Figure 2.14 Cross-layer design for cognitive radio using fuzzy logic.

cognitive radio environment using the concept of cross-layer design. Using a fuzzy logic controller can reduce the complexity of the system. Specifically, using predefined rules for fuzzification and defuzzification processes in a fuzzy logic-based system often reduces the computational complexity significantly when compared to the complexity of solving an optimization problem in an exact manner.

In a fuzzy logic system, the protocol parameters in the different layers can be modeled as fuzzy variables which can be accessed and modified by the fuzzy controller. The values of these fuzzy variables are classified into fuzzy sets with certain probability. The values of input fuzzy variables can be determined by the fuzzification process applied to the measured values. Then, the fuzzy controller applies the inference rules to the input variables and the defuzzification process is used to obtain the values of output variables which are used to control the protocol parameters.

The above concept has been used for cross-layer design in cognitive radio systems. In particular, fuzzy logic was applied to optimize TCP performance over IEEE 802.11 [163] (Figure 2.14). In this case, 802.11 link reliability is represented as a fuzzy variable with two values, i.e. low and high. The input to be fuzzified is the SNR measured from the reception of a frame from an access point. In the transport layer, TCP Veno [164] with a congestion estimation mechanism was used. TCP Veno was modified by adding a fuzzy variable for the congestion level. The input to this fuzzy variable is fuzzified from the congestion indicator available in TCP Veno. This fuzzy variable for congestion has three values, i.e. low, mid, and high. Then, from the set of given rules, the outputs in terms of the increment and decrement steps of the congestion window (cwndIncrement and cwndReductions) used by TCP Veno are defuzzified.

2.3.6 Artificial intelligence approach for designing cognitive radio

Artificial intelligence (AI) techniques for learning and decision making can be applied to design efficient cognitive radio systems [165, 166, 167]. One example was illustrated in [165], where the concept of machine learning was applied to cognitive radio for capacity maximization and dynamic spectrum access. The proposed system architecture is shown in Figure 2.15. Here, the knowledge base maintains the states of the system and the available actions [168]. The reasoning engine uses the knowledge base to select

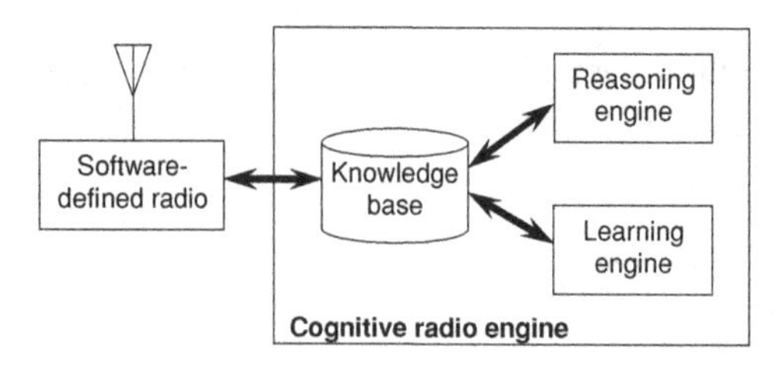

Figure 2.15 Cognitive radio architecture with machine learning.

the best action. The learning engine performs knowledge manipulation based on the observed information (e.g. information on channel availability, channel error rate).

In the knowledge base, two data structures, namely, predicate and action, are defined. The predicate (or inference rule) is used to represent the state of the environment. Based on this state, an action can be performed to change the state so that the system objectives can be achieved. For example, a predicate can be defined as "modulation==QPSK AND SNR==5dB", while the action can be defined as "decrease modulation mode" with precondition "SNR $\leq$ 8dB" and postcondition "modulation==BPSK". Given the input (which is obtained from measurement), the reasoning engine matches the current state (e.g. modulation and SNR in this case) with the predicates and determines the predicate results (i.e. true or false). Then, from the set of predicate results, an appropriate action is taken. In the above example, if the current SNR is equal to 5 dB and current modulation is QPSK, the precondition will be true and the predicate will be active. As a result, the cognitive engine will decide to reduce the modulation mode. In this case, the modulation will be changed to BPSK, as stated in the corresponding postcondition.

A learning algorithm is used to update both the state of the system and the available actions according to the radio environment. This update can be done using an objective function (e.g. minimize the bit error rate) with a goal to determine the best action given the input (e.g. channel quality) and the available knowledge. Different learning algorithms can be used in a cognitive radio network (e.g. based on a hidden Markov model [169], neural network [170], reinforcement learning [171], or genetic algorithm [172]).

Another architecture for knowledge-based cognitive radio was presented in [167]. The two major components in this architecture (as shown in Figure 2.16) are the perception/action and the ontology (rule)/reasoning. The perception/action component is used to observe the input from the environment and the state of the system, and to take the appropriate decisions for the cognitive radio. The ontology/reasoning component stores the knowledge and the rules. The ontology is one of the representations of knowledge in the form of a formal rule language. This is used by the reasoning engine to obtain the best decision according to the defined objective.

2.3.7 Location-aware cognitive radio

Location information can be used to improve the performance of wireless systems. For example, in mobile cellular networks, the location information of the mobile users can be used to optimally reserve the channels for handoff, e.g. using the concept of shadow cluster [173, 174]. Other examples are location-assisted dynamic spectrum

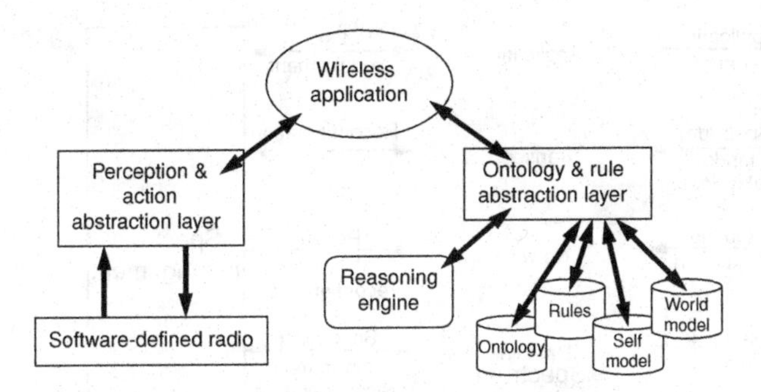

Figure 2.16 Architecture of knowledge-based cognitive radio.

management, transceiver algorithm development and optimization (e.g. location-aided adaptive beamforming), and environment characterization (e.g. location-assisted propagation environment identification) [175, 176]. The location awareness was also used in the context of a cognitive radio system [177, 178]. Two dynamic spectrum access schemes, which were proposed by the FCC to utilize location information for spectrum allocation, are geolocation-database and local beacon [179]. These schemes were proposed for an unlicensed system to access TV bands without interfering with a TV service (i.e. the licensed service).

In a geolocation-database scheme, the licensed services (i.e. TV broadcasters) use location estimation devices (e.g. global positioning system or GPS) to obtain their location information. Then, this information is sent by the licensed service to the FCC central database. This FCC database broadcasts the collected TV band and the location information from the corresponding licensed services to the unlicensed users. Similarly, the unlicensed users also use location estimation devices to obtain their location information. Based on the information from the FCC database and their own location information, the unlicensed users identify available TV bands that they can use without interfering with the licensed services. This step can be performed locally by the unlicensed users. After the unlicensed users determine the TV bands to be used, they submit the request with their location information to the cognitive base station. This cognitive base station allocates the spectrum to the unlicensed users and also reports to the FCC database about the occupancy of the TV bands.

In a similar way to the geolocation-database scheme, a local beacon scheme utilizes the location information to identify the TV bands to be used by the unlicensed users. However, in this scheme, the database is located in a local cell to manage the spectrum allocation. The licensed services report periodically to update their location and TV band information in the local database. The cognitive radio base station broadcasts the geolocation-database to the unlicensed user. The unlicensed user receives this information and uses it to identify available TV bands. Again, its location information is used. The spectrum access plan (e.g. which TV band to access and for how long) is sent to the cognitive radio base station to maintain information on the current spectrum allocation.

Although the proposed dynamic spectrum access can improve the performance of spectrum allocation, there are some issues that need to be addressed. For example, the accuracy of the location information will have an effect on the interference caused to

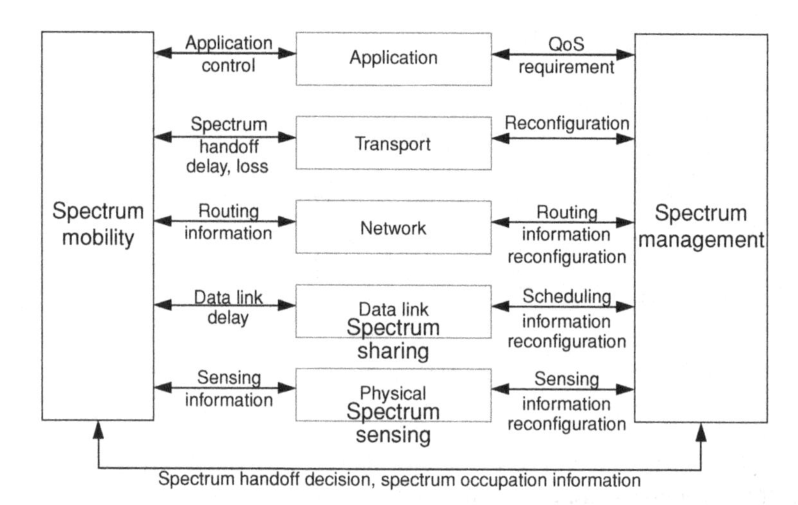

Figure 2.17 Protocol stack for xG networks.

the licensed service, especially when the unlicensed user is located in a place where the geolocation signal cannot be accurately estimated (e.g. indoor). Adaptive position systems based on cognitive radios as proposed in [175] can be applied to mitigate the problem.

2.4 Cognitive radio architectures for NeXt Generation (XG) networks

In order to improve the utilization of radio spectrum, the Defense Advanced Research Projects Agency (DARPA) initiated the NeXt Generation (XG) program [180]. The main idea here is to use the dynamic spectrum access mechanism through cognitive radio for XG networks. In such a network, a cognitive radio user should be able to sense spectrum availability and also detect the presence of incumbent users in a target frequency band so that it is able to release the frequency spectrum when the licensed users are detected. These functions are referred to as spectrum sensing, spectrum management, spectrum sharing, and spectrum mobility, respectively [181]. The locations of these functionalities in the XG network protocol stack are shown in Figure 2.17.

Figure 2.18 shows the general architecture of an XG network. In this architecture, there are two major groups of wireless systems, namely, licensed and unlicensed systems. In a licensed system, the base station or access point provides wireless connections for licensed users. The unlicensed system may or may not use an infrastructure. In the former case (i.e. for an unlicensed system with infrastructure), an XG base station is used to control the spectrum access by XG users. Such a base station may connect to a spectrum broker for synchronization with other XG networks to support coexistence. In the latter case, an ad hoc access mode is used for communication among XG users, and an XG network gateway can be used to connect the unlicensed system to the licensed system.

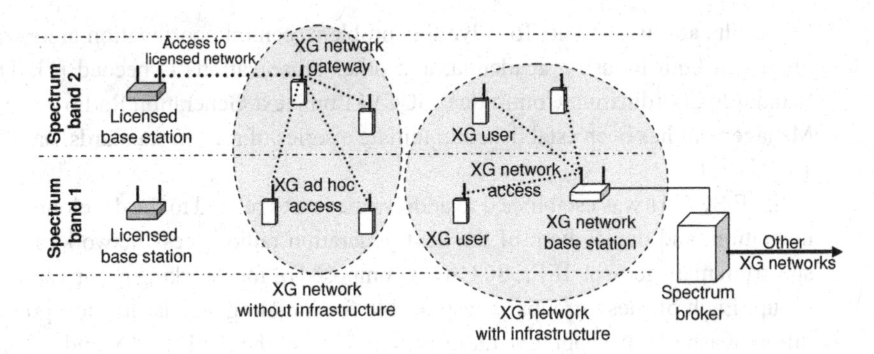

Figure 2.18 XG network architecture.

Several cognitive radio systems based on the XG network architecture have been proposed in the literature. Some of them are described below. Note that a detailed survey on the related works in XG networks can be found in [181].

- *CORVUS:* A cognitive radio approach for usage of virtual unlicensed spectrum (CORVUS) was proposed in [182]. A coordination protocol was proposed to support licensed user detection and spectrum allocation for a group of unlicensed users. There could be multiple groups of unlicensed users, and the coordination among them was also supported in CORVUS.
- *DIMSUMnet:* Dynamic intelligent management of spectrum for ubiquitous mobile network (DIMSUMnet) was proposed in [183]. The concepts of statistically multiplexed access (SMA) and coordinated access band (CAB) were used to improve spectrum utilization, efficiency, and fairness. A spectrum broker was used to allocate CAB to the users.
- *DRiVE/OverDRiVE:* A dynamic radio architecture for IP services in vehicular environments (DRiVE) was proposed to support heterogeneous wireless access [184]. This DRiVE system was extended to support multicast data transmission over dynamic radio networks in vehicular environments (OverDRiVE) [185].
- *OCRA network:* An OFDM-based cognitive radio (OCRA) network was proposed in [186] based on cross-layer design. In this work, the concept of OFDM-based spectrum management in a heterogeneous spectrum environment was presented.

2.5 Cognitive radio standardization

2.5.1 IEEE SCC 41

Since efficient spectrum management using software-defined radio in a cognitive radio network involves many technical and economic aspects, the standardization processes, terms, and other related issues become important. These standardization processes and terms are required for the development and implementation of a cognitive radio network. However, up to now, most individual groups have been working independently and hence the results are incoherent. To solve this problem, central coordination among research groups in both industry, academia, and other organizations is needed [1]. The IEEE Standards Coordinating Committee (SCC) 41 on Next Generation Radio and Spectrum Management has been established to initiate a series of related standards, namely, IEEE 1900 [187].

IEEE SCC 41 was established to address the issues related to the development, implementation, and deployment of the next generation radio access network, and efficient spectrum management. IEEE SCC 41 is composed of four working groups and one study group. Each of these groups is responsible for initiating standardization processes for different aspects of a cognitive radio system. Each of the IEEE 1900 standards in a series will be proposed by each working group. The standardization process works as follows. After finalizing the draft, each working group first submits the resulting document to the IEEE for voting. This voting is arranged by the IEEE Standards Association (SA). IEEE SA members vote as well as submit comments for possible changes. If there is a negative vote, the comments are taken into account by the working group to revise the draft. After revision, the document will be voted on again. The standardization procedure ensures that the standard will be applicable to a variety of products related to cognitive radio. The major components of the IEEE 1900 standards are as follows:

- *IEEE 1900.1:* The major part of this standard is to identify and explain a glossary of terms and concepts related to spectrum management, software-defined radio, adaptive radio, and other relevant technologies. Also, the standard describes the interrelation between the key terms and concepts. The IEEE 1900.1 standard is used as a connection to other IEEE SCC 41 working groups, since all working groups will rely on the definitions provided by this standard.
- *IEEE 1900.2:* This standard relates to the recommended practice for interference and coexistence analysis. In cognitive radio, many wireless devices and services are allowed to coexist in the same location at the same time. Optimization of the operating parameters of these devices and services is a crucial issue to manage and avoid interference. The IEEE 1900.2 standard will recommend the criteria for interference analysis. Also, a framework for measuring, investigating, and analyzing the interference between wireless devices and services will be introduced. This working group will develop a common standard platform to resolve any conflict occurring in cognitive radio (e.g. between the devices working in the same spectrum).
- *IEEE 1900.3:* This standard relates to the recommendation of conformance evaluation of the software modules in software-defined radio. The working group will develop and define the test methods for conformance evaluation of software components in software-defined-radio-based wireless devices. The main objective of this standard is to guarantee coexistence and compliance in the software part. This compliance is required for validation and certification of the final cognitive radio products.
- *IEEE 1900.4:* This standard relates to the coexistence support for the reconfigurable heterogeneous air interface. Heterogeneity will be a key feature in next generation wireless systems in which a mobile device will be able to use multiple wireless

technologies simultaneously (e.g. reconfigurable terminal). Therefore, dynamic spectrum access to support service selection and optimization will be a key component. The working group will define an overall system architecture as well as the major system functionalities. The definitions of the corresponding protocol will also be addressed by this working group.

- *IEEE 1900.A:* This study group's responsibility relates to the certification of dynamic-spectrum-access-based devices. The study group will emphasize the dependability and evaluation of regulatory compliance for wireless devices with dynamic spectrum access. Since a cognitive radio device will have more flexibility than a legacy device, the certification process becomes more challenging. New methodologies and testing procedures need to be developed to ensure that the certified device will not interfere with the transmission of a licensed device. The studies will include a hazard analysis listing the potential causes of interference. The manufacturer will need to avoid these to ensure that the device will conform to the regulations.

There are also other IEEE projects/standards which relate to the next generation wireless systems and cognitive radio, i.e. IEEE 802.18, 19, 21, and 22. IEEE 802.18 is a radio regulatory technical advisory group responsible for participating and monitoring the evolution of the radio regulatory activities in different projects (e.g. for IEEE 802.11 WLAN, IEEE 802.15 WPAN, IEEE 802.16 WMAN, IEEE 802.20 Mobile WMAN, and IEEE 802.22 WRAN). This IEEE 802.18 group may make comments and recommendations to the regulators and other parties to inform any demand for spectrum access. IEEE 802.19 is a technical advisory group for coexistence. This group will address the issues of coexistence between unlicensed wireless networks based on IEEE 802 standards (e.g. IEEE 802.11 and Bluetooth). In this case, when a new standard for an unlicensed wireless network is introduced, this IEEE 802.19 group will review the coexistence assurance of that standard to ensure that the new standard will coexist with existing technologies operating in the same spectrum. IEEE 802.21 is a new standard to support seamless mobility management (e.g. handoff) for both homogeneous and heterogeneous wireless technologies. It will be a fundamental standard for next generation wireless systems in which a mobile user can use multiple wireless technologies concurrently. IEEE 802.22 is a new wireless technology to support data communication with a large coverage area. This technology will operate on TV bands which are largely unoccupied. However, to ensure that the incumbent service will not be interfered with by the IEEE 802.22 devices, dynamic spectrum access will be applied. The relationships among the IEEE SCC 41 components is shown in Figure 2.19.

2.5.2 IEEE 802.22 for wireless regional area networks (WRANs)

Some TV bands have been observed to remain largely unoccupied in many regions. These 6 MHz TV bands can be used for data communication. Since TV bands are mostly in the low-frequency spectrum (e.g. 54–862 MHz in North America and 41–910 MHz internationally), their propagation characteristics are more suitable for long-range transmissions. The IEEE 802.22 standard was proposed for wireless regional

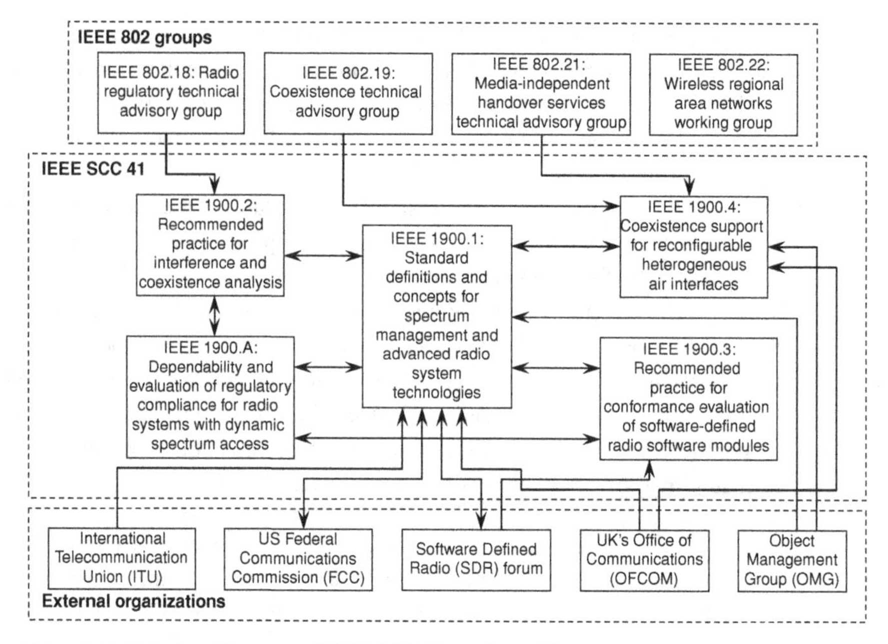

Figure 2.19 Relationship among IEEE SCC 41 projects [1].

area networking (WRAN) technology, which is expected to support mobile users in a cell with a coverage range up to 100 kilometers. Since IEEE 802.22 standard-based systems will reuse TV bands, the cognitive radio concept will be applied to avoid interference to incumbent licensed services (i.e. TV services).

The system architecture of an IEEE 802.22-based WRAN is similar to that of the current broadband wireless access (BWA) network such as IEEE 802.16 WiMAX. In particular, a WRAN is based on point-to-multipoint connections in which the base station (BS) in a cell controls all the connections from consumer premise equipments (CPEs). For example, spectrum access by CPEs and the size of the transmission burst allocated to CPEs in both uplink and downlink are determined by the BS. Also, relay base stations can be deployed to extend the coverage area of a WRAN.

Physical layer in 802.22

In the physical layer, the transmissions of IEEE 802.22-based devices on target frequency bands are based on orthogonal frequency-division multiple access (OFDMA). Multiple TV bands can be used simultaneously to enhance the system throughput by using channel bonding techniques. Channel bonding is a technique used to aggregate multiple bands for transmission of a single data stream. A similar technique is used in IEEE 802.11n-based WLANs. Various modulation and coding modes with spectral efficiency ranging from 0.5 bit/symbol/Hz to 5 bits/symbol/Hz will be supported in the standard. An IEEE 802.22 network is expected to support more than 10 users per cell per band with a downlink and uplink aggregate throughput of 1.5 Mbps and 384 kbps, respectively.

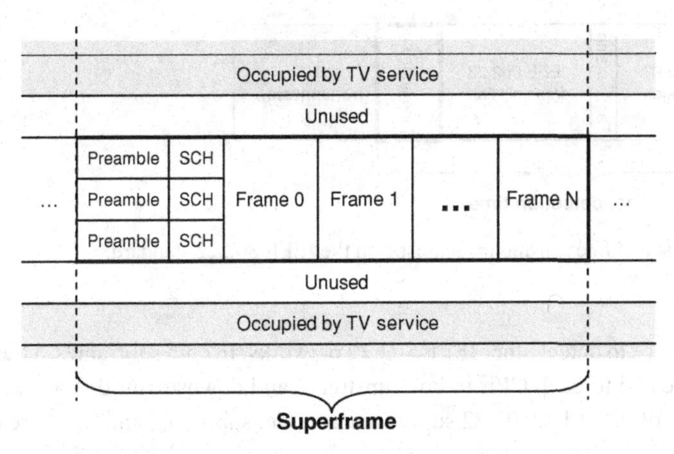

Figure 2.20 Superframe structure in the IEEE 802.22 standard.

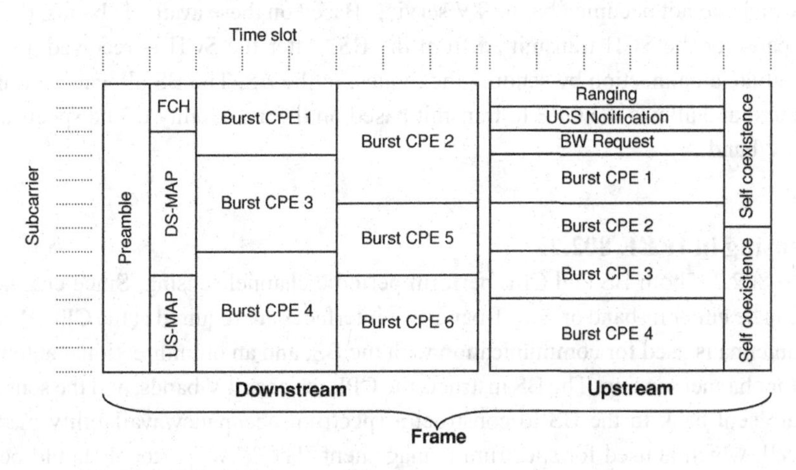

Figure 2.21 IEEE 802.22 frame structure.

MAC layer in 802.22

Figure 2.20 shows the superframe structure of IEEE 802.22 MAC. At the beginning of a superframe, preamble and superframe control header (SCH) information is transmitted by the BS over all available TV bands. The preamble is used to protect incumbent service, while SCH is used by the CPE to synchronize with the BS – it contains all information for a CPE to initiate a connection. Note that, due to the requirement of the FCC, two TV bands remain unused to prevent interference to the TV service.

A superframe consists of multiple transmission frames, and the structure of a transmission frame is shown in Figure 2.21. Transmissions in each frame are time-slotted and are based on OFDM subcarriers. In one frame, there are upstream and downstream subframes. A downstream subframe consists of a preamble to indicate the beginning of the frame. Then, US-MAP and DS-MAP are used to indicate the structure of the upstream and downstream subframes. In the upstream subframe, there is a contention interval for CPE initialization (e.g. ranging), urgent coexistence situation (UCS) notification, and

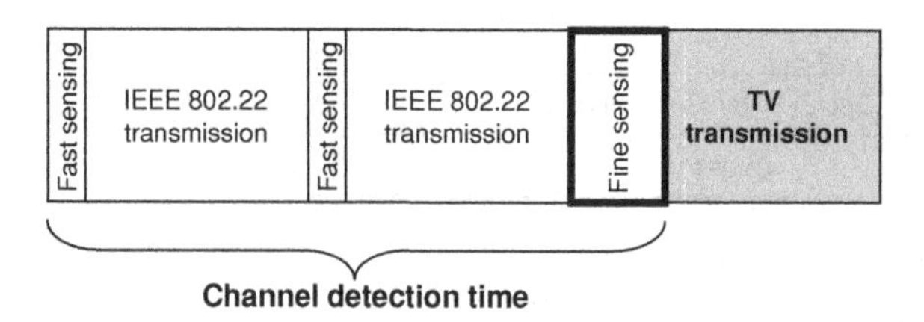

Figure 2.22 Fast and fine sensing mechanisms in the IEEE 802.22 standard.

self-coexistence to detect other IEEE 802.22 networks. In both subframes, a transmission burst is allocated to each CPE in both upstream and downstream directions. Note that the structure of the IEEE 802.22 superframe, frame, subframe, and burst are controlled by the BS.

To initiate a connection with the BS, a CPE scans and identifies the available TV bands which are not occupied by the TV service. Based on these available bands, the CPE then scans for the SCH transmitted from the BS. After the SCH is received, the CPE can initiate a connection by sending the request to the BS. The BS allocates a transmission burst and allows the CPE to transmit based on the availability of the spectrum (i.e. the TV band).

Channel sensing in IEEE 802.22

In 802.22, both BS and CPE perform periodic channel sensing. Since channel sensing can be either in-band or out-of-band, two interfaces are required at the CPE. A directional antenna is used for communication with the BS, and an omnidirectional antenna is used for channel sensing. The BS instructs the CPE to sense TV bands, and the sensing results are sent back to the BS to construct a spectrum occupancy/availability map for each cell, which is used for spectrum management. The MAC protocol should be designed to support the required spectrum management functions including channel switching, channel suspension/resuming, and inclusion/removal of channels from a channel access list.

To detect an incumbent TV service, IEEE 802.22 supports fast and fine in-band sensing mechanisms (Figure 2.22) to improve system efficiency [188]. Fast sensing (e.g. energy detection) is performed over a small time interval to reduce the interruption to data transmission. Results of this sensing are used to analyze the occupancy of channels. If a high transmission energy is detected, fine sensing is performed in the target channel. This fine sensing algorithm takes a longer period of time (e.g. up to 25 ms) to identify the particular signal signature from the incumbent service. These fast and fine sensing mechanisms must ensure the requirements of the IEEE 802.22 standard. Specifically, digital TV signals with a power above the threshold of -116 dBm should be detected with a probability of at least 0.9 (i.e. the maximum probability of a false alarm is 0.1). The incumbent service should be detected within no more than two seconds after the incumbent service starts utilizing the channel.

Open research issues in IEEE 802.22

The major challenges in IEEE 802.22 networks can be summarized as follows:

- *Coexistence with incumbent service:* To avoid interference to the incumbent service, in-band channel sensing is required. A CPE must perform in-band channel sensing and report the sensing result back to the BS. The BS can then construct a spectrum occupancy/availability map, and, subsequently, perform an optimal channel allocation for communications with all CPEs. Efficient channel sensing mechanisms will need to be developed which are able to maximize the exploration of spectrum opportunities with low signaling overhead.
- *Self-coexistence:* Since multiple IEEE 802.22 networks can operate in the same or overlapping areas and the number of unused TV bands is limited, IEEE 802.22 networks require a careful channel access scheme to avoid interference to each other. This issue is important especially in terms of competition and cooperation among IEEE 802.22 network service providers. For self-coexistence, the 802.22 networks may compete or cooperate with each other. Efficient methods for self-coexistence and the corresponding network performance analysis models need to be developed.
- *Economic models and pricing:* Since an IEEE 802.22 WRAN will use TV bands which belong to TV service providers (i.e. TV stations/broadcasters), an IEEE 802.22 WRAN service provider will need to acquire the right to access these TV bands from the spectrum owners (i.e. the TV service providers). Therefore, economic and pricing models need to be developed for spectrum trading between TV band owners and WRAN service providers. From such a model, the optimal number of TV bands to be traded as well as their optimal pricing can be obtained.
- *QoS support:* Development of a QoS framework to support different traffic types in IEEE 802.22 networks is an open research issue. Several types of traffic will need to be supported, including constant-bit-rate, real-time, and best-effort traffic. In this case, traffic scheduling, resource allocation, and admission control in the QoS framework must be designed to support dynamic and opportunistic channel access in 802.22 networks.

2.6 Summary

Due to the command-and-control approach used in traditional spectrum licensing, the radio spectrum cannot be efficiently utilized. Therefore, a new spectrum licensing scheme is being developed which will improve the flexibility of spectrum access. This flexibility will be achieved through the use of cognitive radios implemented as software-defined radios. With cognitive radio, a wireless transceiver can change the transmission parameters dynamically according to the changing environment. A cognitive radio transceiver should have the ability to observe, learn, schedule, plan, and optimize spectrum access to improve the performance of wireless communication systems. With this ability, the frequency spectrum unused by licensed users can be utilized by unlicensed users. However, an unlicensed user must ensure that the interference caused to the licensed user due to its transmission remains below the interference temperature limit. The interference temperature limit is used to quantify the highest transmit power over a particular frequency band.

To detect the presence of licensed users, an unlicensed user has to perform spectrum sensing. This spectrum sensing can be performed either in a non-cooperative or a cooperative manner. In non-cooperative spectrum sensing, each unlicensed user senses the radio spectrum independently. In the case of cooperative spectrum sensing, multiple unlicensed users cooperate by exchanging sensing information/sensing results between each other.

The research issues in different protocol layers of a cognitive radio network have been presented. Also, several approaches in designing cognitive radio, i.e. based on artificial intelligent and cross-layer design, have been discussed.

The major features of the emerging wireless standard for cognitive radio, namely, IEEE 802.22, which will operate on the TV band, have been summarized. IEEE 802.22 will use the cognitive radio concept to access the spectrum allocated to TV services (i.e. licensed systems) to provide WRAN services to unlicensed users.

Part II

Techniques for design, analysis, and optimization of dynamic spectrum access and management

3 Signal processing techniques

Signal processing deals with the analysis, interpretation, and manipulation of signals. Signals of interest include sound, images, biological signals, radar signals, and many others. Processing of such signals includes filtering, storage and reconstruction, separation of information from noise (e.g. aircraft identification by radar), compression (e.g. image compression), and feature extraction (e.g. speech-to-text conversion). In communications systems, signal processing is mostly performed at OSI layer 1, the physical layer (modulation, equalization, multiplexing, radio transmission, etc.), as well as at OSI layer 6, the presentation layer (source coding, including analog-to-digital conversion, and data compression).

In cognitive radio networks, the major task of signal processing is spectrum sensing for detecting the unused spectrum and sharing it without harmful interference to other users. One important requirement in a cognitive radio network is sensing spectrum holes reliably and efficiently. Spectrum sensing techniques can be classified into three categories. First, cognitive radios must be capable of determining if a signal from a primary transmitter is locally present in a certain spectrum. Several approaches are used for transmitter detection, such as matched filter detection, energy detection, cyclostationary feature detection, and wavelet detection. Second, collaborative detection refers to spectrum sensing methods where information from multiple cognitive radio users is exploited for primary user detection. Third, the sensing devices can be separated from the secondary users and can be deployed into the cognitive network by the cognitive radio service provider. By doing this, the cost of the secondary user devices can be reduced and the hidden terminal problem/exposed terminal problem can be mitigated.

Another important task of signal processing is filtering and prediction. The secondary users can utilize these techniques for improved spectrum efficiency. Important filtering and prediction techniques include the autoregressive moving average (ARMA) model, Wiener filter, least mean square filter, recursive least square filter, and Kalman filter.

Finally, most current digital systems use Nyquist sampling theory to obtain the data information. For most applications, the real data space is much less than the sampled space. This creates significant waste in the processing power and transmission requirements. To overcome this problem, compressed sensing has recently been proposed, which uses sampling at much lower rates than the Nyquist sampling rate. The resulting signal can be recovered with a high probability using constrained optimization.

This chapter is organized as follows: first, in Section 3.1, the basics of estimation and detection are described, and then the basic techniques used for spectrum sensing are discussed. In Section 3.2, the concept of collaborative spectrum sensing is illustrated and analyzed. In Section 3.3, the idea of using separate sensing devices is proposed and the resulting improvement of the hidden/exposed terminal problems is investigated. Filtering and prediction techniques are studied in Section 3.4. Finally, the basics of compressed sensing and its applications are discussed in Section 3.5.

3.1 Spectrum sensing

In this section, spectrum estimation techniques are first discussed. Then, the problem of spectrum hole detection is investigated. Finally, practical spectrum sensing techniques are illustrated.

3.1.1 Interference temperature/channel estimation

Estimation theory is a branch of statistics and signal processing that deals with estimating the values of parameters based on measured/empirical data. The parameters describe the physical scenario or object that answers a question posed by the estimator, e.g. the channel estimation and interference temperature estimation for the secondary user. In estimation theory, it is assumed that the desired information is embedded into a noisy signal. Noise adds uncertainty that makes it difficult for correct estimation.

An estimator takes the measured data as input and produces an estimate of the parameters. It is desirable to derive an estimator that exhibits optimality. An optimal estimator would indicate that all available information in the measured data has been extracted. The general steps to design an estimator are as follows:

1. In order to design a desired estimator for estimating a single or multiple parameters, it is first necessary to determine a model for the system. This model should incorporate the process being modeled as well as points of uncertainty and noise.
2. The model describes the physical scenario in which the parameters apply. After deciding upon a model, it is helpful to find the limitations placed upon the estimator. These limitations can be found using the Cramer–Rao bound, for example, which will be discussed later.
3. Next, an estimator needs to be developed or applied (if there is already a known estimator, which is valid for the model). The estimator needs to be tested against the limitations to determine if it is an optimal estimator.
4. Finally, experiments or simulations can be run using the estimator to test its performance.

After developing an estimator, real data might show that the model used to derive the estimator is incorrect, which may require these steps to be repeated to find a new estimator. A non-implementable or infeasible estimator may need to be scrapped and a new estimator found. In summary, the estimator estimates the parameters of a physical model based on the measured data.

To build a model, several statistical components need to be known. These are needed to ensure that the estimator has some mathematical tractability. The first component is a set of statistical samples taken from a random vector (RV) of size N as a vector $\mathbf{x} = [x_1, x_2, \ldots, x_N]^{\mathrm{T}}$. Secondly, from $\mathbf{x}$, we need to estimate M parameters $\theta = [\theta_1, \theta_2, \ldots, \theta_M]^{\mathrm{T}}$, which need to be established with their probability density function (pdf) or probability mass function (pmf) $p(\mathbf{x}|\theta)$. It is also possible for the parameters themselves to have a probability distribution (e.g. Bayesian statistics). It is then necessary to define the epistemic probability $\pi(\theta)$. After the model is formed, the goal is to estimate the parameters, commonly denoted by $\hat{\theta}$. One common estimator is the minimum mean squared error (MMSE) estimator, which utilizes the error between the estimated parameters and the actual values of the parameters $\mathbf{e} = \hat{\theta} - \theta$ as the basis for optimality. This error term is then squared and minimized for the MMSE estimator. If $E[\mathbf{e}] = \mathbf{0}$, the estimator is called an unbiased estimator.

In the following, several typical and commonly used estimators are discussed, namely, the maximum likelihood estimator, maximum a posteriori (MAP)/Bayesian estimator, periodogram, subspace estimator, and blind estimator. Then the fundamental bound for the unbiased estimator, the Cramer–Rao lower bound, is discussed.

Maximum likelihood estimator

Maximum likelihood estimation is a popular statistical method used for fitting a mathematical model to some data. Modeling real world data by estimating maximum likelihood offers a way of tuning the free parameters of the model to provide a good fit. Consider a family D_θ of probability distributions parameterized by an unknown parameter θ (which could be vector-valued), associated with either a known probability density function (continuous distribution) or a known probability mass function (discrete distribution), denoted as f_θ. A sample $x_1, x_2, \ldots, x_N$ of N values is drawn from this distribution, and then using f_θ data, $f_\theta(x_1, \ldots, x_N|\theta)$ is computed. As a function of θ with $x_1, x_2, \ldots, x_N$ fixed, this is the likelihood function

$$L(\theta) = f_\theta(x_1, \ldots, x_N|\theta). \tag{3.1}$$

The maximum likelihood method estimates θ by finding the value of θ that maximizes $L(\theta)$. This is the maximum likelihood estimator (MLE) of θ:

$$\hat{\theta} = \arg\max_\theta L(\theta). \tag{3.2}$$

Commonly, it is assumed that data drawn from a particular distribution are independent, identically distributed (iid) with unknown parameters. This considerably simplifies the problem because the likelihood can then be written as a product of N univariate probability densities:

$$L(\theta) = \prod_{i=1}^{N} f_\theta(x_i|\theta) \tag{3.3}$$

and since maxima are unaffected by monotone transformations, the logarithm of this expression can be taken to turn it into a sum:

$$L(\theta) = \sum_{i=1}^{N} \log f_\theta(x_i|\theta). \tag{3.4}$$

The maximum of this expression can then be found numerically using various optimization algorithms.

Note that the MLE may not be unique, or indeed may not even exist. Moreover, the MLE might not be the unbiased one. In many cases, estimation is performed using a set of independent identically distributed measurements. In such cases, it is of interest to determine the behavior of a given estimator as the number of measurements increases to infinity, which is referred to as the asymptotic behavior. Under certain (fairly weak) regularity conditions, the MLE exhibits several characteristics, which can be interpreted to mean that it is "asymptotically optimal." These include asymptotically unbiased, asymptotically efficient, and asymptotically normal characteristics.

Maximum a posteriori (MAP)/Bayesian estimator

The method of maximum a posteriori (MAP, or posterior mode) estimation can be used to obtain a point estimate of an unobserved quantity on the basis of empirical data. It is closely related to the MLE, but employs an augmented optimization objective that incorporates a prior distribution over the quantity one wants to estimate. MAP estimation can therefore be seen as a regularization of the MLE.

Compared with the MLE, it is assumed that a prior distribution π over θ exists. This allows us to treat θ as a random variable as in Bayesian statistics. Then the posterior distribution of θ is given as follows:

$$\theta \mapsto \frac{f(\mathbf{x}|\theta)\pi(\theta)}{\int_\Theta f(\mathbf{x}|\theta')\pi(\theta')d\theta'} = \arg\max_\theta f(\mathbf{x}|\theta)\pi(\theta), \tag{3.5}$$

where Θ is the domain of π. This is a straightforward application of Bayesian rule.

The method of maximum a posteriori estimation then estimates θ as the mode of the posterior distribution of this random variable:

$$\hat{\theta}_{\mathrm{MAP}}(\mathbf{x}) = \arg\max_\theta f(\mathbf{x}|\theta)\pi(\theta). \tag{3.6}$$

Observe that the MAP estimate of θ coincides with the ML estimate when the prior π is uniform (i.e. a constant function). The MAP estimate is the Bayes estimator under the uniform loss function.

MAP estimates can be computed in several ways:

1. Analytically, when the mode(s) of the posterior distribution can be given in a closed form. This is the case when conjugate priors are used.
2. Via numerical optimization such as the conjugate gradient method or Newton's method. This usually requires first or second derivatives, which have to be evaluated analytically or numerically.

3. Via a modification of an expectation–maximization algorithm. This does not require derivatives of the posterior density.

The expectation–maximization (EM) algorithm alternates between performing an expectation (E) step, which computes an expectation of the likelihood by including the latent variables as if they were observed, and a maximization (M) step, which computes the maximum likelihood estimates of the parameters by maximizing the expected likelihood found on the E step. The parameters found on the M step are then used to begin another E step, and the process is repeated.

- *E-step*: Estimation for unobserved event, conditioned on the observation, using the values from the last maximization step.
- *M-step*: Maximizing the expected log-likelihood of the joint event.

A Bayesian estimator is an estimator or decision rule that maximizes the posterior expected value of a utility function or minimizes the posterior expected value of a loss function (also called posterior expected loss). Suppose an unknown parameter θ is known to have a prior distribution π. Let δ be an estimator of θ (based on some measurements $\mathbf{x}$), and let $R(\theta, \delta)$ be a risk function, such as the mean squared error. The Bayes risk of δ is defined as $E_\pi R(\theta, \delta)$, where the expectation is taken over the probability distribution of θ. An estimator δ is said to be a Bayes estimator if it minimizes the Bayes risk among all estimators.

Periodogram estimator

The periodogram is an estimate of the spectral density of a signal. Let $x_w[n] = w[n]x[n]$ denote a windowed segment of samples from a random process $x[n]$, where the window function $w[n]$ (classically the rectangular window) contains N non-zero samples. Then, the periodogram is defined as the squared-magnitude discrete time Fourier transform (DTFT) divided by N as follows:

$$P(w) = \frac{1}{N}|DTFT(x_w)|^2 = \frac{1}{N}|\sum_{n=0}^{N-1} x_w[n]e^{-jwn}|^2. \tag{3.7}$$

In practice, the periodogram is often computed from a finite-length digital sequence using a fast Fourier transform (FFT). The raw periodogram is not a good spectral estimate because of spectral bias and the fact that the variance at a given frequency does not decrease as the number of samples used in the computation increases.

The spectral bias problem arises from a sharp truncation of the sequence, and can be reduced by first multiplying the finite sequence by a window function which truncates the sequence gracefully rather than abruptly. The variance problem can be reduced by smoothing the periodogram. Spectral estimation deals with various techniques to reduce spectral bias and variance. In the following, two commonly used methods to improve the periodogram are discussed.

Bartlett's method is used for estimating power spectra. Bartlett's method provides a way to reduce the variance of the periodogram at the cost of reduction in resolution, compared to standard periodograms. Common applications of Bartlett's method are frequency response measurements and general spectrum analysis. Bartlett's method consists of the following steps:

1. The original N point data segment is split up into K data segments of length M.
2. For each segment, compute the periodogram by computing the discrete Fourier transform, and then compute the squared magnitude of the result.
3. Average the result of the squared periodograms above for the K data segments. The averaging reduces the variance, compared to the original N point data segment.

The end result is an array of power measurements vs. frequency "bin."

Welch's method is used for estimating the power of a signal vs. frequency, and for reducing noise. Welch's method is based on the concept of using periodograms and is an improvement on the standard periodogram and Barlett's method, in that it reduces noise in the estimated power spectra at the cost of frequency resolution. Due to the noise caused by imperfect and finite data, Welch's method is often desirable due to the reduction in noise.

Welch's method is based on Barlett's method and differs in the two following ways:

1. The signal is split up into overlapping segments. The original data segment is split up into L data segments of length M, with an overlapping of D points. If $D = M/2$, the overlap is said to be 50 percent. If $D = 0$, the overlap is said to be 0 percent, which is the same situation as in Barlett's method.
2. The overlapping segments are then windowed. After the data is split up into overlapping segments, the individual L data segments have a window applied to them (in the time domain). Most window functions have more influence to the data at the center of the set than to data at the edges, which represents a loss of information. To mitigate that loss, the individual data sets are commonly overlapped in time (as in the above step). The windowing of the segments is what makes Welch's method a "modified" periodogram.

After doing the above, the periodogram is calculated by computing the discrete Fourier transform, and then computing the squared magnitude of the result. The individual periodograms are then time-averaged, which reduces the variance of the individual power measurements. The end result is an array of power measurements vs. frequency "bin."

Subspace estimator

Subspace estimator generates a frequency component estimation based on the eigenanalysis or eigendecomposition of the correlation matrix, which is very effective for detection of sinusoids buried in noise, especially when the SNR is low.

Multiple signal classification (MUSIC) is a subspace algorithm widely used for frequency estimation. MUSIC estimates the frequency content of a signal or autocorrelation matrix using an eigenspace method. This method assumes that a signal $\mathbf{x}$ consists of M complex exponentials in the presence of Gaussian white noise. Given an $N \times N$ autocorrelation matrix, $\mathbf{R_x}$, if the eigenvalues are sorted in decreasing order, the M eigenvectors $\mathbf{v}_1, \ldots, \mathbf{v}_M$ correspond to the M largest eigenvalues spanning the signal subspace. Note that $N > M$, so that there are $N - M$ null space for noise and the noise eigenvectors are $\mathbf{v}_{M+1}, \ldots, \mathbf{v}_N$. The frequency estimation function for MUSIC is [189]

$$\hat{w}(e^{jw}) = \frac{1}{\sum_{i=M+1}^{N} |\mathbf{e}^H \mathbf{v}_i|^2}, \tag{3.8}$$

where $\mathbf{e} = [1\ e^{jw}\ e^{2jw} \ldots e^{j(N-1)w}]^T$. In other words, when w is the same to the original frequency in $\mathbf{x}$, the denominator of (3.8) approaches zero due to the orthogonality of the signal space $(\mathbf{v}_1, \ldots, \mathbf{v}_M)$ and noise space $(\mathbf{v}_{M+1}, \ldots, \mathbf{v}_N)$. As a result, $\hat{w}(e^{jw})$ has a peak at the corresponding frequency.

Other popular subspace estimation methods include estimation of signal parameters via rotational invariance techniques (ESPRIT), which uses rotational invariance, and root-MUSIC, which is based on polynomial rooting.

Blind estimator

Blind equalization in communication or blind deconvolution in mathematics are techniques in which the transmitted signal is inferred from the received signal. Unlike other techniques which use training sequence (e.g. known transmitted signals), these techniques make use of only the transmitted signal statistics. Blind equalization is essentially blind deconvolution applied in digital communications. Nonetheless, the emphasis in blind equalization is on online estimation of the equalizer filter, which is the inverse of the channel impulse response, rather than the estimation of the channel impulse response itself. This is due to blind deconvolution's common mode of usage in digital communications systems as a means to extract the continuously transmitted signal from the received signal, with the channel impulse response being of secondary importance.

Many algorithms for the solution of the blind equalization problem have been suggested over the years. For example, Bussgang methods, which make use of the least mean squares filter algorithm; the constant modulus algorithm (CMA), which tries to drive the output signal to one having a constant amplitude; the subspace method; polyspectra techniques, which utilize higher-order statistics in order to compute the equalizer; and techniques that explore the discrete properties of transmitting digital signals.

Cramer–Rao lower bound

The CRB (Cramer–Rao bound) is one of the most important bounding criteria in the estimation theory. In its simplest form, the bound states that the variance of any unbiased estimator is at least as high as the inverse of the Fisher information (defined in (3.10)). An unbiased estimator that achieves this lower bound is said to be efficient. Such a solution achieves the lowest possible mean squared error among all unbiased methods, and is therefore the minimum variance unbiased (MVU) estimator. However, in some cases, no unbiased technique exists which achieves the bound. This may occur even when an MVU estimator exists. The Cramer–Rao bound can also be used to bound the variance of biased estimators. In some cases, a biased approach can result in both a variance and a mean squared error that are below the unbiased Cramer–Rao lower bound. CRB has two forms – one is for the scalar case and one is for the vector case.

In the scalar unbiased case, suppose θ is an unknown deterministic parameter which is to be estimated from measurements $\mathbf{x}$, distributed according to some probability density function $f(\mathbf{x};\theta)$. The variance of any unbiased estimator of $\hat{\theta}$ is then bounded by the inverse of the Fisher information $I(\theta)$ [190]:

$$\mathrm{var}(\hat{\theta}) \geq \frac{1}{I(\theta)}, \tag{3.9}$$

where the Fisher information $I(\theta)$ is defined by [191]

$$I(\theta) = E\left[\left(\frac{\partial l(\mathbf{x};\theta)}{\partial\theta}\right)^2\right] = -E\left[\frac{\partial^2 l(\mathbf{x};\theta)}{\partial\theta^2}\right] \tag{3.10}$$

and $l(\mathbf{x};\theta) = \log f(\mathbf{x};\theta)$ is the natural logarithm of the likelihood function.

The efficiency of an unbiased estimator $\hat{\theta}$ measures how close this estimator's variance comes to this lower bound, which is defined as

$$e(\hat{\theta}) = \frac{I(\theta)^{-1}}{\mathrm{var}(\hat{\theta})} \tag{3.11}$$

or the minimum possible variance for an unbiased estimator divided by its actual variance. The Cramer–Rao lower bound is given by $e(\hat{\theta}) \leq 1$.

In the vector unbiased case, define a parameter column vector $\theta = [\theta_1, \theta_2, \ldots, \theta_M]^{\mathrm{T}}$ with probability density function $f(\mathbf{x};\theta)$, which satisfies the two regularity conditions below:

- The Fisher information is always defined; equivalently, for all $\mathbf{x}$ such that $f(\mathbf{x};\theta) > 0$, $\frac{\partial \ln f(\mathbf{x};\theta)}{\partial\theta}$ exists, and is finite.
- The operations of integration with respect to $\mathbf{x}$ and differentiation with respect to θ can be interchanged in the expectation of T; that is,

$$\frac{\partial}{\partial\theta}\left[\int T(\mathbf{x}) f(\mathbf{x};\theta)\mathrm{d}x\right] = \int T(\mathbf{x})\left[\frac{\partial}{\partial\theta} f(\mathbf{x};\theta)\right]\mathrm{d}x \tag{3.12}$$

whenever the right-hand side is finite. This condition can often be confirmed by using the fact that integration and differentiation can be swapped when either of the following cases hold: (1) the function $f(\mathbf{x};\theta)$ has bounded support in $\mathbf{x}$, and the bounds do not depend on θ; (2) the function $f(\mathbf{x};\theta)$ has infinite support, is continuously differentiable, and the integral converges uniformly for all θ.

The Fisher information matrix is an $M \times M$ matrix with element $\mathbf{I}_{m,k}$ defined as

$$\mathbf{I}_{m,k} = E\left[\frac{\mathrm{d}}{\mathrm{d}\theta_m}\log f(\mathbf{x};\theta)\frac{\mathrm{d}}{\mathrm{d}\theta_k}\log f(\mathbf{x};\theta)\right]. \tag{3.13}$$

Let $\mathbf{T}(\mathbf{x})$ be an estimator of any vector function of parameters, $\mathbf{T}(\mathbf{x}) = [T_1(\mathbf{x}), \ldots, T_n(\mathbf{x})]^{\mathrm{T}}$, and denote its expectation vector by $E[\mathbf{T}(\mathbf{x})]$ by $\psi(\theta)$. The Cramer–Rao bound then states that the covariance matrix of $\mathbf{T}(\mathbf{x})$ satisfies

$$\mathrm{cov}_\theta[\mathbf{T}(\mathbf{x})] \geq \frac{\partial\psi(\theta)}{\partial\theta}[\mathbf{I}(\theta)]^{-1}\left(\frac{\partial\psi(\theta)}{\partial\theta}\right)^{\mathrm{T}}, \tag{3.14}$$

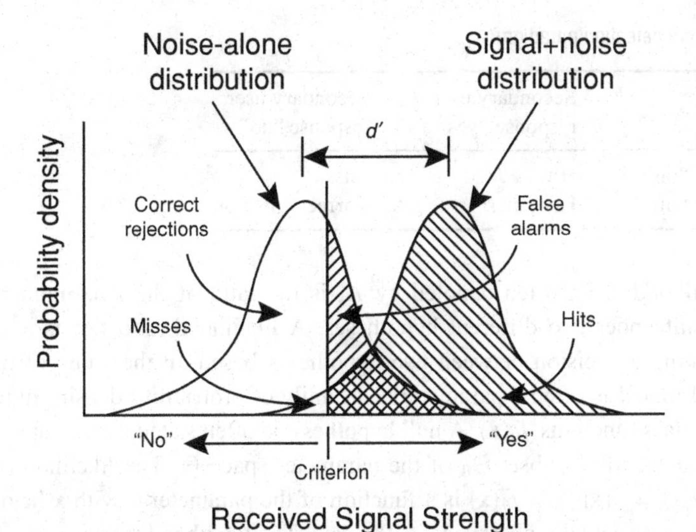

Figure 3.1 Illustration of a typical detection problem.

where the matrix inequality $\mathbf{A} \geq \mathbf{B}$ means that the matrix $\mathbf{A} - \mathbf{B}$ is positive semidefinite, and $\partial\psi(\theta)/\partial\theta$ is a matrix whose ijth element is given by $\partial\psi_i(\theta)/\partial\theta_j$. If $\mathbf{T}(\mathbf{x})$ is an unbiased estimator of θ (i.e. $\psi(\theta) = \theta$), then the Cramer–Rao bound reduces to

$$\text{cov}_\theta(\mathbf{T}(\mathbf{x})) \geq \mathbf{I}(\theta)^{-1}. \tag{3.15}$$

3.1.2 Detection of spectrum holes

The starting point for signal detection theory is that nearly all reasoning and decision making takes place in the presence of some uncertainty. Signal detection theory provides a precise language and graphic notation for analyzing decision making in the presence of uncertainty. The general approach of signal detection theory has direct applications in terms of spectrum sensing for cognitive radios. For instance, the secondary users need to detect whether or not a primary user is present in the network.

As an illustration, the probability distribution functions (pdfs) of the received signals at a secondary user are shown in Figure 3.1. If the primary user is absent, the pdf is a noise-only distribution. If the primary user's signal is being transmitted, the pdf is signal plus noise distribution. According to a certain criterion (or threshold), the secondary user determines if the primary user is present or not. Depending on whether or not the primary user is present and on the secondary user's decision, there are four possibilities as shown in Table 3.1. With the transmission of a primary user, if the secondary user detects the transmission, it is called a "hit"; otherwise, it is called a "miss." In the absence of a primary user, if the secondary user says the primary is "on," the case is called a "false alarm"; otherwise it is the "correct rejection." The false alarm is also called a type-I error and the miss is also called a type-II error. It is evident that the probabilities of all four cases highly depend on the threshold. How to select the optimal threshold using the likelihood ratio test and Neyman–Pearson lemma is discussed next.

Table 3.1 Signal detection paradigm.

	Secondary user response "yes"	Secondary user response "no"
Primary user "on"	Hit	Miss
Primary user "off"	False alarm	Correct rejection

The likelihood ratio, often denoted by Λ, is the ratio of the maximum probability of a result under two different hypotheses. A likelihood-ratio test is a statistical test for making a decision between two hypotheses based on the value of this ratio. A statistical model is often a parameterized family of probability density functions or probability mass functions $f_\theta(\mathbf{x})$. A null hypothesis is often stated by saying the parameter θ is in a specified subset Θ_0 of the parameter space Θ. The likelihood function $L(\theta) = L(\theta|\mathbf{x}) = p(\mathbf{x}|\theta) = f_\theta(\mathbf{x})$ is a function of the parameter θ with $\mathbf{x}$ held fixed at the value that was actually observed, i.e. the data. The likelihood ratio is

$$\Lambda(\mathbf{x}) = \frac{\sup\{L(\theta|\mathbf{x}) : \theta \in \Theta_0\}}{\sup\{L(\theta|\mathbf{x}) : \theta \in \Theta\}}. \tag{3.16}$$

Being a function of the data $\mathbf{x}$, the likelihood ratio is therefore a statistic. The numerator corresponds to the maximum probability of an observed result under the null hypothesis. The denominator corresponds to the maximum probability of an observed result under the alternative hypothesis. Under certain regularity conditions, the numerator of this ratio is less than the denominator. The likelihood ratio under those conditions is between 0 and 1. Lower values of the likelihood ratio mean that the observed result is less likely to occur under the null hypothesis. Higher values mean that the observed result is more likely to occur under the null hypothesis.

The likelihood-ratio test rejects the null hypothesis if the value of this statistic is too small. The threshold value of the statistic depends on the significance level of the test, i.e. on what probability of type-I error (false alarm) is considered tolerable. This is justified by the Neyman–Pearson lemma as follows:

Lemma 3.1 *Neyman–Pearson lemma states that when performing a hypothesis test between two point hypotheses*

$$\begin{cases} H_0 : \theta = \theta_0 \\ H_1 : \theta = \theta_1 \end{cases} \tag{3.17}$$

then the likelihood-ratio test which rejects H_0 in favor of H_1 when

$$\Lambda(x) = \frac{L(\theta_0|x)}{L(\theta_1|x)} \leq \eta \text{ where } P(\Lambda(x) \leq \eta|H_0) = \alpha \tag{3.18}$$

is the most powerful test of size α for a threshold η. If the test is most powerful for all $\theta_1 \in \Theta_1$, it is said to be uniformly most powerful for alternatives in the set Θ_1.

If the distribution of the likelihood ratio corresponding to a particular null and alternative hypothesis can be explicitly determined, then it can directly be used to form

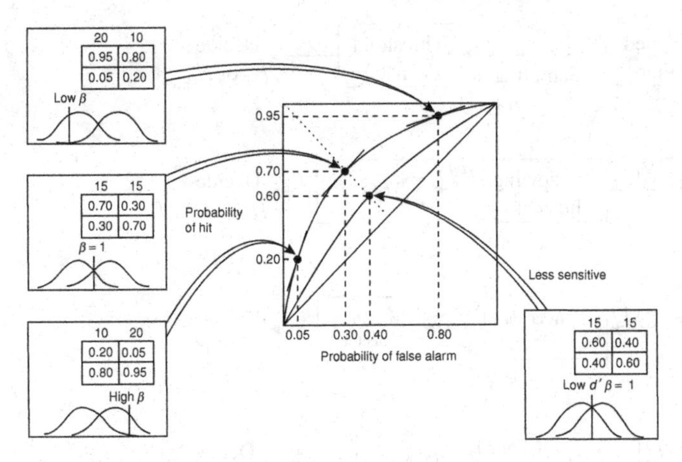

Figure 3.2 Operating characteristic curve for receiver.

decision regions (to accept/reject the null hypothesis). In most cases, however, the exact distribution of the likelihood ratio corresponding to a specific hypothesis is very difficult to determine. A convenient result, though, says that as the sample size approaches infinity, the test statistic $-2\log(\Lambda)$ will be asymptotically χ^2 distributed with degrees of freedom equal to the difference in dimensionality of Θ_1 and Θ_0. This means that for a great variety of hypotheses, a practitioner can take the likelihood ratio Λ, algebraically manipulate Λ into $-2\log(\Lambda)$, compare the value of $-2\log(\Lambda)$ given a particular result to the chi squared value corresponding to a desired statistical significance, and create a reasonable decision based on that comparison.

In signal detection theory, a receiver operating characteristic (ROC), or simply ROC curve, is a graphical plot of the sensitivity vs. (1 – specificity) for a binary classifier system as its discrimination threshold is varied. The ROC can also be represented equivalently by plotting the fraction of true positives (TPR = true positive rate) vs. the fraction of false positives (FPR = false positive rate). It is also known as a relative operating characteristic curve, because it is a comparison of two operating characteristics (TPR and FPR) as the criterion changes. ROC analysis provides tools to select possibly optimal models and to discard suboptimal ones independently from (and prior to specifying) the cost context or the class distribution. ROC analysis is related in a direct and natural way to cost/benefit analysis of diagnostic decision making.

For example, one ROC curve is shown in Figure 3.2. The ideal point is at the left upper corner of the curve where the probability of hit is 1 and the false alarm probability is 0. The 45 degree line from the left lower corner to the right upper corner is the line with the random guess. Any practical estimator should be above this line. The more the curve moves towards left upper, the better the detector is, since with the same false alarm probability the detector gives better hit probability. For each ROC curve of any specific detector, there is a tradeoff between the hit probability and false alarm probability. If the false alarm probability is required to be very small, the hit probability will also be low. All the ROC curves can be drawn by changing the decision threshold.

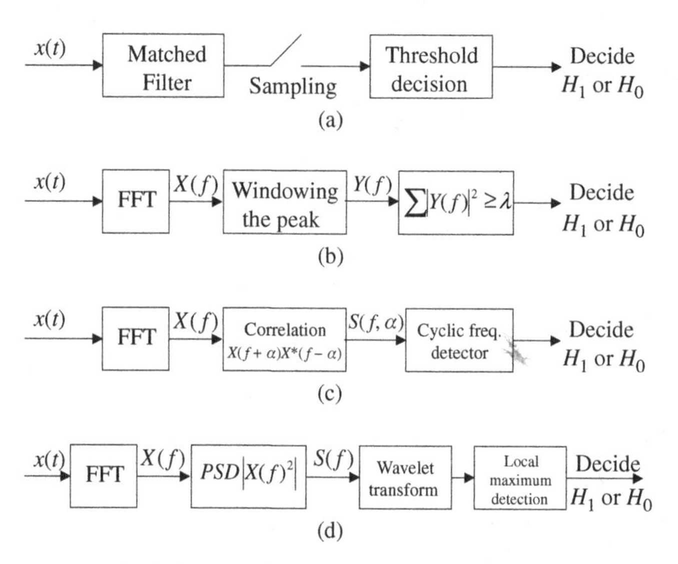

Figure 3.3 Typical approaches for practical spectrum sensing: (a) matched filter, (b) energy detection, (c) cyclostationary detection, (d) wavelet detection.

3.1.3 Practical spectrum sensing approaches

In this subsection, based on the techniques discussed in the previous sections, several popular spectrum sensing solutions are discussed [139]. Suppose the received signal is $x(t)$ and the spectrum detector needs to determine if the primary user's signal is present (hypothesis H_1) or the spectrum is vacant (hypothesis H_0). The four typical practical spectrum sensing approaches are illustrated in Figure 3.3 and described below:

1. *Matched filter*: A matched filter maximizes the signal-to-noise ratio at the sampling time, if the transmission waveform of the primary user is known. This requires the secondary users to have a priori knowledge of the primary user signal such as modulation type and order, pulse shaping, and packet format. The secondary users have to be equipped with carrier synchronization, timing devices, and even an equalizer. This is possible for some types of primary users such as TV signals and OFDM preambles. Due to the coherent nature of the matched filter, detection can be very fast. But if there are multiple types of primary users, the secondary users have to be equipped with multiple dedicated receivers.
2. *Energy detection*: An energy detector is a non-coherent detector that avoids the complicated coherent receivers required by a matched filter, and can be implemented using spectrum analyzing tools such as fast Fourier transform (FFT). Although an energy detector is very simple to implement, there are several drawbacks. First, the spectrum sensing speed is relatively slow. Second, the threshold for detection is very susceptible to the noise level and in-band interference. This is even worse in the frequency-selective and time-varying channels. Third, an energy detector cannot differentiate modulated signals, noise, and interference. As a result, the benefits of detection and interference cancellation techniques cannot be employed. Fourth, the

Table 3.2 Comparison among spectrum sensing solutions.

Sensing approach	Advantage	Disadvantage
Matched filter	Optimal performance and low cost	Prior knowledge of primary user's signal is required
Energy detection	No prior information required and low cost	Cannot work in low SNR; cannot distinguish primary and other secondary users
Cyclostationary detection	Robust in low SNR and interference	Partial information of primary user; high computation cost
Wavelet detection	Effective for wideband signal	Not usable for spread spectrum signals; high computation cost

primary user and the secondary user cannot be distinguished, while only the primary user's transmission should be protected. Finally, energy detection does not work for spread spectrum signals.

3. *Cyclostationary feature detection*: The modulated signals are usually cyclostationary, since built-in periodicity often occurs in training sequence, cyclic prefixes, etc. This periodicity is introduced in the transmitting signal of the primary users so that the receivers can exploit it for timing, channel estimation, etc. Also this periodicity can be used for the detection of a primary user. The basic approach is based on the autocorrelation function and power spectrum density. The spectral correlation function has a cyclic spectrum with cycle frequency. One advantage is that the stationary noise and interference do not exhibit spectral correlation. As a result, the cyclostationary detector can work in a low-SNR region.
4. *Wavelet detection*: Since the spectrum usage can cause irregularity in the power spectrum density, an attractive mathematical tool for analyzing singularities and irregular structure is the wavelet transform, which can investigate the local regularity of signals. The wavelet approach [192] provides advantages for wideband spectrum sensing over the conventional use of multiple narrow band filter banks in terms of implementation cost and flexibility.

The spectrum sensing approaches are compared in Table 3.2. The first tradeoff is the a priori information on the primary user and the performance. Obviously, the a priori information can greatly improve the performance. On the other hand, if the primary users are more "random" in nature, the approaches without requiring the prior information (such as the energy detector) would fit. But the performance can be low especially in the low-SNR area. The second tradeoff is the complexity. The cognitive radio devices for the customers are desired to have low cost. The choice of the spectrum sensing approach is determined by the specific scenario in consideration.

In addition to the above mentioned sensing approaches, cooperative sensing and replacement of sensing devices are attracting increasingly more attentions and will be discussed later in Section 3.2 and Section 3.3, respectively. The basic idea for cooperative sensing is to make the decision based on sensing reports from multiple cognitive radios.

The replacement of sensing devices requires the cognitive radio service provider to deploy separate sensing devices for the secondary users in the network.

3.2 Collaborative sensing

In this section, sensing techniques that involve multiple cognitive radios are discussed. If the collaboration can be performed for spectrum sensing, the performances can be significantly improved. On the other hand, there are many design tradeoffs such as false alarm and missing probability. We will discuss those techniques in two categories. First, we will discuss collaborative sensing involving multiple users exchanging information [135, 193, 194, 195, 196]. Then, we will investigate collaborative sensing using relay transmission [197, 198].

The goal of spectrum sensing is to decide on the hypotheses if the primary user is present (H_1) or not (H_0) as follows:

$$x(t) = \begin{cases} n(t), & H_0 \\ hs(t) + n(t), & H_1, \end{cases} \tag{3.19}$$

where $x(t)$ is the received signal at the cognitive radio, $n(t)$ is the thermal noise, h is the channel gain from the primary transmitter to the secondary user, and $s(t)$ is the transmit symbol from the primary user.

Using the different decision techniques mentioned in the previous section (such as the energy detection), the decision statistic is denoted by Λ. Suppose the decision threshold is λ. The detection probability and false alarm probability are given by

$$P_{\mathrm{d}} = P(\Lambda > \lambda | H_1) \tag{3.20}$$

and

$$P_{\mathrm{f}} = P(\Lambda > \lambda | H_0), \tag{3.21}$$

respectively. Depending on the different statistics of the channel gain h, the values are obtained by expectation.

Figure 3.4 illustrates the collaborative spectrum sensing in a shadowed environment. In this case, only the middle secondary user is able to detect the primary signal. Without collaboration, the two other secondary users would transmit assuming the absence of the primary user. If the collaboration can be performed among the secondary users to jointly detect the primary user, the detection probability can be significantly improved.

Suppose there are N secondary users. An OR-rule or 1-out-of-N decision rule is made for H_1, if any secondary user decides H_1. The probability of detection and probability of false-alarm for collaborative schemes can be written as

$$Q_{\mathrm{d}} = 1 - (1 - P_{\mathrm{d}})^N \tag{3.22}$$

and

$$Q_{\mathrm{f}} = 1 - (1 - P_{\mathrm{f}})^N, \tag{3.23}$$

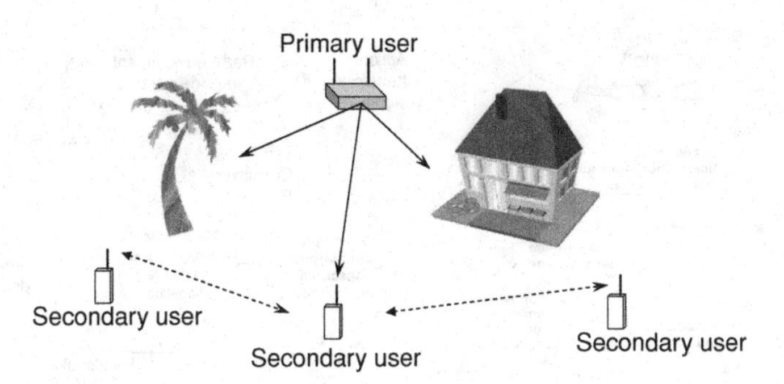

Figure 3.4 Illustration of collaborative sensing.

respectively. Compared with the local sensing for each individual in (3.20) and (3.21), the collaborative scheme increases probability of detection but at the same time increases the probability of false-alarm as well.

There are some variations to analyze collaborative sensing. First, the sensing probabilities for the different secondary users might be different. So equations (3.22) and (3.23) should be modified. Second, if the reporting channels from the secondary users to the decision center are not perfect, there would be tradeoffs [193], such as how many reports should be accepted. Finally, the reports from the secondary users would cost a lot of bandwidth and energy to transmit. A more smart way [195, 196] needs to be constructed if bandwidth and energy are of concern.

In [197, 198], another type of collaborative sensing using relay transmission was proposed. The secondary users operate in a fixed time-division multiple access (TDMA) mode for sending data to some common receiver. Suppose that a primary user starts using the band. The two secondary users need to stop using the band as soon as possible to avoid interference with the primary user. But the detection time might be too long if the secondary user is too far away from the primary user and thus the received signal is too weak. If the two users can help each other using relay (cooperative) transmission protocols such as the amplify-and-forward and decode-and-forward protocols, the spatial diversity inherent in a multiuser scenario can be exploited to achieve better network performance.

Compared with collaborative sensing, the decision can be made at the secondary users and then sent to the decision node, which is usually called the hard decision. Alternatively, the secondary users can send the statistics to the decision node which makes the decision. This is called the soft decision. Collaborative sensing with relay transmission can be viewed as a special kind of distributed soft decision process.

3.3 Replacement of sensing devices in secondary users

In this section, a novel network structure [199] is proposed to separate the spectrum sensing task from the unlicensed users (secondary users). The service provider for the secondary users needs to place sensing devices within the network of licensed users (primary users). These sensing devices sense primary users' activity and also decide

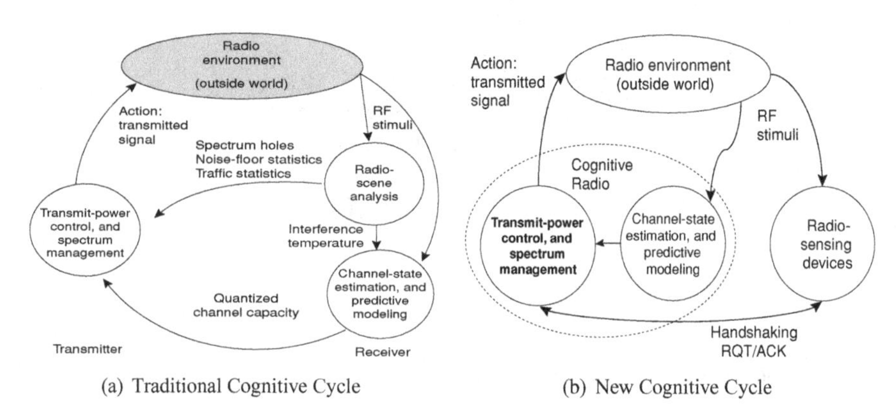

(a) Traditional Cognitive Cycle

(b) New Cognitive Cycle

Figure 3.5 Cognitive cycles.

whether to admit a secondary user's transmission. A new cognitive cycle is proposed accordingly, and the hidden terminal problem/exposed terminal problem are shown to be alleviated. The other advantage of the scheme arises from the business model point of view: the expensive sensing devices will be implemented by the cognitive radio service provider, instead of being built in the secondary user devices which are usually consumer products demanding low implementation cost.

3.3.1 New cognitive cycle with separate sensing devices

Two major objectives of cognitive radio are reliable communication and efficient utilization of the radio spectrum. To achieve these objectives, traditionally, three fundamental cognitive tasks [102] must be fulfilled:

1. Radio-scene analysis and detection of the spectrum holes.
2. Spectrum analysis such as channel-state estimation and prediction of channel capacity.
3. Transmit power control and spectrum management.

In the traditional cognitive cycle as shown in Figure 3.5(a), three tasks interact with each other to handle the outside world so that the best strategy of spectrum access can be calculated and implemented.

As has been mentioned before, the sensing components can be expensive for the secondary users. Moreover, the hidden terminal problem should be solved so that the primary users' performances are not impaired by the hidden secondary users. Next, a new cognitive network structure is developed to overcome these challenges, then a low-temperature handshake mechanism is constructed, and finally the proposed protocols are explained.

A new cognitive network structure is proposed in which the sensing mechanism is not implemented by the secondary users. Instead, the service provider deploys separate sensing devices in the network. Those sensing devices are able to detect primary users' activities and provide admission control for second users for spectrum access. In Figure 3.5(b), the new cognitive cycle is shown.

Compared with the traditional cognitive cycle, within the secondary user, the soft radio part is still maintained. In other words, the cognitive radio devices can still estimate channel state, adjust their transmit power, and manage the spectrum access. The difference is that the sensing part is moved to the sensing devices. The sensing devices perform admission control of secondary users. After admission, it is the secondary user's responsibility to combat the hostile wireless channel and maintain the link quality. The expensive sensing devices are moved out of the secondary users' devices for cheaper implementation.

Since the sensing devices and the cognitive radio users are separated, it is necessary to design a handshaking mechanism between sensing devices and secondary users. The licensed (to the primary users) frequency band is referred to as the *data band*. The handshaking between the sensing devices and the secondary users should not use the data band because the secondary user is not allowed to transmit (its request) until it gets permission. Therefore, the handshaking is traditionally performed at a separate frequency channel, referred to as the *handshake band*. As long as a sensing device detects the activity of primary users, it will deny any transmission request received in the handshake band from the secondary users. However, there are several challenges in the handshaking procedure:

- Data transmissions and handshaking are performed at different frequency bands, with probably different channel gains. It is possible that the sensing device may not be able to detect the transmission request from a secondary user due to deep fading at the handshake band. Note that, unlike a data band, the handshake band is usually a narrow band.
- When there are a number of secondary users, the handshake band may become the bottleneck. In addition, the handshake messages may collide with each other when the secondary users are hidden from each other.
- It is very likely that there are a number of sensing devices in the network. When a secondary user sends a request, more than one sensing devices may receive it. Then the feedback from the sensing devices will collide at the secondary user.

The first challenge implies that the handshake should be carried out at the data band. And the second and third challenges suggest that the traditional request/feedback at the handshake band may not work well. In the following, a handshake (modified from [200]) is proposed to address the above challenges. The handshake consists of two portions: a request from the secondary user and a potential veto from the sensing device, both at the data band with low interference temperature.

If a secondary user has traffic to send, it first sends a request (at the data band) that has a low transmit power level (much lower than power for its normal data transmission) and is spread by a common *request code*, which is known in advance to the sensing devices and the secondary network. No bit-information is carried by the request. The low temperature of the request is to guarantee that the tolerable interference temperatures of primary users are not exceeded by the extra interference from the request signal. The spreading code is to get enough gain such that the sensing devices are able to detect the request.

Table 3.3 Proposed secondary user protocol.

Step 1: Construct a set of channels that secondary users can adapt to.
Step 2: Send the request for the most preferable channel using the low-temperature handshake.
Step 3: Wait for a certain period of time for response.
Step 4: If rejected, remove this channel from the set and go to Step 2.
Step 5: Otherwise, choose this channel while listening to the sensing device. Stop using the channel whenever sensing device sends veto signal.

Table 3.4 Proposed sensing device protocol.

Step 1: Sense the primary user's activities at all times.
Step 2: Veto the secondary user's transmission, using a low-temperature handshake, whenever it conflicts with the primary user.
Step 3: Keep track of the secondary user's admission.
Step 4: Remove the right of spectrum access of secondary users by sending veto signals, as soon as the primary user's transmission is detected again.

Each sensing device continuously detects the received power (from primary users) at the data band. It also scans the common request code for the low-temperature request. If the sensing device detects activity of primary users and also detects a transmission with the common request code, it sends a veto that has a low transmit power level and is spread by a common *veto code*, to notify the secondary user of the rejection of its transmission request. No bit-information is carried by the veto.

After the secondary user sends the low-temperature request, it scans the common veto code. If no veto is detected, the secondary user is allowed to transmit at the data band.

It can be seen that, by the low-temperature request and veto, the challenges in the traditional handshake procedure can be solved effectively, due to the following reasons:

- The request is sent at the data band. Therefore, the different channel gain problem does not exist anymore.
- No bit-information is carried in the request. So when there are a number of secondary users and they all send requests at the same time, the sensing device can detect the energy from all the secondary users using a RAKE receiver. Therefore, the bottleneck effect and collisions of requests are avoided.
- When multiple sensing devices send veto messages, the secondary user can use a RAKE receiver to collect all the energy from the veto messages, since no bit-information is carried by the veto. Thus the secondary user can still be notified not to transmit.

Finally, based on the discussions in the previous two subsections, the protocols for the secondary user and the sensing device are presented in Table 3.3 and Table 3.4, respectively. The secondary user selects the preferred channel from the set of channels that it can adapt to, and then sends the request to the sensing device for spectrum access. The sensing device decides if the secondary user can utilize the channel. If the channel is occupied by the primary user, the sensing device sends the veto to prevent the secondary

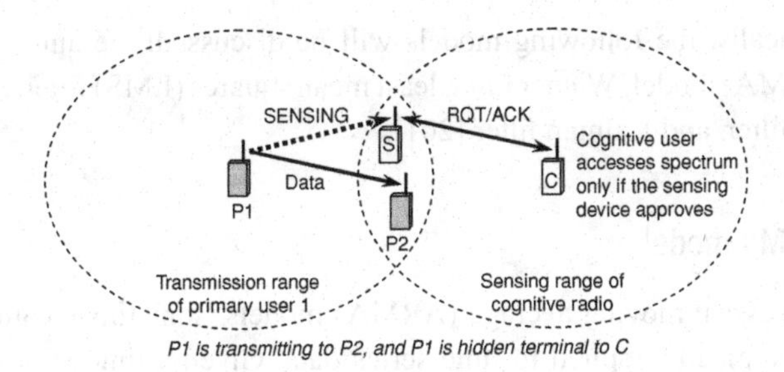

Figure 3.6 Solution of hidden terminal problem by using separate sensing devices.

user's transmission. If the primary user regains the channel, the sensing device sends the veto to stop all secondary users from transmission in this channel.

3.3.2 Remedies for hidden/exposed terminal problems

In a traditional sensing mechanism, a secondary sender senses the activities (of the primary users) in the spectrum band prior to its transmission. However, when a secondary sender is close to a primary receiver but far away from the primary sender, it cannot sense the transmission of the primary sender, and thus transmits its packet. The secondary user's transmission then corrupts the desired reception at the primary receiver. It is worth mentioning that, in wireless LAN, the hidden terminal problem can be solved based on a four-way handshake using request-to-send/clear-to-send messages. However, in many cognitive radio networks, the primary users may not have the provisioning for handshake.

This problem can be solved by the sensing devices. Consider the scenario shown in Figure 3.6. Primary user P1 transmits to primary user P2. If secondary user C is equipped with the sensing ability, it still cannot realize the existence of primary user P1 due to the hidden terminal problem. Instead, let us assume that a sensing device S is located in the network. Before secondary user C tries to access the primary users' spectrum, it sends a request to the nearby sensing device, which senses whether any primary user is utilizing the spectrum. If there is an active primary user, the sensing device sends back a veto message to the secondary user to deny the spectrum access. Otherwise, an acknowledgment message is sent to secondary user C to grant the spectrum access. By employing a sufficient number of sensing devices in the network, the hidden terminal problem is solved and no extra sensing ability is required for the secondary users.

Similarly, the proposed solution can alleviate the exposed terminal problem, because the sensing devices can be placed close to the primary receivers and have better probability of correct sensing than that of cognitive radios with sensing devices.

3.4 Filtering and prediction

In this section, several popular filtering concepts and prediction models are discussed. These concepts can be employed for adapting transmission in a dynamic cognitive radio environment. Specifically, the following models will be discussed: the autoregressive moving average (ARMA) model, Wiener filter, least mean squares (LMS) filter, recursive least squares (RLS) filter, and Kalman filter [201].

3.4.1 AR, MA, and the ARMA model

In statistics, autoregressive moving average (ARMA) models, sometimes called Box–Jenkins models, are typically applied to time series data. Given a time series of data $x[n]$, the ARMA model is a tool for understanding and, perhaps, predicting future values in the series. The model consists of two parts, an autoregressive (AR) part and a moving average (MA) part. The model is usually then referred to as the ARMA(p,q) model, where p is the order of the autoregressive part and q is the order of the moving average part.

The notation AR(p) refers to an autoregressive model of order p. The AR(p) model is given by

$$x[n] = \sum_{i=1}^{p} \varphi_i x[n-i] + \varepsilon[n], \tag{3.24}$$

where φ_i are the parameters of the model and $\varepsilon[n]$ is white noise. An autoregressive model is essentially an infinite impulse response filter with some additional interpretation placed on it. Some constraints are necessary on the values of the parameters of this model in order that the model remains stationary. For example, processes in the AR(1) model with $|\varphi_1| \geq 1$ are not stationary.

The parameters φ_i may be calculated using least squares regression or the Yule–Walker equations:

$$\gamma_m = \sum k = 1^p \varphi_k \gamma_{m-k} + \sigma_\varepsilon^2 \delta_m, \tag{3.25}$$

where $m = 0, \ldots, p$, yielding $p+1$ equations. γ_m is the autocorrelation function of x, σ_ε is the standard deviation of the input noise process, and δ_m is the Kronecker delta function. For $m > 0$, we have

$$\begin{bmatrix} \gamma_1 \\ \gamma_2 \\ \gamma_3 \\ \vdots \end{bmatrix} = \begin{bmatrix} \gamma_0 & \gamma_{-1} & \gamma_{-2} & \cdots \\ \gamma_1 & \gamma_0 & \gamma_{-1} & \cdots \\ \gamma_2 & \gamma_1 & \gamma_0 & \cdots \\ \cdots & \cdots & \cdots & \ddots \end{bmatrix} \begin{bmatrix} \varphi_1 \\ \varphi_2 \\ \varphi_3 \\ \vdots \end{bmatrix}. \tag{3.26}$$

For $m = 0$, we have

$$\gamma_0 = \sum_{k=1}^{p} \varphi_k \gamma_{-k} + \sigma_\varepsilon^2. \tag{3.27}$$

The notation MA(q) refers to the moving average model of order q as

$$x[n] = \varepsilon[n] + \sum_{i=1}^{q} \theta_i \varepsilon[n-i] \tag{3.28}$$

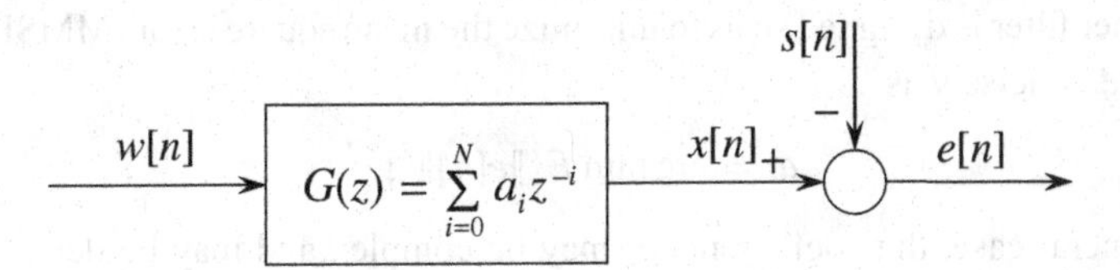

Figure 3.7 Illustration of a Wiener filter.

where θ_i are the parameters of the model and $\varepsilon[n]$ are the error terms. The moving average model is essentially a finite impulse response filter with some additional interpretation placed on it.

The notation ARMA(p,q) refers to the model with p autoregressive terms and q moving average terms. This model contains the AR(p) and MA(q) models as

$$x[n] = \varepsilon[n] + \sum_{i=1}^{p} \varphi_i x[n-i] + \sum_{i=1}^{q} \theta_i \varepsilon[n-i]. \tag{3.29}$$

ARMA models in general can, after choosing p and q, be fitted by least squares regression to find the values of the parameters which minimize the error term. It is generally considered to be a good practice to find the smallest values of p and q which provide an acceptable fit to the data. For a pure AR model the Yule–Walker equations may be used to provide a fit. ARMA is appropriate when a system is a function of a series of unobserved shocks (the MA part) as well as its own behavior.

3.4.2 Wiener filter

The Wiener filter was proposed by Norbert Wiener [202]. Its purpose is to reduce the amount of noise present in a signal by comparing it with an estimate of the desired noiseless signal. Wiener filters are characterized by the following:

- *Assumption*: Signal and (additive) noise are stationary linear stochastic processes with known spectral characteristics or known autocorrelation and cross-correlation.
- *Requirement*: The filter must be physically realizable, i.e. causal (this requirement can be dropped, resulting in a non-causal solution).
- *Performance criterion*: Minimum mean-square error.

A Wiener filter is shown in Figure 3.7. We consider a signal $w[n]$ being fed to a Wiener filter of order N and with coefficients $\{a_i\}, i = 0, \ldots, N$. The output of the filter is denoted as $x[n]$, which is given by

$$x[n] = \sum_{i=0}^{N} a_i w[n-i]. \tag{3.30}$$

The residual error is denoted $e[n]$ and is defined as

$$e[n] = x[n] - s[n], \tag{3.31}$$

where $s[n]$ is the original signal to be estimated/predicted.

The Wiener filter is designed so as to minimize the mean square error (MMSE), which can be stated concisely as

$$a_i = \arg\min E\{\|e[n]\|^2\}. \tag{3.32}$$

In the general case, the coefficients a_i may be complex and may be derived for the case where $w[n]$ and $s[n]$ are complex as well. For simplicity, only the case where all these quantities are real is considered. The mean square error may be rewritten as

$$\begin{aligned} E\{\|e[n]\|^2\} &= E\{(x[n]-s[n])^2\} \\ &= E\{x^2[n]\} + E\{s^2[n]\} - 2E\{x[n]s[n]\} \\ &= E\left\{(\sum_{i=0}^{N} a_i w[n-i])^2\right\} + E\{s^2[n]\} - 2E\left\{\sum_{i=0}^{N} a_i w[n-i]s[n]\right\}. \end{aligned} \tag{3.33}$$

To find the vector $[a_0, \ldots, a_N]$ which minimizes the expression above, the MMSE's derivative w.r.t. a_i is calculated as follows:

$$\begin{aligned} \frac{\partial E\{\|e[n]\|^2\}}{\partial a_i} &= 2E\left\{(\sum_{j=0}^{N} a_j w[n-j]w[n-i])\right\} - 2E\{s[n]w[n-i]\} \\ &= 2\sum_{j=0}^{N} E\{w[n-j]w[n-i]\}\, a_j - 2E\{w[n-i]s[n]\},\ i = 0, \ldots, N. \end{aligned} \tag{3.34}$$

If it is supposed that $w[n]$ and $s[n]$ are stationary, the following sequences $R_w[m]$ and $R_{ws}[m]$, known respectively as the autocorrelation of $w[n]$ and the cross-correlation between $w[n]$ and $s[n]$, can be introduced:

$$R_w[m] = E\{w[n]w[n+m]\} \tag{3.35}$$

$$R_{ws}[m] = E\{w[n]s[n+m]\}. \tag{3.36}$$

The derivative of the MMSE may therefore be rewritten as

$$\frac{\partial E\{\|e[n]\|^2\}}{\partial a_i} = 2\sum_{j=0}^{N} R_w[j-i]a_j - 2R_{sw}[i],\ i = 0, \ldots, N. \tag{3.37}$$

Letting the derivative be equal to zero,

$$\sum_{j=0}^{N} R_w[j-i]a_j = R_{sw}[i],\ i = 0, \ldots, N, \tag{3.38}$$

which can be rewritten in matrix form as

$$\begin{bmatrix} R_w[0] & R_w[1] & \cdots & R_w[N] \\ R_w[1] & R_w[0] & \cdots & R_w[N-1] \\ \vdots & \vdots & \ddots & \vdots \\ R_w[N] & R_w[N-1] & \cdots & R_w[0] \end{bmatrix} \begin{bmatrix} a_0 \\ a_1 \\ \vdots \\ a_N \end{bmatrix} = \begin{bmatrix} R_{sw}[0] \\ R_{sw}[1] \\ \vdots \\ R_{sw}[N] \end{bmatrix}. \tag{3.39}$$

These equations are known as the Wiener–Hopf equations. The matrix appearing in the equation is a symmetric Toeplitz matrix. These matrices are known to be positive

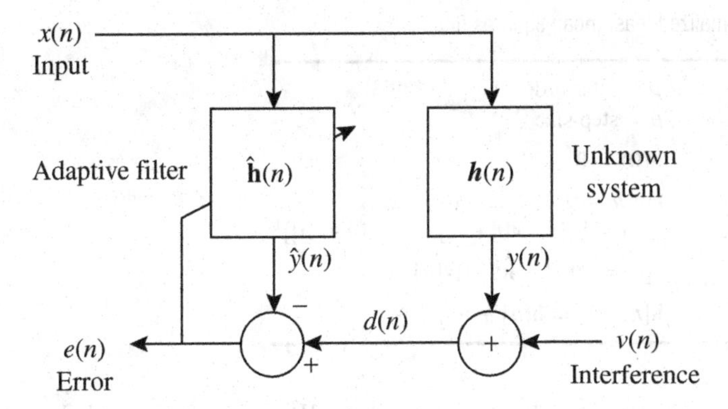

Figure 3.8 Example of an LMS filter.

definite and therefore non-singular yielding a single solution to the determination of the Wiener filter coefficients. Furthermore, there exists an efficient algorithm to solve the Wiener–Hopf equations known as the Levinson–Durbin algorithm.

The Wiener filter is an FIR filter and it is related to the least mean squares (LMS) filter. But for LMS, minimizing the error criterion does not rely on cross-correlations or auto-correlations. Its solution converges to the Wiener filter solution, which will be discussed in the next subsection.

3.4.3 LMS filter

Least mean squares (LMS) algorithms are used in adaptive filters to find the filter coefficients that produce the least mean squares of the error signal. It is a stochastic gradient descent method in that the filter is only adapted based on the error at the current time. Assuming that the unknown system is linear time-invariant and the noise is stationary, the LMS filter solution converges to the Wiener filter solution. Both filters can be used to identify the impulse response of an unknown system, knowing only the original input signal and the output of the unknown system. By relaxing the error criterion to reduce current sample error, the LMS algorithm can be derived from the Wiener filter.

Figure 3.8 shows an unknown system $h[n]$ to be identified. The adaptive filter attempts to adapt the filter $\hat{h}[n]$ to make it as close as possible to $h[n]$, using only the observable signals $x[n]$, $d[n]$, and $e[n]$, as $y[n]$, $v[n]$, and $h[n]$ are not directly observable.

The idea behind LMS filters is to use the method of steepest descent to find a coefficient vector, which minimizes a cost function. The cost function $C[n]$ is defined as MMSE, i.e. $C[n] = E\{\|e[n]\|^2\}$. Applying the steepest descent method means taking the partial derivatives with respect to the individual entries of the filter coefficient vector:

$$\nabla C[n] = \nabla E\{e[n]e^*[n]\} = 2E\{\nabla(e[n]e^*[n])\}, \tag{3.40}$$

where ∇ is the gradient operator.

Table 3.5 Normalized least mean squares filter.

Parameters	p = filter order μ = step size
Initialization	$\hat{\mathbf{h}}(0) = \mathbf{0}$
Computation	For $n = 0, 1, 2, \ldots$ $\mathbf{x}[n] = [x[n], x[n-1], \ldots, x[n-p]]^{\mathrm{T}}$ $e[n] = d[n] - \hat{\mathbf{h}}^H[n]\mathbf{x}[n]$ $\hat{\mathbf{h}}[n+1] = \hat{\mathbf{h}}[n] + \frac{\mu e^*[n]\mathbf{x}[n]}{\mathbf{x}^H[n]\mathbf{x}[n]}$

With $\mathbf{x}[n] = [x[n], x[n-1], \ldots, x[n-p+1]]^T$ and $\nabla e[n] = -\mathbf{x}[n]$, we have

$$\nabla C[n] = -2E\{\mathbf{x}[n]e^*[n]\}. \tag{3.41}$$

$\nabla C[n]$ is a vector which points towards the steepest ascent of the cost function. To find the minimum of the cost function, we need to take a step in the opposite direction of $\nabla C[n]$. Expressed in mathematical terms:

$$\hat{\mathbf{h}}[n+1] = \hat{\mathbf{h}}[n] - \frac{\mu}{2}\nabla C[n] = \hat{h}[n] + \mu E\{\mathbf{x}[n]e^*[n]\}, \tag{3.42}$$

where $\mu/2$ is the step size. This means that we have found a sequential update algorithm which minimizes the cost function.

For most systems the expectation function $E\{\mathbf{x}[n]e^*[n]\}$ must be approximated. This can be done with the following unbiased estimator:

$$\hat{E}\{\mathbf{x}[n]e^*[n]\} = \frac{1}{N}\sum_{i=0}^{N-1}\mathbf{x}[n-i]e^*[n-i], \tag{3.43}$$

where N indicates the number of samples we use for the estimate. The simplest case is $N = 1$ and the updated algorithm is as follows:

$$\hat{\mathbf{h}}[n+1] = \hat{\mathbf{h}}[n] + \mu\mathbf{x}[n]e^*[n]. \tag{3.44}$$

The main drawback of the "pure" LMS algorithm is that it is sensitive to the scaling of its input $x[n]$. This makes it very hard (if not impossible) to choose a learning rate μ that guarantees stability of the algorithm. The normalized least mean squares filter (NLMS) is a variant of the LMS algorithm that solves this problem by normalizing with the power of the input. The NLMS algorithm is summarized in Table 3.5.

It can be shown that if there is no interference (i.e. $v[n] = 0$), then the optimal learning rate for the NLMS algorithm is $\mu_{\text{opt}} = 1$. In the general case with interference ($v[n] \neq 0$), the optimal learning rate is

$$\mu_{\text{opt}} = \frac{E[\|y[n] - \hat{y}[n]\|^2]}{E[\|e[n]\|^2]}. \tag{3.45}$$

The results above assume that the signals $v[n]$ and $x[n]$ are uncorrelated with each other, which is generally the case in practice.

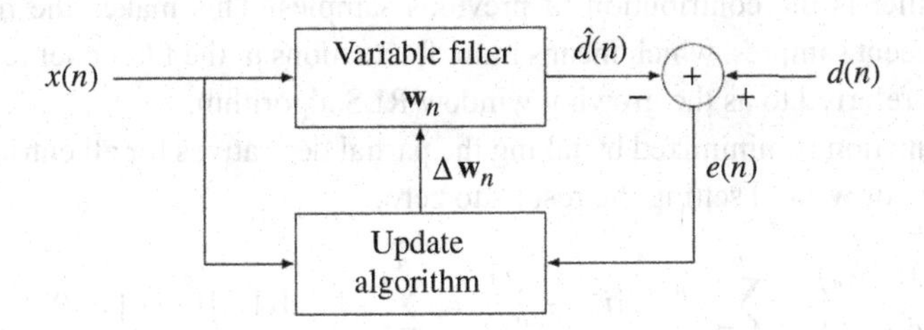

Figure 3.9 Example of an RLS filter.

3.4.4 RLS filter

A recursive least squares (RLS) algorithm is used in adaptive filters to find the filter coefficients that produce the least squares (minimum of the sum of the absolute squared) of the error signal (i.e. difference between the desired and the actual signal). The block diagram of an RLS filter is shown in Figure 3.9. Suppose a signal d is transmitted through a channel. The received signal $x[n]$ is given by

$$x[n] = \sum_{k=0}^{q} b_n[k]d[n-k] = v[n], \tag{3.46}$$

where $v[n]$ is the noise and $b_n[k]$ is the channel response.

The desired signal d is recovered by using a finite impulse response (FIR) filter, $\mathbf{w}$ as

$$\hat{d}[n] = \mathbf{w}_n^{\mathrm{T}}\mathbf{x}[n]. \tag{3.47}$$

The goal is to estimate the parameters of the filter $\mathbf{w}$, and at each time n the new least squares estimate is denoted by $\mathbf{w}_n$. As time evolves, we would like to avoid completely redoing the least squares algorithm to find the new estimate for $\mathbf{w}_{n+1}$ in terms of $\mathbf{w}_n$. The benefit of the RLS algorithm is that there is no need to invert matrices, thereby saving computational power. Another advantage is that it provides intuition behind such results as the Kalman filter.

The idea behind RLS filters is to minimize a cost function C by appropriately selecting the filter coefficients $\mathbf{w}_n$, updating the filter as new data arrives. The error signal $e[n]$ and desired signal $d[n]$ are defined in the negative feedback diagram as shown in Figure 3.9. The error implicitly depends on the filter coefficients through the estimate $\hat{d}[n]$ as

$$e[n] = d[n] - \hat{d}[n]. \tag{3.48}$$

The weighted least squares error function C (the cost function we desire to minimize) being a function of $e[n]$ is therefore also dependent on the filter coefficients:

$$C(\mathbf{w}_n) = \sum_{i=0}^{n} \lambda^{n-i} \|e[i]\|^2, \tag{3.49}$$

where $0 < \lambda \leq 1$ is an exponential weighting factor which effectively limits the number of input samples based on which the cost function is minimized. It also has the interpretation of forgetting factor, i.e. how importantly the past affects the present. The smaller

λ is, the smaller is the contribution of previous samples. This makes the filter more sensitive to recent samples, which means more fluctuations in the filter coefficients. The $\lambda = 1$ case is referred to as the growing window RLS algorithm.

The cost function is minimized by taking the partial derivatives for all entries k of the coefficient vector $\mathbf{w}_n$ and setting the results to zero:

$$\frac{\partial C(\mathbf{w}_n)}{\partial w_n^*[k]} = \sum_{i=0}^{n} \lambda^{n-i} e[i] \frac{\partial e^*[i]}{\partial w_n^*[k]} = \sum_{i=0}^{n} \lambda^{n-i} e[i] x^*[i-k] = 0. \tag{3.50}$$

Next, $e[n]$ is replaced with the definition of the error signal:

$$\sum_{i=0}^{n} \lambda^{n-i} \left[d[i] - \sum_{l=0}^{p} w_n[l] x[i-l] \right] x^*[i-k] = 0. \tag{3.51}$$

Rearranging the equation yields

$$\sum_{l=0}^{p} w_n[l] \left[\sum_{i=0}^{n} \lambda^{n-i} x[i-l] x^*[i-k] \right] = \sum_{i=0}^{n} \lambda^{n-i} d[i] x^*[i-k]. \tag{3.52}$$

This form can be expressed in terms of matrices

$$\mathbf{R}_x[n]\mathbf{w}_n = \mathbf{r}_{dx}[n], \tag{3.53}$$

where $\mathbf{R}_x[n]$ is the weighted autocorrelation matrix for $x[n]$ and $\mathbf{r}_{dx}[n]$ is the cross-correlation between $d[n]$ and $x[n]$. Based on this expression, the coefficients which minimize the cost function can be found as

$$\mathbf{w}_n = \mathbf{R}_x^{-1}[n]\mathbf{r}_{dx}[n]. \tag{3.54}$$

Thus we have obtained a single equation to determine a coefficient vector which minimizes the cost function. Next, we want to derive a recursive solution of the form

$$\mathbf{w}_n = \mathbf{w}_{n-1} + \Delta\mathbf{w}_{n-1}, \tag{3.55}$$

where $\Delta\mathbf{w}_{n-1}$ is a correction factor at time $n-1$. We start the derivation of the recursive algorithm by expressing the cross correlation $\mathbf{r}_{dx}[n]$ in terms of $\mathbf{r}_{dx}[n-1]$ as

$$\mathbf{r}_{dx}[n] = \sum_{i=0}^{n} \lambda^{n-i} d[i]\mathbf{x}^*[i] \tag{3.56}$$

$$= \sum_{i=0}^{n-1} \lambda^{n-i} d[i]\mathbf{x}^*[i] + \lambda d[n]\mathbf{x}^*[n] \tag{3.57}$$

$$= \lambda\mathbf{r}_{dx}[n-1] + d[n]\mathbf{x}^*[n], \tag{3.58}$$

where $\mathbf{x}[i]$ is the $p+1$ dimensional data vector

$$\mathbf{x}[i] = [x[i], x[i-1], \ldots, x[i-p]]^{\mathrm{T}}. \tag{3.59}$$

Table 3.6 Recursive least squares filter.

Parameters	$p =$ filter order $\lambda =$ forgetting factor $\delta =$ value to initialize $\mathbf{P}[0]$
Initialization	$\mathbf{w}_n = 0$ $\mathbf{P}[0] = \delta^{-1}\mathbf{I}$ where $\mathbf{I}$ is the $(p+1)\times(p+1)$ identity matrix
Computation	For $n = 0, 1, 2, \ldots$ $\mathbf{x}[n] = [x[n], x[n-1], \ldots, x[n-p]]^{\mathrm{T}}$ $\alpha[n] = d[n] - \mathbf{w}[n-1]^{\mathrm{T}}\mathbf{x}[n]$ $\mathbf{g}[n] = \mathbf{P}[n-1]\mathbf{x}^*[n]\left\{\lambda + \mathbf{x}^{\mathrm{T}}[n]\mathbf{P}[n-1]\mathbf{x}^*[n]\right\}^{-1}$ $\mathbf{P}[n] = \lambda^{-1}\mathbf{P}[n-1] - \mathbf{g}[n]\mathbf{x}^{\mathrm{T}}[n]\lambda^{-1}\mathbf{P}[n-1]$ $\mathbf{w}_n = \mathbf{P}[n]\mathbf{r}_{dx}[n] = \mathbf{w}_{n-1} + \mathbf{g}[n]\alpha[n]$

Similarly, $\mathbf{R}_x[n]$ can be expressed in terms of $\mathbf{R}_x[n-1]$ as follows:

$$\mathbf{R}_x[n] = \sum_{i=0}^{n} \lambda^{n-i}\mathbf{x}^*[i]\mathbf{x}^{\mathrm{T}}[i] \tag{3.60}$$

$$= \lambda\mathbf{R}_x[n-1] + \mathbf{x}^*[n]\mathbf{x}^T[n]. \tag{3.61}$$

Using the Woodbury matrix identity, we have

$$\begin{aligned}\mathbf{P}[n] = \mathbf{R}_x^{-1}[n] &= \lambda^{-1}\mathbf{R}_x^{-1}[n-1] - \lambda^{-1}\mathbf{R}_x^{-1}[n-1]\mathbf{x}^*[n] \\ &\left\{1 + \mathbf{x}^T[n]\lambda^{-1}\mathbf{R}_x^{-1}[n-1]\mathbf{x}^*[n]\right\}^{-1}\mathbf{x}^T[n]\lambda^{-1}\mathbf{R}_x^{-1}[n-1] \\ &= \lambda^{-1}\mathbf{P}[n-1] - \mathbf{g}[n]\mathbf{x}^T[n]\lambda^{-1}\mathbf{P}[n-1],\end{aligned} \tag{3.62}$$

where

$$\mathbf{g}[n] = \mathbf{P}[n-1]\mathbf{x}^*[n]\left\{\lambda + \mathbf{x}^{\mathrm{T}}[n]\mathbf{P}[n-1]\mathbf{x}^*[n]\right\}^{-1} \tag{3.63}$$

$$= \lambda^{-1}\left\{\mathbf{P}[n-1] - \mathbf{g}[n]\mathbf{x}^{\mathrm{T}}[n]\mathbf{P}[n-1]\right\}\mathbf{x}^*[n] \tag{3.64}$$

$$= \mathbf{P}[n]\mathbf{x}^*[n]. \tag{3.65}$$

By some manipulations, the complete recursion is given by

$$\mathbf{w}_n = \mathbf{P}[n]\mathbf{r}_{dx}[n] = \mathbf{w}_{n-1} + \mathbf{g}[n]\alpha[n], \tag{3.66}$$

where $\alpha[n] = d[n] - \mathbf{x}^{\mathrm{T}}[n]\mathbf{w}_{n-1}$ is the a priori error. Comparing this with the a posteriori error, the error after the filter is updated is

$$e[n] = d[n] - \mathbf{x}^T[n]\mathbf{w}_n \tag{3.67}$$

and the correction factor can be written as

$$\Delta\mathbf{w}_{n-1} = \mathbf{g}[n]\alpha[n]. \tag{3.68}$$

This intuitively satisfying result indicates that the correction factor is directly proportional to both the error and the gain vector, which controls how much sensitivity is desired, through the weighting factor, λ. The RLS algorithm for a p-th order RLS filter is shown in Table 3.6.

3.4.5 Kalman filter

The Kalman filter is an efficient recursive filter that estimates the state of a dynamic system from a series of incomplete and noisy measurements. Together with the linear-quadratic regulator (LQR), the Kalman filter solves the linear-quadratic-Gaussian control problem (LQG). The Kalman filter, the linear-quadratic regulator, and the linear-quadratic-Gaussian controller are solutions to what probably are the most fundamental problems in control theory. An example application would be providing accurate and continuously updated information about the position and velocity of an object given only a sequence of observations about its position, each of which includes some error.

Kalman filters are based on linear dynamical systems discretized in the time domain. They are modeled by Markov chains built on linear operators perturbed by Gaussian noise. The state of the system is represented as a vector of real numbers. At each discrete time increment, a linear operator is applied to the state to generate the new state, with some noise mixed in, and optionally some information from the controls on the system if they are known. Then, another linear operator mixed with more noise generates the visible outputs from the hidden state. The Kalman filter may be regarded as analogous to the hidden Markov model, with the key difference that the hidden state variables take values in a continuous space (as opposed to a discrete state space as in the hidden Markov model). Additionally, the hidden Markov model can represent an arbitrary distribution for the next value of the state variables, in contrast to the Gaussian noise model that is used for the Kalman filter. There is a strong duality between the equations of the Kalman filter and those of the hidden Markov model.

In order to use the Kalman filter to estimate the internal state of a process given only a sequence of noisy observations, one must model the process in accordance with the framework of the Kalman filter as described below:

$$\mathbf{x}_k = \mathbf{F}_k \mathbf{x}_{k-1} + \mathbf{B}_k \mathbf{u}_{k-1} + \mathbf{w}_{k-1}, \tag{3.69}$$

where $\mathbf{F}_k$ is the state transition model applied to the previous state $\mathbf{x}_{k-1}$, $\mathbf{B}_k$ is the control-input model applied to the control vector $\mathbf{u}_k$, and $\mathbf{w}_k$ is the process noise, which is assumed to be drawn from a zero mean multivariate normal distribution with covariance $\mathbf{Q}_k$ i.e. $\mathbf{w}_k \sim N(\mathbf{0}, \mathbf{Q}_k)$.

At time k an observation (or measurement) $\mathbf{z}_k$ of the true state $\mathbf{x}_k$ is made according to

$$\mathbf{z}_k = \mathbf{H}_k \mathbf{x}_k + \mathbf{v}_k, \tag{3.70}$$

where $\mathbf{H}_k$ is the observation model, which maps the true state space into the observed space, and $\mathbf{v}_k$ is the observation noise, which is assumed to be zero mean Gaussian white noise with covariance $\mathbf{R}_k$, i.e. $\mathbf{v}_k \sim N(\mathbf{0}, \mathbf{R}_k)$. The initial state, and the noise vectors at each step $\{\mathbf{x}_0, \mathbf{w}_1, \ldots, \mathbf{w}_k, \mathbf{v}_1, \ldots, \mathbf{v}_k\}$ are all assumed to be mutually independent.

The Kalman filter is a recursive estimator. This means that only the estimated state from the previous time step and the current measurement are needed to compute the estimate for the current state. In contrast to batch estimation techniques, no history of observations and/or estimates is required. It is purely a time domain filter; most filters (for example, a low-pass filter) are formulated in the frequency domain and then transformed back to the time domain for implementation. In what follows, the notation $\hat{\mathbf{x}}_{n|m}$ represents the estimate of $\mathbf{x}$ at time n given observations up to, and including time m.

The state of the filter is represented by two variables: $\hat{\mathbf{x}}_{k|k}$, the estimate of the state at time k; $\mathbf{P}_{k|k} = \mathrm{cov}(\mathbf{x}_k \hat{\mathbf{x}}_{k|k-1})$, the error covariance matrix (a measure of the estimated accuracy of the state estimate). The Kalman filter has two distinct phases: predict and update. The predict phase uses the state estimate from the previous time step to produce an estimate of the state at the current time step. In the update phase, the measured information at the current time step is used to refine this prediction to arrive at a new, (hopefully) more accurate state estimate, again for the current time step. The phases are summarized below:

- *Prediction*

  Predicted state:

$$\hat{\mathbf{x}}_{k|k-1} = \mathbf{F}_k \hat{\mathbf{x}}_{k-1|k-1} + \mathbf{B}_k \mathbf{u}_{k-1}. \tag{3.71}$$

  Predicted estimate covariance:

$$\mathbf{P}_{k|k} = \mathbf{F}_k \mathbf{P}_{k-1|k-1} \mathbf{F}_k^T + \mathbf{Q}_{k-1}. \tag{3.72}$$

- *Update*

  Innovation or measurement residual:

$$\tilde{\mathbf{y}}_k = \mathbf{z}_k + \mathbf{H}_k \hat{\mathbf{x}}_{k|k-1}. \tag{3.73}$$

  Innovation (or residual) covariance:

$$\mathbf{S}_k = \mathbf{H}_k \mathbf{P}_{k|k-1} \mathbf{H}_k^{\mathrm{T}} + \mathbf{R}_k. \tag{3.74}$$

  Optimal Kalman gain:

$$\mathbf{K}_k = \mathbf{P}_{k|k-1} \mathbf{H}_k^{\mathrm{T}} \mathbf{S}_k^{-1}. \tag{3.75}$$

  Updated state estimate:

$$\hat{\mathbf{x}}_{k|k} = \hat{\mathbf{x}}_{k|k-1} + \mathbf{K}_k \tilde{\mathbf{y}}_k. \tag{3.76}$$

  Updated estimate covariance:

$$\mathbf{P}_{k|k} = (\mathbf{I} - \mathbf{K}_k \mathbf{H}_k) \mathbf{P}_{k|k-1}. \tag{3.77}$$

Notice that the formula for the updated estimate covariance above is only valid for the optimal Kalman gain. If the model is accurate, and the values for $\hat{\mathbf{x}}_{0|0}$ and $\mathbf{P}_{0|0}$ accurately reflect the distribution of the initial state values, then the following invariants are preserved: all estimates have mean error zero, i.e.

$$E[\mathbf{x}_k - \hat{\mathbf{x}}_{k|k}] = E[\mathbf{x}_k - \hat{\mathbf{x}}_{k|k-1}] = \mathbf{0} \tag{3.78}$$

$$E[\tilde{\mathbf{y}}_k] = \mathbf{0}. \tag{3.79}$$

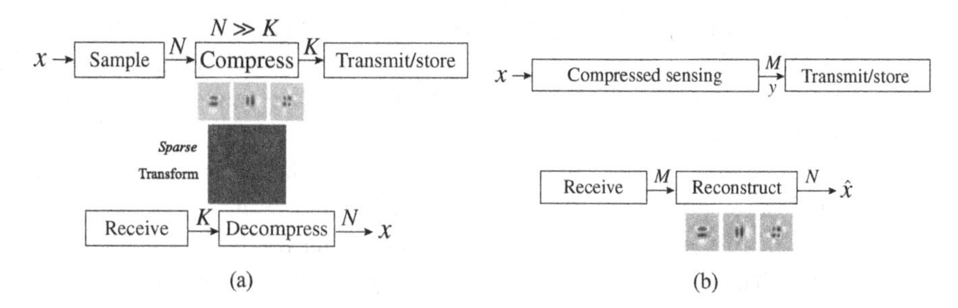

Figure 3.10 Comparison between (a) traditional sampling and (b) compressed sensing.

The Kalman filter can be regarded as an adaptive low-pass infinite impulse response digital filter, with cut-off frequency depending on the ratio between the process and measurement (or observation) noise, as well as the estimate covariance. Frequency response is, however, rarely of interest when designing state estimators such as the Kalman filter, whereas for digital filters such as IIR and FIR filters it is usually of primary concern. For the Kalman filter, the important goal is how accurate the filter is, and this is mostly evaluated based on empirical Monte Carlo simulations, where the "truth" (the true state) is known.

3.5 Compressed sensing

The Shannon/Nyquist theorem states that the original signal can be perfectly reconstructed if the sampling frequency is at least as high as the Nyquist rate (twice the bandwidth of the original signal). As shown in Figure 3.10(a), in the traditional digital data transmission system, the data x is uniformly sampled at Nyquist rate. Then data is compressed from N to K for transmission to save the bandwidth. In the receiver, the data is decompressed.

However, the Shannon/Nyquist theorem is pessimistic in the sense that a $2\times$ oversampling Nyquist rate is a worst-case bound for any bandlimited data. Compressed sensing [203] is a new sampling theory that makes use of compressibility to directly acquire compressed data. It is based on the new uncertainty principles, and randomness plays a key role. As shown in Figure 3.10(b), the compressed sensing samples by more general measurements at a speed M, which is greater than K but much less than N. In the following two subsections, the basic theories of compressed sensing and its possible applications are discussed.

3.5.1 Basic technology

In this subsection, the basic problem statement, solutions and theorems for compressed sensing are studied. For the problem statement, a real-valued, finite-length, one-dimensional, discrete-time signal $\mathbf{x}$ is considered, which is an $N \times 1$ vector in $\mathbb{R}^N$ with element $x[n]$, $n = 1, 2, \ldots, N$. Suppose we have an orthonormal basis of $N \times 1$

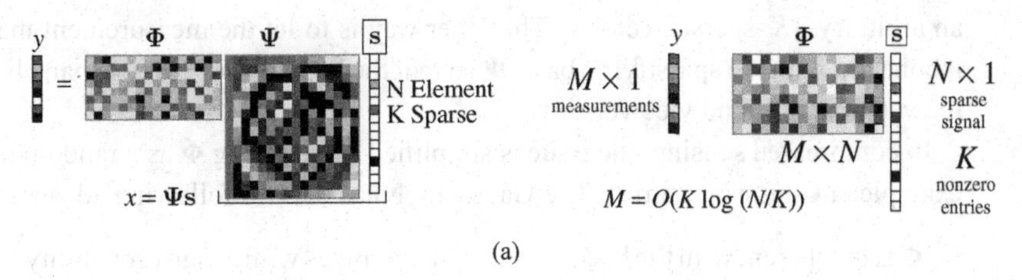

Figure 3.11 Mathematical comparison between (a) traditional sampling and (b) compressed sensing.

vector $\{\psi_i\}_1^N$, and any signal **x** can be expressed as

$$\mathbf{x} = \sum_{i=1}^{N} s_i \psi_i = \Psi \mathbf{s}, \tag{3.80}$$

where **s** is the $N \times 1$ vector of weighting coefficients.

Most signals have sparsity in which **x** is a linear combination of just K basis vectors, with $K \ll N$. For example, for image compression, a JPEG coding can significantly reduce the number of bits needed to represent an image. However, the transform coding techniques require the full N samples of signal **x**, then compute using the full dimension of N, and discard most of the $N - K$ bits.

Compressed sensing is a technique that directly acquires data through a compressed representation. Consider the linear measurement process with dimension $M < N$. The measurement vector **y** has dimension of $M \times 1$ and consider the measurement vectors ϕ_j^{T} as rows of an $M \times N$ matrix $\mathbf{\Phi}$. We have

$$\mathbf{y} = \mathbf{\Phi x} = \mathbf{\Phi \Psi s} = \mathbf{\Theta s}, \tag{3.81}$$

where $\mathbf{\Theta} = \mathbf{\Phi\Psi}$ is an $M \times N$ matrix as described in Figure 3.11(a) and (b). The goal is to design a measurement matrix $\mathbf{\Phi}$ and a reconstruction algorithm for K-space and compressible signals that require only $M \approx K$ or a slightly larger number of measurements.

The solution has two steps: the design of a stable measurement matrix $\mathbf{\Phi}$ which ensures that the information is not damaged due to the reduction of dimension; and a reconstruction algorithm to recover **x** from the measurement **y**.

For a stable measurement matrix $\mathbf{\Phi}$, the number of equations M equals or exceeds the number of unknowns K. A necessary and sufficient condition to ensure that the $M \times K$ system is well conditioned is that for any vector **v** sharing the same K non-zero entries as **s**, we have

$$1 - \epsilon \le \frac{\|\mathbf{\Theta} b\|_2}{\|\mathbf{v}\|_2} \le 1 + \epsilon \tag{3.82}$$

for some $\epsilon > 0$. In other words, $\mathbf{\Theta}$ must preserve the lengths of K-sparse vectors.

In practice, the locations of the K non-zero entries of **s** are unknown, but it has been shown by the restricted isometry property (RIP) [204] that a sufficient condition for a stable inverse for both K-sparse and compressible signals is for $\mathbf{\Theta}$ to satisfy (3.82) for an arbitrary $3K$-sparse vector $\mathbf{v}$. The other way is to let the measurement matrix $\mathbf{\Phi}$ be incoherent with the sparsifying basis $\mathbf{\Psi}$ so that the vectors $\{\phi_j\}$ cannot sparsely represent the vectors $\{\psi_j\}$, and vice versa.

In compressed sensing, the issue is simplified by selecting $\mathbf{\Phi}$ as a random matrix (for example, a Gaussian matrix). The Gaussian $\mathbf{\Phi}$ has the two following advantages:

1. $\mathbf{\Phi}$ is incoherent with the basis $\mathbf{\Psi} = \mathbf{I}$ of delta spikes with a high probability. Using concentration of measure arguments, an $M \times N$ iid Gaussian matrix $\mathbf{\Theta} = \mathbf{\Phi}\mathbf{I} = \mathbf{\Phi}$ can be shown to have the RIP with high probability if $M \geq cK \log(N/K)$, with c as a small constant [204, 205, 203]. As a result, a compressible N-length signal of K-sparse elements can be recovered with high probability as long as $M \geq cK \log(N/K) \ll N$.
2. $\mathbf{\Theta} = \mathbf{\Phi\Psi}$ is also iid Gaussian regardless of the choice of $\mathbf{\Psi}$. So the random Gaussian measurement $\mathbf{\Phi}$ is universal since $\mathbf{\Theta} = \mathbf{\Phi\Psi}$ has the RIP with high probability for every possible $\mathbf{\Psi}$.

The second step is the signal reconstruction step to recover the original N-dimensional signal. Since $M < N$, there are infinitely many $\mathbf{s}'$ satisfying $\mathbf{\Theta s}' = \mathbf{y}$, which lies in the null space. Therefore, the goal is to find the signal's sparse coefficient vector $\mathbf{s}$ as the following formulation:

$$\hat{\mathbf{s}} = \arg\min \|\mathbf{s}'\|_p \quad \text{subject to: } \mathbf{\Theta s}' = \mathbf{y}. \tag{3.83}$$

Usually $p = 0, 1, 2$, and $p = 1$ is mostly used due to the simplicity of the reconstruction algorithm.

3.5.2 Applications

Compressed sensing exploits a priori signal sparsity information and changes the rules of data acquisition. The same principle can be used for any compressible signal class. Each measurement carries the same amount of information so that it is robust to measurement loss and quantization. The encoding is simple and most of the processing is performed at the decoder. Also the random projections can be viewed as weakly encrypted since the data appears as noise. There are many possible applications, and some of them are summarized below:

- *Sensor network*: Sensor networks measure, monitor, and track distributed physical phenomena using wireless embedded sensors, which have limited computation power and energy supply. Moreover, the measured information is very sparse and correlated. As a result, it would be very inefficient for the fusion center to collect data from each sensor. This is similar to the sampling theory, but here it happens over space. Traditionally, the sensors collaborate for sensing, while compressed sensing can reduce the frequency to collect the data and still maintain the sensing quality. The related work can be found in [206, 207, 208, 209, 210].
- *Information scalability*: If we can reconstruct a signal from compressive measurements, then we should be able to perform other kinds of statistical signal processing,

e.g. detection, classification, estimation, and learning, etc. One example is a compressive matched filter, which is also called a "smashed filter." For a traditional matched filter, joint parameter estimation and detection/classification are performed to compute a sufficient statistic for each potential target and articulation, and then to compare "best" statistics to detect/classify. In other words, matched filter classification is to have a closest manifold search. For example, let us assume there are K unknown parameters from N data. Using the compressed sensing, we can only have $M = O(K \log N)$ random measurements to preserve the manifold structure, which enables parameter estimation and matched filter detection/classification directly on compressive measurements. The related work can be found in [211, 212, 213, 214, 215, 216, 217, 218].

- *Image processing*: Some compressed sensing cameras have been implemented at Rice University and Georgia Tech, USA. Compressed sensing can also be applied to bio-medical imaging such as MRI; compressive camera arrays for sampling rate scales logarithmically in both number of pixels and number of cameras; compressive radar and sonar with greatly simplified receivers to explore radar/sonar networks; and compressive DNA microarrays with smaller, more agile arrays for bio-sensing to exploit sparsity in presentation of organisms to array.
- *Other applications related to communications*: First, an analog-to-information (A2I) converter takes an analog input signal and creates discrete (digital) measurements. Compressed sensing can extend the analog-to-digital converter so that it samples at the signal's "information rate" rather than its Nyquist rate.

  Second, the random demodulator can recover the sparse Fourier signals with local sampling rate. This can be employed in frequency hopping spread spectrum, in which the number of channels is much larger than the number of signals.

  For cognitive radio, the compressed sensing techniques were tailored for the coarse sensing of spectrum hole identification [219]. Sub-Nyquist rate samples were utilized to detect and classify frequency bands via a wavelet-based edge detector. Because spectrum location estimation takes priority over fine-scale signal reconstruction, the proposed sensing algorithms are robust to noise and can afford reduced sampling rates.

3.6 Summary

Signal processing techniques have been used widely in spectrum sensing to detect the signature of the signal from a licensed user. A number of algorithms for detecting spectrum holes have been used in the literature. These include matched filter detection, energy detection, cyclostationary feature detection, and wavelet detection. An overview of these signal processing techniques used in spectrum sensing has been presented. The concept of collaborative sensing has been described. An estimation of interference temperature and channel gains based on the maximum likelihood estimator, maximum a posteriori (MAP) estimator, Bayesian estimator, periodogram estimator, and blind estimator have been discussed. To this end, several popular filter concepts and prediction models, namely the autoregressive moving average (ARMA) filter, Wiener filter, least mean squares (LMS) filter, and Kalman filter, have been presented.

4 Optimization techniques

In a wireless network (more specifically, in a cognitive wireless network), the available radio resources such as bandwidth are very limited. On the other hand, the demands for the wireless services are exponentially increasing. Not only are the number of users booming, but also more bandwidth is required for new services such as video telephony, TV on demand, wireless Internet, and wireless gaming. Finding a way to accommodate all these requirements has become an emergent research issue in wireless networking. Resource allocation and its optimization are general methods to improve network performance, but there are tradeoffs for resource usage. One of the major research goals is to present these tradeoffs so that better implementations can be put into practice.

This chapter will focus on how to formulate cognitive wireless networking problems as optimization problems from the perspective of resource allocation. Specifically, this chapter discusses what the resources, parameters, practical constraints, and optimized performances are across the different layers. In addition, it addresses how to perform resource allocation in multiuser scenarios under the presence of the primary users. The tradeoffs between the different optimization goals and different users' interests are also investigated. The goal is to provide a new perspective of wireless networking and resource allocation problems from the optimization point of view.

This chapter is organized as follows: Section 4.1 discusses the basic formulation of the cognitive radio resource allocation as a constrained optimization problem. Section 4.2 studies linear programming and the simplex algorithm as its solution. Section 4.3 investigates how to define a convex optimization problem and some variations. Then the solutions are discussed. In Section 4.4, non-linear programming is explained and some popular solution techniques are illustrated. Section 4.5 first formulates the general integer programming problem, and then general algorithms are explained. In Section 4.6, dynamic programming is explored with some special topics such as Markov decision process. Examples are also shown. In Section 4.7, stochastic programming is investigated. Finally, a chapter summary is provided in Section 4.8.

4.1 Constrained optimization

Many radio resource cognitive resource allocation problems in cognitive radio networks can be formulated as constrained optimization problems. The general formulation can be written as:

$$\min_{\mathbf{x}\in\Omega} f(\mathbf{x}) \tag{4.1}$$

$$\text{subject to:} \begin{cases} g_i(\mathbf{x}) \le 0, & \text{for } i = 1, \ldots, m \\ h_j(\mathbf{x}) = 0, & \text{for } j = 1, \ldots, l, \end{cases}$$

where $\mathbf{x}$ is the parameter vector for optimizing the resource allocation, Ω is the feasible range for the parameter vector, and $f(\mathbf{x})$ is the optimization goal matrix, objective (or goal), or utility function that represents the performance or cost. Here, $g_i(\mathbf{x})$ and $h_j(\mathbf{x})$ are the inequality and equality constraints for the parameter vector, respectively. The optimization process finds the solution $\bar{\mathbf{x}} \in \Omega$ that satisfies all inequality and equality constraints. For an optimal solution, $f(\bar{\mathbf{x}}) \le f(\mathbf{x}), \forall \mathbf{x} \in \Omega$.

4.1.1 Basic definition

If the optimization goal, inequality constraints, and equality constraints are all linear functions of the parameter function $\mathbf{x}$, then the problem in (4.1) is called a *linear program*. One important characteristic of a linear program problem is that there is a global optimal point that is very easy to obtain by linear programming. On the other hand, one major drawback of linear programming is that most of the practical problems in wireless networking and resource allocation are non-linear. Therefore, it is hard to model these practical problems as linear programs. If either the optimization goal or the constraint functions are non-linear, the problem in (4.1) is called a *non-linear program*. In general, there are multiple local optima in a non-linear program, and to find the global optimum is not easy. Furthermore, if the feasible set Ω contains some integer sets, the problem in (4.1) is called an *integer program*. Most integer programs are NP-hard problems which cannot be solved by polynomial time.

A special kind of non-linear program is *convex optimization*, in which the feasible set Ω is a convex set, and the optimization goal and the constraints are convex/concave/linear functions. A convex set is defined as follows:

Definition 4.1 *A set Ω is* convex *if for any $\mathbf{x}_1, \mathbf{x}_2 \in \Omega$ and any θ with $0 \le \theta \le 1$, we have*

$$\theta\mathbf{x}_1 + (1-\theta)\mathbf{x}_2 \in \Omega.$$

A convex function f is defined as:

Definition 4.2 *A function f is* convex *over $\mathbf{x}$, if the feasible range Ω of parameter vector $\mathbf{x}$ is a convex set, and if for all $\mathbf{x}_1, \mathbf{x}_2 \in \Omega$ and $0 \le \theta \le 1$,*

$$f(\theta\mathbf{x}_1 + (1-\theta)\mathbf{x}_2) \le \theta f(\mathbf{x}_1) + (1-\theta) f(\mathbf{x}_2).$$

A function is strictly convex if the strict inequality is held whenever $\mathbf{x}_1 \ne \mathbf{x}_2$ and $0 < \theta < 1$. A function is called concave if $-f$ is convex.

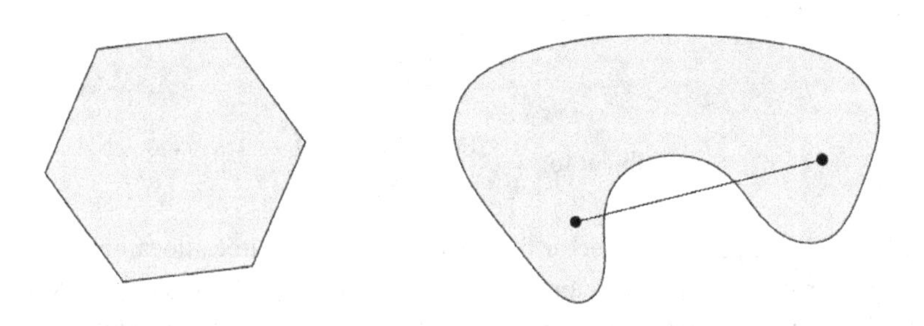

Figure 4.1 Examples: convex set (left) and non-convex set (right).

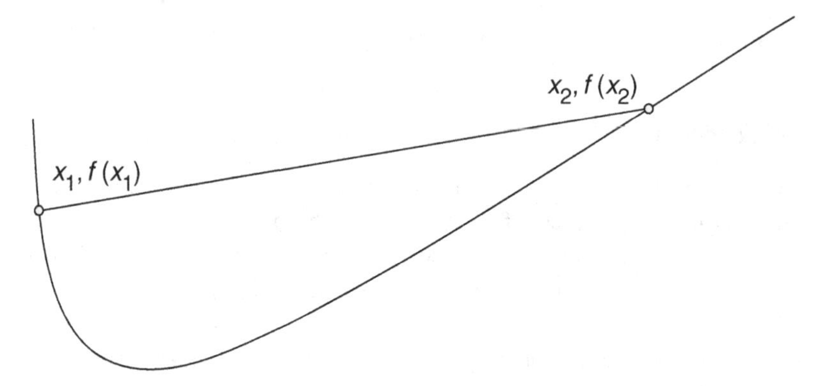

Figure 4.2 Example of convex function.

In Figure 4.1, we show examples of a convex set and a non-convex set. In Figure 4.2, we show the example of a convex function. If the function is differentiable, and if either the following two conditions hold, the function is a convex function:

$$\text{First-order condition: } f(\mathbf{x}_2) \geq f(\mathbf{x}_1) + \nabla f(\mathbf{x}_1)^{\mathrm{T}}(\mathbf{x}_2 - \mathbf{x}_1), \tag{4.2}$$

$$\text{Second-order condition: } \nabla^2 f(\mathbf{x}) \succeq 0. \tag{4.3}$$

One important application of the convex function is Jensen's inequality. Suppose function f is convex and the parameter $\mathbf{x}$ has any arbitrary random distribution over Ω. Then the following equality holds:

$$f(E\mathbf{x}) \leq Ef(\mathbf{x}), \tag{4.4}$$

where E is the expectation.

The advantages of convex optimization technique are as follows:

- There are a variety of applications such as automatic control systems, estimation and signal processing, communications and networks, electronic circuit design, data analysis and modeling, statistics, and finance.
- Computation time is usually quadratic. Problems can then be solved very reliably and efficiently using interior-point methods or other special methods for convex optimization.

- Solution methods are reliable enough to be embedded in a computer-aided design or analysis tool, or even a real-time reactive or automatic control system.
- There are also theoretical or conceptual advantages in formulating a problem as a convex optimization problem.

The challenges of convex optimization are to recognize and model the problem as a convex optimization. Moreover, there are many tricks for transforming problems into convex forms.

How the convex optimization problems can be formulated is discussed next. In a cognitive radio network, the radio resource allocation problem can be formulated as a constrained optimization problem with the parameters, objective functions, and constraints having the following physical meanings:

- *Examples of parameters*:
  Physical layer: Transmitted power, modulation level, channel coding rate, and channel/code selection.
  MAC layer: Transmission time/frequency, service rate, and priorities for transmission.
  Network layer: Route selection and routing cost.
  Application layer: Source-coding rate, buffer priority, and packet arrival rate.
- *Examples of objective functions*:
  Physical layer: Minimal overall power, maximal throughput, maximal rate per joule, and minimal bit error rate.
  MAC layer: Maximal overall throughput, minimal buffer overflow probability, and minimal delay.
  Network layer: Minimal cost and maximal profit.
  Application layer: Minimal distortion, and minimal delay.
- *Examples of constraints*:
  Physical layer: Maximal mobile transmitted power, available modulation constellation, available channel coding rate, and limited energy.
  MAC layer: Contention for channel access, limited time/frequency slot, and limited information about other mobiles.
  Network layer: Maximum number of hops, security constraints.
  Application layer: The base layer transmission, limited source rate, strict delay, and security.
- *Primary user constraints*: Primary users' channel occupancy and interference level; sensing device's accuracy and speed; hidden terminal and exposed terminal problem.

4.1.2 Lagrangian method

After formulating the constrained optimization problem (e.g. for resource allocation in a cognitive wireless network), we need to find solutions. Next, we explain how to obtain a closed-form solution for such a problem. For cases where a closed-form solution cannot be obtained, other methods such as convex programming can be used, which are discussed later in this chapter. One of the most important methods used to find a closed-form solution for constrained optimization is the Lagrangian method, which has the following steps:

1. Rewrite (4.1) as a Lagrangian multiplier function J as
$$J = f(\mathbf{x}) + \sum_{i=1}^{m} \lambda_i g_i(\mathbf{x}) + \sum_{j=1}^{l} \mu_j h_j(\mathbf{x}), \tag{4.5}$$
where λ_i and μ_j are Lagrangian multipliers.
2. Differentiate J over $\mathbf{x}$ and set to zero as
$$\frac{\partial J}{\partial \mathbf{x}} = 0. \tag{4.6}$$
3. From (4.6), solve λ_i and μ_j.
4. Replace λ_i and μ_j in the constraints to obtain optimal $\mathbf{x}$.

Notice that the difficulty arises in the Lagrangian method due to Steps 3 and 4, where the closed-form solution is obtained for the Lagrangian multipliers. Some approximations and mathematical tricks are necessary to obtain the closed form solutions.

Obtaining the analytical results from (4.1) is difficult because of the non-linearity and non-covexity of the constraints and the optimization goal. This makes the Lagrangian multiplier function in (4.5) hard to differentiate in order to obtain the optimal points. If some approximations for the constraints or the optimization goal functions can be obtained under some conditions, (4.1) can be solved by differentiating (4.5) and putting the results back into the constraints to obtain the optimal Lagrangian multiplier. There are several methods to do that, which can be classified as follows:

1. *Parameterized approximation*: In this method, the non-linear or non-convex function is approximated by the parameterized function. The goal is to obtain the optimal parameters such that the approximation errors can be minimized.

   The most common approximation is the *linear approximation*. Suppose the original function is $f(x)$. Then, the linear approximation can be written as
$$\min_{a,b} \int_c^d \|f(x) - (ax + b)\|^2 \mathrm{d}x, \tag{4.7}$$
where $[c, d]$ denotes the region of interest where the approximation needs to be accurate.

   Another type of approximation is *polynomial or Taylor expansion*. The original function can be expanded as:
$$f(x) = c_0 + c_1(x - x_0) + c_2(x - x_0)^2 + \cdots, \tag{4.8}$$
where x_0 is the point where the series are expanded, and $c_0, c_1, c_2, \ldots$ are constants.

   In general, for any convex function $f'(x; \mathbf{a})$ of x with parameter vector $\mathbf{a}$, the approximation within region $[c, d]$ can be written as:
$$\min_{\mathbf{a}} \int_c^d \|f(x) - f'(x; \mathbf{a})\|^2 \mathrm{d}x. \tag{4.9}$$

2. *Omitting the unimportant part*: The basic idea for this type of approximation is that even though the function itself is not convex, within a certain range, some components can be omitted. Consequently, the approximation is convex.

   One good example is the channel capacity function given as

$$C = W \log_2(1 + \text{SNR}), \tag{4.10}$$

   where C is the capacity, W is the bandwidth, SNR is the signal to noise ratio. However, when the SNR is high, i.e. SNR >> 1, we can simplify the capacity formula as

$$C' = W \log_2(\text{SNR}). \tag{4.11}$$

   C' is a convex function. This approximation has been employed to some networking optimizations such as in [220].

4.1.3 Optimality

In this subsection, the optimality of the solution will be discussed, i.e. how to determine if the solution is optimal or not, and in what sense the solution is optimal. The organization of this section is as follows: first, the optimality for the unconstrained problem is discussed. Then, the Fritz John and Karush–Kuhn–Tucker (KKT) conditions are explained. Finally, the second-order conditions are illustrated.

Before unconstrained optimality is discussed, the definitions of global and local optimum will be provided as follows:

Definition 4.3 *Consider* $\min f(x)$ *over* Ω *and let* $\bar{x} \in \Omega$. *If* $f(\bar{x}) \leq f(x)$, $\forall x \in \Omega$, $\bar{x}$ *is called a global minimum. If* $f(\bar{x})$ *is no larger than the neighbor of* $\bar{x}$, $\bar{x}$ *is called a local optimum.*

Some necessary and sufficient conditions for the optimality are as follows:

- *Necessary conditions*:
  1. First-order necessary condition: If $f(\mathbf{x})$ is differentiable at $\bar{\mathbf{x}}$ and if $\bar{\mathbf{x}}$ is a local minimum, then $\nabla f(\bar{\mathbf{x}}) = 0$.
  2. Second-order necessary condition: If $f(\mathbf{x})$ is twice differentiable at $\bar{\mathbf{x}}$ and if $\bar{\mathbf{x}}$ is a local minimum, then $\nabla f(\bar{\mathbf{x}}) = 0$ and Hessian matrix $\mathbf{H}(\bar{\mathbf{x}})$ is positive semi-definite.
- *Sufficient conditions*:
  1. First-order necessary and sufficient condition: If $f(\mathbf{x})$ is pseudo-convex at $\bar{\mathbf{x}}$, then $\bar{\mathbf{x}}$ is a global minimum if and only if $\nabla f(\bar{\mathbf{x}}) = 0$.
  2. Second-order sufficient condition: If $f(\mathbf{x})$ is twice differentiable at $\bar{\mathbf{x}}$, if $\nabla f(\bar{\mathbf{x}}) = 0$ and Hessian matrix $\mathbf{H}(\bar{\mathbf{x}})$ is positive definite, then $\bar{\mathbf{x}}$ is a strict local minimum.

It is worth mentioning that some points on the curves satisfying the necessary conditions might not be the true optimum. For example, in Figure 4.3, the plot for $z = x^2 - y^2$ is shown. It is observed that the necessary condition, i.e. the first-order differential evaluated at $x = 0$ and $y = 0$ is zero, is satisfied for the saddle point at $(0, 0)$. However, the saddle point is not a local optimum.

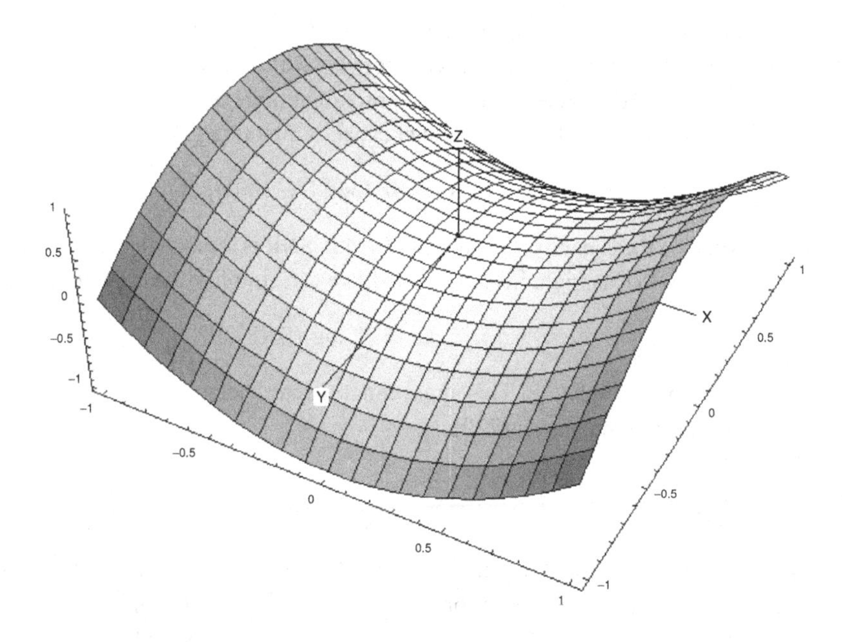

Figure 4.3 Example of a saddle point.

Next a few theorems for the optimality of (4.1) are listed as follows:

- *Fritz John necessary conditions*: Let $\bar{\mathbf{x}}$ be a feasible solution, and let $I = \{i : g_i(\bar{\mathbf{x}}) = 0\}$. Suppose $g_i, i \in I$ is continuous at $\bar{\mathbf{x}}$. f and $g_i, i \in I$ are differentiable at $\bar{\mathbf{x}}$, and $h_j, \forall j$ is continuous and differentiable at $\bar{\mathbf{x}}$. If $\bar{\mathbf{x}}$ is a local minimum, then there exist scalars u_0, u_i, for $i \in I$ and v_j for $j = 1, \ldots, l$ such that

$$u_0 \nabla f(\bar{\mathbf{x}}) + \sum_{i \in I} u_i \nabla g_i(\bar{\mathbf{x}}) + \sum_{j=1}^{l} v_j \nabla h_j(\bar{\mathbf{x}}) = \mathbf{0} \tag{4.12}$$
$$u_0, u_i \geq 0, \forall i \in I, (u_0, \mathbf{u}_I, \mathbf{v}) \neq (0, \mathbf{0}, \mathbf{0}),$$

where $\mathbf{u}_I$ is the vector whose component is u_i and $\mathbf{v} = (v_1, \ldots, v_l)^{\mathrm{T}}$.
- *Fritz John sufficient conditions*: Define $S = \{\mathbf{x} : g_i(\mathbf{x}) \leq 0, \quad \text{for } i \in I, h_j(\mathbf{x}) = 0, \text{ for } j = 1, \ldots, l\}$. If h_j for $j = 1, \ldots, l$ are affine and $\nabla h_j(\bar{\mathbf{x}})$, $j = 1, \ldots, l$ are linearly independent, and if there exists an ε-neighborhood $N_\varepsilon(\bar{\mathbf{x}})$ of $\bar{\mathbf{x}}$, $\varepsilon > 0$, such that f is pseudoconvex on $S \cap N_\varepsilon(\bar{\mathbf{x}})$, then $\bar{\mathbf{x}}$ is a local minimum for (4.1).
- *KKT necessary conditions*: Suppose f and $g_i, i \in I$ are differentiable at $\bar{\mathbf{x}}$, $g_i, i \in I$ is continuous at $\bar{\mathbf{x}}$, and $h_j, \forall j$ is continuous and differentiable at $\bar{\mathbf{x}}$. Suppose $\nabla g_i(\bar{\mathbf{x}}), \forall i \in I$ and $\nabla h_j(\bar{\mathbf{x}})$, $j = 1, \ldots, l$ are linearly independent. If $\bar{\mathbf{x}}$ is a local optimum, then unique scalar $u_i, i \in I$ and v_j for $j = 1, \ldots, l$ exist such that

$$\nabla f(\bar{\mathbf{x}}) + \sum_{i \in I} u_i \nabla g_i(\bar{\mathbf{x}}) + \sum_{j=1}^{l} v_j \nabla h_j(\bar{\mathbf{x}}) = \mathbf{0} \tag{4.13}$$
$$u_i \geq 0, \forall i \in I.$$

- *KKT sufficient conditions*:
  Suppose KKT conditions hold at $\bar{\mathbf{x}}$, i.e. there exist scalars $\bar{u}_i \geq 0$ for $i \in I$ and $\bar{v}_j$ for $j = 1, \ldots, l$ such that

$$\nabla f(\bar{\mathbf{x}}) + \sum_{i \in I} \bar{u}_i \nabla g_i(\bar{\mathbf{x}}) + \sum_{j=1}^{l} \bar{v}_j \nabla h_j(\bar{\mathbf{x}}) = \mathbf{0}. \tag{4.14}$$

  Let $J = \{j : \bar{v}_j > 0\}$ and $K = \{j : \bar{v}_j < 0\}$. Suppose f is pseudoconvex at $\bar{\mathbf{x}}$, g_i is quasiconvex at $\bar{\mathbf{x}}$ for $i \in I$, h_j is quasiconvex at $\bar{\mathbf{x}}$ for $j \in J$ and h_j is quasiconcave at $\bar{\mathbf{x}}$ for $j \in K$. Then $\bar{\mathbf{x}}$ is a global optimal solution of (4.1). If the generalized convexity assumptions on the objective and constraint functions are restricted to a domain $N_\varepsilon(\bar{\mathbf{x}})$ for some $\varepsilon > 0$, then $\bar{\mathbf{x}}$ is a local minimum for (4.1).
- *KKT second-order necessary conditions*:
  Suppose the objective and constraints defined in (4.1) are twice differentiable, and Ω is non-empty. If $\bar{\mathbf{x}}$ is a local optimum, define the restricted Lagrangian function $L(\mathbf{x})$ as

$$L(\mathbf{x}) = \phi(\mathbf{x}, \bar{\mathbf{u}}, \bar{\mathbf{v}}) = f(\mathbf{x}) + \sum_{i \in I} \bar{u}_i g_i(\mathbf{x}) + \sum_{j=1}^{l} \bar{v}_j h_j(\mathbf{x}). \tag{4.15}$$

  Denote its Hessian at $\bar{\mathbf{x}}$ by

$$\nabla^2 L(\bar{\mathbf{x}}) = \nabla^2 f(\bar{\mathbf{x}}) + \sum_{i \in I} \bar{u}_i \nabla^2 g_i(\bar{\mathbf{x}}) + \sum_{j=1}^{l} \bar{v}_j \nabla^2 h_j(\bar{\mathbf{x}}), \tag{4.16}$$

  where $\nabla^2 f(\bar{\mathbf{x}})$, $\nabla^2 g_i(\bar{\mathbf{x}})$ for $i \in I$ and $\nabla^2 h_j(\bar{\mathbf{x}})$ for $j = 1, \ldots, l$ are the Hessians of f, g_i for $i \in I$, and h_j for $j = 1, \ldots, l$, respectively. Assume that $\nabla g_i(\bar{\mathbf{x}})$ for $i \in I$ and $\nabla h_j(\bar{\mathbf{x}})$ for $j = 1, \ldots, l$ are linearly independent. Then $\bar{\mathbf{x}}$ is a KKT point and

$$\mathbf{d}^{\mathrm{T}} \nabla^2 L(\bar{\mathbf{x}}) \mathbf{d} \geq 0 \tag{4.17}$$

  for all $\mathbf{d} \in \{\mathbf{d} \neq \mathbf{0} : \nabla g_i(\bar{\mathbf{x}})^{\mathrm{T}} \mathbf{d} \leq 0\}$ for all $i \in I$, $\nabla h_j(\bar{\mathbf{x}})^{\mathrm{T}} \mathbf{d} = 0$ for all $j = 1, \ldots, l$.
- *KKT second-order sufficient conditions*:
  Suppose the objective and constraints defined in (4.1) are twice differentiable, and Ω is non-empty. Let $\bar{\mathbf{x}}$ be a KKT point for (4.1), with Lagrangian multiplier $\bar{u}$ and $\bar{v}$ associated with the inequality and equality constraints, respectively. Denote $I^+ = \{i \in I : \bar{u}_i > 0\}$ and $I^0 = \{i \in I : \bar{u}_i = 0\}$. Define the restricted Lagrangian function $L(\mathbf{x})$ and its Hessian. Define the cone

$$C = \{\mathbf{d} \neq \mathbf{0}\} : \begin{array}{ll} \nabla g_i(\bar{\mathbf{x}})^{\mathrm{T}} \mathbf{d} = 0, & \text{for } i \in I^+ \\ \nabla g_i(\bar{\mathbf{x}})^{\mathrm{T}} \mathbf{d} \leq 0, & \text{for } i \in I^0 \\ \nabla h_j(\bar{\mathbf{x}})^{\mathrm{T}} \mathbf{d} = 0, & \text{for } j = 1, \ldots, l. \end{array} \tag{4.18}$$

  Then if $\mathbf{d} \in \{\mathbf{d} > \mathbf{0}\}$ for all $\mathbf{d} \in C$, we have that $\bar{\mathbf{x}}$ is a strict local minimum for (4.1).

The optimality discussed in this subsection can be used for different scenarios. For example, it can prove the optimality of a certain solution; it can determine the termination criterion for the adaptive algorithm; and it can be used for convergence analysis.

4.1.4 Primal-dual algorithm

In this subsection, the duality concept for the constrained optimization is defined. Under some convexity assumptions and constraint properties, the primal and dual problems have the same optimal objective values, so it is possible to solve the primal problem by considering the dual problem and developing very efficient algorithms. The dual problem will be defined first, and then the duality theorem will be introduced. After some properties are discussed, the primal-dual algorithm will be introduced.

Consider (4.1) as the primal problem. The Lagrangian dual problem of (4.1) can be defined as:

$$\max \theta(\mathbf{u}, \mathbf{v}) \qquad (4.19)$$
$$\text{subject to: } \mathbf{u} \geq \mathbf{0},$$

where $\theta(\mathbf{u}, \mathbf{v}) = \inf\{f(\mathbf{x}) + \sum_{i=1}^{m} u_i g_i(\mathbf{x}) + \sum_{j=1}^{l} v_j h_j(\mathbf{x}) : \mathbf{x} \in \Omega\}$.

For example, the primal problem for linear programming in standard form can be written as

$$\min_{\mathbf{x}} \mathbf{c}^{\mathrm{T}}\mathbf{x} \qquad (4.20)$$
$$\text{subject to: } \begin{cases} \mathbf{Ax} = \mathbf{b} \\ \mathbf{x} \geq 0, \end{cases}$$

where $\mathbf{x}$ is the optimized vector with dimension $N \times 1$, $\mathbf{c}$ and $\mathbf{b}$ are constant vectors, and $\mathbf{A}$ is a constant matrix. The dual function is

$$\begin{aligned} \theta(\mathbf{u}, \mathbf{v}) &= \inf_{\mathbf{x}}(\mathbf{c}^{\mathrm{T}}\mathbf{x} - \sum_{i=1}^{N} \lambda_i \mathbf{x}_i + \mathbf{v}^{\mathrm{T}}(\mathbf{Ax} - \mathbf{b})) \\ &= -\mathbf{b}^{\mathrm{T}}\mathbf{v} + \inf_{\mathbf{x}}(\mathbf{c} + \mathbf{A}^{\mathrm{T}}\mathbf{v} - \lambda)^{\mathrm{T}}\mathbf{x}, \end{aligned} \qquad (4.21)$$

where $\lambda = [\lambda_1 \ldots \lambda_N]^{\mathrm{T}}$. Since a linear function $(\mathbf{c} + \mathbf{A}^{\mathrm{T}}\mathbf{v} - \lambda)$ is unbounded below, $\theta(\mathbf{u}, \mathbf{v}) = -\infty$ only when $\mathbf{c} + \mathbf{A}^{\mathrm{T}}\mathbf{v} - \lambda = 0$. We have the dual problem as follows:

$$\max\ \theta(\mathbf{u}, \mathbf{v}) = \begin{cases} -\mathbf{b}^{\mathrm{T}}\mathbf{v}, \ \mathbf{c} + \mathbf{A}^{\mathrm{T}}\mathbf{v} - \lambda = 0 \\ -\infty, \ \text{otherwise}. \end{cases} \qquad (4.22)$$
$$\text{subject to: } \lambda \geq \mathbf{0}.$$

One of the major applications of the dual problem is the duality theorem. The following weak duality theorem states that the objective value of any feasible solution to the dual problem is a lower bound on the objective value of any feasible solution to the primal problem:

Theorem 4.1 *Weak duality theorem: If $\boldsymbol{x}$ is a feasible solution to the primal problem in (4.1) and $(\boldsymbol{u}, \boldsymbol{v})$ are the feasible solution to the dual problem in (4.19), then $f(\boldsymbol{x}) \geq \theta(\boldsymbol{u}, \boldsymbol{v})$. The duality gap is defined as $f(\boldsymbol{x}) - \theta(\boldsymbol{u}, \boldsymbol{v})$.*

Under convexity assumptions and constraint qualification, the duality gap is zero under the following strong duality theorem:

Theorem 4.2 *Strong duality theorem: Let Ω be a convex set, g_i be convex, and h_j be affine (i.e. $\mathbf{h}(\mathbf{x}) = \mathbf{A}\mathbf{x} + \mathbf{b}$ with a matrix $\mathbf{A}$ and a vector $\mathbf{b}$). Suppose there exists an $\mathbf{x} \in \Omega$ that $\mathbf{g}(\mathbf{x}) \le \mathbf{0}$ and $\mathbf{h}(\mathbf{x}) = \mathbf{0}$, and $\mathbf{0}$ is the interior point for $\mathbf{h}$ (Slater's (interiority) condition). Then the optimal point of a primal problem is the same as the optimal point of the dual problem, i.e.*

$$\inf\{(4.1)\} = \sup\{(4.19)\}. \tag{4.23}$$

The necessary and sufficient condition for the zero duality gap is the existence of a saddle point, defined by the following:

Definition 4.4 *$(\bar{\mathbf{x}}, \bar{\mathbf{u}}, \bar{\mathbf{v}})$ with $\bar{\mathbf{x}} \in \Omega$ and $\bar{\mathbf{u}} \ge \mathbf{0}$ if and only if*

1. $\phi(\bar{\mathbf{x}}, \bar{\mathbf{u}}, \bar{\mathbf{v}}) = \min(f(\mathbf{x}) + \sum_{i \in I} \bar{u}_i g_i(\mathbf{x}) + \sum_{j=1}^{l} \bar{v}_j h_j(\mathbf{x}))$;
2. $\mathbf{g}(\bar{\mathbf{x}}) \le \mathbf{0}$, $\mathbf{h}(\bar{\mathbf{x}}) = \mathbf{0}$;
3. $\bar{\mathbf{u}}^T \mathbf{g}(\bar{\mathbf{x}}) = 0$.

The other similar interpretation is the complementary slackness [220].

If the primal and dual problems are the same, it is possible to solve the primal problem indirectly by solving the dual problem. For non-linear non-convex problems, the duality gap might not be zero. Sometimes in practice, the dual problem will need less information to optimize and consequently it is easier to solve. In addition, primal and dual problems can be solved iteratively to find the optimal solution. One solution is called the cutting plane or outer-linearization method, as follows:

1. Initialization.
2. Solve primal optimization.
3. Solve dual optimization.
4. Go back to Step 1, until the optimal condition, such as the KKT condition, is satisfied.

4.2 Linear programming and the simplex algorithm

Linear programming (LP) is the problem of maximizing/minimizing a linear function over a convex polyhedron. Linear programming is extensively used in engineering. Linear programming can be solved using the simplex method [221] [222], which runs along polytope edges of the visualization solid to find the best answer. Some other algorithms will also be discussed.

An LP problem can be expressed in *standard form* as follows:

$$\min \mathbf{c}^{\mathrm{T}} \mathbf{x} \tag{4.24}$$

$$\text{subject to: } \begin{cases} \mathbf{A}\mathbf{x} = \mathbf{b} \\ \mathbf{x} \ge \mathbf{0}, \end{cases}$$

where $\mathbf{x}$ is the vector of variables to solve, $\mathbf{A}$ is a matrix of known coefficients, and $\mathbf{c}$ and $\mathbf{b}$ are vectors of known coefficients. The expression $\mathbf{c}^T\mathbf{x}$ is called the objective function, and the equations $\mathbf{Ax} = \mathbf{b}$ are called the constraints. The matrix $\mathbf{A}$ is generally not square and usually $\mathbf{A}$ has more columns than rows, and $\mathbf{Ax} = \mathbf{b}$ is therefore quite likely to be under-determined, leaving great latitude in the choice of $\mathbf{x}$ which minimizes $\mathbf{c}^T\mathbf{x}$.

Two families of solution techniques are widely used today. Both the techniques visit a progressively improving series of trial solutions until a solution is reached that satisfies the conditions for an optimum:

- The *simplex methods*, introduced by Dantzig about 50 years ago, visit "basic" solutions computed by fixing enough of the variables at their bounds to reduce the constraints $\mathbf{Ax} = \mathbf{b}$ to a square system, which can be solved for unique values of the remaining variables. Basic solutions represent extreme boundary points of the feasible region defined by $\mathbf{Ax} = \mathbf{b}$, $\mathbf{x} \geq \mathbf{0}$, and the simplex method can be viewed as moving from one such point to another along the edges of the boundary.
- *Barrier or interior-point methods*,[1] by contrast, visit points within the interior of the feasible region. These methods derive from techniques for non-linear programming that were developed and popularized in the 1960s by Fiacco and McCormick, but their application to linear programming dates back only to Karmarkar's innovative analysis in 1984.

First, the solution of an LP problem falls on the boundary of the feasible region by the following theorem:

Theorem 4.3 *Extreme point (or simplex filter) theorem: If the maximum or minimum value of a linear function defined over a polygonal convex region exists, then it is found at the boundary of the region.*

This theorem implies that a finite number of extreme points indicate a finite number of solutions. Hence, the search is reduced to a finite set of points. However, a finite set can still be too large for practical problems. The simplex method provides an efficient systematic search guaranteed to converge in a finite number of steps.

LP problems must be converted into augmented form before solving by the simplex algorithm. This form introduces non-negative slack variables to replace non-equalities with equalities in the constraints. The problem can then be written in the following augmented form:

$$\begin{pmatrix} 1 & -\mathbf{c}^T & 0 \\ \mathbf{0} & \mathbf{A} & \mathbf{I} \end{pmatrix} \begin{pmatrix} Z \\ \mathbf{x} \\ \mathbf{x}_S \end{pmatrix} = \begin{pmatrix} 0 \\ \mathbf{b} \end{pmatrix}, \tag{4.25}$$

where $\mathbf{x}$ is the variable vector from the standard form in (4.24), $\mathbf{x}_S$ is the introduced slack variable vector from the augmentation process, c contains the optimization coefficients, $\mathbf{A}$ and $\mathbf{b}$ describe the system of constraint equations, and Z is the variable to be maximized.

[1] Barrier or interior-point methods will be discussed in detail in Section 4.4.

Generally, the number of variables exceeds the number of equations. The difference between the number of variables and the number of equations gives us the degrees of freedom associated with the problem. Any solution, optimal or not, will therefore include a number of variables of arbitrary value. The simplex algorithm uses zero as this arbitrary value, and the number of variables with value zero equals the degrees of freedom.

Variables of non-zero values are called basic variables, and variables of zero values are called non-basic variables in the simplex algorithm. The augmented form simplifies finding the initial basic feasible solution. The simplex method provides an efficient systematic search guaranteed to converge in a finite number of steps. The algorithm is as follows:

1. Begin the search at an extreme point (i.e. a basic feasible solution).
2. Determine if the movement to an adjacent extreme can improve on the optimization of the objective function. If not, the current solution is optimal. If, however, improvement is possible, then proceed to the next step.
3. Move to the adjacent extreme point which offers (or, perhaps, appears to offer) the most improvement in the objective function.
4. Continue Steps 2 and 3 until the optimal solution is found or it can be shown that the problem is either unbounded or infeasible.

In 1972, Klee and Minty gave an example of a linear programming problem in which the polytope P is a distortion of an n-dimensional cube. They showed that the simplex method as formulated by Dantzig visits all $2n$ vertices before arriving at the optimal vertex. This shows that in the worst-case the algorithm has exponential time complexity. Similar examples have been found for other pivot rules. It remains an open question whether there is a pivot rule with polynomial time worst-case complexity. Nevertheless, the simplex method is remarkably efficient in practice. Attempts to explain this efficiency employ the notion of average complexity or (recently) smoothed complexity.

The importance of linear programming derives in part from its many applications and in part from the existence of good general-purpose techniques for finding optimal solutions. For example, Khachian [223] found a $O(x^5)$ polynomial time algorithm. A much more efficient polynomial time algorithm was found by Karmarkar [224]. This method goes through the middle of the solid (making it a so-called interior-point method), and then transforms and wraps. Interior-point methods were known as early as the 1960s in the form of the barrier function methods. These techniques take as input only an LP in the standard form, and determine a solution without utilizing any special structure of the original LP problem. These techniques are fast and reliable over a substantial range of problem sizes and applications.

On the other hand, there are some limitations of LP. In practice, the objective and the constraints are rarely linear functions of the optimization parameters. Under the non-linear conditions, there might be many local optima and the simplex algorithm cannot find good solutions. In the next few sections, we will discuss more generalized programming methods that deal with the more complicated problems.

4.3 Convex programming

A convex optimization problem can be defined as follows:

$$\min f_0(x) \tag{4.26}$$
$$\text{subject to: } \begin{cases} f_i(x) \leq 0, i = 1, \ldots, m \\ \mathbf{a}_i^{\mathrm{T}}\mathbf{x} = \mathbf{b}_i, i = 1, \ldots, p, \end{cases}$$

where the objective function f_0 is convex, the inequality constraint functions $f_1, \ldots, f_m$ are convex, and the equality constraints $g_i(\mathbf{x}) = \mathbf{a}_i^{\mathrm{T}}\mathbf{x} - \mathbf{b}_i$ are affine.

A fundamental property of convex optimization problems is that any locally optimal point is also globally optimal. Moreover, by using duality theory studied in the previous section, the optimality conditions can be easily identified. Since linearity is a special kind of convexity, linear programming is a special kind of convex optimization. Some typical convex programming problems are discussed in the following subsections.

4.3.1 Quadratic, geometric, and semi-definite programming

A problem is called a *quadratic program* if the objective function is quadratic and the constraint functions are affine as

$$\min \mathbf{x}^{\mathrm{T}}\mathbf{P}\mathbf{x} + 2\mathbf{q}^{\mathrm{T}}\mathbf{x} \tag{4.27}$$
$$\text{subject to: } \begin{cases} \mathbf{G}\mathbf{x} \leq \mathbf{h} \\ \mathbf{A}\mathbf{x} = \mathbf{b}, \end{cases}$$

where $\mathbf{P} = \mathbf{P}^{\mathrm{T}}$. If $\mathbf{P}$ is a positive semi-definite matrix, then the problem in (4.27) is a convex function. In this case, the quadratic program has a global optimal point, if there exists at least one vector $\mathbf{x}$ satisfying the constraints and the optimization goal is bounded on the feasible region. If $\mathbf{P}$ is positive definite, then the global optimal point is unique. If $\mathbf{P}$ is zero, the problem becomes a linear program. The KKT condition is sufficient when the problem is convex.

The dual of a quadratic program is also a quadratic program. For example, if we ignore the equality constraint in (4.27), the dual function is given by

$$q(\mathbf{u}) = \inf_{\mathbf{x}} \left(\frac{1}{2}\mathbf{x}^{\mathrm{T}}\mathbf{P}\mathbf{x} + \mathbf{q}^{\mathrm{T}}\mathbf{x} + \mathbf{u}^{\mathrm{T}}(\mathbf{G}\mathbf{x} - \mathbf{h}) \right). \tag{4.28}$$

The infimum is attained for $\mathbf{x} = -(\mathbf{P})^{-1}(\mathbf{q} + \mathbf{G}^{\mathrm{T}}\mathbf{u})$, and we have the dual problem as

$$\min \frac{1}{2}\mathbf{u}^{\mathrm{T}}\mathbf{G}(\mathbf{P})^{-1}\mathbf{G}^{\mathrm{T}}\mathbf{u} + (\mathbf{h}^{\mathrm{T}} + \mathbf{G}(\mathbf{P})^{-1}\mathbf{q})\mathbf{u} \tag{4.29}$$
$$\text{subject to: } \mathbf{u} \geq \mathbf{0}.$$

For a positive definite $\mathbf{P}$, the ellipsoid method solves the problem in polynomial time [220]. If $\mathbf{P}$ is negative definite (even if Q has only one negative eigenvalue), then the problem is NP-hard [220].

For *geometric programming*, let us define the following two functions.

Definition 4.5 *A function f is called monomial if*

$$f(x) = cx_1^{a_1}x_2^{a_2}\dots x_n^{a_n}, \tag{4.30}$$

where $c \geq 0$ and $a_i \in \boldsymbol{R}$.

Definition 4.6 *A function f is called posynomial if*

$$f(x) = \sum_{k=1}^{K} c_k x_1^{a_{1k}} x_2^{a_{2k}} \dots x_n^{a_{nk}}, \tag{4.31}$$

where $c_k \geq 0$ and $a_{ik} \in \boldsymbol{R}$.

An optimization problem is called geometric programming if

$$\min f_0(x) \tag{4.32}$$

$$\text{subject to: } \begin{cases} f_i(x) \leq 1, i = 1, \dots, m \\ h_i(x) = 1, i = 1, \dots, p, \end{cases}$$

where $f_0, \dots, f_m$ are posynomials and $h_1, \dots, h_p$ are monomials. Geometric programming has numerous applications, such as circuit sizing and parameter estimation via logistic regression in statistics. The maximum likelihood estimation in logistic regression is a geometric programming.

Geometric programs are not (in general) convex optimization problems, but they can be transformed to convex problems by a change of variables and a transformation of the objective and constraint functions. For example, by changing $y = \log x$, a posynomial becomes a sum of exponentials of affine functions.

The field of *semidefinite programming* (SDP) or semidefinite optimization (SDO) deals with optimization problems over symmetric positive semidefinite matrix variables with linear cost function and linear constraints. Popular special cases are linear programming and convex quadratic programming with convex quadratic constraints. A semidefinite programming problem can be expressed as

$$\min \mathbf{c}^{\mathrm{T}}\mathbf{x} \tag{4.33}$$

$$\text{subject to: } F(\mathbf{x}) \geq 0,$$

where

$$F(\mathbf{x}) = \mathbf{F}_0 + \sum_{i=1}^{m} x_i \mathbf{F}_i \tag{4.34}$$

and $F(\mathbf{x})$ is positive semidefinite.

The dual of a semidefinite program is given by

$$\min \mathbf{F}_0\mathbf{y} \tag{4.35}$$

$$\text{subject to: } \begin{cases} \mathbf{F}_i\mathbf{y} = 0, \ \forall i \\ \mathbf{y} \geq 0. \end{cases}$$

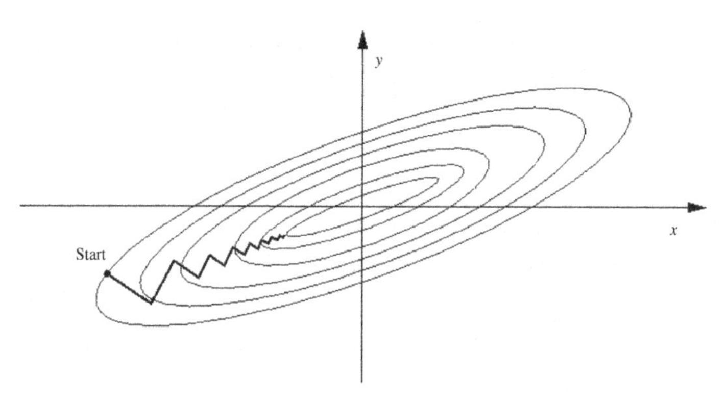

Figure 4.4 Example of gradient descent method.

Many practical problems in operations research and combinatorial optimization can be modeled or approximated as semidefinite programming (SDP) problems, such as the solution of the max cut problem with an approximation ratio of 0.878 56. There are two types of algorithms for solving SDP problems. One is the interior-point method, and the other is the specialized general convex optimization algorithm.

4.3.2 Gradient method, Newton method, and their variations

The other advantage of convex optimization is the fact that there are many simple methods that can be used to find the global optimum. The solutions of unconstrained optimization will be discussed first, and then the solutions will be generalized to constrained optimizations. For the unconstrained cases, some methods are as follows:

- *Gradient method*: The gradient method finds the nearest local minimum of a function with the assumption that the gradient of the function can be computed. The method of steepest descent, also called the gradient descent method, starts at a point $\mathbf{x}_0$ and, as many times as needed, moves from $\mathbf{x}_i$ to $\mathbf{x}_{i+1}$ by minimizing along the line extending in the direction of $-\nabla f(\mathbf{x})$, the local downhill gradient. The pseudocode is given as follows:

  Gradient descent method
  - Take a feasible starting point $\mathbf{x}_0$.
  - Repeat:
    1. Calculate the gradient $\nabla f(\mathbf{x})$.
    2. Line search: choose the step size t that optimizes $f(\mathbf{x}_i - t\nabla f(\mathbf{x}_i))$.
    3. Update: $\mathbf{x}_{i+1} = \mathbf{x}_i - t\nabla f(\mathbf{x}_i)$.
  - Until the stopping criteria, such as the KKT condition (defined in the previous section) or the desired accuracy, are satisfied.

  Figure 4.4 shows an example of the gradient method. The method of steepest descent is simple, easy to apply, and each iteration is fast. It is also very stable; if the minimum points exist, the method is guaranteed to locate them. But, even with all these advantages, the method has one very important drawback: it generally has slow convergence. For badly scaled systems, i.e. if the eigenvalues of the Hessian matrix at the solution point are different by several orders of magnitude, the method can end up spending a large number of iterations before locating a minimum point. It starts out with a reasonable convergence, but the progress gets slower and slower.

- *Newton's method*: This is a well-known method for finding roots of equations in one or more dimensions. It can also be used to find local maxima and local minima of functions. Since a real number x^* is a minima or a maxima if the first-order gradient of the function $f(x)$ is equal to zero at that point, x^* can be obtained by applying Newton's method to $f'(x)$.

Definition 4.7 *Given that the second-order differential exists, the Newton step is defined as:*

$$\nu_{\text{nt}} = -\nabla^2 f(x)^{-1} \nabla f(x). \tag{4.36}$$

Definition 4.7 has the following interpretations:

- $\mathbf{v}_{\text{nt}}$ is the minimizer of the second-order approximation

$$\hat{f}(\mathbf{x}+\mathbf{v}) = f(\mathbf{x}) + \nabla f(\mathbf{x})^{\mathrm{T}}\mathbf{v} + \frac{1}{2}\mathbf{v}^{\mathrm{T}}\nabla^2 f(\mathbf{x})\mathbf{v}. \tag{4.37}$$

- $\mathbf{v}_{\text{nt}}$ is the steepest descent direction in the Hessian norm.
- $\mathbf{v}_{\text{nt}}$ is the solution of linearized optimality condition for the first-order differential $\nabla f(\mathbf{x}^*) = 0$.
- The Newton step is independent of the linear changes of the parameters.

Definition 4.8 *The Newton decrement is defined as:*

$$\lambda(x) = (\nabla f(x)^{\mathrm{T}} \nabla^2 f(x)^{-1} \nabla f(x))^{\frac{1}{2}}. \tag{4.38}$$

$\lambda^2/2$ is an estimate of $f(\mathbf{x}) - f^*$, based on the quadratic approximation of f at $\mathbf{x}$ [220].

The geometric interpretation for Newton's method is that at each iteration one approximates $f(\mathbf{x})$ by a quadratic function around $\mathbf{x}$, and then takes a step towards the maximum/minimum of that quadratic function. The detailed steps are given as follows:

Newton's method

- Start with a feasible starting point $\mathbf{x}_0$, tolerance $\epsilon > 0$.
- repeat
  1. Calculate the Newton step and decrement.
  2. Quit if $\frac{\lambda^2}{2} \leq \epsilon$.
  3. Line search: choose the step size t.
  4. Update: $\mathbf{x}_{i+1} = \mathbf{x}_i + t\mathbf{v}_{\text{nt}}$.

Newton's method converges much faster towards a local maximum or minimum than the gradient descent method. However, to use Newton's method one needs to know the Hessian of $f(\mathbf{x})$, which sometimes can be difficult to compute. There exist various quasi-Newton methods, where an approximation for the Hessian is used instead.

Another drawback of Newton's method is that finding the inverse of the Hessian can be an expensive operation.

There are some other methods for unconstrained optimization:

- *Conjugate gradients*: The method attempts to solve this problem by "learning" from experience. It uses conjugate directions instead of the local gradient for going downhill. The method proceeds by generating vector sequences of iterates (i.e. successive approximations to the solution), residuals corresponding to the iterates, and search directions used in updating the iterates and residuals. Although the length of these sequences can become large, only a small number of vectors need to be stored in memory. In every iteration of the method, two inner products are performed in order to compute update scalars that are defined to make the sequences satisfy certain orthogonality conditions. If the vicinity of the minimum has the shape of a long and narrow valley, the minimum is reached in far fewer steps than the case using the method of steepest descent.
- *Secant method*: For Newton's method, the Hessian matrix can be both laborious to calculate and to invert for systems with a large number of dimensions. The idea of the secant method is not to use the Hessian matrix directly, but rather to start the procedure with an approximation to the matrix, which, as one gets closer to the solution, gradually approaches the Hessian.
- *Stochastic (or "on-line") gradient descent*: The true gradient is approximated by the gradient of the cost function only evaluated on a single training example. The parameters are then adjusted by an amount proportional to this approximate gradient. Therefore, the parameters of the model are updated after each training example. For large data sets, on-line gradient descent can be much faster than batch gradient descent.
- *Broyden–Fletcher–Goldfarb–Shanno (BFGS) method*: Quasi-Newton or variable metric methods can be used when the Hessian matrix is difficult or time-consuming to evaluate. Instead of obtaining an estimate of the Hessian matrix at a single point, these methods gradually build up an approximate Hessian matrix by using gradient information from some or all of the previous interactions visited by the algorithm.

The rest of this section will discuss how to solve constrained optimization problems. First, by using the KKT conditions, sometimes it can be proven that the constraints can be eliminated without affecting the final solutions. Moreover, by utilizing the dual problem, some constraints can be removed without a loss of optimality. If the constraints cannot be reduced, the following methods are the general approaches to solve the problems:

- *Projected gradient method/Newton method*: If there is no inequality constraint, the gradient or search direction can be projected according to the equality constraint. Suppose the optimization problem is given by

$$\begin{aligned} &\text{minimize: } f(\mathbf{x}) \\ &\text{subject to: } \mathbf{A}\mathbf{x} = \mathbf{b}. \end{aligned} \tag{4.39}$$

For the projected gradient method, the project gradient $\mathbf{v}_{pg}$ can be written as:

$$\begin{pmatrix} \mathbf{I} & \mathbf{A}^T \\ \mathbf{A} & \mathbf{0} \end{pmatrix} \begin{pmatrix} \mathbf{v}_{pg} \\ \mathbf{w} \end{pmatrix} = \begin{pmatrix} -\nabla f(\mathbf{x}) \\ 0 \end{pmatrix}, \tag{4.40}$$

where $\mathbf{w}$ is an estimate of the Lagrangian multiplier.

For the projected Newton method, the projected Newton step $\mathbf{v}_{nt}$ can be expressed as:

$$\begin{pmatrix} \nabla^2 f(\mathbf{x}) & \mathbf{A}^T \\ \mathbf{A} & \mathbf{0} \end{pmatrix} \begin{pmatrix} \mathbf{v}_{nt} \\ \mathbf{w} \end{pmatrix} = \begin{pmatrix} -\nabla f(\mathbf{x}) \\ 0 \end{pmatrix}. \tag{4.41}$$

- *Interior-point method/barrier method*: When the searching point is approaching the boundary of the feasibility, the approach is to add the penalty to the objective, so that the solution always satisfies the inequality constraints. These approaches will be discussed in detail in Section 4.4 on non-linear programming.
- *Cutting planes method*: The basic idea of this approach is to find a hyperplane so that the searching space for the optimal solution can be greatly reduced. We will discuss this method in more detail in Section 4.5.

4.4 Non-linear programming

A non-linear programmming (NLP) problem can be expressed in the following form:

$$\text{minimize: } F(\mathbf{x}) \tag{4.42}$$

$$\text{subject to: } \begin{cases} g_i(\mathbf{x}) = 0, & \text{for } i = 1, \ldots, m_1, \text{ where } m_1 >= 0 \\ h_j(\mathbf{x}) \geq 0, & \text{for } j = m_1 + 1, \ldots, m, \text{ where } m \geq m_1, \end{cases}$$

where F is one scalar-valued function of variable vector $\mathbf{x}$. We seek to minimize F subject to one or more other such functions that limit or define the values of the variable vector. F is called the "objective function," and the various other functions are called the "constraints." Since the objective function or the constraints can be non-linear, the optimization in (4.42) is called non-linear programming.

One of the greatest challenges in NLP is that some problems exhibit "local optima," i.e. the solutions that merely satisfy the requirements are not necessarily good. These situations are similar to the multiple peaks. It is difficult for an algorithm that tries to move from point to point only by climbing uphill, since the peak it achieves might not be the highest. Algorithms that propose to overcome this difficulty are termed as "global optimization" algorithms. How to find a local optimum from an initialization will be discussed next. Specifically, the barrier method/interior point method will be discussed. Then, the techniques to find the global optimum will be discussed.

The idea of encoding the feasible set using a barrier and designing barrier methods was studied in the early 1960s by Fiacco–McCormick and others. These ideas were mainly developed for general non-linear programming. Nesterov and Nemirovskii introduced a special class of such barriers that can be used to encode any convex set. They

Table 4.1 Barrier method.

Given a feasible initialization $\mathbf{x}_0$, tolerance $\epsilon > 0$, and $t > 0$, $\mu > 1$.
Repeat: 1) Calculate $\mathbf{x}^*$ in (4.45). 2) $\mathbf{x} = \mathbf{x}^*$. 3) If the error is smaller than the tolerance ϵ, return $\mathbf{x}$. 4) $t = \mu t$.

guarantee that the number of iterations of the algorithm is bounded by a polynomial in the dimension and accuracy of the solution.

4.4.1 Barrier/interior-point method

In constrained optimization, a barrier function is a continuous function whose value on a point increases to infinity as the point approaches the boundary of the feasible area. It is used as a penalizing term for violations of constraints. The barrier function will be also convex and smooth. The two most common types of barrier function are inverse barrier functions and logarithmic barrier functions, given by

$$I_{\text{inv}} = \begin{cases} \sum_{j=m_1}^{m} \frac{1}{h_j(\mathbf{x})}, & \text{if } h_j \geq 0,\ j = m_1, \ldots, m \\ +\infty, & \text{otherwise} \end{cases} \tag{4.43}$$

and

$$I_{\log} = \begin{cases} -\sum_{j=m_1}^{m} \log(h_j(\mathbf{x})), & \text{if } h_j \geq 0,\ j = m_1, \ldots, m \\ +\infty, & \text{otherwise,} \end{cases} \tag{4.44}$$

respectively.

By adding the barrier function to the objective function $F(\mathbf{x})$, the problem in (4.42) becomes

$$\begin{aligned} &\text{minimize: } tF(\mathbf{x}) + I(\mathbf{x}) \\ &\text{subject to: } g_i(\mathbf{x}) = 0, \quad \text{for } i = 1, \ldots, m_1. \end{aligned} \tag{4.45}$$

In the extreme case, where t is large enough, the barrier function I becomes an ideal barrier function, and the problem in (4.45) becomes the problem in (4.42).

The barrier method (path-following algorithm) tries to solve the problem in (4.42) by solving a sequence of the simplified problem in (4.45). In other words, the method computes the optimal $\mathbf{x}^*$ for a sequence of increasing values of t until the solution is close enough to the original problem. The details of the barrier method are shown in Table 4.1.

There is a tradeoff in the choice of μ. If μ is small, the complexity in solving the problem in (4.45) for each iteration is small, and the iterations closely follow the central path within the feasible range, which can avoid possible local optima. However, it needs more iterations. On the other hand, if μ is large, the barrier function converges fast to

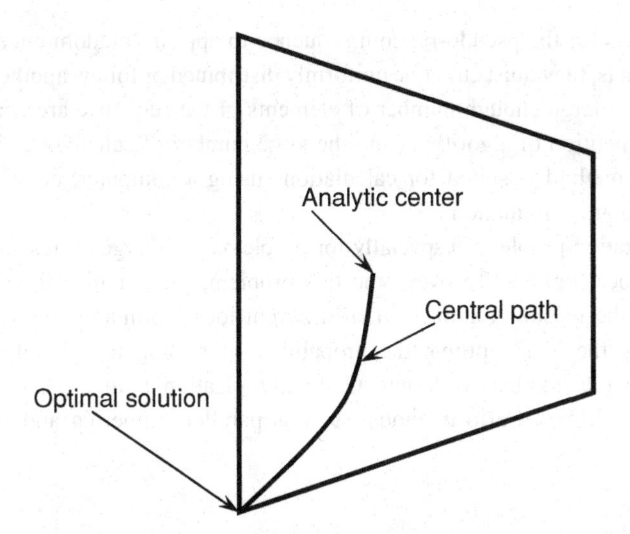

Figure 4.5 Example of the interior-point method.

the ideal solution, but the complexity for solving (4.45) increases and a certain possible local optimum might appear.

Interior-point methods are a certain class of algorithms used to solve linear and non-linear convex optimization problems. These algorithms are inspired by the algorithms of Narendra Karmarkar for linear programming, developed in 1984. The basic elements of the method consist of a self-concordant barrier function used to encode the convex set. Mehrotra's predictor–corrector algorithm is a common implementation of an interior-point method. The primal-dual interior-point method proposed by Kojima, Mizuno, and Yoshise is also widely used. The interior-point method starts from the analytic center (which is proposed by Sonnevend and Megiddo), then follows the central path, and finally converges to an optimal solution. One example is shown in Figure 4.5.

To achieve the global optimum, the initial starting point for the algorithms plays an important role. One popular approach starts with some heuristics with good solutions, and then uses a certain algorithm to converge to a better solution. In most cases, some good solutions can be obtained. However, the global optimum solution is not guaranteed. In order to obtain the global optimum, other methods are widely employed.

4.4.2 Monte Carlo method

Monte Carlo methods are a class of computational algorithms for simulating the behavior of various physical and mathematical systems. They are distinguished from other simulation methods (such as molecular dynamics) by being stochastic, which are non-deterministic in a certain manner. The algorithm usually uses random numbers (or more often pseudo-random numbers) as opposed to deterministic algorithms.

Interestingly, the Monte Carlo method does not require truly random numbers to be useful. Most of the useful techniques use deterministic, pseudo-random sequences, making it easy to test and re-run simulations. The only quality usually necessary to make good simulations is for the pseudo-random sequence to appear "random enough" in a certain sense. That is, they must either be uniformly distributed or follow another desired distribution when a large enough number of elements of the sequence are considered. Because of the repetition of algorithms and the large number of calculations involved, the Monte Carlo method is suited for calculations using a computer, utilizing many techniques of computer simulation.

For the optimization problems, especially for problems with large dimensions, there might be many local optima. To overcome this problem, we initialize the algorithm randomly within the feasible region, so that different local optima can be converged to. By comparing the local optima, the probability of finding the global optimum is increased with the number of Monte Carlo initialization points. There are also some variations of Monte Carlo methods, such as parallel tempering and stochastic tunneling.

4.4.3 Simulated annealing

Simulated annealing (SA) is a generic probabilistic meta-algorithm for the global optimization problem. It was independently invented by S. Kirkpatrick, C. D. Gelatt and M. P. Vecchi in 1987, and by V. Cerny in 1985.

The name and inspiration come from annealing in metallurgy, a technique involving the heating and controlled cooling of a material to increase the size of its crystals and reduce their defects. The heat causes the atoms to become unstuck from their initial positions (a local minimum of the internal energy) and wander randomly through states of higher energy; the slow cooling gives them more chances of finding configurations with lower internal energy than the initial one.

By analogy with this physical process, each step of the SA algorithm replaces the current solution by a random "nearby" solution. This solution is chosen with a probability that depends on the difference between the corresponding function values and on a global parameter T (called the temperature). The temperature is gradually decreased during the process. The dependency is such that the current solution changes almost randomly when T is large, but increases "downhill" as T goes to zero. The allowance for "uphill" moves reduces the chance that the method gets stuck at a local minimum.

At each step, the SA heuristic considers a certain neighbor of the current state, and probabilistically decides between moving the system to a neighboring state or staying in the current state. The probabilities are chosen so that the system ultimately tends to move to states of lower energy. The probability is large when the temperature is high so that the algorithm will not be stuck in a certain local optimum. On the other hand, the probability is low since the probability of local optima is low. When the temperature is zero, the algorithm reduces to the greedy algorithm. Typically this step is repeated until the system reaches a state that is good enough for the application, or until a given computation budget has been exhausted. It can be shown that, for any given finite problem, the probability that the simulated annealing algorithm terminates with the global optimal solution approaches 1 as the annealing schedule is extended.

Table 4.2 Genetic algorithm.

Choose the population of random initializations.
Repeat: 1. Evaluate the individual performance of a certain proportion of the population. 2. Select pairs of best-ranking individuals to reproduce. 3. Apply the crossover operator, which determines the probabilities that the two selected individuals will be actually combined together for the offspring. 4. Apply the mutation operator, in which a small probability of mutation is added to the offspring. Until terminating condition is met.

4.4.4 Genetic algorithm

A genetic algorithm (GA) is a search technique used in computer science to find approximate solutions for optimization. Genetic algorithms are a particular class of evolutionary algorithms that use techniques inspired by evolutionary biology such as inheritance, mutation, natural selection, and recombination (or crossover).

Genetic algorithms are typically implemented as a computer simulation in which a population of abstract representations (called chromosomes) of candidate solutions (called individuals) to an optimization problem evolves toward better solutions. Traditionally, solutions are represented in binary as strings of 0s and 1s, but different encodings are also possible. The evolution starts from a population of completely random individuals and happens in generations. In each generation, the fitness of the whole population is evaluated. Then, multiple individuals are stochastically selected from the current population (based on their fitness), and modified (mutated or recombined) to form a new population, which becomes the current population in the next iteration of the algorithm. The skeleton of a genetic algorithm is shown in Table 4.2. A more detailed discussion on the genetic algorithm and its application to wireless networking problems will be provided in Chapter 6 (Section 6.4).

4.5 Integer programming

In discrete optimization problems the decision variables assume discrete values from a specified set. The combinatorial optimization problems, on the other hand, are problems of choosing the best combination out of all possible combinations. Most combinatorial problems can be formulated as integer programs.

In wireless cognitive networks, integer/combinatorial optimization problem formulations can be used to obtain efficient resource allocation methods which meet the desired objectives when the values of some or all of the decision variables are restricted to be integers. Constraints on basic resources, such as modulation, channel allocation, and coding rate, restrict the possible alternatives that are considered. For example, channel allocation, modulation level, channel coding rate, and even power are discrete in a practical system. To design the future wireless cognitive networks, it is important to study these integer optimizations, especially from a practical implementation point of view.

The versatility of the integer/combinatorial optimization model stems from the fact that, in many practical problems, activities and resources, such as channel, user, or time slot, are indivisible. Also, many problems have only a finite number of alternative choices (such as modulation) and consequently can appropriately be formulated as combinatorial optimization problems (the word combinatorial refers to the fact that only a finite number of alternative feasible solutions exist). Combinatorial optimization models are often referred to as integer programming models, where programming refers to "planning" where some or all of the decisions can take on only a finite number of alternative possibilities. Integer optimization is the process of finding one or more best (optimal) solutions in a well-defined discrete problem space. In this section, we study how to use integer optimization for wireless networking and resource allocation problems.

The major difficulty with these problems is that we do not have any optimality conditions to check if a given (feasible) solution is optimal or not. For example, in linear programming we do have an optimality condition: when a candidate solution is given, we check to see if there exists an "improving feasible direction" to move; if there isn't, then the solution is optimal. If we can find a direction to move that results in a better solution, then the solution is not optimal. There are no such global optimality conditions in discrete or combinatorial optimization problems. In order to guarantee that a given feasible solution is optimal, the solution is to be compared with every other feasible solution. Doing this explicitly amounts to total enumeration of all possible alternatives that are computationally prohibitive due to the NP-completeness of integer programming problems. Therefore, this comparison must be done implicitly, resulting in partial enumeration of all possible alternatives.

There are three different approaches for solving integer programming problems, although they are frequently combined into "hybrid" solution procedures in computational practice. These are as follows:

- relaxation and decomposition techniques;
- enumerative techniques;
- cutting planes approaches based on polyhedral combinatorics.

Before these techniques are studied, the general problem and one simple example, namely, the knapsack problem, are investigated.

4.5.1 General formulation

In this subsection, the general problem formulation for integer optimization is discussed first. Then the potential applications of integer optimization models for wireless networking and resource allocation problems are discussed.

Most of the integer optimization research to date covers only the linear case. A survey of non-linear integer programming approaches is given in [225]. The general problem formulation can be written as follows:

$$\min_{\mathbf{x,y,z}} f(\mathbf{x,y,z}) \tag{4.46}$$

$$\text{subject to: } \begin{cases} g_i(\mathbf{x,y,z}) \leq 0, \text{ for } i = 1, \ldots, m \\ h_j(\mathbf{x,y,z}) = 0, \text{ for } j = 1, \ldots, l \\ \mathbf{x} \in \mathcal{R}, \mathbf{y} \in \{0, 1\}, \text{ and } \mathbf{z} \in \mathcal{I}, \end{cases}$$

where function f is the objective function, function g_i is the equality constraint function, function h_j is the inequality constraint function, the component of vector $\mathbf{x}$ is a real-valued variable, the component of vector $\mathbf{y}$ is a variable which can assume a value of either 0 or 1, and the component of vector $\mathbf{z}$ is a integer value in a space $\mathcal{I}$. If $\mathbf{y} = \mathbf{0}$ and $\mathbf{z} = \mathbf{0}$, (4.46) becomes a non-linear optimization case. If $\mathbf{z} = \mathbf{0}$, the problem in (4.46) is referred to as a pure 0–1 integer-programming problem; if $\mathbf{y} = \mathbf{0}$, the problem in (4.46) is called a pure integer programming problem. Otherwise, the problem is a mixed integer programming problem.

For wireless networking and resource allocation problems, there are many potential applications of integer optimization models. Some representative examples are described next.

Network, routing, and graph problems

Many optimization problems can be represented by a network, where the network (or graph) is defined by nodes and by arcs connecting those nodes. Many practical problems arise around physical networks such as communication networks. In addition, there are many problems that can be modeled as networks even when there is no underlying physical network. For example, one can think of the assignment problem where one wishes to assign a set of users to a certain set of jobs in a way that minimizes the cost of the assignment. Here one set of nodes represents the users to assign, another set of nodes represents the possible jobs, and there is an arc connecting a user to a job if that user is capable of performing that job.

In addition, there are many graph-theoretic problems that examine the properties of the underlying graph or network. Such problems include the Chinese postman problem where one wishes to find a path (a connected sequence of edges) through the graph that starts and ends at the same node, which covers every edge of the graph at least once, and that has the shortest length possible. If one adds the restriction that each node must be visited exactly one time and drops the requirement that each edge be traversed, the problem becomes the notoriously difficult traveling salesman problem. Other graph problems include the vertex coloring problem, whose objective is to determine the minimum number of colors needed to color each vertex of the graph in order that no pair of adjacent nodes (nodes connected by an edge) share the same color; the edge coloring problem, whose objective is to find a minimum total weight collection of edges such that each node is incident to at least one edge; the maximum clique problem, whose objective is to find the largest subgraph of the original graph such that every node is connected to every other node in the subgraph; and the minimum cut problem, whose objective is to find a minimum weight collection of edges that (if removed) would disconnect a set of nodes s from a set of nodes t.

Although these combinatorial optimization problems on graphs might appear, at first glance, to be interesting mathematically but having little application to the decision making in management or engineering, their domain of applicability is extraordinarily broad. The traveling salesman problem has applications in routing and scheduling, in large-scale circuitry design and in strategic defense. The four-color problem (can a map be colored in four colors or less?) is a special case of the vertex coloring problem. Both the clique problem and the minimum cut problem have important implications for the reliability of large systems.

Scheduling problem

Space-time networks are often used in scheduling applications. Here one wishes to meet specific demands at different points in time. To model this problem, different nodes represent the same entity at different points in time. An example of the many scheduling problems that can be represented as a space-time network is the channel assignment problem. This problem requires that users are assigned with channels so that the system performance is maximized. Each time, a channel must have one and only one user assigned to it, and a user can be assigned to a channel according to its instantaneous channel conditions, buffer state, as well as the quality-of-service (QoS) requirements.

Assignment problem

The assignment problem is one of the fundamental combinatorial optimization problems. In its most general form, the problem is as follows: There are a number of agents and a number of tasks. Any agent can be assigned to perform any task, incurring a certain cost that may vary depending on the assignment. It is required to perform all tasks by assigning exactly one agent to each task in such a way that the total cost of the assignment is minimized. If the number of agents and tasks are equal and the total cost of the assignment for all tasks is equal to the sum of the costs for each agent (or equivalently the sum of the costs for each task), then the problem is called a linear assignment problem. Commonly, an assignment problem without any additional qualification refers to a linear assignment problem. Other types of assignment problem include quadratic assignment problem and bottleneck assignment problem.

The assignment problem is a special case of another optimization problem known as the transportation problem, which is a special case of the maximal flow problem, which in turn is a special case of a linear program. While it is possible to solve any of these problems using the simplex algorithm, each problem has more efficient algorithms designed to take advantage of its special structure. Algorithms are known which solve the linear assignment problem within time bounded by a polynomial expression in terms of the number of agents.

The restrictions on agents, tasks, and cost in the (linear) assignment problem can be relaxed, as shown in the example below. Suppose that a taxi firm has three taxis (the agents) available, and three customers (the tasks) wishing to be picked up as soon as possible. The firm prides itself on speedy pickups, so for each taxi the "cost" of picking up a particular customer will depend on the time taken for the taxi to reach the pickup point. The solution to the assignment problem will be whichever combination of taxis and customers results in the least total cost.

However, the assignment problem can be made rather more flexible than it first appears. In the above example, suppose that there are four taxis available, but still only three customers. Then a fourth task can be invented, perhaps called "sitting still doing nothing," with a cost of 0 for the taxi assigned to it. The assignment problem can then be solved in the usual way and can still give the best solution to the problem.

Similar tricks can be played in order to allow more tasks than agents, tasks to which multiple agents must be assigned (for instance, a group with number of customers more than that which will fit in one taxi), or maximizing profit rather than minimizing cost.

The formal definition of the assignment problem (or linear assignment problem) can be written as follows: each of n tasks can be performed by any of n agents. The cost of task i being accomplished by agent j is c_{ij}. Assign one agent to each task to minimize the total cost as:

$$\text{minimize: } \sum_{i=1}^{n}\sum_{j=1}^{n} c_{ij}x_{ij} \tag{4.47}$$

$$\text{subject to: } \begin{cases} \sum_{j=1}^{n} x_{ij} = 1, & \text{for all } i \\ \sum_{i=1}^{n} x_{ij} = 1, & \text{for all } j \\ x_{ij} \in \{0, 1\}. \end{cases}$$

Next we discuss the formulation considerations. The versatility of the integer programming formulation, as illustrated by the above examples, shall provide sufficient explanation for the high activity in the field of combinatorial optimization that involves development of solution procedures for such problems. Since there are often different ways of mathematically representing the same problem, and since obtaining an optimal solution to a large integer programming problem in a reasonable amount of computer time may well depend on the way it is "formulated," significant amounts of recent research have been directed towards the reformulation of integer programming problems. In this regard, it is sometimes advantageous to increase (rather than decrease) the number of integer variables, the number of constraints, or both.

Once the problem has been formulated (or reformulated) into an integer programming problem, solution approaches for obtaining optimal or at least near optimal solutions must be found. In the next subsection, we discuss a special case of integer programming, namely, the knapsack problem, and describe its applications and variations.

4.5.2 Knapsack problem

Suppose one wants to fill a knapsack that can hold a total weight of c with a certain combination of items from a list of n possible items, each with weight w_i and value p_i, so that the value of the items packed into the knapsack is maximized. This problem has a single linear constraint (i.e. the weight of the items in the knapsack should not exceed c), a linear objective function which sums the values of the items in the knapsack, and the added restriction that each item either be in the knapsack or not (i.e. a fractional amount of the item is not possible). Define a vector of binary variable x_j having the following meaning:

$$x_j = \begin{cases} 1, & \text{if item } j \text{ is selected} \\ 0, & \text{otherwise.} \end{cases} \tag{4.48}$$

The simplest problem is how to pack the knapsack with as many valuable items as possible, which has the following formulation:

$$\max_{x_j} \sum_{j=1}^{n} p_j x_j \tag{4.49}$$

$$\text{subject to: } \sum_{j=1}^{n} w_j x_j \leq c.$$

In general, the knapsack problems are NP-hard. For solution approaches specific to the knapsack problem, see [226].

Although this problem might seem almost too simple to have much applicability, the knapsack problem is important to cryptographers and to those interested in protecting computer files, electronic transfers of funds, and electronic mail. These applications use a "key" to ensure information security. Often the keys are designed based on linear combinations of a certain collection of data items that must equal a certain value. This problem is also structurally important in that most integer programming problems are generalizations of this problem (i.e. there are many knapsack constraints that together compose the problem). Approaches for the solution of multiple knapsack problems are often based on examining each constraint separately.

Next, all types of knapsack problem formulations are listed. For wireless networking and resource allocation problems, the best fits for the specific problem can be selected.

- *0–1 knapsack problem*: The problem formulation is the same as (4.49). The problem itself has attracted a lot of attention due to the following facts: first, it can be viewed as the simplest integer optimization problem. Second, it appears as a subproblem of many complex problems. Third, it is applicable to many practical situations.
- *Bounded knapsack problem*: The bounded knapsack problem is a generalized 0–1 knapsack problem since x_j can be any integer and bounded by b_j rather than being binary only. The problem is formulated as:

$$\max_{x_j} \sum_{j=1}^{n} p_j x_j \tag{4.50}$$

$$\text{subject to: } \begin{cases} \sum_{j=1}^{n} w_j x_j \leq c \\ 0 \leq x_j \leq b_j \text{ and } b_j \text{ is an integer, } j \in N = \{1, \ldots, n\}. \end{cases}$$

- *Subset–sum problem*: The subset–sum problem is also called the value-independent knapsack problem or stick-stacking problem. It is a special case of the 0–1 knapsack problem with $p_j = w_j, \ \forall j$. The problem can be written as follows:

$$\max_{x_j} \sum_{j=1}^{n} w_j x_j \tag{4.51}$$

$$\text{subject to: } \begin{cases} \sum_{j=1}^{n} w_j x_j \leq c \\ x_j = 1, \ \text{if item j is selected; } x_j = 0, \ \text{otherwise.} \end{cases}$$

- *Change-making problem*: The problem is to make changes for a fixed amount of money and minimize the number of changes. The unit for each change is denoted by w_j and the overall amount of money is c. The problem can be formulated as follows:

$$\min_{x_j} \sum_{j=1}^{n} x_j \tag{4.52}$$

$$\text{subject to: } \begin{cases} \sum_{j=1}^{n} w_j x_j = c \\ 0 \leq x_j \text{ and is an integer, } j \in N = \{1, \ldots, n\}. \end{cases}$$

- *Multiple knapsack problem*: If there are more than one knapsack, the problem is a multiple knapsack problem. Suppose there are a total of m knapsacks and n items. Then, the 0–1 multiple knapsack problem is as follows:

$$\max \sum_{i=1}^{m} \sum_{j=1}^{n} p_j x_{ij} \tag{4.53}$$

$$\text{subject to: } \begin{cases} \sum_{j=1}^{n} w_j x_{ij} \leq c_i, \ \forall i \in M = \{1, \ldots, m\} \\ \sum_{i=1}^{m} x_{ij} \leq 1, \ \forall j \in N = \{1, \ldots, n\} \\ x_{ij} = 1, \ \text{if item } j \text{ is selected for knapsack } i; \ x_{ij} = 0, \ \text{otherwise.} \end{cases}$$

- *Generalized assignment problem*: In the generalized assignment problem, the profit and weight are different for different knapsacks, i.e. we have p_{ij} and w_{ij}. The problem formulation is given by:

$$\max \sum_{i=1}^{m} \sum_{j=1}^{n} p_{ij} x_{ij} \tag{4.54}$$

$$\text{subject to: } \begin{cases} \sum_{j=1}^{n} w_{ij} x_{ij} \leq c_i, \ \forall i \in M = \{1, \ldots, m\} \\ \sum_{i=1}^{m} x_{ij} = 1, \ \forall j \in N = \{1, \ldots, n\} \\ x_{ij} = 1, \ \text{if item } j \text{ is selected for knapsack } i; \ x_{ij} = 0, \ \text{otherwise.} \end{cases}$$

- *Bin-packing problem*: The bin-packing problem is to select the minimal number of knapsacks with capacity c to pack all the items. Suppose $y_i = 1$ if the ith knapsack is occupied; otherwise $y_i = 0$. Then, the problem formulation is given by:

$$\min \sum_{i=1}^{n} y_i \tag{4.55}$$

$$\text{subject to: } \begin{cases} \sum_{j=1}^{n} w_j x_{ij} \leq c y_i, \ \forall i \in N = \{1, \ldots n\} \\ \sum_{i=1}^{m} x_{ij} = 1, \ \forall j \in N = \{1, \ldots, n\} \\ x_{ij} = 1, \ \text{if item } j \text{ is selected for knapsack } i; \ x_{ij} = 0, \ \text{otherwise.} \end{cases}$$

In the next three subsections, three different ways of solving the integer/combinatorial problem will be discussed. Examples are given to clarify each approach.

4.5.3 Relaxation and decomposition

One way of solving integer programming problems is to take a set of "complicated" constraints into the objective function in a Lagrangian fashion (with fixed multipliers that are changed iteratively). This approach is known as Lagrangian relaxation. By removing the complicated constraints from the constraint set, the resulting subproblem usually becomes considerably easier to solve. The latter is a necessity for the approach to work because the subproblems must be solved repetitively until the optimal values for the multipliers are found. The bound found by Lagrangian relaxation can be tighter than that found by linear programming, but only at the expense of solving subproblems in integers, i.e. only if the subproblems do not have the integrality property. A problem has the integrality property if the solution to the Lagrangian problem is unchanged when the integrality restriction is removed. Before applying Lagrangian relaxation, the structure of the problem needs to be identified in order to relax the constraints that are "complicated" [227]. A related approach that attempts to strengthen the bounds of Lagrangian relaxation, is called Lagrangian decomposition [228]. This approach consists of isolating sets of constraints so as to obtain separate, easy problems to solve over each of the subsets. The dimension of the problem is increased by creating linking variables that link the subsets. All Lagrangian approaches are problem dependent and no underlying general theory has evolved.

Consider the following example of constrained combinatorial optimization:

$$\max \sum_{j=1}^{n} p_j x_j \tag{4.56}$$

$$\text{subject to: } \begin{cases} \sum_{j=1}^{n} w_j x_j = c \\ x_j = \{0, 1\}, \forall j. \end{cases}$$

The Lagrangian relaxation relaxes the complicated constraint to the objective function, which is given by

$$\max \sum_{j=1}^{n} p_j x_j + \lambda \left(c - \sum_{j=1}^{n} w_j x_j \right) \tag{4.57}$$

$$\text{subject to: } x_j = \{0, 1\}, \forall j$$

where λ is the Lagrangian multiplier. The goal is to obtain a simple solution of (4.57) with a fixed λ. Consequently, the complexity can be reduced significantly. By adjusting λ, the feasibility and slackness of the complicated constraints are improved. This method is called the sub-gradient method and can be summarized as follows:

1. Begin with each λ at 0. Let the step size be a certain (problem-dependent value) k.
2. Solve (4.57) to get current solution x.
3. For every constraint violated by x, increase the corresponding λ by k.

4. For every constraint with positive slack relative to x, decrease the corresponding λ by k.
5. If m iterations have passed since the best relaxation value has decreased, cut k in half.
6. Go to 2.

Next, we give an example for Lagrangian relaxation. The integer optimization problem is

$$\max_{x_i \in \{0,1\}} \; 4x_1 + 5x_2 + 6x_3 + 7x_4 \tag{4.58}$$

$$\text{subject to: } \begin{cases} 2x_1 + 2x_2 + 3x_3 + 4x_4 \leq 7 \\ x_1 - x_2 + x_3 - x_4 \leq 0. \end{cases}$$

The Lagrangian relaxation is given by

$$\max_{x_i \in \{0,1\}} \; 4x_1 + 5x_2 + 6x_3 + 7x_4 + \lambda_1(7 - 2x_1 - 2x_2 - 3x_3 - 4x_4) + \lambda_2(-x_1 + x_2 - x_3 + x_4). \tag{4.59}$$

We select the initial values as $\lambda_1 = \lambda_2 = 0$ and step size to be 0.5. The solution is $x_j = 1, \forall j$. However, this violates the first constraint. When we set $\lambda_1 = 0.5$ and $\lambda_2 = 0$, the solution is obtained, but again the constraint is violated. We set the new values as $\lambda_1 = 2$ and $\lambda_2 = 0$. There is slackness in the first constraint, but the second constraint is satisfied. Since we cannot go back to $\lambda_1 = 1.5$, which causes a violation of the constraint, we reduce the step size by setting $\lambda_1 = 1.75$ and $\lambda_2 = 0$. The process stops when $\lambda_1 = 1.83$ and $\lambda_2 = 0.33$ with optimal solution obtained as follows: $x_1 = x_2 = x_3 = 1$ and $x_4 = 0$.

Most Lagrangian-based strategies provide approaches that deal with special row structures. Other problems may possess a special column structure, such that when some subsets of the variables are assigned specific values, the problem reduces to one that is easy to solve. Benders' decomposition algorithm fixes the complicated variables, and solves the resulting problem iteratively [229]. Based on the problem's associated dual, the algorithm must then find a cutting plane (i.e. a linear inequality) which "cuts off" the current solution point but no integer feasible points. This cut is added to the collection of inequalities and the problem is re-solved.

Since each of the decomposition approaches described above provides a bound on the integer solution, it can be incorporated into a branch-and-bound algorithm (discussed in the next subsection), instead of the more commonly used linear programming relaxation. However, these algorithms are special-purpose algorithms in that they exploit the "constraint pattern" or special structure of the problem.

4.5.4 Enumeration technique: branch-and-bound

The simplest approach to solving a pure integer programming problem is to enumerate all possibilities. However, due to the "combinatorial explosion" resulting from the parameter "size," only the smallest instances can be solved by such an approach. Sometimes one can implicitly eliminate many possibilities by domination or feasibility arguments. Besides straightforward or implicit enumeration, the most commonly used enumerative approach is called branch and bound, where the "branching" refers to the enumeration part of the solution technique and "bounding" refers to the seeking of possible solutions by comparing to a known upper or lower bound on the solution value. Next, we will discuss the branch-and-bound approach in detail and illustrate some examples.

The general idea may be described in terms of finding the minimal value of a function $f(x)$ over a set of admissible values of the argument x called the feasible region. Both f and x could be arbitrary in nature. A branch-and-bound procedure requires two tools.

The first tool is a smart way of covering the feasible region by several smaller feasible subregions (ideally, splitting into subregions). This is called branching, since the procedure is repeated recursively to each of the subregions and all produced subregions naturally form a tree structure, called a search tree or a branch-and-bound-tree. Its nodes are the constructed subregions. Another tool is bounding, which is a fast way of finding the upper and lower bounds for the optimal solution within a feasible subregion.

The core of the approach is a simple observation that (for a minimization task) if the lower bound for a subregion A from the search tree is greater than the upper bound for any other (previously examined) subregion B, then A may be safely discarded from the search. This step is called pruning. It is usually implemented by maintaining a global variable m that records the minimum upper bound seen among all subregions examined so far; any node whose lower bound is greater than m can be discarded.

It may happen that the upper bound for a node matches its lower bound; that value is then the minimum of the function within the corresponding subregion. Sometimes there is a direct way of finding such a minimum. In both these cases it is said that the node is solved. Note that this node may still be pruned as the algorithm progresses.

Ideally the procedure stops when all nodes of the search tree are either pruned or solved. At that point, all non-pruned subregions will have their upper and lower bounds equal to the global minimum of the function. In practice, the procedure is often terminated after a given time; at that point, the optimal lower bound and the optimal upper bound, among all non-pruned sections, define a range of values that contains the global minimum.

The efficiency of the method depends critically on the effectiveness of the branching and bounding algorithms used; bad choices can lead to repeated branching, without any pruning, until the subregions become very small. In that case the method will be reduced to an exhaustive enumeration of the domain, which is often impractically large. There is no universal bounding algorithm that works for all problems, and there is little hope that one will ever be found; therefore, the general paradigm needs to be implemented separately for each application, with branch-and-bound algorithms that are specially designed for it.

In the following, an example branch-and-bound algorithm will be given to maximize the following constrained integer optimization:

$$\max Z = 21x_1 + 11x_2 \tag{4.60}$$

$$\text{subject to: } \begin{cases} 7x_1 + 4x_2 \leq 13 \\ x_1 \geq 0, x_2 \geq 0 \\ x_1, x_2 \text{ are integers.} \end{cases}$$

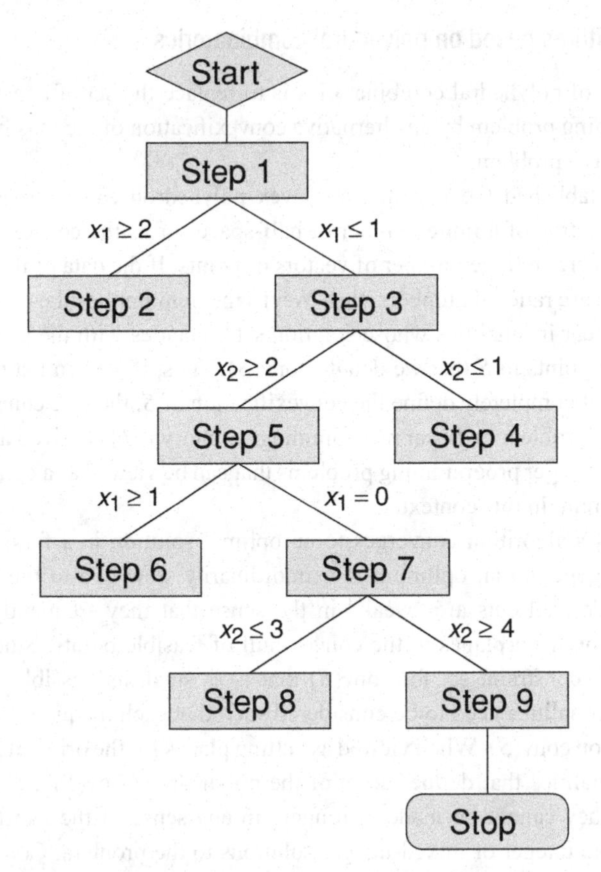

Figure 4.6 An example of branch-and-bound.

1. The first step relaxes (4.60) assuming that x_1 and x_2 are continuous variables. We have the solution $Z = 39$, $x_1 = 1.86$, and $x_2 = 0$. Then we try to branch on x_1 to Steps 2 and 3, since it is not an integer value.
2. $x_1 \geq 2$. In this case, there is no feasible solution.
3. $0 \leq x_1 \leq 1$. The solution is $Z = 37.5$, $x_1 = 1$, and $x_2 = 1.5$. Then we try to branch on x_2 to Steps 4 and 5, since it is not an integer value.
4. $0 \leq x_1 \leq 1$ and $0 \leq x_2 \leq 1$. The solution is $Z = 32$, $x_1 = 1$, and $x_2 = 1$. Since all variables are integers, stop branching. Return one of the possible solutions.
5. $0 \leq x_1 \leq 1$ and $x_2 \geq 2$. The solution is $Z = 37$, $x_1 = 0.71$, and $x_2 = 2$. Then we try to branch on x_1 to Steps 6 and 7, since it is not an integer value.
6. $x_1 = 1$ and $x_2 \geq 2$. No feasible solution.
7. $x_1 = 0$ and $x_2 \geq 2$. The solution is $Z = 35.75$, $x_1 = 0$, and $x_2 = 3.25$. Then we try to branch on x_2 to Steps 8 and 9, since it is not an integer value.
8. $x_1 = 0$ and $2 \leq x_2 \leq 3$. The solution is $Z = 33$, $x_1 = 0$, and $x_2 = 3$. Since all variables are integer, stop branching. Return one of the possible solutions.
9. $x_1 = 0$ and $x_2 \geq 4$. No feasible solution.

The steps are shown in Figure 4.6. Compared with the results in Steps 4 and 8, the optimal solution is $Z = 32$, $x_1 = 1$, and $x_2 = 1$.

4.5.5 Cutting plane algorithms based on polyhedral combinatorics

The underlying idea of polyhedral combinatorics is to replace the set of constraints in an integer programming problem by an alternative convexification of the feasible points and extreme rays of the problem.

H. Weyl [230] established the fact that a convex polyhedron can alternatively be defined as the intersection of a finite number of half-spaces or as the convex hull plus the conical hull of a certain finite number of vectors or points. If the data of the original problem formulation are rational numbers, then Weyl's theorem implies the existence of a finite system of linear inequalities whose solution set coincides with the convex hull of the mixed-integer points in S that we denote conv(S). Thus, if we can list the set of linear inequalities that completely define the convexification of S, then we can solve the integer programming problem by linear programming. Gomory [231] derived a "cutting plane" algorithm for integer programming problems that can be viewed as a constructive proof of Weyl's theorem, in this context.

Although Gomory's algorithm converges to an optimal solution in a finite number of steps, the convergence to an optimum is extraordinarily slow due to the fact that these algebraically-derived cuts are "weak" in the sense that they frequently do not even define supporting hyperplanes to the convex hull of feasible points. Since one is interested in a linear constraint set for conv(S) that is as small as possible, minimal systems of linear inequalities need to be considered such that each inequality defines a facet of the polyhedron conv(S). When viewed as cutting planes for the original problem, then the linear inequalities that define facets of the polyhedron conv(S) are the "best possible" cuts, i.e. they cannot be made "stronger" in any sense of the word without losing certain feasible integer or mixed-integer solutions to the problem. Considerable research activity has been focused on identifying parts (or all) of those linear inequalities for specific combinatorial optimization problems that are however derived from an underlying general theme due to Weyl's theorem that applies generally. Since for most interesting integer programming problems the minimal number of inequalities necessary to describe this polyhedron is exponential in the number of variables, one is led to wonder whether such an approach can ever be computationally practical. It is remarkable that the implementation of cutting plane algorithms based on polyhedral theory has been successful in solving problems of sizes previously believed to be intractable. The numerical success of the approach can be explained, in part, by the fact that we are interested in proving optimality of a single extreme point of conv(S). We therefore do not require the complete description of S, but rather only a partial description of S in the neighborhood of the optimal solution.

As a first step, the general cutting plane approach relaxes the integrality restrictions on the variables and solves the resulting linear program over the set S. If the linear program is unbounded or infeasible, so is the integer program. If the solution to the linear program is an integer, then one has solved the integer program. If not, then one solves a facet-identification problem whose objective is to find a linear inequality that "cuts off" the fractional linear programming solution while assuring that all feasible integer points satisfy the inequality, i.e. an inequality that "separates" the fractional point from the polyhedron conv(S). The algorithm continues until: (1) an integer solution is found (the problem is solved); (2) the linear program is infeasible and therefore the integer problem is infeasible; or (3) no cut is identified by the facet-identification procedures either because a full description of the facial structure is unknown or because the facet-identification procedures are inexact, i.e. one is unable to algorithmically generate cuts of a known form. If we terminate the cutting plane procedure because of the third possibility, then, in general, the process has "tightened" the linear programming formulation so that the resulting linear programming solution value is much closer to the integer solution value. In general, the cutting plane method can be summarized as follows:

1. Solve the linear programming relaxation.
2. If the solution to the relaxation is feasible in the integer programming problem, the algorithm stops and the solution is optimal.
3. Otherwise, find one or more cutting planes that separate the optimal solution to the relaxation from the convex hull of feasible integral points, and add a subset of these constraints to the relaxation.
4. Return to the first step.

Typically, the first relaxation is solved using the primal simplex algorithm. After the addition of cutting planes, the current primal solution is no longer feasible. However, since the dual problem is only modified by the addition of some variables, if these extra dual variables are given the value of zero, the current dual solution is still feasible. Therefore, subsequent relaxations are solved using the dual simplex method. Notice that the values of the relaxations provide lower bounds on the optimal value of the integer program. These lower bounds can be used to measure progress towards optimality, and to give performance guarantees on integral solutions.

Consider the following example of an integer programming problem:

$$\text{minimize:} \quad -2x_1 - x_2 \tag{4.61}$$

$$\text{subject to:} \quad \begin{cases} x_1 + 2x_2 \leq 7 \\ 2x_1 - x_2 \leq 3 \\ x_1, x_2 \geq 0, \text{ integer.} \end{cases}$$

The feasible integer points are indicated in Figure 4.7. The linear programming relaxation (or LP relaxation) is obtained by ignoring the integrality restrictions; this is given by the polyhedron contained in the solid lines. The boundary of the convex hull of the feasible integer points is indicated by dashed lines.

If a cutting plane algorithm is used to solve this problem, the linear programming relaxation will first be solved, giving the point $x_1 = 2.6$, $x_2 = 2.2$, which has value -7.4. The inequalities $x_1 + x_2 \leq 4$ and $x_1 \leq 2$ are satisfied by all the feasible integer points but they are violated by the point (2.6, 2.2). Thus, these two inequalities are valid cutting planes. These two constraints can then be added to the relaxation, and when the relaxation is solved again, the point $x_1 = 2$, $x_2 = 2$ results, with value -6. Notice that this point is feasible in the original integer program, so it must actually be optimal for that problem, since it is optimal for a relaxation of the integer program.

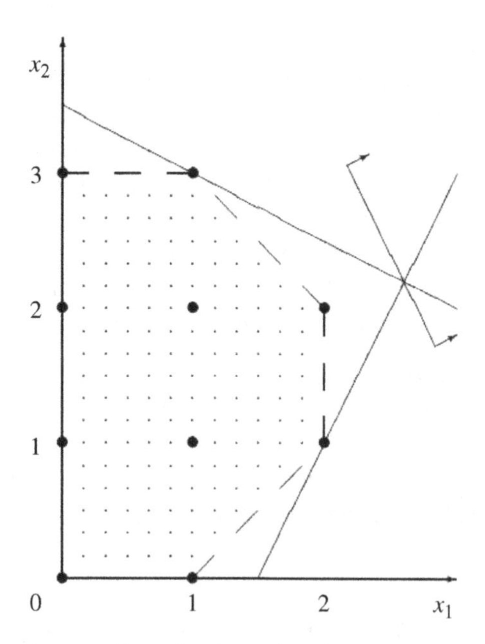

Figure 4.7 Example of a cutting plane.

If, instead of adding both inequalities, only the inequality $x_1 \leq 2$ is added, the optimal solution to the new relaxation becomes $x_1 = 2, x_2 = 2.5$, with value -6.5. The relaxation can then be modified by adding a cutting plane that separates this point from the convex hull, for example $x_1 + x_2 \leq 4$. Solving this new relaxation will again result in the optimal solution to the integer program.

4.6 Dynamic programming and the Markov decision process

The basics of dynamic programming are first studied. Then, one major application, the Markov decision process (MDP), is introduced and solved by using the Bellman's equation. Then, for the cases when the transition probability is not random or not known for the Markov process, partially observable MDP and reinforcement learning methods are investigated. Finally, two examples of using MDP models for wireless networking problems are illustrated.

4.6.1 General definition of dynamic programming

One of the major categories of programming methods is dynamic programming. Basically, the dynamic programming approach solves a high complexity problem by combining the solutions of a series of low complexity subproblems. Dynamic programming relies on the principle of optimality, which states that in an optimal sequence of decisions or choices, each subsequence must also be optimal. To understand the basic problem formulation, we define the following concepts:

Table 4.3 Classifications of dynamic programming.

Discrete	Continuous
Finite horizon	Infinite horizon
Deterministic	Stochastic
Constrained	Unconstrained
Perfect state information	Imperfect state information
Single objective	Multiple objective

- *State:* A state is a configuration of a system and is identified by a label which indicates the properties corresponding to that state.
- *Stage:* A stage is a single step in the system undergoing a certain process and corresponds to the transition from one state to an adjacent state.
- *Action:* At each state there is a set of actions available from which a choice must be made.
- *Policy:* A policy is a set of actions, one for each of a number of states. An optimal policy is the best set of actions in accordance with a given objective.
- *Return:* A return is an outcome which a system generates over one stage of a process. A return is usually something like a profit, a cost, a distance, a yield, or the consumption of a product.
- *Value of a state:* The value of a state is a function of returns (sum) generated when the system starts in that state and a particular policy is followed. The value of a state under an optimal policy is the optimal value.

Depending on the different characteristics of the above definitions, the dynamic programming problems can be classified as shown in Table 4.3. Dynamic programming generally takes one of two approaches:

- *Top-down approach:* The problem is broken into subproblems, and these subproblems are solved and the solutions are stored (i.e. remembered), in case they need to be solved again. This approach combines recursion and memorization together.
- *Bottom-up approach:* All subproblems that might be needed are solved in advance and then used to build up solutions to larger problems. This approach is slightly better in stack space usage and the number of function calls, but it is sometimes not intuitive to determine all the subproblems needed for solving the given problem.

In practice, to formulate the problem, the following steps are required:

1. Characterize the structure of the optimal solution.
2. Define the value of the optimal solution recursively.
3. Compute the optimal solution values using either top-down with caching or bottom-up in a table.
4. Construct the optimal solution from the computed values.

Some typical applications are:

- Problems which involve a sequence of decisions in time: e.g. production planning, stock control, investment decision making, or replacement policy.
- Problems in which a sequence of decisions is not directly related to time: e.g. sequential production processes, optimal path problems, or optimal search problems.
- Allocation problems: e.g. deciding a sequence of allocations of one or more limited resources.
- Combinatorial and graph theoretic problems: e.g. scheduling, sequencing, or set partitioning.

4.6.2 Markov decision process

One of the major applications of dynamic programming is in Markov decision processes (MDPs), which provide a mathematical framework for modeling decision-making in situations where outcomes are partly random and partly under the control of the decision maker. In MDP, the Markov property is assumed, i.e. the effects of an action taken in a state depend only on that state and not on the prior history. MDPs are useful for studying a wide range of optimization problems for cognitive radios [232].

An MDP model contains the action at time n as α_n, state as s_n, reward function as $R(s_n, \alpha_n)$, and the forgetting factor or discounting factor as β. The discounted time average reward for an infinite time scale can be expressed as:

$$V = \sum_{n=1}^{\infty} \beta^n R(s_n, \alpha_n). \tag{4.62}$$

Here the action can be deterministic or stochastic. The deterministic action is $\alpha_n = \pi(s_n)$, where π is the policy. The stochastic action is characterized by $P(s'|s, \pi(s))$, which denotes the probability of going to the next state s' given the current state s and policy $\pi(s)$.

Bellman's equation, which is also called an optimality equation or a dynamic programming equation, is given by

$$V^{\pi}(s) = R(s) + \beta \sum_{s'} P(s'|s, \pi(s)) V^{\pi}(s'). \tag{4.63}$$

The equation for the optimal policy is known as Bellman's optimality equation:

$$V^{*}(s) = R(s) + \max_{\alpha} \beta \sum_{s'} P(s'|s, \alpha) V^{*}(s'). \tag{4.64}$$

In other words, Bellman's equation calculates the action that gives the best expected return.

One major disadvantage of MDP is the *curse of dimensionality*. The state space is typically astronomically large. To overcome this curse, factored MDPs are one approach used to represent large, structured MDPs compactly, in which a state is implicitly described by assigning it to some set of state variables.

Table 4.4 Types of planning problems.

	State	Action model
Classical planning	Observable	Deterministic
MDP	Observable	Stochastic
POMDP	Partial observable	Stochastic

The solution above assumes that the state s is known when action is to be taken; otherwise $\pi(s)$ cannot be calculated. When this assumption is not true, the problem is called a partially observable Markov decision process (POMDP). MDPs are a special case of POMDPs in which the observations always uniquely identify the true state (i.e. the states are directly observed or can be directly deduced from the observation). In Table 4.4, the classical planning, MDP, and POMDP problems are compared.

Since the true state of the world cannot be uniquely identified, a POMDP reasoner must maintain a probability distribution, called the belief state, which describes the probabilities for each true state of the world. Maintenance of the belief states is Markovian, in that it only requires knowledge of the previous belief state, the action taken, and the observation seen.

Each belief is a probability distribution, thus, each value in a POMDP is a function of an entire probability distribution. This is problematic, since probability distributions are continuous. Additionally, we have to deal with the huge complexity of belief spaces. POMDPs so far have only been applied successfully to very small state spaces with small numbers of possible observations and actions.

If the probabilities of MDP are unknown, the problem is one of reinforcement learning, which is a subarea of machine learning concerned with how an agent ought to take actions in an environment so as to maximize some notion of long-term reward. Reinforcement learning algorithms attempt to find a policy that maps states of the world to the actions the agent ought to take in those states. The environment is typically formulated as a finite-state MDP, and reinforcement learning algorithms for this context are highly related to dynamic programming techniques. State transition probabilities and reward probabilities in the MDP are typically stochastic but stationary over the course of the problem.

Q-learning is the most popular model-free reinforcement learning technique that works by learning an action-value function that gives the expected utility of taking a given action in a given state and following a fixed policy thereafter. A strength with Q-learning is that it is able to compare the expected utility of the available actions without requiring a model of the environment. The core of the algorithm is a simple value iteration update. For each state, s, from the state set S, and for each action, a, from the action set A, the following update to its expected discounted reward can be calculated:

$$Q_{t+1}(s_t, a_t) \leftarrow Q_t(s_t, a_t) + \alpha_t(s_t, a_t)[R_t + \beta \max_a Q_t(s_{t+1}, a_{t+1}) - Q_t(s_t, a_t)], \tag{4.65}$$

where R_t is an observed real reward at time t, $\alpha_t(s, a)$ are the learning rates such that $0 \leq \alpha_t(s, a) \leq 1$, and β is the discount factor such that $0 \leq \beta < 1$.

See Chapter 6 (Sections 6.2 and 6.3) for more discussions on MDPs and Q-learning and their applications in the context of cognitive wireless networks.

4.6.3 Examples in wireless networks

Two of the most important dynamic programming examples which are used in wireless communications are given below:

- *Shortest path for routing*

  Problem definition: Given a connected graph $G = (V, E)$, a weight $d : E \rightarrow R^+$ and two vertices s and v in V, find the shortest path between these two vertices such that the sum of the weight of its constituent edges is minimized. Dijkstra's algorithm is known to be one of the best algorithms to find the shortest path. The time required by Dijkstra's algorithm is $O(|V|^2)$. It can be reduced to $O(|V| \log |V|)$. The detailed pseudo code for the algorithm is as follows:

  1. Set iteration $i = 0$, initial point $u_0 = s$, route set $S_0 = u_0 = s$, cost function $L(u_0) = 0$, and $L(v) = \infty$ for $v \neq u_0$. If $|V| = 1$ then stop, otherwise go to Step 2.
  2. For each v in $V \backslash S_i$, replace $L(v)$ with $\min L(v), L(u_i) + d_v^{u_i}$. If $L(v)$ is replaced, put a label $(L(v), u_i)$ on v.
  3. Find a vertex v which minimizes $\{L(v) : v \in V \backslash S_i\}$, say $u_i + 1$.
  4. Let $S_{i+1} = S_i \cup \{u_i + 1\}$.
  5. Replace i by $i + 1$. If $i = |V| - 1$ then stop, otherwise go to Step 2.

- The *Viterbi algorithm* is a dynamic programming algorithm for finding the most likely sequence of hidden states, known as the Viterbi path, that result in a sequence of observed events, especially in the context of hidden Markov models. The forward algorithm is a closely related algorithm for computing the probability of a sequence of observed events. These algorithms form a subset of information theory.

  The Viterbi algorithm was originally conceived as an error-correction scheme for noisy digital communication links, finding universal applications in decoding the convolutional codes used in both CDMA and GSM digital cellular, dial-up modems, satellite, deep-space communications, and 802.11 wireless LANs. The basic idea is to store a buffer of received signals and determine the most likely sequence of the original transmitted sequence. It is now also commonly used in speech recognition, keyword spotting, computational linguistics, and bioinformatics. For example, in speech-to-text speech recognition, the acoustic signal is treated as the observed sequence of events, and a string of text is considered to be the "hidden cause" of the acoustic signal. The Viterbi algorithm finds the most likely string of text given the acoustic signal.

  The algorithm is not general; it makes a number of assumptions. First, both the observed events and hidden events must be in a sequence. This sequence often corresponds to time. Second, these two sequences need to align, and an observed event needs to correspond to exactly one hidden event. Third, computing the most likely hidden sequence up to a certain point t must depend only on the observed event at

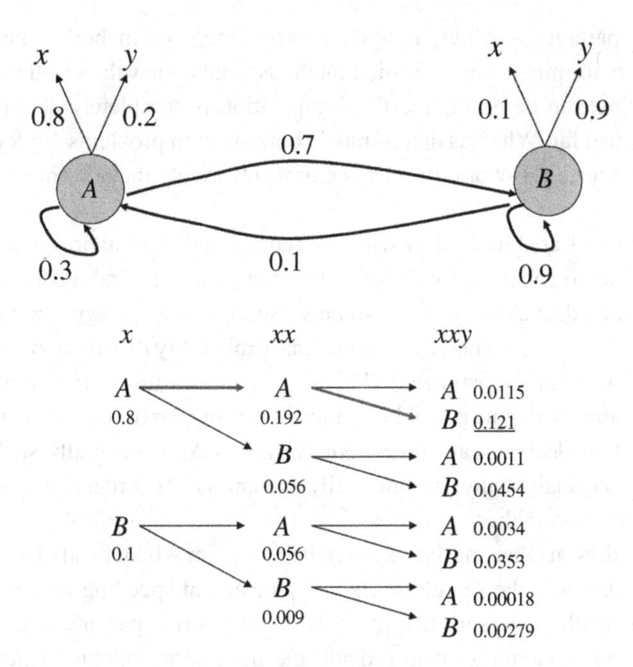

Figure 4.8 Example of a Viterbi algorithm.

point t, and the most likely sequence at point $t-1$. These assumptions are all satisfied in a first-order hidden Markov model.

The terms "Viterbi path" and "Viterbi algorithm" are also employed to related dynamic programming algorithms that discover the single most likely explanation for an observation. For example, in stochastic parsing a dynamic programming algorithm can be used to discover the single most likely context-free derivation (parse) of a string, which is sometimes called the "Viterbi parse."

In Figure 4.8, an example of the Viterbi decoding is provided. There are two states A and B with the transition probability as shown. In each state, two different observations x and y can be detected. The observed sequence is xxy. In the Viterbi algorithm, there is a decoding depth which represents the number of symbols that are jointly decoded. The deeper the decoding depth, the higher the error correction capability, but also the higher the computation complexity and the latency. The question is: what is the most likely sequence of states? From the example, it can be observed that the sequence AAB has the largest probability of $0.8 \times 0.3 \times 0.8 \times 0.7 \times 0.9 = 0.121$.

4.7 Stochastic programming

Stochastic optimization refers to the minimization (or maximization) of a function in the presence of randomness in the optimization process. Stochastic optimization methods are optimization algorithms which incorporate probabilistic (random) elements in the problem data (the objective function, the constraints, etc.), in the algorithm itself (through random parameter values, random choices, etc.), or in both. The concept contrasts with the deterministic optimization methods, where the values of the objective functions are assumed to be exact, and the computation is completely determined by the values sampled so far. Whereas deterministic optimization problems are formulated with known parameters, real-world problems almost invariably include some unknown parameters.

When the parameters are known only within certain bounds, one approach to tackling such problems is called robust optimization. Here the goal is to find a solution which is feasible for all such data and optimal in some sense. Stochastic programming models are similar in style but take advantage of the fact that probability distributions governing the data are known or can be estimated. The goal here is to find some policy that is feasible for all (or almost all) the possible data instances and maximizes the expectation of some function of the decisions and the random variables. More generally, such models are formulated, solved analytically or numerically, and analyzed in order to provide useful information to a decision-maker.

On the other hand, even when the data is exact, it is sometimes beneficial to deliberately introduce randomness into the search process as a means of speeding up convergence and making the algorithm less sensitive to modeling errors (i.e. perturbation analysis). Further, the injected randomness may provide the necessary impetus to move away from a local solution when searching for a global optimum. Indeed, this randomization principle is known to be a simple and effective way to obtain algorithms with good performance uniformly across all data sets, for all sorts of problems.

In this section, the general problem formulation is first studied. Then several categories of solutions, such as the distribution problem, chance constraint and penalize shortage, and recourse are investigated. Simple examples are also given.

4.7.1 Problem definition

To understand the stochastic programming, let us consider the following simple linear programming example:

$$\text{minimize: } x_1 + x_2 \tag{4.66}$$
$$\text{subject to: } \begin{cases} w_1 x_1 + x_2 \geq 7 \\ w_2 x_1 + x_2 \geq 4 \\ x_1 \geq 0 \\ x2 \geq 0, \end{cases}$$

where w_1 is uniformly distributed from 1 to 4 (i.e. $U(1, 4)$), and $w_2 \sim U(1/3, 1)$.

The problem in (4.66) is shown in Figure 4.9. The random variables w_1 and w_2 present the gradients of the constraints. The solution can be any point within the highlighted area. However, the solution point can be optimal with a specific value of the random variable, while it will not be optimal or even feasible with the other values. The question now is how to solve the problem. Or more precisely, under what criteria should the problem be solved?

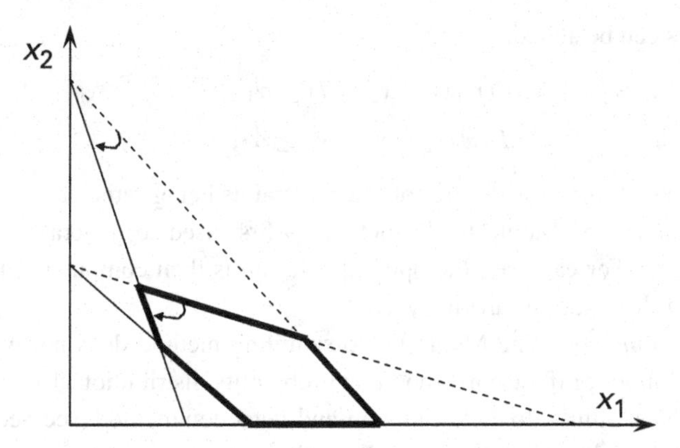

Figure 4.9 An example of random linear programming.

The general form of the problem formulation can be written as

$$\text{minimize: } g_0(x,\xi) \tag{4.67}$$
$$\text{subject to: } \begin{cases} g_i(x,\xi) \le 0, i = 1,\ldots,m \\ x \in X \subset \mathbb{R}^n, \end{cases}$$

where ξ is a random vector and g_i are deterministic functions.

If it is possible to decide how to solve the problem after observing the value of the random variable, we simply solve it as a deterministic program problem – referred to as the *wait-and-see* approach. Unfortunately, this approach is not generally appropriate since we need to decide what to do without knowing the exact values of the random variables. Next, we discuss several types of solutions.

The first category of solution tries to explore the distribution of the random variable. The simplest method is to guess the uncertainty. For the example in (4.66), we show the following three possible guesses, each of which tells us something about the level of "risk":

1. *Unbiased:* Choose mean values for each random variable.
   Since the mean $E(w) = (2.5, 1.5)$, we can take $w_1 = 2.5$ and $w_2 = 1.5$ into (4.66). The resulting solution is $(\hat{x}_1, \hat{x}_2) = (18/11, 32/11)$.
2. *Pessimistic:* Choose the worst-case value for random variables.
   The worst case $(w_1, w_2) = (1, 1/3)$, and the resulting solution is $(\hat{x}_1, \hat{x}_2) = (0, 7)$.
3. *Optimistic:* Choose best-case values for random variables.
   The best case $(w_1, w_2) = (4, 1)$, and the resulting solution is $(\hat{x}_1, \hat{x}_2) = (4, 0)$.

The advantage of the guessing method above is its simplicity. However, since only rough information about the randomness is utilized, the results can be very inaccurate or even infeasible. To overcome this problem, the following methods are usually employed:

1. *Chance constraint:* In this method, we enforce that the probability of a constraint being true is sufficiently large. For example, for the problem in (4.66), the following

two constraints can be added:

$$P(w_1 x_1 + x_2 \geq 7) \geq \alpha_1, \tag{4.68}$$

$$P(w_2 x_1 + x_2 \geq 4) \geq \alpha_2, \tag{4.69}$$

where α_1 and α_2 are the probabilities of the constraints being satisfied.

2. *Sampling method:* The Monte Carlo method is first used to generate the sets of random variables. For each set, the optimal solution is then computed. Finally, the distributions of the solutions are analyzed.
3. *Likelihood ratio method:* The Monte Carlo sampling method does not work if the goal is discontinuous or if the corresponding probability distribution also depends on decision variables. Similar to the estimation-and-detection method (see Section 3.1), likelihood ratios can be defined to solve the problem.
4. *Genetic algorithm:* A search technique used to find exact or approximate solutions to optimization and search problems, which was discussed in Section 4.4.
5. *Simulated annealing:* A generic probabilistic meta-algorithm for the global optimization problem, namely locating a good approximation to the global optimum of a given function in a large search space. It was also discussed in Section 4.4.

4.7.2 Recourse

The most widely applied and studied stochastic programming models are two-stage recourse programs. Here the decision maker takes some action in the first stage, after which a random event occurs which affects the outcome of the first-stage decision. A recourse decision can then be made in the second stage which compensates for any bad effects that might have been experienced as a result of the first-stage decision. The optimal policy from such a model is a single first-stage policy and a collection of recourse decisions (a decision rule) defining which second-stage action should be taken in response to each random outcome.

Similar to the barrier method discussed in Section 4.4, the infeasibility can be accepted but with a penalty of the expected shortage. Define $x^- = \max(0, -z)$ as the negative part of z. For the constraint $w_1 x_1 + x_2 \geq 7$ in (4.66), the shortfall is $(w_1 x_1 + x_2 - 7)^-$. For each constraint, the shortfall cost q_i is assigned. Therefore, the optimization problem in (4.66) becomes

$$\min_{x \in \mathbb{R}_+^2} = \left\{x_1 + x_2 + q_1 E_{w_1}[(w_1 x_1 + x_2 - 7)^-] + q_2 E_{w_2}[(w_2 x_1 + x_2 - 4)^-]\right\}. \tag{4.70}$$

The above problem is convex, and is equivalent to the following problem:

$$\min_{x \in \mathbb{R}_+^2} = \{x_1 + x_2 + Q(x_1, x_2)\}, \tag{4.71}$$

where

$$Q(x_1, x_2) = E_w \left[\min_{y \in \mathbb{R}_+^2} \{q_1 y_1 + q_2 y_2\} \right] \tag{4.72}$$

$$\text{subject to: } \begin{cases} w_1 x_1 + x_2 + y_1 = 7 \\ w_2 x_1 + x_2 + y_2 = 4. \end{cases}$$

Here $Q(x_1, x_2)$ is called the recourse function. Using the above equations, the actions can be corrected by the following sequence of events:

1. First period decision is made.
2. Nature makes a random decision.
3. A second period decision is made to repair the havoc wrought by nature in Step 2.

The idea of two-stage recourse can be easily extended to multistage recourse, which fits the situation where decisions are made periodically based on currently known realizations of some of the random variables. It has some learning flavor in the multistage recourse. An H-stage stochastic linear program with fixed recourse can be written as:

$$\min c^1 x^1 + E\{\min x^2(w) x^2(w) + \cdots + \min x^H(w) x^H(w)\} \tag{4.73}$$

$$\text{subject to: } \begin{cases} W^1 x^1 = h^1, \\ T^1(w) x^1 + W^2 x^2(w) = h^2(w) \\ \cdots\cdots \\ T^{H-1}(w) x^{H-1} + W^H x^H(w) = h^H(w) \\ x^1 \geq 0, \ldots, x^H(w) \geq 0, \end{cases}$$

where the decision variables $x^2(w), \ldots, x^H(w)$ are functions of the random variable w. $T(w)$ and $H(w)$ are technology and recourse matrices, respectively.

4.8 Summary

In this chapter, an overview of different optimization techniques has been provided. These techniques can be classified into the following categories:

- *Closed-form solution:* Using the Lagrangian method, some approximations and mathematical manipulations are necessary to solve the Lagrangian equations in closed form.
- *Mathematical programming:* Many real-world and theoretical problems can be modeled in the general framework of mathematical programming. There are five major subfields of mathematical programming: linear, convex, non-linear, dynamic, and stochastic programming.
- *Integer/combinatorial optimization:* In cognitive radio-specific resource allocation problems, many parameters have only integer values, such as the modulation rate. Other parameters, such as the channel allocation, have a combinatorial nature. There are several possible solutions, such as relaxation and decomposition, enumeration, cutting planes, and solutions of the knapsack problem.

Compared to the resource allocation in a general wireless network, a cognitive radio network has more design constraints due to the presence of primary users. Moreover, spectrum sensing brings additional considerations to the problem.

The majority of the approaches discussed in this chapter need a centralized control to gather all the information needed for optimization. Among all resource allocation schemes, the centralized scheme has the best performance but incurs considerable signaling overhead to update information. This requirement can significantly reduce system efficiency. The centralized scheme can fit the network scenarios in which the topology is simple, mobility is low, and computation power is high. It can serve as a performance upper bound against which other more practical schemes can be compared.

5 Game theory

The ideas underlying game theory have appeared throughout history, in the Bible, the Talmud, the works of Descartes and Sun Tzu, and the writings of Chales Darwin. Modern game theory, however, can be considered as an outgrowth of three seminal works:

- Augustin Cournot's *Research into the Mathematical Principles of the Theory of Wealth* in 1838 gives an intuitive explanation of what would eventually be formalized as the Nash equilibrium, as well as provides an evolutionary, or dynamic notion of best-response to the actions of others.
- Francis Ysidro Edgeworth's *Mathematical Psychics* demonstrated the notion of competitive equilibria in a two-person (as well as two-type) economy; Emile Borel, in "Algèbre et calcul des probabilités," *Comptes Rendus Académie des Sciences*, vol. 184, 1927, provided the first insight into mixed strategies that randomization may support a stable outcome.
- While many other contributors hold a place in the history of game theory, it is widely accepted that modern analysis began with John von Neumann and Oskar Morgenstern's book, *Theory of Games and Economic Behavior*. Then building on von Neumann and Morgenstern's results John Nash developed the modern framework for methodological analysis.

Depending on the nature of the different approaches, there are different possible applications of game theory. If the information is strictly limited to local information, the non-cooperative game might be the only choice for each individual to play. However, such a game might have a very low-efficiency outcome. To overcome this problem, pricing or referee approaches have been proposed. If the users care about long-term benefits, the repeated game can be employed to enforce cooperation by the threat of future punishment from others. Example schemes are the tit-for-tat and cartel maintenance approaches. If signaling is allowed and the problem is about the contract on efficient usage of limited resources, cooperative game theory can be employed to have mutual benefits. Moreover, if the outcome is binary and signaling is allowed, auction theory and mechanism design can be used for resource allocation.

There are other game theory approaches. For example, if the information is not deterministic but statistical, the stochastic game [233, 234] can be employed to solve the problem and analyze the stability. Even if the distributed individual users have an incentive to cooperate, they might not know what the best point for cooperation is. Under this condition, reinforcement learning or artificial intelligence approaches [171] can also be employed to find the best cooperation point. The game with incomplete information is usually investigated using the Bayesian game. Depending on the different applications, various game theory approaches can be employed to optimize system performance. Game theory approaches bring a new perspective to traditional wireless resource allocation and networking optimization.

In this chapter, different types of games will be discussed, basic game theory concepts will be described and simple examples illustrated. The goal is to allow readers to understand basic game theory problems and approaches so that they can formulate game theory problems in their own research areas. Since the purpose is not to teach game theory, mathematical details are limited only to what are necessary. For more details of game theory, readers are referred to [235, 236, 237]. This chapter is organized as follows: in Section 5.1, we describe the basic definitions used in game theory. In Section 5.2, we discuss non-cooperative game models. In Section 5.3, dynamic and repeated game models are described. In Section 5.4, the bargaining game model is introduced, which provides benefits to all game players. In Section 5.5, the coalition game concept is discussed. In Section 5.6, games with imperfect information are explained. Finally, in Section 5.7, the basic concepts of some special types of games are reviewed.

5.1 Basics of game theory

In this section, the basic concepts and elements of game theory are discussed. To explain the players' behavior in a game, prisoner's dilemma, a simple two-player non-cooperative game, is illustrated. Next, the games are classified according to different criteria. The use of utility functions in wireless networking and resource allocation problems is discussed. Utility functions are used to represent users' interests. One well-known wireless resource allocation problem [238] is then discussed. Finally, the cognitive cycle of a cognitive radio is mapped to a game.

A game can be roughly defined as a method of interaction between users (or players), where each user adjusts its strategy to optimize its own utility while competing with the others. Strategy and utility can be defined as:

Definition 5.1 *A **strategy** r is a complete contingent plan, or a decision rule, that defines the action an agent will select in every distinguishable state Ω of the world.*

Definition 5.2 *In any game, **utility** (payoff) u represents the motivations of players. A utility function for a given player assigns a number for every possible outcome of the game with the property that a higher (or lower) number implies that the outcome is more preferred.*

One of the most common assumptions made in game theory is rationality. In its mildest form, rationality implies that all players are motivated by maximizing their own utilities. In a stricter sense, it implies that every player always maximizes his or her utility, thus being able to perfectly calculate the probabilistic result of every action.

Table 5.1 The prisoner's dilemma.

	Cooperate	Defect
Cooperate	$(3,3)$	$(0,5)$
Defect	$(5,0)$	$(1,1)$

A game can be defined as follows:

Definition 5.3 *A* ***game*** *G in the strategic form has three elements: the set of players $i \in \mathcal{I}$, which is a finite set $\{1, 2, ..., N\}$; the strategy space Ω_i for each player i; and utility function u_i, which measures the outcome of user i for each strategy profile $\mathbf{r} = (r_1, r_2, ..., r_N)$. We define $\mathbf{r}_{-i}$ as the strategies of player i's opponents, i.e. $\mathbf{r}_{-i} = (r_1, ..., r_{i-1}, r_{i+1}, ..., r_N)$. In static games, the interaction between users occurs only once, while in dynamic games the interaction occurs several times.*

A very famous yet simple example of a game is "the prisoner's dilemma." The name comes from a hypothetical situation: two criminals are arrested for committing the same crime, but the police do not have enough proof to convict either. Thus, the police separate the two and offer a deal: if one testifies to convict the other (i.e. defects), he or she will get a reduced sentence or go free. Here the prisoners do not have information about the other's "move." The payoff, if they both remain quiet (and thus cooperate with each other) is high, since neither can be convicted without further proof. If one betrays the other, but the other remains silent, then the first benefits because he or she goes free while the second is imprisoned, because there is sufficient proof to convict the silent one. If they both betray each other, they both get a reduced sentence, which can be described as a null result. The obvious dilemma is the choice between two options, where a good decision cannot be made without information.

The prisoner's dilemma is a two-player game. One player plays as the row player and one plays as the column player (Table 5.1). Both can choose to cooperate (C) or defect (D). Thus, there are four possible outcomes of the game: $\{(C, C), (D, D), (C, D), (D, C)\}$. Under mutual cooperation, $\{(C, C)\}$, both players will receive the reward payoff, 3. Under mutual defection, $\{(D, D)\}$, both players receive the punishment of defection, 1. When one player cooperates while the other defects, $\{(C, D), (D, C)\}$, the cooperating player receives the payoff, 0, while the defecting player receives the payoff, 5.

In the prisoner's dilemma, if one player cooperates, the other will have a better payoff (5 instead of 3) if he/she defects; if one player defects, the other player will still have a better payoff (1 instead of 0) if he/she defects. Regardless of the other player's strategy, the first player always selects defect and $\{(D, D)\}$ is an equilibrium. Although cooperation will give each player a better payoff of 3, greediness leads to an inefficient outcome. That is, the outcome of the non-cooperative game can be less efficient.

Games can be separated into three basic classifications:

- *Non-cooperative vs. cooperative game*: In non-cooperative games individual players act to maximize their own payoff, while in cooperative games, coalitions of players are formed and players have joint actions so as to gain mutual benefits.

- *Strategic (or normal form) games and extensive (form) games*: The normal form is a matrix representation of a simultaneous game. For two players, one is the "row" player, and the other is the "column" player. Each row or column represents a strategy and each box represents the payoff of each player for every combination of strategies.

  The extensive form (also called a game tree) is a graphical representation of a sequential game. It provides information about the players, payoffs, strategies, and the order of moves. The game tree consists of nodes (or vertices), which are points at which players can take actions, connected by edges, which represent the actions that may be taken at that node. An initial (or root) node represents the first decision to be made. Every set of edges from the first node through the tree eventually arrives at a terminal node, which represents an end to the game. Each terminal node is labeled with the payoff earned by each player if the game ends at that node.
- *Games with complete or incomplete information*: A game with complete information is one in which all factors of the game are common knowledge. Specifically, each player is aware of all other players, the timing of the game, and the set of strategies and payoffs for each player.

  A sequential game with imperfect information is one in which a player does not know exactly what actions other players took up to that point. Technically, there exists at least one information set with more than one node. If every information set contains exactly one node, the game is one with perfect information. Intuitively, if it is my turn to move, I may not know what every other player has done up to now. Therefore, I have to infer from the likelihood of their actions and from Bayesian rule to decide on my current decision.

In the rest of this section, the formulation of games in wireless networking and resource allocation problems is discussed. In such formulations, the utility function represents the interests of the mobile users. For example, the utility can be the communication rate, power, energy, or a combination of them. The challenge is how to formulate the utility function so that it has a physical meaning and the game outcome is not trivial. Typically, if the utility function is monotonic, the game outcome occurs in the boundary situation, which is trivial. Therefore, the utility function shall be designed so that the utility is either quasiconcave or quasiconvex. In addition, the optimal point is selected to be somewhere within the practical parameter range. Moreover, the location of this optimal point also depends on the behavior of other users. If the utility function has the above properties, the analysis and improvement of the game outcomes can be further studied. A commonly used utility function in wireless networking and resource allocation problems proposed in [238] will be studied next.

Users access a wireless system through the air interface that is a common resource and they transmit information using battery energy. Since the air interface is a shared medium, each user's transmission is a source of interference for others. The signal-to-interference ratio (SIR) is a measure of the quality of signal reception for the wireless user. Typically, a user intends to achieve a high quality of reception (high SIR) while at the same time expending a small amount of energy. Thus, it is possible to view both SIR and battery energy (or equivalently transmitted power) as commodities that a wireless user desires. There exists a tradeoff between high SIR and low energy consumption. Finding a good balance between the two conflicting objectives is the primary focus of the power control component of radio resource management in a wireless network.

An optimum power control algorithm for wireless voice systems maximizes the number of conversations that can simultaneously achieve a certain QoS objective. Typically, the QoS objective for a voice terminal is to achieve a minimum acceptable SIR. However, this approach is not appropriate for the efficient operation of a wireless data system. This is because the QoS objective for data services differs from the QoS objective for voice service. In a data-oriented system, error-free communication has the highest priority. The SIR is an important quantity since there is a direct relationship between the SIR and the probability of transmission error.

Suppose the frame length is M bits with L information bits and $M - L$ overhead bits (e.g. for protection against channel errors). The rate is R bits/s and power per bit is P. Let P_c denote the probability of correct reception of a frame at the receiver which is a function of the signal-to-interference-plus-noise ratio (SINR). The utility function proposed in [238] is given by

$$u = \frac{\mathrm{LRP_c}}{MP} \text{ bits/joule.} \tag{5.1}$$

The utility has the physical meaning of the number of successfully received information bits per joule of energy cost. To avoid the problem of infinite utility with $P = 0$, an efficiency function P_e is proposed to approximate P_c as

$$f(\mathrm{SNR}) = (1 - 2P_e(\mathrm{SNR}))^M, \tag{5.2}$$

where P_e depends on the modulation and coding scheme of the system.

Suppose $M = L = 10$, $R = 1$, and $P = SNR$. The utility function is shown in Figure 5.1 for the DPSK scheme, where $P_e = 1/2\mathrm{e}^{-\mathrm{SNR}}$. This is a quasiconcave function of SNR. When the SNR is small, the frame error rate in (5.2) is large so that the utility in (5.1) is small. On the other hand, when SNR is high, increasing the power cannot improve the frame error rate, but causes only a wastage of transmission energy. Therefore, the utility decreases.

For the cognitive cycle, a cognitive radio observes the environment and sets the intelligent move, which can be mapped into a game. The cognitive radio users are the players in the game, the payoff is the utility, the observations are the arguments of the utility function, and the outside world can be interpreted as the outcome space of the game.

5.2 Non-cooperative static game

In this section, first the non-cooperative static game is defined. Then, two approaches to present a game (i.e. normal and extensive forms) are discussed. Some properties of a game, such as dominance, Nash equilibrium, Pareto optimality, and mix strategies are studied next. Finally, the low efficiency of the outcome for a non-cooperative static game

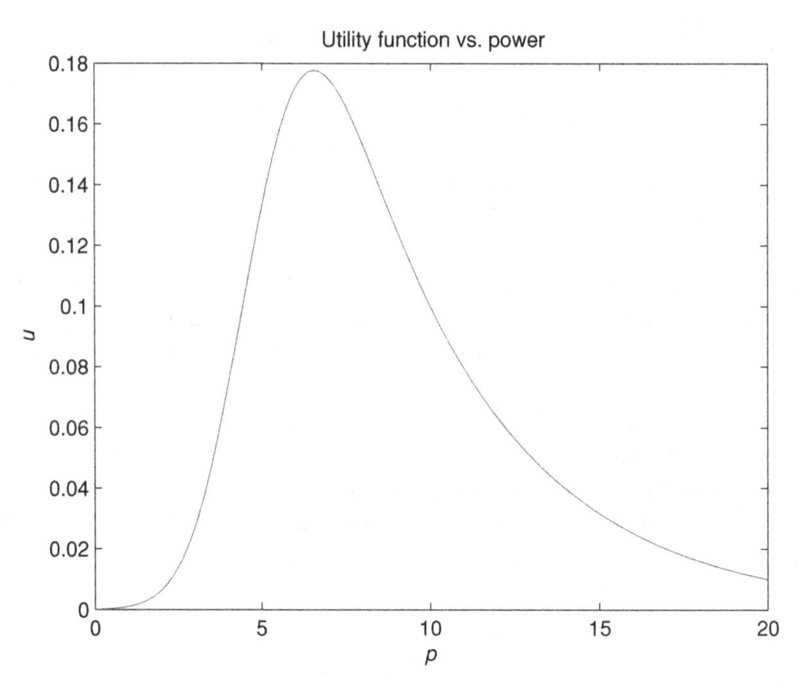

Figure 5.1 Example of a utility function.

model, which is employed in wireless networking and resource allocation problems, is discussed. To improve the game outcomes, two methods, namely, pricing method and referee method, are briefly discussed.

Definition 5.4 *A* ***non-cooperative game*** *is one in which players are unable to make enforceable contracts outside of those specifically modeled in the game. Hence, it is not defined as games in which players do not cooperate, but as games in which any cooperation must be self-enforcing.*

Definition 5.5 *A* ***static game*** *is one in which all players make decisions (or select a strategy) simultaneously, without knowledge of the strategies that are being chosen by other players. Even though the decisions may be made at different points in time, the game is simultaneous because each player has no information about the decisions of others; thus, it is as if the decisions are made simultaneously.*

5.2.1 Normal form of static game

Static games can be represented by the normal form, which is defined as follows:

Definition 5.6 *The* ***strategic (or normal) form*** *is a matrix representation of a simultaneous game. For two players, one is the "row" player, and the other is the "column" player. Each row or column represents a strategy and each box represents the payoff of each player for every combination of strategies.*

Table 5.2 Battle of the sexes (wife, husband).

	Boxing	Opera
Boxing	(1, 2)	(0, 0)
Opera	(0, 0)	(2, 1)

One example is the "battle of the sexes" shown in Table 5.2 with the following scenario: a husband and a wife have agreed to attend a rare entertainment event in the evening. Unfortunately, neither remembers which of the two special events in town they had agreed on: the boxing match or the opera. The husband prefers the boxing match while the wife prefers the opera; yet, both prefer being together to being apart. They must decide simultaneously and without communication which event to attend. There are two pure strategy equilibria. Different pure strategy equilibria are preferred by each player. However, either equilibrium is preferred by both players to any of the non-equilibrium outcomes.

For each user, there is a strategy space. Some of the strategies are superior to the others for the user's interests. To define such superiority, we have the following two definitions:

Definition 5.7 ***Dominant strategies***: *A strategy is dominant if, regardless of what other players do, the strategy earns a player a larger payoff than any other. Hence, a strategy is dominant if it is always better than any other strategy, regardless of what opponents may do. If a player has a dominant strategy then he or she will always play it in equilibrium. Also, if one strategy is dominant, then all others are dominated.*

Definition 5.8 ***Dominated strategies***: *A strategy is dominated if, regardless of what any other players do, the strategy earns a player a smaller payoff than some other strategy. Hence, a strategy is dominated if it is always better to play some other strategy, regardless of what opponents may do. A dominated strategy is never played in equilibrium.*

For example, in the prisoner's dilemma in Table 5.1, each player has a dominated strategy of cooperation and a dominant strategy of defection. This is because no matter what other player's strategy is, the defection for each player always yields a higher utility.

5.2.2 Nash equilibrium, Pareto optimality, and mixed strategy

To analyze the outcome of the game, the Nash equilibrium (named after John Nash) is a well-known concept which states that in the equilibrium every agent will select a utility-maximizing strategy given the strategies of every other agent:

Definition 5.9 *Define a strategy vector* $\mathbf{r} = [r_1 \ldots r_K]$ *and define the strategy vector of the ith player's opponents as* $\mathbf{r}_i^{-1} = [r_1 \ldots r_{i-1}\ r_{i+1} \ldots r_K]$*, where* K *is the number of users and* r_i *is the strategy of user* i*. If* u_i *is the utility of user* i*, the* ***Nash equilibrium*** *point* $\mathbf{r}$ *is defined as:*

$$u_i(r_i, \mathbf{r}_i^{-1}) \geq u_i(\tilde{r}_i, \mathbf{r}_i^{-1}), \ \forall i, \ \forall \tilde{r}_i \in \Omega, \ \mathbf{r}_i^{-1} \in \Omega^{K-1}. \tag{5.3}$$

Table 5.3 Game of chicken (driver1, driver2).

	Stay	Swerve
Stay	(−100,−100)	(1,−1)
Swerve	(−1,1)	(0,0)

In other words, a Nash equilibrium is a set of strategies, one for each player, such that no player has incentive to unilaterally change his/her action. Players are in an equilibrium if a change in strategies by any one of them will lead that player to earn less than if it remained with its current strategy.

The existence of Nash equilibrium is difficult to prove. There are some theorems which state the conditions of existence of Nash equilibrium. It is only necessary to prove that the proposed game satisfies the requirements of the theorems. For example, in [235], it was shown that a Nash equilibrium point (NEP) exists if $\forall\ i$:

1. Ω, the support domain of $u_i(\mathbf{r}_i)$, is a non-empty, convex, and compact subset of a certain Euclidean space $\Re^L$.
2. $u_i(\mathbf{r}_i)$ is continuous in $\mathbf{r}_i$ and quasiconvex in $\mathbf{r}_i$.

There might be an infinite number of Nash equilibria. Among all these equilibria, the optimal one needs to be selected. There are many criteria by which to judge whether the equilibrium is optimal or not. Among these criteria, the Pareto optimality is one of the most important definitions:

Definition 5.10 ***Pareto optimality***: *Named after Vilfredo Pareto, the Pareto optimality is a measure of efficiency. An outcome of a game is Pareto optimal if there is no other outcome that makes every player at least as well off and at least one player strictly better off. That is, a Pareto optimal outcome cannot be improved upon without hurting at least one player. Often, a Nash equilibrium is not Pareto optimal, which implies that the players' payoffs can all be increased.*

Until now, only deterministic or pure strategies have been discussed. A pure strategy defines a specific move or action that a player will follow in every possible attainable situation in a game. Such moves may not be random, or drawn from a distribution, as in the case of mixed strategies:

Definition 5.11 ***Mixed strategy***: *A strategy consists of possible moves and a probability distribution (collection of weights) which corresponds to how frequently each move is about to play. A player will only use a mixed strategy when he/she is indifferent to several pure strategies. Moreover, if the opponent can benefit from knowing the next move, the mixed strategy is preferred since it is desirable to keep the opponent guessing.*

To illustrate the idea of Pareto optimality and mixed strategy, an example of the "game of chicken" is shown in Table 5.3 (it is also called the hawk–dove game). The scenario is as follows: two people drive directly at each other along a narrow road. The first

Table 5.4 Matching pennies (different, same).

	Heads	Tails
Heads	(−1,1)	(1,−1)
Tails	(1,−1)	(−1,1)

to swerve loses face among his/her peers. If neither swerves, however, both are killed. There are two pure strategy equilibria. A different pure strategy equilibrium is preferred by each player. Both equilibria – (1,−1) and (−1,1) – are Pareto optimal. A mixed strategy equilibrium also exists, where each user plays swerve with probability 0.99 and plays stay with probability 0.01.

A zero-sum game is a special case of a constant sum game in which the sum of all players' payoffs is 0. Hence, a gain for one participant is always at the expense of another. Therefore, one strategy used to gain personal benefit is to suppress the other's play. Given the conflicting interests, the equilibrium of such game is often a mixed strategy.

One example of a zero-sum game is the game of "matching pennies" shown in Table 5.4. Between two children, the scenario determines who is required to do the nightly chores. The two children first select who will be represented by "same" and who will be represented by "different." Then, each child conceals in his/her palm a penny which is either face up or face down. Both coins are revealed simultaneously. If they match (both are heads or both are tails), the child who claims "same" wins. If they are different (one heads and one tails), the child who claims "different" wins. The game is equivalent to "odds or evens" and quite similar to a three-strategy version – rock, paper, scissors. The game is a zero-sum game. The only equilibrium is in mixed strategies. Each player plays each strategy with equal probability, resulting in an expected payoff of zero for each player.

Unfortunately, the Nash equilibria of non-cooperative static games often have low efficiency. For example, in the example of the prisoner's dilemma in Table 5.1, if one player cooperates, the other player will have better payoff if he/she defects; if one player defects, the other player will still have better payoff if he/she defects. Therefore, regardless of the other strategy, each player always selects defect and $\{(D, D)\}$ is a Nash equilibrium. Although cooperation will give each player a better payoff of 3, greediness leads to an inefficient outcome, so the outcome of the non-cooperative static game can be less efficient. To overcome this low efficiency problem, there are many schemes in the literature. In the following, two examples, namely, the pricing and referee-based approaches, are discussed.

5.2.3 Social optimum: price of anarchy and referee-based game

Since the individual user has no incentive to cooperate with the other users in the system and imposes harm to the other users' resources, the outcome of a game might not be optimal from the system's point of view. Traditionally, the social optimum is calculated by assuming there is a Genie that can tell the complete information of the system. Then the social optimum is calculated using the optimization techniques that were discussed in the previous chapter. The resulting social optimal solutions are compared with the game outcomes. Next, two techniques which can improve the game outcomes to approach social optima are studied.

To overcome the social optimal problem, pricing (or taxation) has been used as an effective tool by both economists and researchers in computer networks. The pricing technique is motivated by the following two objectives:

- System revenue is optimized.
- Cooperation is encouraged for resource usage.

An efficient pricing mechanism can make the distributed decisions compatible with the system efficiency obtained by centralized control. A pricing policy is called *incentive compatible* if pricing enforces a Nash equilibrium that achieves the system optimum. In [238, 239], a usage-based pricing policy was proposed, where the price that a user pays for using the resources is proportional to the amount of resources consumed by the user. Specifically, for the utility function in (5.1), the new utility with pricing becomes

$$u' = \frac{\mathrm{LRP_c}}{MP} - \alpha p, \tag{5.4}$$

where α is the price for transmitted power P and the price can be different for different users. It was shown in [238] that the above utility function can achieve Pareto optimality.

Second, the basic idea for the referee-based scheme is to introduce a referee for a non-cooperative game. The game may have multiple Nash equilibria. A referee is in charge of detecting the less efficient Nash equilibria and then it changes the game rules to prevent the players from falling into undesirable game outcomes. It is worth mentioning that the non-cooperative game is still played in a distributive way. The referee intervenes only when it is necessary. In [240, 241], the above idea was employed in a multicell OFDMA network to find an efficient distributed resource allocation.

5.3 Dynamic/repeated game

When players interact by playing a similar game a number of times, the game is called a dynamic, or repeated game. Unlike simultaneous games, players have at least some information about the strategies chosen by others and thus their play may be contingent on past moves. The cooperation among autonomous players can be encouraged by considering the long-term benefit and threat from others. In this section, representation of a dynamic game is first discussed. Then, the information available to the players and the properties of the dynamic games are discussed. Next, two practical strategies, tit-for-tat and cartel maintenance, are proposed using a repeated game. Finally, a special kind of game, the stochastic game (Markov game), is described.

5.3.1 Sequential game and extensive form

First, let us define the concept of a sequential game:

Definition 5.12 *A* ***sequential game*** *is a game in which players make decisions (or select a strategy) following a certain predefined order, and in which at least some players can observe the moves of players who preceded them. If no players observe the moves of previous players, then the game is simultaneous. If every player observes the moves of every other player who has gone before her, the game is one of perfect information. If some (but not all) players observe prior moves, when others move simultaneously, the game is one of imperfect information.*

As far as the information available to a game is concerned, there are three options:

Definition 5.13 *A game is one of* ***complete information*** *if all parameters of the game are common knowledge. Specifically, each player is aware of all other players, the timing of the game, and the set of strategies and payoffs for each player.*

Definition 5.14 *A sequential game is one of* ***perfect information*** *if only one player moves at a time and if each player knows every action of the players that moved before him/her at every point. Technically, every information set contains exactly one node. Intuitively, if it is my turn to move, I always know what every other player has done up to this point. All other games are games of imperfect information.*

Definition 5.15 *A sequential game is one of* ***imperfect information*** *if a player does not know exactly what actions other players took up to this point. Technically, there exists at least one information set with more than one node. Intuitively, if it is my turn to move, I may not know what every other player has done up to now. Therefore, I have to infer from the likelihood of actions of other players and from Bayesian rule to decide on my current decision.*

Sequential games are represented by game trees (the extensive form):

Definition 5.16 *A* ***game tree*** *(also called the* ***extensive form****) is a graphical representation of a sequential game. It provides information about the players, payoffs, strategies, and the order of moves. The game tree consists of nodes, which are points at which players can take actions, connected by vertices, which represent the actions that may be taken at that node. An initial (or root) node represents the first decision to be made. Every set of vertices from the first node through the tree eventually arrives at a terminal node, representing an end to the game. Each terminal node is labeled with the payoffs earned by each player if the game ends at that node.*

Figure 5.2 illustrates this idea. The game has two players: 1 and 2. The numbers by every non-terminal node indicate to which player that decision node belongs. The numbers by every terminal node represent the payoffs to the players (e.g. 2,1 represents a payoff of 2 to player 1 and a payoff of 1 to player 2). The labels by every edge of the graph are the name of the action represented by that edge.

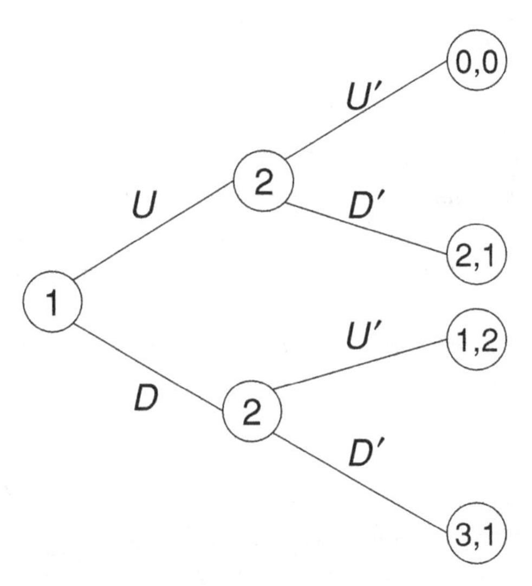

Figure 5.2 Example of a game in extensive form (game tree).

The initial node belongs to player 1, indicating that that player moves first. According to the tree, the game is as follows: player 1 chooses between U and D; player 2 observes player 1's choice and then chooses between U' and D'. The payoffs are as specified in the tree. There are four outcomes represented by the four terminal nodes of the tree: (U, U'), (U, D'), (D, U'), and (D, D). The payoffs associated with each outcome, respectively, are as follows: (0, 0), (2,1), (1,2), and (3,1).

A sequential game can be solved using the concept of subgame perfect equilibrium, as follows:

Definition 5.17 *A* ***subgame perfect*** *Nash equilibrium is an equilibrium such that players' strategies constitute a Nash equilibrium in every subgame of the original game. It may be found by backward induction, an iterative process for solving finite extensive form or sequential games. First, one determines the optimal strategy of the player who makes the last move of the game. Then, the optimal action of the next-to-last moving player is determined taking the last player's action as given. The process continues in this way backwards in time until all players' actions have been determined.*

In the example shown in Figure 5.2, if player 1 plays D, player 2 will play U' to maximize his/her payoff and so player 1 will only receive 1. However, if player 1 plays U, player 2 maximizes his payoff by playing D' and player 1 receives 2. Player 1 prefers the payoff of value 2 to 1 and so player 1 will play U and player 2 will play D'. This is the subgame perfect equilibrium.

Subgame perfect equilibria eliminate non-credible threats. A non-credible threat is a threat made by a player in a sequential game that will not be in the best interest for that player to carry out. The player hopes that the threat is believed, in which case there is no need to carry it out. Even though Nash equilibria may depend on non-credible threats, backward induction eliminates them.

The technical definition of a repeated game can be given as follows:

Definition 5.18 *Let G be a static game and β be a discount factor. The $\mathcal{T}$-period* ***repeated game****, denoted as $G(\mathcal{T}, \beta)$, consists of game G repeated $\mathcal{T}$ times. The payoff for such a game is given by*

$$V_i = \sum_{t=1}^{\mathcal{T}} \beta^{t-1} \pi_i^t, \tag{5.5}$$

where π_i^t denotes the payoff to player i in period t. If $\mathcal{T}$ approaches infinity, then $G(\infty, \beta)$ is referred to as the infinitely-repeated game.

According to the Folk theorem, stated below, in an infinitely repeated game, any feasible outcome that gives each player better payoff than the Nash equilibrium can be obtained:

Theorem 5.1 ***Folk theorem****: Let $(e_1, \ldots, e_n)$ be the payoffs from a Nash equilibrium of G and let $(x_1, \ldots, x_n)$ be any feasible payoffs from G. If $x_i > e_i$ for every player i, then there exists an equilibrium (it is also subgame-perfect) of $G(\infty, \beta)$ that attains $(x_1, \ldots, x_n)$ as the average payoff, provided that β is sufficiently close to 1.*

5.3.2 Tit-for-tat and trigger-price strategy

By using the repeated game, any feasible payoffs better than the Nash equilibrium can be obtained. The reason is that, with enough patience, a player's non-cooperative behavior will be punished by the future revenge from other cooperative players, or on the other hand, a player's cooperation can be rewarded in the future by others' cooperation. The remaining problem is how to define a good rule to achieve these better payoffs by enforcing cooperation among players. In the following, two approaches, namely, tit-for-tat and cartel maintenance, are proposed.

Tit-for-tat is a type of trigger strategy in which a player responds in one period with the same action his/her opponent used in the last period. Many research works used this method [242, 243]. The advantage of tit-for-tat is its implementation simplicity. However, there are some potential problems. The best response of a player is not the same action of the other opponent. The information of others' actions is hard to obtain. All these limit the possible applications for tit-for-tat. There are many less harsh and more optimal variation of trigger strategies than tit-for-tat. One of the optimal design criteria is cartel maintenance [244].

The basic idea for the cartel maintenance in repeated game framework is to provide enough threat to greedy players so as to prevent them from deviating from cooperation. First, the cooperative point is obtained so that all players have better performances than those of non-cooperative NEPs. However, if any player deviates from cooperation while others still play cooperatively, this deviating player has a better utility, while others have relatively worse utilities. If no rule is employed, the cooperative players will also have incentives to deviate. Consequently, the network performance deteriorates due to the non-cooperation. The proposed framework provides a mechanism so that the current defecting gains of the selfish player will be outweighed by future punishment strategies from other players. For any rational player, this threat of punishment prevents him/her from deviation, so the cooperation is enforced.

The proposed trigger strategy is a strategy to introduce punishment on the defecting players. In the trigger strategy, the players start with cooperation. Assume each player can observe the public information (e.g. the outcome of the game), P_t at time t. Examples of this public information can be the successful transmission rate and network throughput. Note that such public information is mostly imperfect or simply partial information about the players' strategies. Here, it is assumed that a larger P_t stands for a higher cooperation level, resulting in higher performances for all players. Let the cooperation strategies be $\bar{\lambda} = [\lambda_1, \lambda_2, \ldots, \lambda_N]^{\mathrm{T}}$ and the non-cooperative strategies be $\bar{s} = [s_1, s_2, \ldots, s_N]^{\mathrm{T}}$, respectively. The trigger–punishment game rule is characterized by three parameters: the optimal punishment time T, the trigger threshold P^*, and the cooperation strategy $\bar{\lambda}$. The trigger–punishment strategy $(\bar{\lambda}, P^*, T)$ for a distributed player i is given as follows:

1. Player i plays the strategy of the cooperation phase, $\bar{\lambda}$, in period 0.
2. If the cooperation phase is played in period t and $P_t > P^*$, player i plays the cooperation phase in period $t + 1$.
3. If the cooperation phase is played in period t and $P_t < P^*$, player i switches to a punishment phase for $T - 1$ periods, in which the players play a static Nash equilibrium $\bar{s}$ regardless of the realized outcomes. At the Tth period, play returns to the cooperative phase.

Note that $\bar{s}$ generates the non-cooperation outcome, which is much worse than that generated by the cooperation strategy $\bar{\lambda}$. Therefore, the selfish players who deviate will have much lower utilities in the punishment phase. Moreover, the punishment time T is designed to be long enough to let all defecting gains of the selfish players be outweighed by the punishment. Therefore, the players have no incentive to deviate from cooperation, since the players aim to maximize the long-run payoffs over time.

The rest of the problem is how to calculate the optimal parameters of $(\bar{\lambda}, P^*, T)$, i.e. how to construct a cartel so that the benefit is optimized and the incentive for deviation is eliminated. Let us define $\mathcal{P}_{\text{trig}} = Pr(P_t < P^*)$, which is the trigger probability that the realization of public information is less than the trigger threshold. Based on (5.5), the expected payoff V_i is given as follows:

$$V_i(\bar{\lambda}, P^*, T) = \pi_i(\bar{\lambda}) + (1 - \mathcal{P}_{\text{trig}})\beta V_i(\bar{\lambda}, P^*, T) + \mathcal{P}_{\text{trig}}[\sum_{t=1}^{T-1} \beta^t \pi_i(\bar{s}) + \beta^T V_i(\bar{\lambda}, P^*, T)], \forall i, \tag{5.6}$$

where $\pi_i(\bar{\lambda})$ and $\pi_i(\bar{s})$ are the cooperation and non-cooperation payoffs, respectively. The first term at the right-hand side of (5.6) is the current expected payoff if cooperation is played, the second term and third term are the payoffs for two different results depending on whether or not the punishment is triggered, respectively. Note that V_i is a function of not only players' strategies $\bar{\lambda}$ but also the game parameters P^* and T. The objective is to maximize the expected payoff V_i for each user while the optimal strategy yields an NEP for the proposed algorithm. In order to achieve the NEP given P^* and T, the optimal strategies of the repeated games can also be characterized by the following first-order necessary conditions:

$$\frac{\partial V_i(\lambda_i, \lambda_{-i})}{\partial \lambda_i} = 0, \forall i. \tag{5.7}$$

If all players have the same utility and the game outcome is symmetric for all players, the solution λ^* of the first-order conditions is the same for all players. This solution is also a function of parameter P^* and T. In order to obtain the optimal P^* and T for maximizing the expected payoff V_i, we have the following differential equations:

$$\frac{\partial V_i(P^*, T)}{\partial P^*} = 0, \quad \frac{\partial V_i(P^*, T)}{\partial T} = 0, \forall i. \tag{5.8}$$

In general, (5.7) and (5.8) need to be solved via numerical methods. For a certain structure of the payoff function, the closed-form optimal configuration $\{\bar{\lambda}, P^*, T\}$ can be derived.

5.3.3 Stochastic game

In game theory, a stochastic game is a dynamic, competitive game with probabilistic transitions played by one or more players. The game is played in a sequence of stages. At the beginning of each stage the game is in some state. The players select actions and each player receives a payoff that depends on the current state and the chosen actions. The game then moves to a new random state whose distribution depends on the previous state and the actions chosen by the players. The procedure is repeated at the new state and the game continues for a finite or infinite number of stages. The total payoff to a player is often taken to be the discounted sum of the stage payoffs or the limit inferior of the averages of the stage payoffs. Note that a one-state stochastic game is equivalent to an (infinitely) repeated game, and a one-agent stochastic game is equivalent to a Markov decision process (MDP). The formal definition is as follows:

Definition 5.19 *An N-player stochastic game G consists of a finite, non-empty set of state S, N players, a finite set of actions A_i for the players, a conditional probability distribution p on $S \times (A_1 \times A_2 \times \cdots \times A_N)$ called the law of motion, and bounded real valued payoff functions u_i defined on the history space $H = S \times A \times S \times A \cdots$, where $A = A_1 \times A_2 \times \cdots \times A_N$. The game is called an N-player deterministic game if for all state $s \in S$ and action choice $a = (a^1, a^2, \ldots, a^N)$ there is a unique state s' such that $p(s'|s, a) = 1$.*

If there is a finite number of players and the action sets and the set of states are finite, then a stochastic game with a finite number of stages always has a Nash equilibrium. The same is true for a game with infinitely many stages if the total payoff is the discounted sum. Shapley proposed an algorithm [233] for two-player zero-sum stochastic games using value iteration.

Stochastic games were invented by Lloyd Shapley [233] in the early 1950s. Readers are referred to the book edited by Neyman and Sorin [246]. The book by Filar and Vrieze [247] is more elementary in nature, and provides a unified rigorous treatment of the theories of MDP and two-person stochastic games. They coin the term *competitive MDP* to encompass both one- and two-player stochastic games. Stochastic games have applications in economics and evolutionary biology. In wireless networking, stochastic games have been used in areas such as flow control, routing, and scheduling [248, 249].

5.4 Bargaining game

A cooperative game is one in which players are able to make enforceable contracts. Hence, it is not defined as the game in which players actually do cooperate, but as games in which any cooperation is enforceable by an outside party (e.g. a judge and police). There are two major concepts in cooperative game theory: the bargaining solution and coalition concepts. The bargaining game is discussed in this section. Coalition concepts will be discussed in Section 5.5.

The bargaining problem of cooperative game theory can be described as follows [239, 250, 236]:

Definition 5.20 *Let $\mathbf{N} = \{1, 2, \ldots, N\}$ be the set of players. Let $\boldsymbol{S}$ be a closed and convex subset of $\Re^N$ to represent the set of feasible payoff allocations that the players can obtain if they all work together. Let $u^i_{\min}$ be the minimal payoff that player i will expect; otherwise, he/she will not cooperate. Suppose $\{u_i \in \boldsymbol{S} | u_i \geq u^i_{\min}, \forall i \in \mathbf{N}\}$ is a non-empty bounded set. Define $\boldsymbol{u}_{\min} = (u^1_{\min}, \ldots, u^N_{\min})$, then the pair $(\boldsymbol{S}, \boldsymbol{u}_{\min})$ is called an N-person bargaining problem.*

Within the feasible set $\mathbf{S}$, the notion of Pareto optimality is defined as a selection criterion for the bargaining solutions:

Definition 5.21 *The point $(u_1, \ldots, u_N)$ is said to be* ***Pareto optimal****, if and only if there is no other allocation u'_i such that $u'_i \geq u_i, \forall i$, and $u'_i > u_i, \exists i$, i.e. there exists no other allocation that leads to superior performance for some users without deteriorating the performance for some other users.*

Next, some bargaining solutions are discussed and several examples are illustrated.

5.4.1 Solution of bargaining games

Since there might be an infinite number of Pareto optimal points, a selection criterion is required to choose a bargaining result. A possible criterion is fairness. One commonly used fairness criterion for wireless resource allocation is max-min fairness [251], where the performance of the user with the worst channel conditions is maximized. This criterion penalizes the users with good channels and as a result deteriorates overall system performance. Now, let us introduce the criterion of fairness used for Nash Bargaining Solution (NBS) [236]. The intuitive idea is that, after the minimal requirements are satisfied for all users, the rest of the resources are allocated to users in proportion to their conditions. Many types of bargaining solutions [236] exist; among them, the NBS provides a unique and fair Pareto optimal operation point under the following conditions.

Definition 5.22 *$\bar{\boldsymbol{u}}$ is said to be a **Nash bargaining solution** in $\boldsymbol{S}$ for $\boldsymbol{u}_{\min}$, i.e. $\bar{\boldsymbol{u}} = \phi(\boldsymbol{S}, \boldsymbol{u}_{\min})$, if the following axioms are satisfied:*

1. *Individual rationality: $\bar{u}_i \geq u^i_{\min}, \forall i$.*
2. *Feasibility: $\bar{\boldsymbol{u}} \in \boldsymbol{S}$.*
3. *Pareto optimality: For every $\hat{\boldsymbol{u}} \in \boldsymbol{S}$, if $u_i \geq \bar{u}_i, \forall i$, then $\hat{u}_i = \bar{u}_i, \forall i$.*
4. *Independence of irrelevant alternatives: If $\bar{\boldsymbol{u}} \in \boldsymbol{S}' \subset \boldsymbol{S}$, $\bar{\boldsymbol{u}} = \phi(\boldsymbol{S}, \boldsymbol{u}_{\min})$, then $\bar{\boldsymbol{u}} = \phi(\boldsymbol{S}', \boldsymbol{u}_{\min})$.*
5. *Independence of linear transformations: For any linear scale transformation ψ, $\psi(\phi(\boldsymbol{S}, \boldsymbol{u}_{\min})) = \phi(\psi(\boldsymbol{S}), \psi(\boldsymbol{u}_{\min}))$.*
6. *Symmetry: If $\boldsymbol{S}$ is invariant under all exchanges of agents, $\phi_j(\boldsymbol{S}, \boldsymbol{u}_{\min}) = \phi_{j'}(\boldsymbol{S}, \boldsymbol{u}_{\min}), \forall j, j'$.*

Axioms 4-6 are called axioms of fairness. Axiom 4 asserts that eliminating the feasible solutions that would not have been chosen shall not affect the NBS solution. Axiom 5 asserts that the bargaining solution is scale invariant. Axiom 6 asserts that if the feasible ranges for all users are completely symmetric, then all users have the same solution.

The following optimization has been shown to have the NBS that satisfies the above axioms [236]:

Theorem 5.2 ***(Existence of NBS)**: There is a solution function $\phi(\boldsymbol{S}, \boldsymbol{u}_{\min})$ that satisfies all six axioms in Definition 5.22, and this solution satisfies*

$$\phi(\boldsymbol{S}, \boldsymbol{u}_{\min}) \in \arg \max_{\bar{\boldsymbol{u}} \in \boldsymbol{S}, \bar{u}_i \geq u^i_{\min}, \forall i} \prod_{i=1}^{N} \left(\bar{u}_i - u^i_{\min}\right). \tag{5.9}$$

Two other bargaining solutions which are alternatives to the NBS are – the Kalai–Smorodinsky solution (KSS) [235] and the egalitarian solution (ES). To define these solutions, let us introduce the following definitions:

Definition 5.23 *(Restricted monotonicity): If $\mathcal{U} \subset \mathcal{V}$ and $H(\mathcal{U}, \boldsymbol{u}_{\min}) = H(\mathcal{V}, \boldsymbol{u}_{\min})$ then $\phi(\mathcal{U}, \boldsymbol{u}_{\min}) \geq \phi(\mathcal{V}, \boldsymbol{u}_{\min})$, where $H(\mathcal{U}, \boldsymbol{u}_{\min})$, called the* utopia point, *is defined as:*

$$H(\mathcal{U}, \boldsymbol{u}_{\min}) = \left[\max_{\boldsymbol{u} > \boldsymbol{u}_{\min}} u_1(\boldsymbol{u}) \;\; \max_{\boldsymbol{u} > \boldsymbol{u}_{\min}} u_2(\boldsymbol{u}) \right]. \tag{5.10}$$

Now the KSS is defined as follows:

Definition 5.24 ***(Kalai–Smorodinsky solution)**: Let Λ be a set of points on the line containing $\boldsymbol{u}_{\min}$ and $H(\mathcal{U}, \boldsymbol{u}_{\min})$. $\phi(\mathcal{U}, \boldsymbol{u}_{\min})$ is the KSS, which can be expressed as*

$$\phi(\mathcal{U}, \boldsymbol{u}_{\min}) = \max \left\{ \boldsymbol{u} > \boldsymbol{u}_{\min} \left| \frac{1}{\theta_1}(u_1 - u^1_{\min}) = \frac{1}{\theta_2}(u_2 - u^2_{\min}) \right. \right\}, \tag{5.11}$$

where $\theta_i = H_i(\mathcal{U}, \boldsymbol{u}_{\min}) - u^i_{\min}$. The solution is in Λ.

Definition 5.25 ***(Egalitarian solution)**: $\phi(\mathcal{U}, \boldsymbol{u}_{\min})$ is the ES, which can be expressed as follows:*

$$\phi(\mathcal{U}, \boldsymbol{u}_{\min}) = \max \left\{ \boldsymbol{u} > \boldsymbol{u}_{\min} | u_1 - u_{\min}^1 = u_2 - u_{\min}^2 \right\}. \tag{5.12}$$

Next, the ultimatum game (sequential bargaining) is studied in which players interact anonymously and only once. For example, for the two-player case, the first player proposes how to divide a sum of money between themselves, and the second player can either accept or reject this proposal. If the second player rejects, neither player receives anything. If the second player accepts, the money is split according to the proposal. Since the game is played only once, and anonymously, reciprocation is not an issue.

For illustration, suppose there is a smallest division of the good available (say 1 cent). Suppose that the total amount of money available is x. The first player chooses some amount in the interval $[0, x]$. The second player chooses some function $f : [0, x] \to \{\text{"}accept\text{,"}\ \text{"}reject\text{"}\}$ (i.e. the second player chooses which divisions to accept and which to reject). The strategy profile is represented as (p, f), where p is the proposal and f is the function. If $f(p) =$ "*accept*" the first receives p and the second $x - p$, otherwise both get zero. (p, f) is a Nash equilibrium of the Ultimatum game if $f(p) =$ "*accept*" and there is no $y > p$ such that $f(y) =$ "*accept*" (i.e. p is the largest amount the second will accept). The first player would not want to unilaterally increase his/her demand since the second will reject any higher demand. The second would not want to reject the demand, since he/she would then get nothing.

There is one other Nash equilibrium where $p = x$ and $f(y) =$ "*reject*" for all $y > 0$ (i.e. the second player rejects all demands that gives the first player any non-zero amount). Here both players receive nothing, but neither could receive more by unilaterally changing his/her strategy.

However, only one of these Nash equilibria satisfies a more restrictive equilibrium concept, namely, subgame perfection. Suppose that the first player demands a large amount that gives the second player some (small) amount of money. By rejecting the demand, the second player is choosing nothing rather than something. So, it would be better for the second player to choose to accept any demand that gives him/her any amount whatsoever. If the first player knows this, he/she will give the second the smallest (non-zero) amount possible.

5.4.2 Applications of bargaining games

Bargaining game models can be used for many wireless networking problems. Several example applications are discussed below:

- *Resource allocation in OFDMA networks* [252]: For multiuser single cell orthogonal frequency-division multiple-access (OFDMA) systems, the resource allocation problem is to allocate subcarrier, rate, and power to maximize the overall system rate, under each user's maximal power and minimal rate constraints, while considering the fairness among users. The approach in [252] considers a new fairness criterion, which is a generalized proportional fairness based on NBS and coalitions. First, a two-user algorithm was developed to bargain subcarrier usage between two users. Then a multiuser bargaining algorithm was developed based on optimal coalition pairs among users. The simulation results showed that the proposed algorithm not only provides fair resource allocation among users, but also achieves overall system rates comparable with that for a scheme maximizing the total rate without considering fairness. Also, a much higher rate is achieved compared to a max-min fairness-based scheme. Moreover, the iterative fast implementation of the scheme has the complexity for each iteration of only $O(K^2 N \log N + K^4)$, where N is the number of subcarriers and K is the number of users.

- *Spectrum allocation in dynamic spectrum access networks* [253]: Dynamic spectrum access in a wireless system can achieve high utilization by allowing devices to sense and utilize the available spectrum opportunistically. However, a naive distributed spectrum assignment can lead to significant interference between devices. A general framework was defined for the spectrum access problem for several definitions of overall system utility. By reducing the allocation problem to a variant of the graph coloring problem, the global optimization problem was shown to be NP-hard, and a general approximation methodology through vertex labeling was proposed. A centralized strategy, where a central server calculates an allocation assignment based on global knowledge, and a distributed approach, where devices collaborate to negotiate local channel assignments towards global optimization, were investigated in [253]. The experimental results showed that the allocation algorithms can dramatically reduce interference and improve throughput (as much as 12-fold). It was also shown that the distributed algorithms generate allocation assignments similar in performance to the centralized algorithms using global knowledge, while incurring substantially less computational complexity in the process.
- *Spectrum allocation in multi-hop wireless networks* [254]: There is a need for new spectrum access protocols that are opportunistic, flexible, and efficient, yet fair. Game theory provides an ideal framework for analyzing spectrum access, a problem that involves complex distributed decisions by independent spectrum users. A cooperative game theory model was developed in [254] to analyze a scenario where nodes in a multi-hop wireless network need to agree on a fair allocation of spectrum. In high interference environments, the utility space of the game was shown to be non-convex, which may make some optimal allocations unachievable with pure strategies. However, as the number of available channels increases, the utility space becomes close to convex and thus optimal allocations become achievable with pure strategies. NBS was shown to achieve a good compromise between fairness and efficiency, using a small number of channels. Finally, a distributed algorithm was proposed for spectrum sharing and it was shown that this distributed algorithm achieves allocations reasonably close to the NBS.
- *Scheduling in wireless mesh networks* [255]: A fair scheduling scheme was proposed for OFDMA-based wireless mesh networks (WMNs), which fairly allocate subcarriers and power to mesh routers (MRs) and mesh clients to maximize the NBS fairness criterion. In WMNs, since not all the information necessary for scheduling is available at a central scheduler (e.g. the MR), it is advantageous to involve the MR and as many

mesh clients as possible in distributed scheduling based on the limited information that is available locally at each node. Instead of solving a single global control problem, the subcarrier and power allocation problem was decoupled into two subproblems hierarchically, where the MR allocates groups of subcarriers to the mesh clients, and each mesh client allocates transmit power among its subcarriers to each of its outgoing links. The two subproblems were formulated based on non-linear integer programming and non-linear mixed integer programming, respectively. A simple and efficient solution algorithm was developed for the MR's problem. Also, a closed-form solution was obtained by transforming the mesh client's problem into a time-division scheduling problem. Extensive simulation results demonstrated that the proposed scheme provides fair opportunities to the respective users (mesh clients) and a comparable overall end-to-end rate when the number of mesh clients increases.

- *Multimedia transmission* [256]: Multiuser multimedia applications such as enterprise streaming, surveillance, and gaming are recently emerging, and they are often deployed over bandwidth-constrained network infrastructures. To meet the QoS requirments of the delay-sensitive and bandwidth intensive multimedia data for these applications, efficient resource (bandwidth) management becomes paramount. The well-known game theoretic concept of bargaining was used in [256] to allocate bandwidth fairly and optimally among multiple collaborative users. Specifically, two bargaining solutions, namely, NBS and KSBS, were considered for the resource management problem. Interpretations were provided for the two investigated bargaining solutions for multiuser resource allocation – the NBS can be used to maximize the system utility, while the KSBS ensures that all users incur the same utility penalty relative to the maximum achievable utility. The bargaining strategies and solutions were implemented in the network using a resource manager, which explicitly considers the application-specific distortion for the bandwidth allocation. The bargaining solutions exhibit important properties (axioms) that can be used for effective multimedia resource allocation. Moreover, several criteria were proposed for determining bargaining powers for these solutions, which provide additional flexibility in choosing the solution by taking into consideration the visual quality impact, the deployed spatiotemporal resolutions, etc.

5.5 Coalition game

In the previous section, the cooperative game by bargaining has been discussed. In addition to bargaining, other analysis tools for cooperative games include coalition, core, Shapley function, and nucleolus. In the following, these concepts will be explained with examples. Also the fairness issue will be discussed. Finally, a merge/split algorithm will be shown in order to achieve the coalition in a distributed manner.

5.5.1 Characteristic function and core

First, we discuss how to divide the benefits among users, and the properties and stability of the division by introducing the following concepts:

Definition 5.26 *A **coalition** S is defined to be a subset of the total set of players N (i.e. $S \in N$). The users in a coalition try to cooperate with each other. The **coalition form** of a game is given by the pair (N, v), where v is a real-valued function, called the **characteristic function**. $v(S)$ is the value of the cooperation for coalition S with the following properties:*

1. *$v(\emptyset) = 0$.*
2. *(Superadditivity) if S and T are disjoint coalitions ($S \bigcap T = \emptyset$), then $v(S) + v(T) \leq v(S \bigcup T)$.*

The coalition states the benefit obtained via a cooperation agreement. But we still need to study how to divide the benefit to the cooperative players. One of the possible properties of an agreement on a fair division is stability – that is, no coalition shall have the incentive and power to upset the cooperative agreement. The set of such divisions of v is called the core, which is defined as follows:

Definition 5.27 *A payoff vector $\mathbf{x} = (x_1, \ldots, x_N)$ is said to be **group rational** or **efficient** if $\sum_{i=1}^{N} x_i = v(N)$. A payoff vector x is said to be **individually rational** if the player can obtain the benefit no less than acting alone, i.e. $x_i \geq v(\{i\}),\ \forall i$. An **imputation** is a payoff vector satisfying the above two conditions.*

Definition 5.28 *An imputation $\mathbf{x}$ is said to be unstable through a coalition S if $v(S) > \sum_{i \in S} x_i$, i.e. the players have incentive for coalition S and upset the proposed $\mathbf{x}$. The set C of a stable imputation is called the **core**, i.e.*

$$C = \{\mathbf{x} : \sum_{i \in N} x_i = v(N) \text{ and } \sum_{i \in S} x_i \geq v(S),\ \forall S \subset N\}. \tag{5.13}$$

The core gives a reasonable set of possible shares. A combination of shares is in a core if there exists no subcoalition in which its members may gain a higher total outcome than the share of concern. If the share is not in a core, some members may become frustrated and may leave the whole group in order to form a smaller group.

To illustrate the idea of the core, let us consider the following example. Assume a game with the following characteristic functions:

$$v(\emptyset) = 0,\ v(\{1\}) = 1,\ v(\{2\}) = 0,\ v(\{3\}) = 1 \tag{5.14}$$
$$v(\{1,2\}) = 4,\ v(\{1,3\}) = 3,\ v(\{2,3\}) = 5,\ v(\{1,2,3\}) = 8.$$

By using $v(\{2,3\}) = 5$, payoff vector such as (4, 3, 1) can be eliminated, since player 2 and player 3 can achieve a better payoff by forming a coalition themselves. Using the same analysis, the final core of the game is (3,4,1), (3,3,2), (3,2,3), (3,1,4), (2,5,1), (2,4,2), (2,3,3), (2,2,4), (1,5,2), (1,4,3), and (1,3,4).

5.5.2 Fairness

The concept of the core defines the stability of an allocation of payoff. Next, we investigate how to divide the benefits among players according to different fairness criteria.

First, we study each individual player's power in the coalition by defining a value called Shapley function.

Definition 5.29 *A **Shapley function** ϕ is a function that assigns to each possible characteristic function v a real number, i.e.*

$$\phi(v) = (\phi_1(v), \phi_2(v), \ldots, \phi_N(v)) \tag{5.15}$$

where $\phi_i(v)$ represents the worth or value of player i in the game. The Shapley axioms for $\phi(v)$ are:

1. ***Efficiency**: $\sum_{i \in N} \Phi_i(v) = v(N)$.*
2. ***Symmetry**: If i and j are such that $v(S \bigcup \{i\}) = v(S \bigcup \{j\})$ for every coalition S not containing i and j, then $\phi_i(v) = \phi_j(v)$.*
3. ***Dummy axiom**: If i is such that $v(S) = v(S \bigcup \{i\})$ for every coalition S not containing i, then $\phi_i(v) = 0$.*
4. ***Additivity**: If u and v are characteristic functions, then $\phi(u + v) = \phi(v + u) = \phi(u) + \phi(v)$.*

It can be proven that there exists a unique function ϕ that satisfies the Shapley axioms. To calculate the Shapley function, suppose we form the grand coalition by entering the players into this coalition one at a time. As each player enters the coalition, he/she receives the amount by which his/her entry increases the value of the coalition he/she has entered. The amount a player receives by this scheme depends on the order in which the players enter. The Shapley value is just the average payoff to the players if the players enter in a completely random order, i.e.

$$\phi_i(v) = \sum_{S \subset N, i \in S} \frac{(|S| - 1)!(N - |S|)!}{N!}[v(S) - v(S - \{i\})]. \tag{5.16}$$

For the example in (5.14), it can be shown that the Shapley value is $\phi = (14/6, 17/6, 17/6)$.

Another concept in multiple cooperative games is the nucleolus. For a fixed characteristic function, an imputation $\mathbf{x}$ is found such that the worst inequity is minimized, i.e. for each coalition S and its associated dissatisfaction, an optimal imputation is calculated to minimize the maximum dissatisfaction. First we define the concept of excess, which measures the dissatisfaction:

Definition 5.30 *The measure of the inequity of an imputation $\mathbf{x}$ for a coalition S is defined as the **excess**:*

$$e(\mathbf{x}, S) = v(S) - \sum_{j \in S} x_j. \tag{5.17}$$

Obviously, any imputation $\boldsymbol{x}$ is in the core, if and only if all its excesses are negative or zero.

Among all allocations, kernel is a fair allocation, which is defined as follows:

Definition 5.31 *A **kernel** of v is the set of all allocations **x** such that*

$$\max_{S \subseteq N-j, i \in S} e(\mathbf{x}, S) = \max_{T \subseteq N-i, j \in T} e(\mathbf{x}, T). \tag{5.18}$$

If players i and j are in the same coalition, then the highest excess that player i can make in a coalition without player j is equal to the highest excess that player j can make in a coalition without player i.

Finally, the nucleolus is defined as follows:

Definition 5.32 *The **nucleolus** is the allocation **x** which minimizes the maximum excess:*

$$\mathbf{x} = \arg\min_{\mathbf{x}}(\max e(\mathbf{x}, S), \ \forall S). \tag{5.19}$$

The nucleolus has the following property: the nucleolus of a game in coalitional form exists and is unique. The nucleolus is group and individually rational, and satisfies both the symmetry and dummy axioms. If the core is not empty, the nucleolus is in the core and kernel. In other words, the nucleolus is the best allocation with the min-max criterion.

5.5.3 Merge/split algorithm

Using the coalition formation concepts described in the previous section, a coalition formation algorithm for self-organization can be introduced. This algorithm will be based on simple rules of merge-and-split that allow to modify a partition T of N nodes as follows [257]:

- *Merge rule*: Merge any set of coalitions $\{S_1, \ldots, S_l\}$ where $\{\bigcup_{j=1}^{l} S_j\} \rhd \{S_1, \ldots, S_l\}$, therefore, $\{S_1, \ldots, S_l\} \to \{\bigcup_{j=1}^{l} S_j\}$. (Each S_i denotes a coalition).
- *Split rule*: Split any coalition $\bigcup_{j=1}^{l} S_j$ where $\{S_1, \ldots, S_l\} \rhd \{\bigcup_{j=1}^{l} S_j\}$, thus, $\{\bigcup_{j=1}^{l} S_j\} \to \{S_1, \ldots, S_l\}$.

In brief, multiple coalitions will merge (split) if merging (splitting) yields a preferred collection based on a chosen $\rhd$. In [257], it was shown that any arbitrary iteration of merge-and-split operations *terminates*. Therefore, it will be suitable to devise a coalition formation algorithm by means of merge-and-split. The Pareto order is highly appealing as a comparison relation $\rhd$ for the merge-and-split rules. With the Pareto order, coalitions will merge only if at least one user can enhance its individual payoff through this merge without decreasing the other users' payoffs. Similarly, a coalition will split only if at least one user in that coalition is able to strictly improve its individual payoff through the split without hurting other users. A decision to merge or split is, thus, tied to the fact that all users must benefit from merge or split, thus, any merged (or split) form is reached only if it allows all involved users to maintain their payoffs with at least one user improving. In Figure 5.3, an example of forming a virtual multiple-input multiple-output (MIMO) system in wireless networks is shown which is based on coalition games.

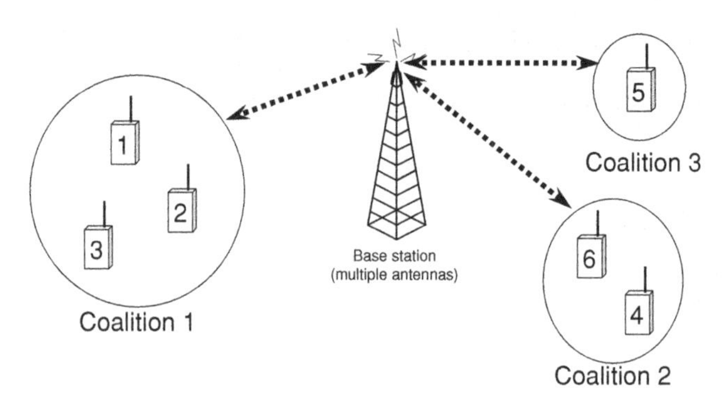

Figure 5.3 Coalition game and merge/split algorithm for virtual MIMO.

Two important defection functions can be pinpointed [258, 259]. First, the $\mathbb{D}_{\mathrm{hp}}(T)$ function (denoted $\mathbb{D}_{\mathrm{hp}}$), which associates with each partition T of N the family of all partitions of N that the players can form through merge-and-split operations applied to T. This function allows any group of players to leave the partition T of N through merge-and-split operations to create another *partition* in N. Second, the $\mathbb{D}_{\mathrm{c}}(T)$ function (denoted $\mathbb{D}_{\mathrm{c}}$), which associates with each partition T of N the family of all collections in N. This function allows any group of players to leave the partition T of N through *any* operation and create an arbitrary *collection* in N. Two forms of stability stem from these definitions: $\mathbb{D}_{\mathrm{hp}}$ stability and a stronger $\mathbb{D}_{\mathrm{c}}$ stability. A partition T is $\mathbb{D}_{\mathrm{hp}}$-stable if no players in T are interested in leaving T through merge-and-split to form other partitions in N; while a partition T is $\mathbb{D}_{\mathrm{c}}$-stable if no players in T are interested in leaving T through *any* operation (not necessarily merge or split) to form other collections in N.

5.6 Game with imperfect information

In game theory, a Bayesian game is one in which information about characteristics of the other players (i.e. payoffs) is incomplete. Following John C. Harsanyi's framework, a Bayesian game can be modeled by introducing nature as a player in the game. Nature assigns a random variable to each player. This random variable with given probability density function takes values of types for each player. (In the course of the game, nature randomly chooses a type for each player according to the probability distribution across each player's type space.) Harsanyi's approach to modeling a Bayesian game in such a way allows games of incomplete information to become games of imperfect information (in which the history of the game is not available to all players). The type of a player determines that player's payoff function and the probability associated with the type is the probability that the player for whom the type is specified is that type. In a Bayesian game, the incompleteness of information means that at least one player is unsure of the type (and so the payoff function) of another player.

Such games are called Bayesian because of the probabilistic analysis inherent in the game. Players have initial beliefs about the type of each player (where a belief is a probability distribution over the possible types for a player) and can update their beliefs according to Bayesian rule as play takes place in the game, i.e. the belief a player holds about another player's type might change on the basis of the actions they have played. The lack of information held by players and modeling of beliefs mean that such games are also used to analyze imperfect information scenarios.

The impacts of imperfect information on the normal and extensive forms of game are discussed in the following two subsections.

5.6.1 Bayesian game in normal form

The normal form representation of a non-Bayesian game with perfect information is a specification of the strategy spaces and payoff functions of players. A strategy for a player is a complete plan of action that covers every contingency of the game, even if that contingency can never arise. The strategy space of a player is thus the set of all strategies available to a player. A payoff function is a function from the set of strategy profiles to the set of payoffs (normally the set of real numbers), where a strategy profile is a vector specifying a strategy for every player.

In a Bayesian game, it is necessary to specify the strategy spaces, type spaces, payoff functions, and beliefs for every player. A strategy for a player is a complete plan of actions that covers every contingency that might arise for every type that player might be. A strategy must not only specify the actions of the player given the type that he/she is, but must also specify the actions that would be taken if he/she were of another type. Strategy spaces are defined accordingly. A type space for a player is just the set of all possible types of that player. The beliefs of a player describe the uncertainty of that player about the types of the other players. Each belief is the probability of the other players having particular types, given the type of the player with that belief (i.e. the belief is Prob(types of other players|type of this player)). A payoff function is a two-place function of strategy profiles and types. If a player has a payoff function $u(x, y)$ and he/she has type t, the payoff received is $u(x^*, t)$, where x^* is the strategy profile played in the game (i.e. the vector of strategies played). The formal definition of a Bayesian game is given as follows:

Definition 5.33 *A* ***Bayesian game*** *of incomplete information is defined as*

1. *Set of players:* $i \in \{1, 2, \ldots, N\}$.
2. *Actions available to player* i*:* A_i *for* $i \in \{1, 2, \ldots, N\}$. *Let* $a_i \in A_i$ *denote a typical action for player* i.
3. *Sets of possible types for all players:* T_i *for* $i \in \{1, 2, \ldots, N\}$. *Let* $t_i \in T_i$ *denote a typical type of player* i.
4. *Let* $\boldsymbol{a} = (a_1, \ldots, a_N)$, $\boldsymbol{t} = (t_1, \ldots, t_N)$, $\boldsymbol{a}_{-i} = (a_1, \ldots, a_{i-1}, a_{i+1}, \ldots, a_N)$, $\boldsymbol{t}_{-i} = (t_1, \ldots, t_{i-1}, t_{i+1}, \ldots, t_N)$.
5. *Nature's move:* $\boldsymbol{t}$ *is selected according to a joint probability distribution* $p(\boldsymbol{t})$ *on* $T = T_1 \times \cdots \times T_N$.

6. *Strategies:* $s_i : T_i \to A_i$, *for* $i \in \{1, 2, \ldots, N\}$. $s_i(t_i) \in A_i$ *is then the action that type* t_i *of player* i *takes.*
7. *Payoff:* $u_i(a_1, \ldots, a_N; t_1, \ldots, t_N)$.

The game proceeds as follows: first, nature chooses $\mathbf{t}$ according to probability $p(\mathbf{t})$. Then each player i observes realized type $\hat{t}_i$ and updates its beliefs: each player calculates with conditional probability of remaining types conditioned on $t_i = \hat{t}_i$; denote distribution of $\mathbf{t}_{-i}$ conditioned on $\hat{t}_i$ by $p_i(\mathbf{t}_{-i}|\hat{t}_i)$. Finally, players take actions simultaneously.

The expected payoff is calculated as follows: given strategy s_i, type t_i of player i plays action $s_i(t_i)$. With vector of type $\mathbf{t} = (t_1, \ldots, t_N)$ and strategies $(s_1, \ldots, s_N)$, the realized action profile is $(s_1(t_1), \ldots, s_N(t_N))$. Player i of type $\hat{t}_i$ has beliefs about types of other players given by conditional probability distribution $p_i(\mathbf{t}_{-i}|\hat{t}_i)$. The expected payoff of action s_i is

$$\sum_{\mathbf{t}:t_i=\hat{t}_i} u_i(s_i, \mathbf{s}_{-i}(\mathbf{t}_{-i}), \mathbf{t}) p_i(\mathbf{t}_{-i}|\hat{t}_i). \tag{5.20}$$

The action $s_i(\hat{t}_i)$ for player i is a best response to $\mathbf{s}_{-i}(\mathbf{t}_{-i})$ if and only if for all $s_i' \in A_i$,

$$\sum_{\mathbf{t}:t_i=\hat{t}_i} u_i(s_i(\hat{t}_i), \mathbf{s}_{-i}(\mathbf{t}_{-i}), \mathbf{t}) p_i(\mathbf{t}_{-i}|\hat{t}_i) \geq \sum_{\mathbf{t}:t_i=\hat{t}_i} u_i(s_i', \mathbf{s}_{-i}(\mathbf{t}_{-i}), \mathbf{t}) p_i(\mathbf{t}_{-i}|\hat{t}_i). \tag{5.21}$$

Using the above equation, the solution concept of the Bayesian game, namely, the Bayesian Nash equilibrium, is introduced as follows:

Definition 5.34 *A strategy profile* $(s_1(t_1), \ldots, s_N(t_N))$ *is a* ***Bayesian Nash equilibrium*** *if* $s_i(t_i)$ *is a best response to* $\mathbf{s}_{-i}(\mathbf{t}_{-i}$ *for all* $t_i) \in T_i$ *and for all* i. *In other words, an action specified by the strategy of any given player has to be optimal, given strategies of all other players and beliefs of players.*

In a non-Bayesian game, a strategy profile is a Nash equilibrium if every strategy in that profile is a best response to every other strategy in the profile, i.e. there is no strategy that a player could play that would yield a higher payoff, given all the strategies played by the other players. In a Bayesian game (where players are modeled as risk-neutral), rational players are seeking to maximize their expected payoff, given their beliefs about the other players (in the general case, where players may be risk averse or risk-loving, the assumption is that the players aim to maximize the expected utility). A Bayesian Nash equilibrium is defined as a strategy profile and beliefs specified for each player about the types of the other players that maximizes the expected payoff for each player given their beliefs about the other players' types and given the strategies played by the other players.

Next, an example is given on Cournot competition (which will be described in more detail in Chapter 11) with privately known cost. Suppose two firms compete by deciding simultaneously quantities q_1 and q_2. One firm has a known marginal cost of 2; the other firm may have either a high or a low cost: $t_2 = t_{\mathrm{H}} = 3$ with probability 0.5 and $t_2 = t_{\mathrm{L}} = 1$ with probability 0.5. The price is a function of the quantities q_1 and q_2 and is given by: $P = 4 - q_1 - q_2$. The question is what is the Bayesian Nash equilibrium? Now $q_2^{\mathrm{H}}, q_2^{\mathrm{L}}, q_1$ is an equilibrium if

$$q_2^{\mathrm{H}} = \arg\max_{q_2^{\mathrm{H}}}(4 - q_1 - q_2^{\mathrm{H}})q_2^{\mathrm{H}} - 3q_2^{\mathrm{H}}, \tag{5.22}$$

$$q_2^{\mathrm{L}} = \arg\max_{q_2^{\mathrm{L}}}(4 - q_1 - q_2^{\mathrm{L}})q_2^{\mathrm{L}} - q_2^{\mathrm{L}}, \tag{5.23}$$

$$q_1 = \arg\max_{q_1} \frac{1}{2}[(4 - q_1 - q_2^{\mathrm{H}})q_1 - 2q_1] + \frac{1}{2}[(4 - q_1 - q_2^{\mathrm{L}})q_1 - 2q_1]. \tag{5.24}$$

Using the first-order condition, we get the Bayesian Nash equilibrium solution: $q_1 = 2/3$, $q_2^{\mathrm{L}} = 7/6$, and $q_2^{\mathrm{H}} = 1/6$.

5.6.2 Bayesian game in extensive form

The solution concept of Bayesian Nash equilibrium yields an abundance of equilibria in dynamic games, when no further restrictions are placed on players' beliefs. This makes Bayesian Nash equilibrium an incomplete tool to analyze dynamic games of incomplete information.

Bayesian Nash equilibrium results in some implausible equilibria in dynamic games, where players take turns sequentially rather than simultaneously. Similarly, implausible equilibria might arise in games with perfect and complete information, such as incredible threats and promises. Such equilibria might be eliminated in perfect and complete information games by applying subgame perfect Nash equilibrium. However, it is not always possible to avail oneself of this solution concept in incomplete information games because such games contain non-singleton information sets and since subgames must contain complete information sets, sometimes there is only one subgame – the entire game – and so every Nash equilibrium is trivially subgame perfect. Even if a game does have more than one subgame, the inability of subgame perfection to cut through information sets can result in implausible equilibria not being eliminated.

To refine the equilibria generated by the Bayesian Nash solution concept or subgame perfection, one can apply the perfect Bayesian equilibrium (PBE) solution concept. PBE is in the spirit of subgame perfection in that it demands that subsequent play be optimal. However, it places player beliefs on decision nodes that enables moves in non-singleton information sets to be dealt with more satisfactorily.

Definition 5.35 *A **perfect Bayesian equilibrium** is a strategy profile and a set of beliefs for each player, $(s^{\mathrm{C}}, i^{\mathrm{C}})$ such that:*

1. *at every information set I_i, player i's strategy maximizes its payoff, given actions of all other players, and player i's beliefs;*
2. *at information sets reached with positive probability when s^{C} is played, beliefs are formed according to s^{C} and Bayesian rule when necessary;*
3. *at information sets that are reached with probability zero when s^{C} is played, beliefs may be arbitrary but must be formed according to Bayesian rule when possible.*

So far in our discussion on Bayesian games, it has been assumed that information is perfect (or, if imperfect, play is simultaneous). In examining dynamic games, however, it might be necessary to have the means to model imperfect information. PBE provides this means: players place beliefs on nodes occurring in their information sets, which means that the information set can be generated by nature (in the case of incomplete information) or by other players (in the case of imperfect information).

The beliefs held by players in Bayesian games can be approached more rigorously in PBE. A belief system is an assignment of probabilities to every node in the game such that the sum of probabilities in any information set is 1. The beliefs of a player are exactly those probabilities of the nodes in all the information sets at which that player has the move (a player's belief might be specified as a function from the union of his/her information sets to [0,1]). A belief system is consistent for a given strategy profile if and only if the probability assigned by the system to every node is computed as the probability of that node being reached given the strategy profile, i.e. by Bayesian rule.

The notion of sequential rationality is what determines the optimality of subsequent play in PBE. A strategy profile is sequentially rational at a particular information set for a particular belief system if and only if the expected payoff of the player whose information set it is (i.e. who has the move at that information set) is maximal given the strategies played by all the other players. A strategy profile is sequentially rational for a particular belief system if it satisfies the above condition for every information set.

Next, let us give an example, shown in Figure 5.2, in which information is imperfect since player 2 does not know what player 1 does when he/she comes to play. If both players are rational and both know that both players are rational and everything that is known by any player is known to be known by every other player (i.e. player 1 knows player 2 knows that player 1 is rational and player 2 knows this, etc.), play in the game will be according to perfect Bayesian equilibrium as follows:

In the game shown in Figure 5.2, player 2 cannot observe player 1's move. Player 1 would like to fool player 2 into thinking he/she has played U when he/she has actually played D, so that player 2 will play D′ and player 1 will receive a payoff of 3. In fact in the second game there is a perfect Bayesian equilibrium where player 1 plays D and player 2 plays U′ and player 2 holds the belief that player 1 will definitely play D (i.e. player 2 places a probability of 1 on the node reached if player 1 plays D). In this equilibrium, every strategy is rational given the beliefs held and every belief is consistent with the strategies played. In this case, the perfect Bayesian equilibrium is the only Nash equilibrium.

Sequential equilibrium is a refinement of Nash equilibrium for extensive form games due to David M. Kreps and Robert Wilson [260]. A sequential equilibrium specifies not only a strategy for each of the players but also a belief for each of the players. A belief gives, for each information set of the game belonging to the player, a probability distribution on the nodes in the information set. A profile of strategies and beliefs is called an assessment for the game. Informally speaking, an assessment is a sequential equilibrium if its strategies are sensible given its beliefs and its beliefs are sensible given its strategies.

The formal definition of a strategy being sensible given a belief is straightforward; the strategy should simply maximize expected payoff in every information set. It is also straightforward to define what a sensible belief should be for those information sets that are reached with positive probability given the strategies; the beliefs should be the conditional probability distribution on the nodes of the information set, given that it is reached.

It is far from straightforward to define what a sensible belief should be for those information sets that are reached with probability zero, given the strategies. Indeed, this is the main conceptual contribution of Kreps and Wilson [260]. Their consistency requirement is the following: the assessment should be a limit point of a sequence of totally mixed strategy profiles and associated sensible beliefs, in the above straightforward sense.

Sequential equilibrium is a further refinement of the subgame perfect equilibrium and even the perfect Bayesian equilibrium. It is itself refined by extensive-form trembling hand perfect equilibrium. Strategies of sequential equilibria (or even extensive-form trembling hand perfect equilibria) are not necessarily admissible. A refinement of sequential equilibrium that guarantees admissibility is quasi-perfect equilibrium.

5.7 Other special types of games

In this section, several important special games are described which are widely used to model wireless networking problems.

5.7.1 Zero-sum game

First, let us discuss the zero-sum game which appears widely in daily life. In game theory and economic theory, zero-sum describes a situation in which a participant's gain or loss is exactly balanced by the losses or gains of the other participant(s). That is, when the total gains of the participants are added up, and the total losses are subtracted, they will sum to zero. "Go" is an example of a zero-sum game: it is impossible for both players to win. Zero-sum can be thought of more generally as a constant sum where the benefits and losses to all players sum to the same value of money. Cutting a cake is zero- or constant-sum because taking a larger piece reduces the amount of cake available for others. In contrast, non-zero-sum describes a situation in which the interacting parties' aggregate gains and losses is either less than or more than zero.

Situations where participants can all gain or suffer together, such as a country with an excess of bananas trading with another country for their excess of apples, where both benefit from the transaction, are referred to as non-zero-sum. Other non-zero-sum games are games in which the sum of gains and losses by the players are always more or less than what they began with. For example, a game of poker, disregarding the house's rake, played in a casino is a zero-sum game unless the pleasure of gambling or the cost of operating a casino is taken into account, making it a non-zero-sum game.

The concept was first developed in game theory and consequently zero-sum situations are often called zero-sum games, though this does not imply that the concept, or game

Table 5.5 Example of a zero-sum game.

	A	B	C
1	30,−30	−10,10	20,−20
2	10,−10	20,−20	−20,20

theory itself, applies only to what are commonly referred to as games. In pure strategies, each outcome is Pareto optimal (generally, any game where all strategies are Pareto optimal is called a conflict game) [1]. Nash equilibria of two-player zero-sum games are exactly pairs of minimax strategies.

In 1944 John von Neumann and Oskar Morgenstern proved that any zero-sum game involving n players is in fact a generalized form of a zero-sum game for two players, and that any non-zero-sum game for n players can be reduced to a zero-sum game for $n + 1$ players; the $(n + 1)$ player representing the global profit or loss.

Next, let us show an example of zero-sum game which is represented by a payoff matrix. Consider for example the two-player zero-sum game as shown in Table 5.5. The order of the play is as follows: the first player chooses in secret one of the two actions 1 or 2; the second player, unaware of the first player's action, chooses in secret one of the three actions A, B, or C. Then, the choices are revealed and each player's points total is affected according to the payoff for those choices. For example, player 1 chooses action 2 and player 2 chooses action B. When the payoff is allocated, player 1 gains 20 points and player 2 loses 20 points.

Now, in this example game, both players know the payoff matrix and attempt to maximize their points (i.e. payoffs). What should they do? Player 1 could reason as follows: "With action 2, I could lose up to 20 points and can win only 20, while with action 1 I can lose only 10 but can win up to 30, so action 1 looks a lot better." With similar reasoning, player 2 would choose action C. If both players take these actions, player 1 will win 20 points. But what happens if player 2 anticipates player 1's reasoning and choice of action 1, and deviously goes for action B, so as to win 10 points? Or if player 1 in turn anticipates this devious trick and goes for action 2, so as to win 20 points after all?

John von Neumann had presented the fundamental and surprising insight that probability provides a way out of this conundrum. Instead of deciding on a definite action to take, the two players assign probabilities to their respective actions, and then use a random device which, according to these probabilities, chooses an action for them. Each player computes the probabilities so as to minimize the maximum expected point-loss independent of the opponent's strategy. This leads to a linear programming problem with a unique solution for each player. This minimax method can compute provably optimal strategies for all two-player zero-sum games.

For the example given above, it turns out that player 1 should choose action 1 with probability 4/7 and action 2 with probability 3/7, while player 2 should assign the probabilities 0, 4/7 and 3/7 to the three actions A, B, and C. Player 1 will then win 20/7 points on average per game.

5.7.2 Potential game

A game is considered to be a potential game if the incentive of all players to change their strategies can be expressed in one global function, the potential function. The concept was proposed by Dov Monderer and Lloyd Shapley [261]. The potential function is a useful tool to analyze equilibrium properties of games, since the incentives of all players are mapped into one function, and the set of pure Nash equilibria can be found by simply locating the local optima of the potential function. A potential game can be formally defined as follows:

Definition 5.36 *Let N be the number of players, A be the set of action profiles over the action sets A_i of each player and u be the payoff function. Then, a game $G = (N, A = A_1 \times \cdots \times A_N, u : A \to \mathcal{R}^N)$ is a (cardinal)* **potential game** *if there is an exact potential function $\Phi : A \to \mathcal{R}$ such that $\forall i$*

$$\Phi(b_i, a_{-i}) - \Phi(a_i, a_{-i}) = u_i(b_i, a_{-i}) - u_i(a_i, a_{-i}), \; \forall a_i, b_i \in A_i, \forall a \in A. \quad (5.25)$$

A game is a general (ordinal) potential game if there is an ordinal potential function $\Phi : A \to \mathcal{R}$ such that

$$\text{sgn}[\Phi(b_i, a_{-i}) - \Phi(a_i, a_{-i})] = \text{sgn}[u_i(b_i, a_{-i}) - u_i(a_i, a_{-i})] \; \forall a_i, b_i \in A_i, \forall a \in A, \quad (5.26)$$

where sgn *denotes the sign function.*

In other words, in cardinal games, *the difference in individual payoffs for each player from individually changing one's strategy with other things the same has to have the same value as the difference in values for the potential function. In* ordinal games, *only the signs of the differences have to be the same.*

Next, a simple example of a two-player, two-strategy game is shown with externalities. Individual players' payoffs are given by the function $u_i(s_i, s_j) = b_i s_i + w s_i s_j$, where s_i is player i's strategy, s_j is the opponent's strategy, and w is a positive externality from choosing the same strategy. The strategy choices are $+1$ and -1, as seen in the payoff matrix in Table 5.6(a). This game has a potential function

$$P(s_1, s_2) = b_1 s_1 + b_2 s_2 + w s_1 s_2. \quad (5.27)$$

If player 1 moves from -1 to $+1$, the payoff difference is $\Delta u_1 = u_1(+1, s_2) - u_1(-1, s_2) = 2b_1 + 2ws_2$. The change in potential is $\Delta P = P(+1, s_2) - P(-1, s_2) = (b_1 + b_2 s_2 + w s_2) - (-b_1 + b_2 s_2 - w s_2) = 2b_1 + 2ws_2 = \Delta u_1$. The solution for player 2 is obtained in a similar way.

Using numerical values $b_1 = 2$, $b_2 = -1$, $w = 3$, this example transforms into a simple "battle of the sexes," as shown in Table 5.6(b). The game has two pure Nash equilibria, $(+1, +1)$ and $(-1, -1)$. These are also the local maxima of the potential function (Table 5.6(c)). The only stochastically stable equilibrium is $(+1, +1)$, the global maximum of the potential function.

There has been a lot of work where wireless networking and resource allocation problems have been modeled using potential games such as those in [262, 263, 264, 265, 266, 267, 268, 263, 269, 270, 271]. The major applications are in power control and

Table 5.6 (a) Example of potential game, (b) battle of the sexes (payoffs), (c) battle of the sexes (potentials).

	+1	−1
+1	(b_1+w, b_2+w)	$(b_1-w, -b_2-w)$
−1	$(-b_1-w, b_2-w)$	$(-b_1+w, -b_2+w)$

(a)

	+1	−1
+1	(5,2)	(−1,−2)
−1	(−5,-4)	(1,4)

(b)

	+1	−1
+1	4	0
−1	−6	2

(c)

waveform adaptation. The design implications are the nice properties of convergence with shared and independent outcomes for each individual.

Let us give an example based on the discussions on [263]. Consider a single-cell CDMA network with $\mathcal{M} = \{1, \cdots, M\}$ users. The received SINR of user $i \in \mathcal{M}$ is

$$\gamma_i(\mathbf{p}) = \frac{P_i h_i}{n_0 + \sum_{j \neq i} P_j h_j}$$

where h_i is the channel gain from user i's transmitter to the base station. For each user $i \in \mathcal{M}$, we want to solve the following power control problem:

$$\begin{aligned} \text{minimize} \quad & P_i \\ \text{subject to:} \quad & f_i(\gamma_i(\mathbf{p})) \geq \gamma_i^{\text{thresh}}, \\ \text{variables} \quad & P_i \in [0, P_i^{\max}]. \end{aligned} \tag{5.28}$$

Here $P_i^{\max}$ is the maximum power, f_i is the QoS function, and γ_i^{thresh} is the QoS threshold. It was shown in [263] that, for all user i, Problem (5.28) can be transformed into an equivalent non-cooperative game $G = \left[\mathcal{M}, \mathcal{A}, \left\{\log(P_i^{\max} - P_i)\right\}_{i \in \mathcal{M}}\right]$, with the *coupled* action set as

$$\mathcal{A} = \left\{\mathbf{p} : f_i(\gamma_i(\mathbf{p})) \geq \gamma_i^{\text{thresh}}, P_i \in [0, P_i^{\max}], \forall i \in \mathcal{M}\right\}.$$

Furthermore, the game G admits a potential function

$$Z(\mathbf{p}) = \sum_{i \in \mathcal{M}} \log(P_i^{\max} - P_i).$$

We can then maximize function $Z(\mathbf{p})$ over set $\mathcal{A}$, and the corresponding maximizer(s) will be the NE(s) of game G, and thus the optimal solution(s) of problem (5.28) for all i.

5.7.3 Super-modular game

A super-modular game is an important class of games that has the nice properties of existence and convergence to Nash equilibrium. In a super-modular game, when one player takes an action with higher value, the others want to do the same. The game has many applications in wireless networking and radio resource allocation, such as the power control game. Moreover, the super-modular game is analytically appealing with many properties and good behaviors for learning algorithms.

First, let us define the super-modular game as follows:

Definition 5.37 *Let A be the action and u be the utility for an N user game. The game $G(A_1, \ldots, A_N; u_1, \ldots, u_N)$ is a* **super-modular game** *if, for all i,*

- *A_i is a compact subset of $\mathbb{R}$;*
- *u_i is upper semi-continuous in (A_i, A_{-i}); and*
- *u_i has increasing differences in (A_i, A_{-i}).*

The major properties of super-modular games are explained in the following theorem and properties:

Theorem 5.3 *Let $G(A, u)$ be a super-modular game. The set of strategies surviving iterated strict dominance has the greatest and the smallest element $\overline{A}$, $\underline{A}$, and $\overline{A}$, $\underline{A}$ are both Nash equilibria. In other words,*

- *a pure strategy of Nash equilibrium exists;*
- *the largest and smallest strategies compatible with iterated deletion, rationalizability, and Nash equilibrium are the same;*
- *if a super-modular game has a unique Nash equilibrium, then it is dominance solvable, and lots of adjustment rules will converge to it, e.g. best-response dynamics.*

Proposition 5.1 *Suppose a super-modular game is indexed by t. The largest and smallest Nash equilibria are increasing in t.*

Proposition 5.2 *Suppose a super-modular game with positive spillovers (for all i, $u_i(A_i, A_{-i})$ is increasing in A_{-i}). Then the largest Nash equilibrium is Pareto-preferred.*

Example applications of super-modular games include the following:

- *Investment game*: Suppose N firms simultaneously make investments $A_i \in \{0, 1\}$ and payoffs are

$$u_i(A_i, A_{-i}) = \begin{cases} \pi(\sum A_i) - k, & A_i = 1 \\ 0, & A_i = 0, \end{cases} \tag{5.29}$$

  where π is increasing in aggregate investment and k is a constant.
- *Bertrand competition*: Suppose N firms simultaneously set prices and that

$$D_i(p_i, p_{-i}) = a_i - b_i p_i + \sum_{j \neq i} d_{ij} p_j, \tag{5.30}$$

  where $b_i, d_{ij} > 0$.
- *Cournot duopoly*: Super-modular if A_1 is firm 1's quantity, and A_2 is the negative of firm 2's quantity.
- *Diamond search model*: N agents who exert efforts looking for trading partners. Let A_i denote the effort of agent i, and $c(A_i)$ cost of this effort, where c is increasing and

continuous. The probability of finding a partner is $A_i \sum_{j \neq i} A_j$. Then

$$u_i(A_i, A_{-i}) = A_i \sum_{j \neq i} A_j - c(A_i) \tag{5.31}$$

has increasing differences in (A_i, A_{-i}). Hence it is super-modular.

5.7.4 Correlated equilibrium

A special kind of equilibrium, called correlated equilibrium, is described in this subsection. In 2006, Robert J. Aumann was awarded the Nobel Prize for introducing the concept of correlated equilibrium [272, 273]. Unlike Nash equilibrium, in which each player only considers his own strategy, correlated equilibrium achieves better performance by allowing each user to consider the joint distribution of players' actions. In other words, each player needs to consider the other players' behaviors to determine whether there are mutual benefits to explore. The idea is that a strategy profile is chosen randomly according to a certain distribution. Given the recommended strategy, it is in the players' best interests to conform with this strategy. The distribution is called the correlated equilibrium. It has been shown that the correlated equilibrium can yield a better payoff than that of the convex hull of the Nash equilibria.

If a player follows an action in every possible attainable situation in a game, the action is called pure strategy, in which the probability of using action v_l, $p(r_i^n = v_l)$, has only one non-zero value 1 for all l. In the case of mixed strategies, the player will follow a probability distribution over different possible actions, i.e. different rate l.

Next, let us define the correlated equilibrium. Let $\mathbb{G} = \{K, (\Omega_i)_{i \in K}, (u_i)_{i \in K}\}$ be a finite K-player game in strategic form, where Ω_i is the strategy space for player i, and u_i is the utility function for player i. Define Ω_{-i} as the strategy space for player i's opponents. Let us denote the action for player i and his opponents as $\mathbf{r}_i$ and $\mathbf{r}_{-i}$, respectively. Then, the correlated equilibrium is defined as:

Definition 5.38 *A probability distribution p is a correlated strategy of game $\mathbb{G}$, if and only if, for all $i \in K$, $\mathbf{r}_i \in \Omega_i$, and $\mathbf{r}_{-i} \in \Omega_{-i}$,*

$$\sum_{\mathbf{r}_{-i} \in \Omega_{-i}} p(\mathbf{r}_i, \mathbf{r}_{-i})[u_i(\mathbf{r}_i', \mathbf{r}_{-i}) - u_i(\mathbf{r}_i, \mathbf{r}_{-i})] \leq 0, \forall \mathbf{r}_i' \in \Omega_i. \tag{5.32}$$

By dividing inequality in (5.32) with $p(\mathbf{r}_i) = \sum_{\mathbf{r}_{-i} \in \Omega_{-i}} p(\mathbf{r}_i, \mathbf{r}_{-i})$, we have

$$\sum_{\mathbf{r}_{-i} \in \Omega_{-i}} p(\mathbf{r}_{-i} | \mathbf{r}_i)[u_i(\mathbf{r}_i', \mathbf{r}_{-i}) - u_i(\mathbf{r}_i, \mathbf{r}_{-i})] \leq 0, \forall \mathbf{r}_i' \in \Omega_i. \tag{5.33}$$

The inequality (5.33) means that when the recommendation to player i is to choose action $\mathbf{r}_i$, then choosing action $\mathbf{r}_i'$ instead of $\mathbf{r}_i$ cannot obtain a higher expected payoff to player i.

Note that the set of correlated equilibria is nonempty, closed, and convex in every finite game. Moreover, it may include the distribution that is not in the convex hull of the Nash equilibrium distributions. In fact, every Nash equilibrium is a correlated equilibrium and Nash equilibria correspond to the special case where $p(\mathbf{r}_i, \mathbf{r}_{-i})$ is a product of

Table 5.7 Two secondary players' game: (a) reward table, (b) Nash equilibrium, (c) mixed Nash equilibrium (lower left), and (d) correlated equilibrium.

	0	1
0	(5,5)	(6,3)
1	(3,6)	(0,0)

(a)

	0	1
0	0	(0 or 1)
1	(1 or 0)	0

(b)

	0	1
0	9/16	3/16
1	3/16	1/16

(c)

	0	1
0	0.6	0.2
1	0.2	0

(d)

each individual player's probability for different actions, i.e. the play of the different players is independent [272, 273, 274]. The correlated equilibrium can be calculated via linear programming. If only local information is available, learning algorithms such as non-regret learning can achieve the correlated equilibrium by the probability of 1 [274].

In Table 5.7, an example is shown for two secondary players with different actions. Table 5.7(a) lists the utility function for two players taking action 0 and 1. It can be observed that when two players take action 0, they have the highest aggregated benefit. We can see this action as a cooperative action, or in our case the secondary players act less aggressively. But if any player plays more aggressively using action 1 while the other still plays action 0, the aggressive player has a better utility, but the other player has a lower utility and the overall benefit is reduced. In this case, the aggressive player can achieve a higher payoff. However, if both players play aggressively using action 1, they both obtain very low utilities. Table 5.7(b) shows two Nash equilibria, where one player dominates the other. The dominating player has a utility of 6 and the dominated player has a utility of 3, which is unfair. Table 5.7(c) shows the mixed Nash equilibrium where two players have a probability of 0.75 for action 0 and 0.25 for action 1, respectively. The utility for each player is 4.5.

In Table 5.7(b) and (c), the Nash equilibria and mixed Nash equilibria are all within the set of correlated equilibria. In Table 5.7(d), we show an example where the correlated equilibrium is outside the convex hull of the Nash equilibrium. Note that the joint distribution is not the product of two players' probability distributions, i.e. the actions of two players are not independent. Moreover, the utility for each player is 4.8, which is higher than that of the mixed strategy.

5.8 Summary

In this chapter, an introduction to game theory has been presented. This serves as a background for developing game-theoretic models for dynamic spectrum access in a cognitive radio environment. In a cognitive radio environment, unlicensed users are rational to maximize their benefits (i.e. payoffs), and they may cooperate or compete with each other in order to achieve their desired objectives. First, the basics of non-cooperative game theory and the concepts of Nash equilibrium and Pareto optimality have been discussed. Then, the basics of a repeated game has been presented. A brief introduction to stochastic game models has been provided. Next, the concept of cooperative games (i.e. the bargaining game and coalition game) has been presented. Several applications of cooperative game models to wireless networking problems have been discussed. Game models with imperfect information such as Bayesian games and several other special types of games including zero-sum game, potential game, and super-modular game have been introduced. To this end, the concept of correlated equilibrium has been discussed.

6 Intelligent algorithms

Cognitive radios need to have the ability to learn and adapt their wireless transmission according to the ambient radio environment. Intelligent algorithms such as those based on machine learning, genetic algorithms, and fuzzy control are therefore key to the implementation of cognitive radio technology. In general, these algorithms are used to observe the state of the wireless environment and build knowledge about the environment. This knowledge is used by a cognitive radio to adapt its decision on spectrum access. For example, a cognitive radio (i.e. an unlicensed user) can observe the transmission activity of primary (i.e. licensed) users on different channels. This enables the cognitive radio to build knowledge about the licensed users' activity on each channel (e.g. how often and how long the channel will be occupied by the licensed user). This knowledge is then used by the cognitive radio to decide which channel to access so that the desired performance objectives can be achieved (e.g. throughput is maximized while the interference or collision caused to the licensed users is maintained below the target level).

Applications of different learning and intelligent algorithms to a cognitive radio system were summarized in [2] (Figure 6.1). These algorithms can be used for learning/reasoning and decision/adaptation in cognitive radio systems. In this chapter, three major categories of intelligent algorithm, namely, machine learning, genetic algorithm, and fuzzy logic-based algorithms, are discussed. Machine learning can be divided into three categories: supervised learning, unsupervised learning, and reinforcement learning. The related algorithms and their applications to cognitive radio are discussed in Sections 6.1–6.3. The basics of genetic algorithms and their application to cognitive radio are presented in Section 6.4. Fuzzy logic control and its application to cognitive radio is discussed in Section 6.5. A summary of the chapter is given in Section 6.6.

6.1 Machine learning

Machine learning is one of the disciplines in artificial intelligence. It deals with the design and development of learning algorithms using sample data or past experience to optimize the performance of a system given certain objectives and constraints. Machine learning utilizes the theories of mathematics and statistics to construct inference models from sample data so that algorithms can be designed based on these models. Two major steps in machine learning are training and making inference/decision. In the training

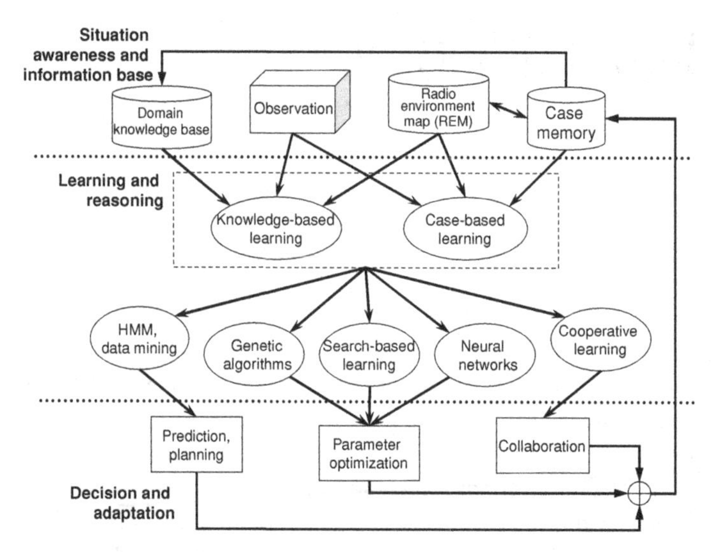

Figure 6.1 Applications of intelligent algorithms in a cognitive radio system [2].

step, sample data or past experience are used to build knowledge about the environment or the system. In this step, efficient algorithms are required to extract useful information from raw data. Once the knowledge is built, an inference/decision is made based on the available knowledge and current state and input data.

Machine learning techniques have been applied to solve problems related to natural language processing, pattern recognition, and robotics. For example, in natural language processing, machine learning is used to understand and also to generate natural human language. In this application, the algorithm analyzes the structure of the input data which is in the form of human language. The meaning is extracted and used so that the system can react properly. In pattern recognition, machine learning is applied to recognize a specific pattern in the input data (e.g. human face or speech recognition). In robotics, machine learning is used by autonomous agents to observe the environment and perform the necessary tasks.

6.1.1 Supervised learning

In supervised learning, algorithms are developed to learn and extract knowledge from a set of training data. The training data is composed of inputs and corresponding outputs. The output can be a continuous value for a regression problem, or can be a discrete value for a classification problem. The objective of supervised learning is to predict the output given any valid input. For example, in the cognitive radio context, the input can be the number of unlicensed users and the output can be the level of interference temperature at the licensed users. Given the historical data on these inputs and outputs, the knowledge about the interference temperature can be constructed. This knowledge can be used to predict the level of interference when a new unlicensed user is allowed to transmit. For this, a mathematical model is developed and the parameters of the model are obtained by using training data (i.e. training the model).

Regression technique

For supervised learning where the output is a continuous value, regression is used to analyze the training data consisting of one or more inputs and one output (i.e. dependent and independent variables, respectively). In this case, the output is modeled as a function of the inputs. This function is referred to as a regression equation. This regression equation consists of the regression parameters and noise. This noise is considered to be a random variable, while the parameters are constants which are estimated so that the output provides a best fit to the numerical data. The fitness of the model to the numerical data can be measured by using different methods (e.g. least square). Then, the regression equation is used to generate the output for any arbitrary input.

The set of training data is denoted by $\mathbb{X} = \{\mathbf{x}_n, r_n\}_{n=1}^{N}$ where $\mathbf{x}_n$ is a vector of inputs and r_n is the output corresponding to input $\mathbf{x}_n$. The number of elements in this set $\mathbb{X}$ is N. The regression equation is given by

$$r_n = \mathscr{F}(\mathbf{x}) + \epsilon, \tag{6.1}$$

where ϵ denotes the noise. Since the noise cannot be observed, the hidden variables $\mathbf{y}$ are added into the equation as follows:

$$r_n = \mathscr{F}(\mathbf{x}_n, \mathbf{y}_n). \tag{6.2}$$

The error which indicates the fitness of the model to a set of training data $\mathbb{X}$ can be obtained from

$$\mathrm{Er}(\mathscr{F}|\mathbb{X}) = \frac{1}{N}\sum_{n=1}^{N}(r_n - \mathscr{F}(\mathbf{x}_n))^2 . \tag{6.3}$$

The objective is to find $\mathscr{F}(\cdot)$ which minimizes the error. This $\mathscr{F}(\cdot)$ can be chosen to be a linear function as follows:

$$\mathscr{F}(\mathbf{x}) = w_0 + w_1 x_1 + \cdots + w_J x_J = \sum_{j=1}^{J} w_j x_j + w_0, \tag{6.4}$$

where w_j is the weight used to construct a function $\mathscr{F}(\cdot)$ and J is the total number of inputs. Let us consider a simple case of one input x and one output r. Then, the error can be expressed as follows:

$$\mathrm{Er}(w_1, w_0|\mathbb{X}) = \sum_{n=1}^{N}(r_n - (w_1 x_n + w_0))^2 . \tag{6.5}$$

We can minimize this error by taking the derivatives of an error function with respect to w_1 and w_0. Then, these derivatives are set to be zeros, and the unknown values of w_1 and w_0 are solved. In this way, we obtain

$$w_1 = \frac{\sum_{n=1}^{N} x_n r_n - \overline{x}\,\overline{r}N}{\sum_{n=1}^{N}(x_n)^2 - \overline{x}^2 N}, \qquad w_0 = \overline{r} - w_1\overline{x}, \tag{6.6}$$

where $\overline{x} = \sum_{n=1}^{N} x_n/N$, and $\overline{r} = \sum_{n=1}^{N} r_n/N$. An example of training data and its estimated regression function are shown in Figure 6.2.

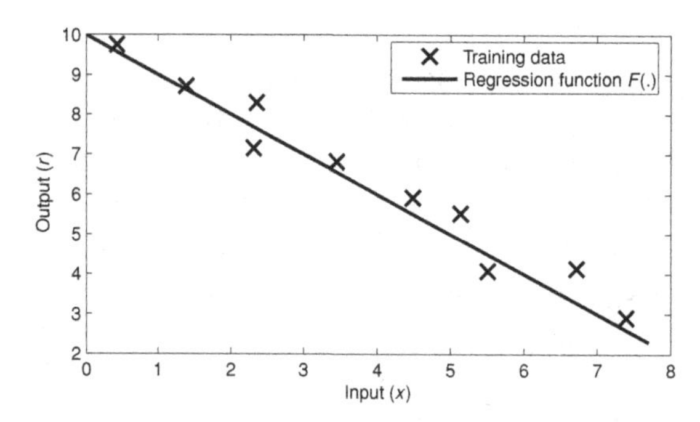

Figure 6.2 Training data and regression function.

Classification technique

For a classification problem, the output of supervised learning is a discrete value. The objective of a classification problem is to assign new inputs to one of the number of discrete classes or categories. The basic technique used for this classification problem is based on Bayesian rule. Given a large set of training data, the a priori probability $P(C_k)$ of the input to be in class C_k can be obtained. Basically, this a priori probability is given by the fraction of the input data in each class. Let y denote the feature variable of the input to be classified. This feature variable can be obtained as a function of input x. Let us assume further that the value of this feature variable is in a discrete set denoted by $\{Y_l\}$. The joint probability $P(C_k, Y_l)$ is the probability that inputs with feature value Y_l belong to class C_k. From a set of training data, this probability is obtained as the fraction of the training data with feature value Y_l and classified into class C_k. Then, the conditional probability $P(Y_l|C_k)$ is used to determine the probability that the input has value Y_l given that the input belongs to class C_k. This conditional probability is the fraction of the training data with class C_k which has a value of Y_l. The relationship between the joint probability, conditional probability, and prior probability can be expressed as follows:

$$P(C_k, Y_l) = P(Y_l|C_k)P(C_k). \tag{6.7}$$

Similarly, we have

$$P(C_k, Y_l) = P(C_k|Y_l)P(Y_l), \tag{6.8}$$

where $P(C_k|Y_l)$ is the probability that the input belongs to class C_k given that the feature value of input is Y_l, and $P(Y_l)$ is the probability that the input has a feature value of Y_l. This probability is the fraction of the total number of training data with feature value Y_l. Therefore, we obtain

$$P(C_k|Y_l) = \frac{P(Y_l, C_k)P(C_k)}{P(Y_l)}, \tag{6.9}$$

which is Bayesian rule. In particular, $P(C_k|Y_l)$ is defined as the *posterior* probability, since it indicates the probability that the input will be in class C_k if the feature value is Y_l. Since we can obtain $P(Y_l, C_k)$, $P(C_k)$, and $P(Y_l)$ from a set of training data, the posterior probability $P(C_k|Y_l)$ can be easily calculated. To classify the new input data,

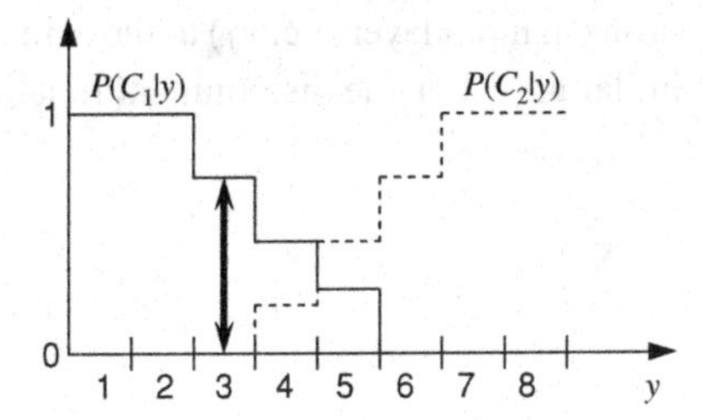

Figure 6.3 Classification example in which y is the feature variable of the input x.

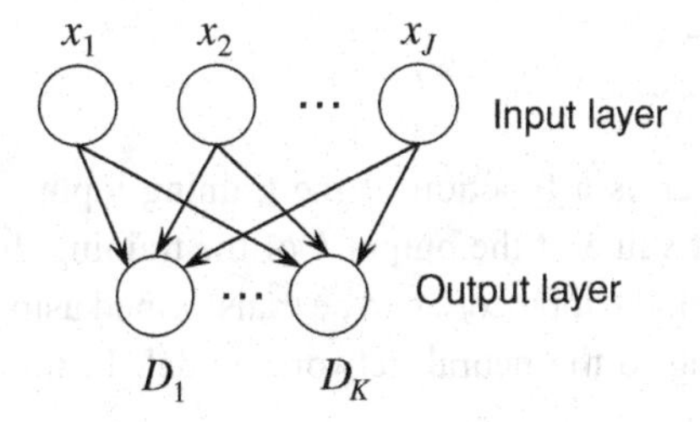

Figure 6.4 A neural network model with J inputs and K outputs.

we can measure its feature value y. The classification will be performed such that the probability of misclassification is minimized. This corresponds to the case that class C_k is assigned if the posterior probability $P(C_k|Y_l)$ is the largest. For example, with two classes (e.g. C_1 and C_2) and one input (e.g. y), the posterior probabilities are shown in Figure 6.3. With these posterior probabilities, which are obtained from a set of training data, if the new feature value of input is 3, the class of this input should be C_1 to minimize the misclassification error.

To classify the input formally, a discriminant function $\mathscr{D}_k(\mathbf{x})$ is introduced. In particular, the input $\mathbf{x}$ will belong to class k if $\mathscr{D}_k(\mathbf{x}) > \mathscr{D}_{k'}(\mathbf{x})$ for all $k' \neq k$. In an above example, the discriminant function can be chosen as $\mathscr{D}_k(\mathbf{x}) = P(C_k|\mathbf{x})$.

Neural network

Neural network is a technique in machine learning which can be used to model complex relationships between inputs and outputs. The neural network model is inspired by biological neural networks (e.g. the human brain), which are composed of a group of connected neurons or perceptrons. An example of a three-layer neural network is shown in Figure 6.4 in which the input and output layers have J and K neurons, respectively. This neural network model corresponds to the discriminant function of class k,

$$\mathscr{D}_k(\mathbf{x}) = \mathbf{w}_k^{\mathrm{T}}\mathbf{x} + w_{k,0}, \tag{6.10}$$

where $\mathbf{w}_k$ is a weight vector and parameter $w_{k,0}$ is a bias. The boundary for separating class k and k' is the hyperplane given by

$$(\mathbf{w}_k - \mathbf{w}_{k'})^{\mathrm{T}}\mathbf{x} + (w_{k,0} - w_{k',0}) = 0. \tag{6.11}$$

In this case, each line connecting the neurons in input layer (i.e. x_j) to those in the output layer (i.e. $\mathscr{D}_k(\cdot)$) corresponds to $w_{k,j}$. Similar to (6.10), the discriminant function can be expressed as follows:

$$\mathscr{D}_k(\mathbf{x}) = \sum_{j=0}^{J} w_{k,j} x_j, \tag{6.12}$$

where $x_1 = 1$.

A training process is applied to obtain the weights $w_{k,j}$. The standard method is to minimize the sum-of-squares error, which can be expressed as follows:

$$\text{Er}(\mathbf{w}) = \frac{1}{2} \sum_{n=1}^{N} \sum_{k=1}^{K} \left(\mathscr{D}_k(\mathbf{x}_n; \mathbf{w}) - r_{n,k} \right)^2, \tag{6.13}$$

where $\mathscr{D}_k(\mathbf{x}_n; \mathbf{w})$ is the output k, which is a function of the training input $\mathbf{x}_n$ and the weight vector $\mathbf{w}$; and $r_{n,k}$ is the target value of the output k of the training data n. The size of the training data is N. In general, the input $x_{j,n}$ can be transformed using function $\phi(\cdot)$ into $\hat{x}_{n,j} = \phi(x_{j,n})$ before feeding to the neural network model. In this case, the sum-of-squares error function becomes

$$\text{Er}(\mathbf{w}) = \frac{1}{2} \sum_{n=1}^{N} \sum_{k=1}^{K} \left(\sum_{j=0}^{J} w_{k,j} \hat{x}_{j,n} - r_{n,k} \right)^2. \tag{6.14}$$

To minimize this error, $\text{Er}(\mathbf{w})$ is differentiated with respect to $w_{k,j}$ and the derivative is set to zero. This gives the following set of equations:

$$\frac{\partial \text{Er}}{\partial w_{k,j}} = \sum_{n=1}^{N} \left(\sum_{j'=0}^{J} w_{k,j'} \hat{x}_{j',n} - r_{n,k} \right) \hat{x}_{j,n} = 0, \quad \forall j, k. \tag{6.15}$$

This set of equations can be written in matrix form as follows:

$$\left(\hat{\mathbf{X}}^{\mathrm{T}} \hat{\mathbf{X}} \right) \mathbf{W}^{\mathrm{T}} = \hat{\mathbf{X}}^{\mathrm{T}} \mathbf{R}, \tag{6.16}$$

where $\hat{\mathbf{X}}$ is a matrix of $\hat{x}_{j,n}$ with size $J \times N$, $\mathbf{W}$ is a matrix of $w_{k,j}$ with size $K \times J$, and $\mathbf{R}$ is a matrix of $r_{n,k}$ with size $N \times K$. In this case, we can obtain

$$\mathbf{W}^{\mathrm{T}} = \hat{\mathbf{X}}^{\dagger} \mathbf{R}, \tag{6.17}$$

where $\hat{\mathbf{X}}^{\dagger}$ is the pseudo-inverse of $\hat{\mathbf{X}}$, which is obtained from

$$\hat{\mathbf{X}}^{\dagger} = (\hat{\mathbf{X}}^{\mathrm{T}} \hat{\mathbf{X}})^{-1} \hat{\mathbf{X}}^{\mathrm{T}}. \tag{6.18}$$

However, in practice $\hat{\mathbf{X}}^{\mathrm{T}} \hat{\mathbf{X}}$ could be singular or nearly singular for which it would be difficult to obtain the inverse. Therefore, the solution (i.e. $w_{k,j}$) can be obtained using the gradient method. In this case, the weights are iteratively computed as follows:

$$w_{k,j}(t+1) = w_{w,j}(t) - \alpha \frac{\partial \text{Er}}{\partial w_{k,j}}, \tag{6.19}$$

where t indicates the index of iteration and α is the learning rate.

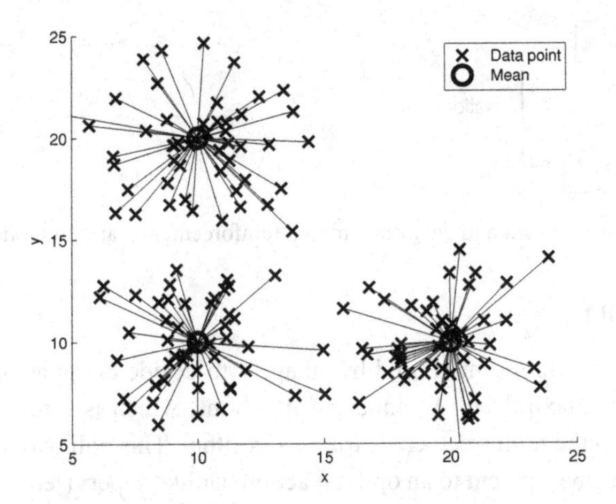

Figure 6.5 Example results obtained from the K-means algorithm.

6.1.2 Unsupervised learning

In supervised learning, there is a set of training data consisting of some inputs and their corresponding known/correct outputs. In contrast, in unsupervised learning there is no such correct output. The objective of unsupervised learning is to identify the structure of the input data. This structure can be identified if certain patterns occur more often than others. The basic method in unsupervised learning is clustering. This clustering is used to find the groups of inputs which have similarity in their characteristics.

One of the clustering algorithms is the K-means algorithm. In this algorithm, the number of groups K is provided in advance. With T data points, the algorithm will partition these data points into K disjoint subsets. Each subset $\mathbb{X}_k$ contains T_k data points. The objective of this clustering algorithm is to minimize the sum-of-squares clustering function, which is defined as follows:

$$S = \sum_{k=1}^{K} \sum_{t \in \mathbb{X}_k} ||\mathbf{x}_t - \boldsymbol{\mu}_k||^2, \tag{6.20}$$

where $\mathbf{x}_t$ is the data point and $\boldsymbol{\mu}_k$ is the mean of the data points in subset $\mathbb{X}_k$. This mean is obtained from

$$\boldsymbol{\mu}_k = \frac{1}{T_k} \sum_{t \in \mathbb{X}_k} \mathbf{x}_t. \tag{6.21}$$

The algorithm works as follows. First, the data points are randomly assigned to subset $\mathbb{X}_k$. Then, the mean is computed as in (6.21). Next, each point is re-assigned to a new subset whose mean is the nearest. The means of all subsets are then re-computed. These steps are repeated until there is no change in the grouping of the data points. An example of the results obtained from the K-means algorithm is shown in Figure 6.5.

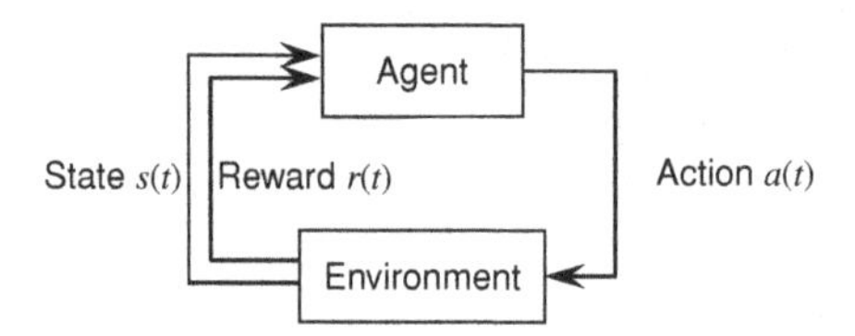

Figure 6.6 Interaction between agent and environment in a reinforcement learning model.

6.1.3 Reinforcement learning

A reinforcement learning algorithm is used by an agent to decide on an action so that its long-term reward is maximized. A sequence of actions, which is referred to as a policy, is obtained using the reinforcement learning algorithm. This policy is a mapping between the state of an environment to an optimal action. Unlike supervised learning, in reinforcement learning an agent will not know the inputs and their corresponding correct output. Also, the objective of reinforcement learning is different from that of supervised and unsupervised learning, because with reinforcement learning an agent focuses on the on-line performance rather than the off-line performance. Since the knowledge about the environment is not known a priori, an agent has to explore all the feasible actions and their consequences (i.e. exploration). Thereby, the agent can build the knowledge and then exploit the knowledge to achieve its objective (i.e. exploitation).

Since reinforcement learning can be used without training data and because it aims to maximize the long-term on-line performance, it is particularly suitable for a cognitive radio network. Specifically, an unlicensed user can use a reinforcement learning algorithm to explore the possible transmission strategies and exploit the knowledge thus obtained to adapt its transmission parameters to achieve the desired objective (e.g. maximize throughput) while the constraints (e.g. on the interference temperature limit) are satisfied. More detailed discussions on reinforcement learning and its applications to distributed wireless communications and networking are presented in the following sections.

6.2 Reinforcement learning models and algorithms

With reinforcement learning, an agent making a decision will map the situation (i.e. state) into the action to maximize the reward (or minimize the cost) [171]. In this case, the agent does not know which action will yield the highest reward, but it will learn by interacting with the environment. The outcome (i.e. reward and change of state) from taking any action is observed and the agent will generalize this outcome into the knowledge and utilize it for making decisions in the future (Figure 6.6). The agent continually learns and makes decisions, while the environment responds to an action from the agent by changing its state.

Specifically, the agent and the environment interact over a sequence of discrete time steps $t = 0, 1, 2, 3, \ldots$. At each time step t, the agent can observe the system state $s(t)$, which can be a composite state of the agent and the environment. State $s(t)$ can be defined as $s(t) \in \mathbb{S}$, where $\mathbb{S}$ is the state space. The agent makes a decision and performs

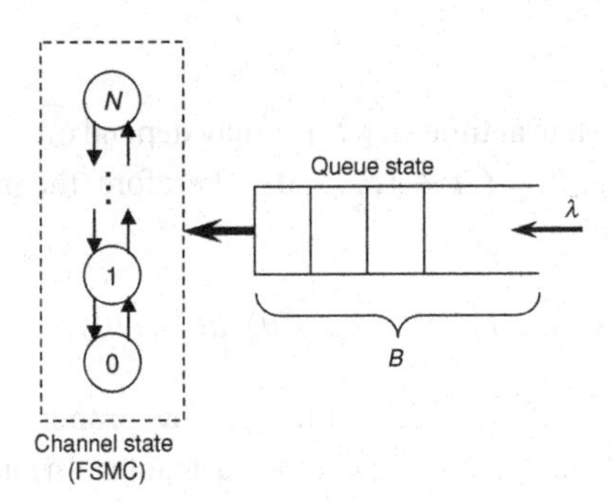

Figure 6.7 Example of wireless transmission.

a selected action $a \in \mathbb{A}(s(t))$, where $\mathbb{A}(s(t))$ is the set of feasible actions at state $s(t)$. In the next time step $t+1$, the agent receives the reward (i.e. the consequence of the action). This reward is defined as $r(t+1) \in \mathbb{R}$. Also, the agent can observe the next state $s(t+1)$. The state transition from $s(t)$ to $s(t+1)$ depends on the action taken by the agent. To make a decision in each time step, the agent uses a policy π. The policy π of the agent is a mapping of situation or state to action. In particular, given the state of the system, a policy will tell the agent what action to take. The policy is considered to be a solution of reinforcement learning since it will affect the long-term reward of the agent. The policy can be deterministic (i.e. given a state, a certain action will be taken) or stochastic (i.e. given a state, any action is taken randomly according to a certain probability distribution). While a reward is determined as a short-term objective of performing an action at a particular state, a value function is a long-term objective achieved by a policy. The value of a state is defined as the total reward that an agent accumulates.

Let us consider a simple time-slot-based wireless transmission scenario where a mobile transmits data to a receiver. A queue is used in the transmitter to buffer data packets from the application in the upper layer of the communication protocol stack. Fading in the wireless channel will result in a time-varying transmission rate (Figure 6.7). Let us consider a slowly fading wireless channel which can be modeled by a finite-state Markov chain (FSMC). Since the energy supply of the mobile is limited, but at the same time the delay of the packets waiting in the queue has to be minimized, the transmitter has to make a decision on whether to transmit or not given the channel and queue state in a time-slot. The objective of the transmitter is to minimize the long-term cost due to energy consumption and delay.

For the above decision problem on wireless transmission, a reinforcement learning problem can be formulated as follows: the state space is defined as $\mathbb{S} = \{(\mathcal{C}, \mathcal{X}), \mathcal{C} \in \{0, 1, \ldots, N\}, \mathcal{X} \in \{0, 1, \ldots, B\}\}$, where $\mathcal{C}$ and $\mathcal{X}$ denote the channel state and the queue state, respectively. N and B are the maximum number of channel states and queue states, respectively. The set of actions is $\mathbb{A}(s) = \{0, 1\}$ $\forall s \in \mathbb{S}$ where 1 and 0 indicate transmitting and not transmitting data, respectively. The reward is defined as $r = -(a\eta + \mathcal{X}/\lambda)$, where a is the action, η is the amount of energy required for transmitting a packet, and λ is the average packet arrival rate.

The methods of obtaining the solution (i.e. the optimal policy) for a reinforcement learning problem formulation will be discussed in the following sections.

6.2.1 Value function

In a general scenario, transition from a state at time step $t+1$ may depend on all previous states and actions (i.e. $s(t')$ and $a(t')$ for $t' = t, t-1, \ldots, 0$). Therefore, the probability of transition can be defined as follows:

$$P\left(s(t+1) = s', r(t+1) = r | s(t), a(t), r(t), \ldots, s(\theta), a(0), r(0)\right). \tag{6.22}$$

However, if the state transitions of a system follow the Markov property, then the transition of a state depends only on the current state, not the past states. In such a case, the probability of transition can be defined as follows:

$$P\left(s(t+1) = s', r(t+1) = r | s(t), a(t)\right). \tag{6.23}$$

With the Markov property of the system states, the state and expected reward in the next time step can be completely predicted by using the knowledge of current state and action. Consequently, the decision and the value can be assumed to be functions only of the current state. With the Markov property, the reinforcement learning model becomes similar to a Markov decision process (MDP). Given a policy π, the expected reward function can be defined as follows:

$$V^{\pi}(s) = E_{\pi}\left(R(t)|s(t) = s\right) = E_{\pi}\left(\sum_{t'=0}^{\infty} \beta^{t'} r(t+t'+1) \middle| s(t) = s\right), \tag{6.24}$$

where $E_{\pi}(\cdot)$ is the expectation operator over the policy π and $\beta^{t'}$ is a discount factor. In reinforcement learning, this V^{π} is called the state-value function for policy π. Alternatively, if an action is taken, the expected reward function can be defined as follows:

$$\begin{aligned} Q^{\pi}(s,a) &= E_{\pi}\left(R(t)|s(t) = s, a(t) = a\right) \\ &= E_{\pi}\left(\sum_{t'=0}^{\infty} \beta^{t'} r(t+t'+1) \middle| s(t) = s, a(t) = a\right), \end{aligned} \tag{6.25}$$

which is called the action-value function for policy π.

The solution of the reinforcement learning model is an optimal policy which maximizes the long-term reward. Based on the partial ordering theory, the policy π is defined to be better than or equal to a policy π' if its expected reward is greater than or equal to that of π' for all states. That is, $\pi \geq \pi'$ if and only if $V^{\pi}(s) \geq V^{\pi'}(s), \forall s \in \mathbb{S}$. Given this partial ordering policy, an optimal policy π^* is defined as the policy (at least one) that is better than or equal to all other policies. This can be defined as

$$V^*(s) = \max_{\pi} V^{\pi}(s), \quad \forall s \in \mathbb{S}, \tag{6.26}$$

$$Q^*(s,a) = \max_{\pi} Q^{\pi}(s,a), \quad \forall s \in \mathbb{S}, \forall a \in \mathbb{A}(s) \tag{6.27}$$

for optimal state-value function, and optimal action-value function, respectively. Based on the Bellman's optimality equation, the optimal state-value function can be expressed as

$$V^*(s) = \max_{a \in \mathbb{A}(s)} Q^{\pi^*}(s, a)$$
$$= \max_a \sum_{s'} P(s'|s, a) \left(R(s'|s, a) + \beta V^*(s') \right) \quad (6.28)$$

where $P(s'|s, a)$ and $R(s'|s, a)$ are the probability and reward given that at current state s, action a is taken, and the system changes to state s'. The optimal action-value function can be expressed as

$$Q^*(s, a) = \sum_{s'} P(s'|s, a) \left(R(s'|s, a) + \beta \max_{a'} Q^*(s', a') \right). \quad (6.29)$$

In this case, given $P(s'|s, a)$ and $R(s'|s, a)$ of the system, $V^*(s)$ and $Q^*(s, a)$ can be obtained by solving a set of non-linear equations. This $V^*(s)$ can be used to obtain the optimal policy. That is, for each state, there will be at least one action for which $V^*(s)$ is maximized. The policy which assigns non-zero probability to those actions will be an optimal policy. This can be achieved by performing a one-step search or a greedy search, since the solution of this search will be optimal given an optimal $V^*(s)$. Note that this one-step search will maximize the long-term reward since the state-value function has already taken the reward in the subsequent step into account. $Q^*(s, a)$ can be obtained in a similar way. However, in this case, a one-step search is not required, since the expression of $\max_{a'} Q^*(s', a')$ is already in the calculation of $Q^*(s, a)$. Again, this $Q^*(s', a')$ provides an optimal long-term reward for the agent.

6.2.2 Dynamic programming

Dynamic programming is one of the approaches that can be used to obtain the optimal policy. This approach uses a set of algorithms which can be applied when perfect knowledge of the system is available.

To find an optimal policy, the value of any policy has to be obtained. This state-value function is defined as follows:

$$V^\pi(s) = \sum_a \pi(s, a) \sum_{s'} P(s'|s, a) \left(R(s'|s, a) + \beta V^\pi(s') \right). \quad (6.30)$$

$V^\pi(s)$ can be obtained by solving the above set of equations (i.e. policy evaluation). Alternatively, it can be obtained iteratively from Algorithm 6.1.

After the value function for a policy is known, a policy improvement can be applied to obtain a better policy than π. In this case, a policy which is better than π can be determined by evaluating the action-state value of taking action a at state s which is defined as

$$Q^\pi(s, a) = \sum_{s'} P(s'|s, a) \left(R(s'|s, a) + \beta V^\pi(s') \right). \quad (6.31)$$

Let $\pi'(s)$ denote the policy corresponding to taking action a at state s. If $Q^\pi(s, \pi'(s)) \geq V^\pi(s)$, then policy π' can provide an identical or better long-term reward than that of π,

Algorithm 6.1 Iterative policy evaluation algorithm.

Input: Policy π to be evaluated

1: $V(s) = 0, \forall s \in \mathbb{S}$ {Initialize}
2: **repeat**
3: $\quad \hat{v} = 0$
4: $\quad$ **for all** $s \in \mathbb{S}$ **do**
5: $\quad\quad v = V(s)$
6: $\quad\quad V(s) = \sum_a \pi(s,a) \sum_{s'} P(s'|s,a)(R(s'|s,a) + \beta V(s'))$
7: $\quad\quad \hat{v} = \max(\hat{v}, |v - V(s)|)$
8: $\quad$ **end for**
9: **until** $\hat{v}$ is smaller than a threshold

Output: V^π

i.e. $V^{\pi'}(s) \geq V^\pi(s)$. This alternative policy π' can be determined as follows:

$$\begin{aligned} \pi'(s) &= Q^\pi(s,a) \\ &= \arg\max_a \sum_{s'} P(s'|s,a)\left(R(s'|s,a) + \beta V^\pi(s')\right), \end{aligned} \tag{6.32}$$

where π' is referred to as a greedy policy.

The policy evaluation and policy improvement can be combined into a policy iteration algorithm. In particular, a policy evaluation is used to determine the value of the policy. Then, policy improvement is applied to find a better policy. These steps are repeated until a better policy cannot be found. This policy iteration is described in Algorithm 6.2. It was shown that this policy iteration algorithm always converges to an optimal policy [171].

Algorithm 6.2 Policy iteration algorithm.

1: Initialize $V(s)$ and $\pi(s), \forall s \in \mathbb{S}$
2: **repeat**
3: $\quad$ **repeat**
4: $\quad\quad \hat{v} = 0$ {Perform policy evaluation}
5: $\quad\quad$ **for all** $s \in \mathbb{S}$ **do**
6: $\quad\quad\quad v = V(s)$
7: $\quad\quad\quad V(s) = \sum_a \pi(s,a) \sum_{s'} P(s'|s,a)(R(s'|s,a) + \beta V(s'))$
8: $\quad\quad\quad \hat{v} = \max(\hat{v}, |v - V(s)|)$
9: $\quad\quad$ **end for**
10: $\quad$ **until** $\hat{v}$ is smaller than a threshold
11: $\quad$ Initialize $\hat{\pi}(s)$ for $\hat{\pi}(s) \neq \pi(s)$ {Perform policy improvement}
12: $\quad$ **while** $\hat{\pi}(s) \neq \pi(s)$ **do**
13: $\quad\quad \hat{\pi}(s) = \pi(s)$
14: $\quad\quad \pi(s) = \arg\max_a \sum_{s'} P(s'|s,a)(R(s'|s,a) + \beta V(s'))$
15: $\quad$ **end while**
16: **until** There is no change in policy π

6.2.3 Monte Carlo methods

When perfect knowledge about the system is not available, the dynamic programming approach cannot be applied to obtain the optimal policy. In such a case, the Monte Carlo method can be applied when the experiences of the system are available to an agent. This experience is defined as the sequences of states, actions, and rewards which can be obtained from simulations of the system. Although in this simulation-based approach the model of the system (i.e. agent and environment) is required, only sample transitions can be used to obtain an optimal policy. This sample transition is easier to obtain than the transition probability as in the dynamic programming approach. The Monte Carlo method is an algorithm used to solve the reinforcement learning problem by using average sample return. To average the sample return, the experience is divided into episodes. Each episode finishes regardless of the taken action. The average return in the episode is used to compute the value function and adjust the policy.

Similar to the dynamic programming approach, Monte Carlo policy evaluation is first used to determine the value functions of V^{π} and Q^{π}. Then, the policy improvement is applied. Monte Carlo policy evaluation and policy improvement are combined into the policy iteration algorithm. In Monte Carlo policy evaluation, the state-value function for a given policy is evaluated to obtain the average return. This average return is obtained by averaging the observed return when visiting a particular state in the episodes. To estimate $V^{\pi}(s)$, every-visit and first-visit methods can be applied. In the every-visit method, this state-value function $V^{\pi}(s)$ is estimated as the average of the return of all visits to state s. In contrast, in the first-visit method, the state-value function is estimated by using only the first visits to state s. The first-visit method is shown in Algorithm 6.3.

Algorithm 6.3 Monte Carlo policy evaluation algorithm.

```
1: Initialize V and π, and Ret(s), which is a return of state s
2: loop
3:    Simulate the episode using π
4:    for all First visit to s ∈ S do
5:       Observe immediate return of the first visit to state s
6:       Compute average return and store in Ret(s)
7:    end for
8: end loop
```

However, if a model of the system is not available, the state value is not sufficient to determine a policy. In this case, the action-value function $Q^{\pi}(s, a)$ is estimated by observing the return after action a is taken in state s according to policy π. A method similar to that in Algorithm 6.3 can be used to estimate this action-value function.

Policy improvement is applied after policy evaluation is performed. This policy evaluation and improvement is referred to as a Monte Carlo control algorithm, and is shown in Algorithm 6.4. Again, this algorithm is based on episodes. After each episode, the observed return is used by policy evaluation to estimate the action-value function, and

Algorithm 6.4 Monte Carlo control algorithm.

1: Initialize V and π, and Ret(s) which is a return of state s
2: **loop**
3: Simulate the episode using π
4: **for** Each pair of visiting s and a in the episode **do**
5: Observe immediate return of taking action a at state s
6: Compute average return and store in $Q(s, a)$
7: **end for**
8: **for** Each state s in the episode **do**
9: $\pi(s) = \arg\max_a Q(s, a)$
10: **end for**
11: **end loop**

it is used to improve the policy of all states. Note that this algorithm uses an estimation of the action-value function instead of state-value function as in Algorithm 6.3.

Although the Monte Carlo control algorithm usually converges to the optimal policy, its convergence property has not been proven.

6.2.4 Temporal-difference learning

Temporal-difference learning is a combination of the Monte Carlo and dynamic programming approaches. It does not require a model of the system. However, it can update the estimate of value and policy without waiting for the final outcome as in the Monte Carlo approach. The main idea of temporal-difference (TD) learning is the prediction. In each time step, TD learning (i.e. the TD(0) algorithm) updates and predicts the value function by an immediate reward for the current state as follows:

$$V(s(t)) = V(s(t)) + \alpha \left(R(t+1) + \beta V(s(t+1)) - V(s(t))\right), \tag{6.33}$$

where α is a weight. The corresponding algorithm to implement this prediction is shown in Algorithm 6.5. Again, the episode is defined as the sequence of states and state-action pairs. This value prediction is then used for optimizing the policy.

Algorithm 6.5 TD(0) algorithm.

1: Initialize $V(s)$
2: **loop**
3: Initialize state s {for the current episode}
4: **loop**
5: Choose action a according to policy π
6: Take action a and observe reward R and the next state s'
7: $V(s) = V(s) + \alpha\left(R + \beta V(s') - V(s)\right)$
8: $s = s'$
9: **end loop**
10: **end loop**

The basic of TD learning is the Sarsa algorithm. This algorithm requires an exploration step to explore possible actions and their corresponding returns. Then, an exploitation step is applied to take an optimal action. The Sarsa algorithm first learns an action-value function. The following prediction method is used:

$$Q(s(t), a(t)) = Q(s(t), a(t)) + \alpha \left(R(t+1) + \beta Q(s(t+1), a(t+1)) - Q(s(t), a(t))\right). \tag{6.34}$$

The Sarsa algorithm is shown in Algorithm 6.6. For an ϵ-greedy policy, the algorithm will choose the action which will maximize the return. However, to explore all possible actions, with a small probability ϵ, the algorithm chooses an action randomly and observes any change of the system. Then, each action will be chosen with the probability $1/|\mathbb{A}|$, where $|\mathbb{A}|$ is the number of elements in set $\mathbb{A}$. This Sarsa algorithm converges to an optimal policy with probability one if all state-action pairs are visited an infinite number of times. With the greedy policy (e.g. ϵ-greedy), the Sarsa algorithm was proven to converge within the limit.

Algorithm 6.6 Sarsa algorithm.

1: Initialize $Q(s, a)$
2: **loop**
3: Initialize state s {for the current episode}
4: **loop**
5: Choose action a at state s to explore and exploit $Q(s, a)$ (e.g. ϵ-greedy)
6: Take action a and observe reward R and the next state s'
7: Choose a' from s' to explore and exploit $Q(s, a)$ (e.g. ϵ-greedy)
8: $Q(s, a) = Q(s, a) + \alpha \left(R + \beta Q(s', a') - Q(s, a)\right)$
9: $s = s', a = a'$
10: **end loop**
11: **end loop**

Another widely adopted TD learning technique is the Q-learning algorithm [275]. In Q-learning, the action-value is estimated from

$$Q(s(t), a(t)) = Q(s(t), a(t)) + \alpha \left(R(t+1) + \beta \max_{a} Q(s(t+1), a) - Q(s(t), a(t))\right). \tag{6.35}$$

The Q-learning algorithm is shown in Algorithm 6.7.

This algorithm converges to an optimal policy with probability one as long as all state-action pairs are continually updated.

6.2.5 Learning in games

While a Markov decision process formulation can be applied to obtain the solution for a system with a single agent and a single objective, game theory can be applied for a system with multiple agents. In such systems, different agents have their own objectives

Algorithm 6.7 Q-learning algorithm.

1: Initialize $Q(s, a)$
2: **loop**
3: Initialize state s {for the current episode}
4: **loop**
5: Choose action a at state s to explore and exploit $Q(s, a)$ (e.g. ϵ-greedy)
6: Take action a and observe reward R and the next state s'
7: $Q(s, a) = Q(s, a) + \alpha (R + \beta \max_{a'} Q(s', a') - Q(s, a))$
8: $s = s', a = a'$
9: **end loop**
10: **end loop**

which could be conflicting with each other. Again, reinforcement learning can also be used by the agents to learn and perform the best actions. Two basic approaches of reinforcement learning in game theory are based on fictitious play and Q-learning with an extension to multi-agent settings.

Fictitious play

The main idea of fictitious play is that the players in a game will choose their best strategies in each period to maximize expected payoff. This strategy selection is based on the knowledge which is the prediction of the strategy that each opponent player will choose in that period. Let $q_{-i}(i, t)$ denote a weight (i.e. propensity) for the strategies of other players maintained by player i at time t. $\mathbf{s}_{-i}$ is a vector of strategies chosen by all players except player i. This weight is updated as follows:

$$q_{\mathbf{s}_{-i}}(i, t) = \alpha q_{\mathbf{s}_{-i}}(i, t-1) + \begin{cases} 1, & \text{if } \mathbf{s}_{-i,t-1} = \mathbf{s}_{-i} \\ 0, & \text{if } \mathbf{s}_{-i,t-1} \neq \mathbf{s}_{-i} \end{cases} \tag{6.36}$$

where α is a discount factor. In particular, the weight is increased by one when the opponent players choose a certain strategy. Then, the probability that the opponent players will choose strategy $\mathbf{s}_{-i}$, which is estimated by player i, can be obtained from

$$\rho_{(\mathbf{s}_{-i})}(i, t) = \frac{q_{\mathbf{s}_{-i}}(i, t)}{\sum_{\mathbf{s}'_{-i}} q_{\mathbf{s}'_{-i}}(i, t)}. \tag{6.37}$$

Then, the fictitious play is defined as the rule $\sigma_{(\rho(i,t))}(i, t)$ for which $\sigma_{(\rho(i,t))}(i, t) \in \mathscr{B}_i(\rho(i, t))$ where $\mathscr{B}_i(\rho(i, t))$ is the best response of player i given the probability that strategy $\mathbf{s}_{-i}$ is adopted by all other players. This fictitious play is similar to the Bayesian inference. In this case, player i has a belief $\rho_{(\mathbf{s}_{-i})}(i, t)$ of the strategies from other players.

Q-learning and game

Although Q-learning was originally developed for decision making by a single agent, it was extended for multi-agent decision [276] based on the concept of stochastic game [247]. An n-player stochastic game can be defined as $(\mathbb{S}, \mathbb{A}_1, \ldots, \mathbb{A}_n,$

$R_1, \ldots, R_n, P)$, where $\mathbb{S}$ is the state space, $\mathbb{A}_i$ is the set of actions for agent i (i.e. player i), R_i is the reward, and P is the transition probability. The objective of each agent is to maximize the long-term reward. This long-term reward is discounted where β is a discount factor. Let π_i denote the strategy of player i. The objective of player i is to maximize

$$J_i(s, \pi_1, \ldots, \pi_n) = \sum_{t=0}^{\infty} \beta^t E(R_i(t), \pi_1, \ldots, \pi_n, s(0) = s), \tag{6.38}$$

where $s(0)$ is an initial state at time step $t = 0$ and $E(\cdot)$ is an expectation over the strategies of all players. The solution of this stochastic game is the Nash equilibrium which is a tuple of n strategies $(\pi_1^*, \ldots, \pi_n^*)$. This Nash equilibrium has the following property:

$$J_i(s, \pi_1^*, \ldots, \pi_i^*, \ldots, \pi_n^*) \geq J_i(s, \pi_1^*, \ldots, \pi_i, \ldots, \pi_n^*), \quad \forall \pi_i. \tag{6.39}$$

It was shown that every n-player discounted stochastic game has at least one Nash equilibrium in stationary strategies [277].

Based on this stochastic game, the concept of multi-agent Q-learning was introduced. In this case, the action-value function is extended to have both states and actions of all players, i.e. $Q(s, a_1, \ldots, a_n)$. The Nash Q-value is defined as the expected sum of discounted rewards given that all players adopt the Nash equilibrium strategies:

$$Q_i^*(s, a_1, \ldots, a_n) = R_i(s, a_1, \ldots, a_n) + \beta \sum_{s' \in \mathbb{S}} P(s'|s, a_1, \ldots, a_n) J_i(s', \pi_1^*, \ldots, \pi_n^*). \tag{6.40}$$

However, this stochastic game may have many Nash equilibria, and the Nash Q-values will be different.

The Nash Q-learning algorithm was proposed by extending the Q-learning algorithm for a single agent environment. In this multi-agent Q-learning algorithm, the Q-value is updated with the future payoff so that each agent can observe and estimate the return for using a particular strategy. However, in this case, each agent must be able to observe not only its payoff, but also payoffs of other players. In each time step, a player observes the current state s, and takes action a. An immediate reward R and the next state s' are observed. The Nash equilibrium $NQ = (\pi_1(s'), \ldots, \pi_n(s'))$ is computed from

$$NQ_i(s', t) = (\pi_1(s'), \ldots, \pi_n(s')) Q_i(s', t) \tag{6.41}$$

where the Q-value is updated as follows:

$$Q_i(s, a_1, \ldots, a_n, t) = (1 - \alpha(t)) Q_i(s, a_1, \ldots, a_n, t) + \alpha(t) \left(R_i(t) + \beta N Q_i(s', t) \right) \tag{6.42}$$

in which $\alpha(t)$ is the learning rate at time step t. The Nash Q-learning algorithm for each player is shown in Algorithm 6.8.

Algorithm 6.8 Nash Q-learning algorithm.

1: Initialize $Q(s, a_1, \ldots, a_n)$
2: **loop**
3: Choose action $a_i(t)$ (e.g. using ϵ-policy)
4: Take action $a_i(t)$ and observe reward $R_i(t)$ of all players (i.e. $i = 1, \ldots, n$), and actions $a_j(t)$ of other players (i.e. $j \neq i$), and the next state $s(t+1) = s'$
5: **for all** $j = 1, \ldots, n$ **do**
6: $Q_j(s, a_1, \ldots, a_n, t+1) = (1 - \alpha(t))Q_j(s, a_1, \ldots, a_n, t) + \alpha(t)\left(R_j(t) + \beta N Q_j(s', t)\right)$
7: **end for**
8: $s = s'$
9: **end loop**

6.3 Applications of machine-learning techniques in wireless communications and networking

6.3.1 Neural network and cognitive radio

Applications of neural networks in a cognitive radio network can be found in [2, 278, 279, 280]. The neural network models in these works use training data for estimating network performances. For example, in [278], a multilayered feedforward neural network [281] was used to model and estimate the performances of IEEE 802.11 networks. This neural network provides a black-box model for the non-linear relationship between the inputs (i.e. network parameters) and the outputs (i.e. network performances). This neural network model can learn from training data which can be obtained in an on-line manner when the real-time measurement data are available. Although training a neural network model requires a large amount of computational resources, the computation of output is much simpler and it only incurs a small overhead. Therefore, this model is suitable for a cognitive radio network for which a prompt response to the changing radio environment is required from an unlicensed user. For example, the unlicensed user must stop transmission as soon as the licensed user's activity on the same channel is detected.

The multilayered feedforward neural network presented in [278] is shown in Figure 6.8. This neural network model is composed of hidden and output layers. The inputs of the model are channel quality, the number of successfully received frames, the number of erroneous frames, and the fraction of time in which a channel is sensed to be idle. The outputs of the model are estimated throughput, delay, and reliability of the network. When training this neural network model, all measured inputs are used to adjust the weights to minimize the error compared with the known outputs. This adjustment is repeated until the error falls below a certain threshold. To adjust the weight, the back-propagation algorithm [282] is used. The training data consists of network parameters and the corresponding performance measures.

This neural network model was evaluated and it was shown that the estimated performances obtained from this neural network model are close to those from the simulation

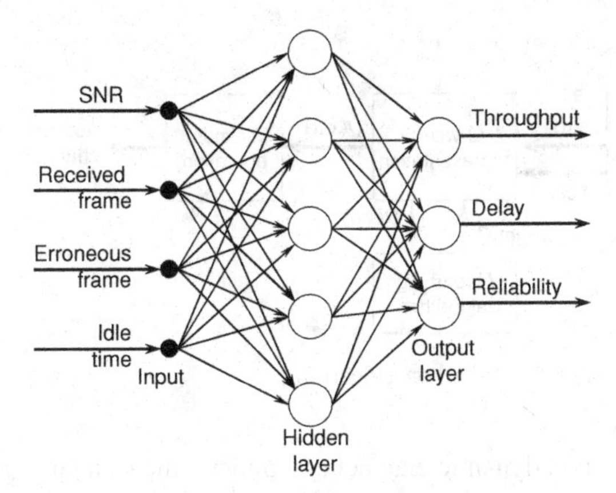

Figure 6.8 A neural network for modeling the performance of IEEE 802.11-based channel access.

results. This neural network model can estimate the network performances as accurately as the analytical models (e.g. Markov chain model in [283]) without requiring information about the number of users in the network. In presence of channel errors, this neural network model is superior to that in [283]. When the multirate transmission of IEEE 802.11 using different modulation and coding rates is considered, this neural network model outperforms the performances of those of auto rate fallback (ARF) [284] and MPDU-based link adaptation scheme (MBLAS) [285].

6.3.2 Q-learning and dynamic channel assignment

In a cellular network, the radio spectrum/channels have to be assigned to a group of cells. This assignment should be made such that it meets users' demand as well as minimizing inter-cell interference. Since a static channel assignment scheme may fail to adapt to the time-varying users' demand (i.e. load), a dynamic channel assignment strategy will be more desirable. A special type of reinforcement learning technique, namely, Q-learning, was applied to solve the dynamic channel assignment problem [286]. In the system model under consideration, there are A cells and N channels. In each cell, the users' call arrival is assumed to follow a Poisson process, while the channel holding time is exponentially distributed. The state of the system is defined as $x(t) = (i, N_v(i))_t$ where i is the cell with call arrival and departure events, and $N_v(i)$ is the number of channels available to cell i at time t. The action is the assignment of a channel k to the current call request in cell i. The immediate cost is calculated from

$$R(x, k) = n_1(k)r_1 + n_2(k)r_2 + n_3(k)r_3, \tag{6.43}$$

where $n_1(k)$ is the number of compact cells with interference to cell i when channel k is being used, $n_2(k)$ is the number of cochannel cells, $n_3(k)$ is the number of other cochannel cells using channel k, r_1, r_2, and r_3 are the weights. This immediate cost indicates the inter-cell interference [287]. The state transitions are determined by call arrivals and departures.

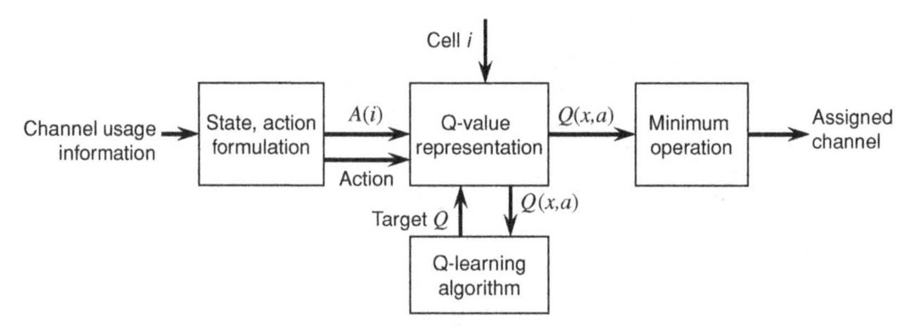

Figure 6.9 Dynamic channel assignment based on Q-learning.

To obtain the solution of this dynamic channel assignment, the standard Q-learning technique is applied. The structure of this Q-learning-based dynamic channel assignment algorithm is shown in Figure 6.9. In the first step, the state-action is determined (i.e. the current assigned and available channels). This channel usage information is then passed to the next step to determine the Q-value. In this step, the Q-value is recalculated based on the Q-learning algorithm. Then, the Q-value is used to determine the best action to minimize the cost. The best action is defined as the channel to be assigned to an arriving call and can be obtained as follows:

$$k^* = \min_k \left(Q(x(t), k)\right). \tag{6.44}$$

Simulation results showed that this Q-learning-based dynamic channel assignment, when compared with the MAXAVAIL algorithm [288], can achieve performance (e.g. in terms of call blocking probability) close to the MAXAVAIL scheme but with much lower implementation and computational complexity.

6.3.3 Q-learning and pricing in cognitive radio

Q-learning was applied to the market-based resource management in a cognitive radio network [289]. A single-hop communications model based on TDMA in an ad hoc unlicensed network was considered. An unlicensed user from node j sends the request with transmission rate $\hat{R}_{i,j}$ to the source node (e.g. the base station) i. Node i computes the size of transmission slot and power for node j. The minimum and maximum rate of transmission between node i and j can be determined from

$$R_{\text{min},i} = 0, \quad R_{\text{max},i} = \frac{\tau_{\text{max},i}}{T} W \log\left(1 + \frac{P_{\text{max},i} h_{i,j}}{kWT_I}\right), \tag{6.45}$$

where $\tau_{\text{max},i}$ is the maximum size of transmission slot, T is the size of slot duration, W is the channel bandwidth, $P_{\text{max},i}$ is the maximum transmit power of node i, $h_{i,j}$ is the channel gain between nodes i and j, k is the Boltzmann constant, and T_I is the total interference temperature at node i. If the transmission rate requirement $\hat{R}_{i,j}$ can be satisfied (i.e. $\hat{R}_{i,j} \leq R_{\text{max},i}$), then node i allocates rate $\hat{R}_{i,j}$ to node j. Otherwise, node i finds the rate $\tilde{R}_{i,j}$ that maximizes its differential utility $\Delta U_{i,j}$. This rate can be obtained by considering the revenue of node i as follows:

$$\tilde{R}_{i,j} = \arg\max_{R_{i,j}} \Delta U_{i,j} \tag{6.46}$$
$$= \arg\max_{R_{i,j}} \left(w_1 \log\left(1 - \frac{R_{i,j} + R_{\text{tot},i}}{C_i}\right) - w_1 \log\left(1 - \frac{R_{\text{tot},i}}{C_i}\right) + p_i R_{i,j} \right),$$

where w_1 is the weight of the utility function, $R_{\text{tot},i}$ is the total downstream rate of node i, C_i is total capacity of node i, and p_i is the price per unit of bit rate charged by node i to node j.

To solve for an optimal unit price p_i, the Q-learning algorithm is applied. In this case, the state of the system is the average of the downstream rate, which can be obtained by dividing the total transmission rate by the total number of connections. The action is the price p_i. The action a at state s is chosen with a probability $\phi(a, s)$, which can be obtained from

$$\phi(a, s) = \frac{\epsilon e^{Q(s,a)}}{\sum_{a'} \epsilon e^{Q(s,a')}}, \tag{6.47}$$

where parameter ϵ is used to control the fraction of time for exploration or exploitation processes, and $Q(s, a)$ is the Q-value of taking action a at state s. This Q-value is updated from

$$Q(s(t+1), a) = (1 - \alpha)Q(s(t), a) + \alpha(r + \beta \max_{a'} Q(s(t), a')), \tag{6.48}$$

where $s(t)$ is the state at time step t. The immediate reward is obtained from

$$r = p_i \frac{R_{\text{tot},i}}{w_2} + p_i \log\left(1 - \frac{R_{\text{tot},i}}{C_i}\right), \tag{6.49}$$

where w_2 is the weight.

Performance evaluation results showed that the Q-learning algorithm provides higher network throughput compared to the static case (i.e. when the price is fixed). Therefore, the probability of rejecting a new connection is lower, and also the interference temperature becomes lower.

6.3.4 Radio environment map and learning algorithms

In [290], architecture of a cognitive engine for IEEE 802.22-based WRAN was proposed. This cognitive engine is designed to use a radio environment map (REM) database [291, 292] and a learning algorithm to optimize the spectrum usage. The major tasks of this cognitive engine are as follows: map the current radio scenario into a state vector which will be used by the optimizer; learn the solution of the current radio scenario using case-based and knowledge-based learning algorithm [293]; and optimize the solution into the actual action, e.g. using a local search or a genetic algorithm.

The architecture of this cognitive engine is shown in Figure 6.10. In this case, the radio scenario is defined as the spectrum availability and transmit power mask between the WRAN base station (BS) and the customer premises equipment (CPE). The objective

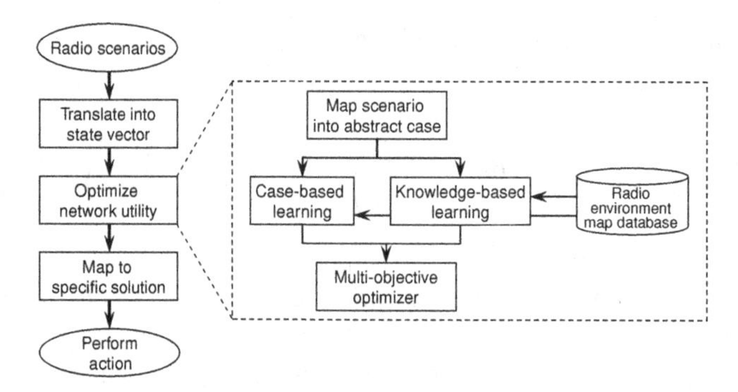

Figure 6.10 Architecture of a cognitive engine for IEEE 802.22-based WRAN.

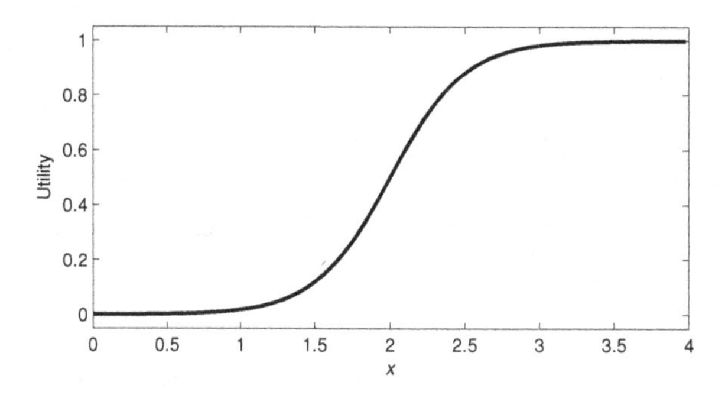

Figure 6.11 A plot of the function $\mathscr{V}(\cdot)$.

of this cognitive engine is to maximize the global utility function, which is defined as follows:

$$\mathscr{U} = \prod_k (u_k)^{w_k}, \tag{6.50}$$

where u_k is the utility metric k and w_k is its weight. The utility metric function for a CPE is a function of bit error rate P_b, data rate R_b, and transmit power P_t. This utility function can be defined as follows:

$$u_{\mathrm{cpe}} = \mathscr{V}\left(1/P_\mathrm{b}, 1/\hat{P}_\mathrm{b}\right)^2 \mathscr{V}\left(1/R_\mathrm{b}, 1/\hat{R}_\mathrm{b}\right)^2 \mathscr{V}\left(1/P_\mathrm{t}, 1/\hat{P}_\mathrm{t}\right)^2, \tag{6.51}$$

where $\hat{P}_\mathrm{b}$, $\hat{R}_\mathrm{b}$, and $\hat{P}_\mathrm{t}$ are the target bit error rate, target data rate, and target transmit power, respectively. The function $\mathscr{V}(\cdot,\cdot)$ is defined as follows:

$$\mathscr{V}(x,\hat{x}) = \frac{1}{2}\left(\tanh\left(\log\left(\frac{x}{\hat{x}}\right) - \eta\right)\sigma + 1\right), \tag{6.52}$$

where η and σ are the adjustable parameters of this function. The utility of the base station u_{bs} is defined as the normalized spectral efficiency (i.e. the average number of available subcarriers per channel) of the base station. An example of the function $\mathscr{V}(\cdot,\cdot)$ with $\hat{x} = 2$, $\eta = 4$, and $\sigma = 1$ is shown in Figure 6.11.

This cognitive engine works as follows. First, it collects current radio scenario information from the radio environment map database and also from the spectrum sensing

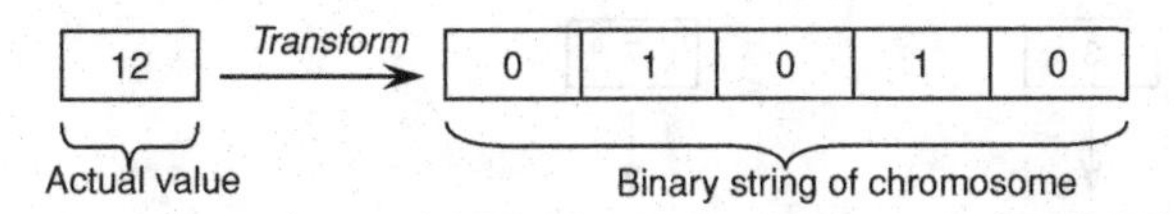

Figure 6.12 Binary encoding of a single decision variable in genetic algorithm.

module. This information is processed and mapped into the state parameters. Then, the cognitive engine searches the case library and knowledge base to determine the feasible solution. This case library and the knowledge base are also updated with the estimated performances. Then, an optimizer is applied to obtain the solution which maximizes the global utility.

Performance evaluation results showed that the cognitive engine can achieve high utility and can adapt to the changes in the radio environment quickly. However, in some extreme cases (e.g. when the required spectrum from the CPE reaches the capacity of the base station), this cognitive engine cannot obtain the feasible solution.

6.4 Genetic algorithms

Genetic algorithms were invented to solve optimization problems using Darwin's *natural evolution* principle. Genetic algorithms are typically implemented by computer simulations. In a genetic algorithm, a population is a collection of individuals (i.e. chromosomes). This individual represents the candidate solution to an optimization problem. The population evolves so that the better solution can be reached. The evolution of the population in a genetic algorithm uses probabilistic rules rather than deterministic rules. Also, a genetic algorithm relies on many chromosomes, which represent different points in the problem. As a result, a genetic algorithm can avoid obtaining locally optimal solutions, which are common in a non-convex problem. In addition, since a genetic algorithm requires only the evaluation of the objective function, it can perform without knowing the exact expression of the objective function. With these advantages, genetic algorithms are used to solve many optimization problems, especially those with non-convex objective functions.

In a genetic algorithm, the decision variables to be optimized are encoded into a chromosome. While there are different choices of encoding schemes, the most common scheme is a binary string. In this case, the decision variable of an optimization is transformed into a sequence of 0s and 1s. In a problem with a single decision variable, e.g. x, the value of this variable is transformed into binary data (e.g. as shown in Figure 6.12). In the case of multiple decision variables, multiple concatenation method will be used. In this case, the encoded binary strings of the decision variables are concatenated into a single string as shown in Figure 6.13.

This population of chromosomes evolves and a new generation is created. The major operations in the evolution are as follows:

1. *Reproduction:* In this operation, the chromosomes in the current population are selected with a certain probability based on their goodness. These selected

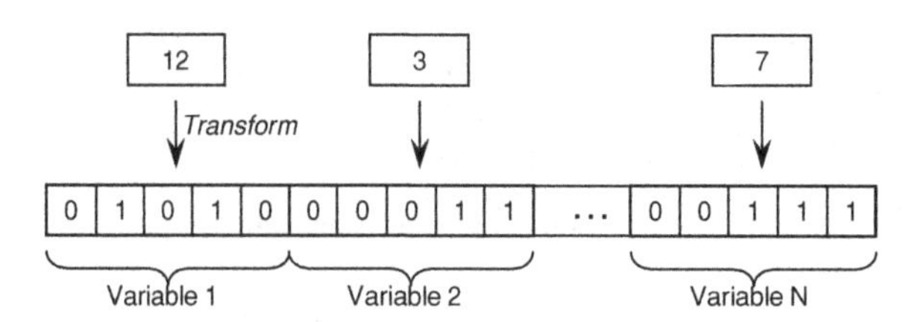

Figure 6.13 Binary encoding of multiple decision variables in a genetic algorithm.

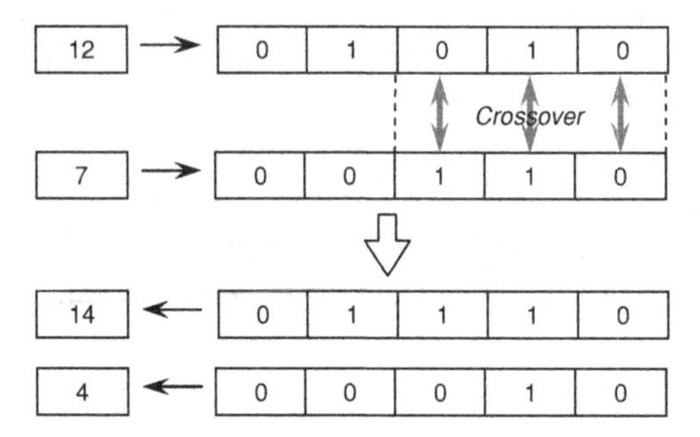

Figure 6.14 Crossover operation in a genetic algorithm.

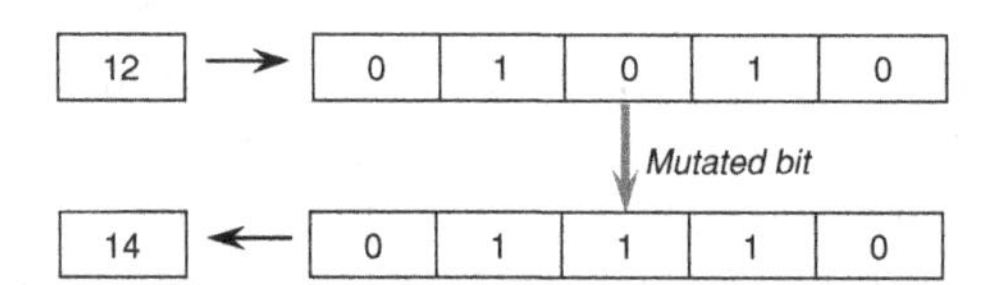

Figure 6.15 Mutation operation in a genetic algorithm.

chromosomes are re-copied to produce a new population. The new population is defined as the successive generation.

2. *Crossover:* In this operation, two chromosomes are randomly selected according to their goodness and then their sub-strings (whose locations in the original strings are randomly chosen) are interchanged, for example, as shown in Figure 6.14.
3. *Mutation:* In this operation, a chromosome is randomly selected according to its goodness. Then, some bits which are randomly chosen in the selected chromosome are altered (e.g. complemented from 0 to 1 and from 1 to 0) as shown in Figure 6.15. The resulting chromosome, after its bits are altered, is included in the new generation.

Each chromosome is characterized by its goodness (i.e. fitness), which is referred to as the fitness function. The fitness of a chromosome is a direct function of the objective in the optimization problem (but not necessarily the same). In a genetic algorithm, the population is evolved so that its chromosomes contain the highest fitness function in the case of maximization. In particular, the chromosome with the highest fitness will be reproduced most often. As a result, the globally optimal solution will be reached. Note that this fitness function is defined to be different from the objective function of the problem so that it can be suitably used by the reproduction operation.

Algorithm 6.9 Genetic algorithm.

1: Choose the population of random initializations
2: **repeat**
3: Evaluate the fitness of each chromosome
4: Select pairs of best-ranking chromosomes to reproduce
5: Apply crossover operation
6: Apply mutation operation
7: **until** Stop criteria are met

The outline of a genetic algorithm is shown in Algorithm 6.9. The algorithm will be terminated if the stop criteria are met. The stop criterion can be defined as the maximum number of iterations. Alternatively, the mean and the maximum fitness values of the population can be observed. If these two values reach a predefined threshold, the algorithm will stop. In this case, it is likely that the chromosomes in the population will encode the points which are close to the optimal solution.

A genetic algorithm was applied to optimize the performance of cognitive radio [150]. The system model considers autonomous vehicular communications. Due to the geographically varying environment, the transmission parameters are dynamically adjusted to optimize the fitness given the system parameters of signal bandwidth and bit error rate. The fitness of this genetic algorithm is defined as a function of key performance parameters including link margin LM, spectral efficiency SE, data rate DR, and received carrier power/interference level CI. The fitness measure is expressed as follows:

$$\mathrm{FM} = w_1\mathrm{LM} + w_2\mathrm{SE} + w_3\mathrm{DR} + w_4\mathrm{CI}, \tag{6.53}$$

where w_1, w_2, w_3, and w_4 are the weights corresponding to the performance parameters. In this case, the spectral efficiency is computed from

$$\mathrm{SE} = \frac{W}{\mathrm{CG} \times \mathrm{DR}}, \tag{6.54}$$

where W is the bandwidth, CG is the coding gain, and DR is the data rate. The received carrier power/interference level can be obtained from

$$\mathrm{CI} = \mathrm{EIRP} + G_{\mathrm{r}} - L_{\mathrm{s}}, \tag{6.55}$$

where EIRP is the equivalent isotropic radiated power, which is obtained by multiplying the transmit power and transmit antenna gain; G_{r} is the receive antenna gain; and L_{s} is the loss from the antenna configuration. The link margin is defined as the difference between required and actual signal energy normalized to thermal noise.

In a genetic algorithm, the chromosomes are used to maintain the values of the decision variables. In this case, the chromosomes are defined for modulation and coding scheme, system noise figure, transmit power level, transmit antenna gain, receive antenna gain, antenna parameters, and operating frequency. With these many adjustable transmission parameters, formulating and solving an optimization problem by standard method is complex and practically infeasible. However, with a genetic algorithm, near-optimal transmission parameters can be obtained while the algorithm converges within 30 iterations (i.e. 30 generations of chromosome evolution).

6.5 Fuzzy logic

The theory of fuzzy logic was developed to present approximate knowledge which may not be suitably expressed by the conventional crisp method (i.e. bivalent set theory). In the conventional method, *true* and *false* are represented by values 1 and 0, respectively. By using membership functions, fuzzy logic extends and generalizes these true and false values to any value between 0 to 1. A membership function indicates the degree by which the value belongs to a fuzzy set.

Fuzzy logic can be used to make decisions by using incomplete, approximate, and vague information. Therefore, it is suitable for a complex system for which it is difficult to compute/express the parameters precisely (e.g. by using a mathematical model). Fuzzy logic also enables us to make inferences by using the approximate information in order to decide on an appropriate action for given inputs. In short, instead of using complicated mathematical formulations, fuzzy logic uses human-understandable fuzzy sets and inference rules (e.g. IF, THEN, ELSE, AND, OR, NOT) to obtain the solution that satisfies the desired system objectives. The main advantage of fuzzy logic is its low complexity. Therefore, fuzzy logic is suitable for real-time cognitive radio applications in which the response time is critical to system performance.

6.5.1 Fuzzy set

The fuzzy set theory is similar to traditional bivalent set theory. However, a fuzzy set has no clear or crisp boundaries. Fuzzy set is any set that allows its members to have different degrees of membership (through the membership function) in the interval [0,1]. The membership function is defined as the degree of membership of a particular element within the set. Also, since the fuzzy set is represented by membership functions, it is possible for one element to belong to more than one set (to a different degree). For example, there are two sets to represent vehicle speed (e.g. "slow" and "fast"). For a measured vehicle speed (which could be inaccurate) of 130 km/h, it will be either "slow" or "fast" (e.g. if the threshold for "fast" is 140 km/h, this measured vehicle speed is precisely "slow"). On the other hand, in fuzzy logic, this measured vehicle speed could be in set "fast" with a membership value of 80 percent and in set "slow" with a membership value of 20 percent (i.e. not quite slow, but relatively fast). Note that here the vehicle speed is called a fuzzy variable, while "slow" and "fast" are called linguistic variables.

The membership function assigns a value in the interval [0,1] to a fuzzy variable. This membership function represents the possibility function (i.e. not the probability function) of an element being in a particular fuzzy set. The membership function can be expressed as $\mu_A(a)$, where A is a fuzzy set, and a is a fuzzy variable. An example of membership functions is shown in Figure 6.16, where fuzzy sets A and B are for "slow" and "fast," respectively.

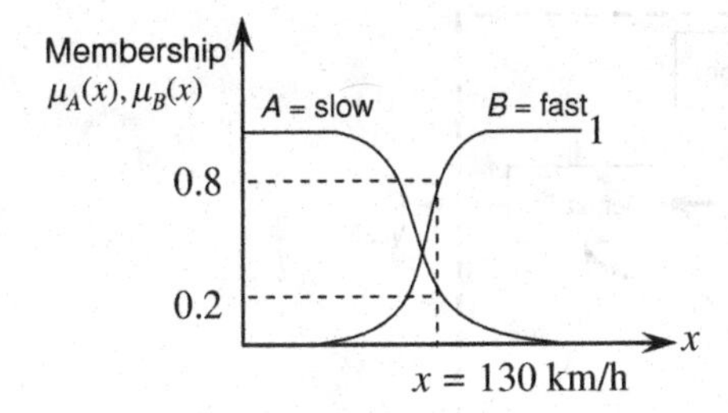

Figure 6.16 Membership functions for vehicle speed.

6.5.2 Fuzzy operation

The operations on fuzzy sets are similar to those in bivalent set theory (i.e. NOT, OR, AND).

- *NOT* denotes the complement of a fuzzy set A (i.e. A') and the corresponding operation on membership function is given by $\mu_{A'}(a) = 1 - \mu_A(a)$.
- *OR* denotes the union of fuzzy sets $A \cup B$ and the corresponding operation on membership function is given by $\mu_{A \cup B}(a) = \max(\mu_A(a), \mu_B(a))$. Since this OR operation is a union of multiple fuzzy sets, membership function of this operation is the largest membership from all fuzzy sets.
- *AND* denotes the intersection of fuzzy set $A \cap B$ and the corresponding operation on membership function is given by $\mu_{A \cap B}(a) = \min(\mu_A(a), \mu_B(a))$. Since this AND operation is the intersection of multiple fuzzy sets, the membership function is the lowest membership from all fuzzy sets.

6.5.3 Fuzzy rule

Knowledge in fuzzy logic is represented in the form of linguistic rules. This rule is based on cause and effect (e.g. in IF–THEN format). The knowledge base is composed of several fuzzy rules. To determine the outcome (or decision), these rules are evaluated and the outcomes are aggregated to the final solution. If A and C denote the fuzzy sets, the rule representing IF A THEN C can be expressed as $A \rightarrow C$. In this rule, A is called the cause, condition, or antecedent of the rule, while C is called the effect, action, or consequence of the fuzzy rule. For this IF–THEN rule, the membership function of the outcome c for a given input a can be obtained using many different approaches. The most widely used implication rule is the Mamdani implication, which can be expressed as follows:

$$\mu_{A \rightarrow C}(a, c) = \min(\mu_A(a), \mu_C(c)). \tag{6.56}$$

6.5.4 Fuzzy logic control

A fuzzy logic control system can be used to obtain the solution to a problem given imprecise, noisy, and incomplete input information. In general, there are three major components in a fuzzy logic control system: fuzzifier, fuzzy logic processor, and defuzzifier (Figure 6.17). While the fuzzifier is used to map the crisp inputs into fuzzy sets, the

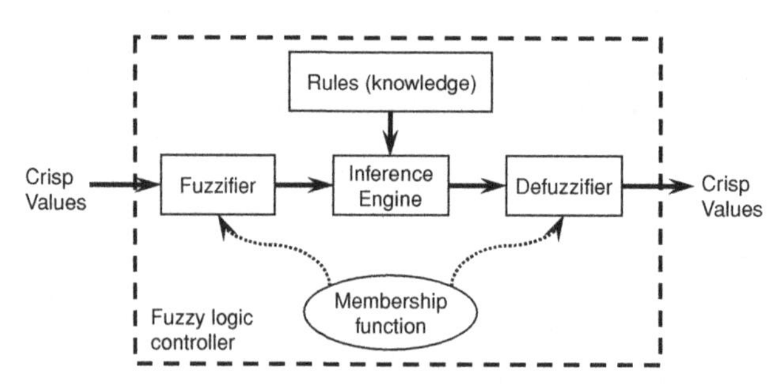

Figure 6.17 Fuzzification, inference engine, and defuzzification.

fuzzy logic processor implements an inference engine to obtain the solution based on predefined sets of rules. Then, the defuzzifier is applied to transform the solution to the crisp output.

In the fuzzification process, input values are fuzzified to determine the membership functions. Then, these fuzzified inputs are used by the inference rules to determine an outcome or decision. For example, let A_i and B_i denote the fuzzified inputs, and C_i denote an output. A set of rules can be defined as follows:

Rule 1: A_1 AND $B_1 \rightarrow C_1$
Rule 2: A_2 AND $B_2 \rightarrow C_2$
Rule n: A_n AND $B_n \rightarrow C_n$

Since these rules are related through the IF–THEN implication, the membership function of an output of a particular rule can be expressed as $\mu_{R_i}(a,b,c) = \min(\mu_A(a), \mu_B(b), \mu_C(c))$. Then, the outcomes of all rules are combined using the maximum function of each rule as follows:

$$\mu_R(a,b,c) = \max_i \min\left(\mu_{A_i}(a), \mu_{B_i}(b), \mu_{C_i}(c)\right). \tag{6.57}$$

However, this outcome from all rules determines the membership function, not the crisp value. Therefore, we need the defuzzification process to obtain the final output of the controller. The most widely used method is centroid, in which the output is determined from a center of gravity of the membership function from the outcome of the set of rules. Let $\mathbb{K} = \{c|\mu_C(c) > 0\}$ denote a set of outputs c with membership value larger than zero. Then, the defuzzified output can be obtained from $\hat{c} = \int_{c\in\mathbb{K}} c\mu_C(c)\mathrm{d}c / \int_{c\in\mathbb{K}} \mu_C(c)\mathrm{d}c$ if $\mathbb{K}$ is a continuous set, and $\hat{c} = \sum_{c\in\mathbb{K}} c\mu_C(c) / \sum_{c\in\mathbb{K}} \mu_C(c)$ if $\mathbb{K}$ is a discrete set.

6.5.5 Applications of fuzzy logic control to cognitive radio

Network access selection

Fuzzy logic control has been used in cognitive radio design [294, 166, 295]. For example, in [295], fuzzy logic control is used to model the distributed decision of a cognitive radio user to choose from multiple available wireless access options (e.g. multiple IEEE 802.11 access points). In this case, a cognitive radio user will choose the wireless access based on the performances in terms of throughput, delay, and reliability (transmission error rate). The knowledge of the system is represented by fuzzy rules which are constructed by using feedback information from all users in the network. A new user initiating a connection will utilize this knowledge to make the best decision on choosing network access. This fuzzy logic decision making process considers the application requirements (i.e. QoS requirements such as throughput, delay, and reliability, which are denoted by t, d, r, respectively).

For the fuzzy logic decision making, the performances of each network access are evaluated and the best network access will be chosen by a cognitive radio for data transmission. First, for network access i, the estimated network-layer performance is computed by combining the radio link level and core network performances as follows:

- *Throughput:* $\tilde{t}_e(i) = \min(\tilde{t}_l(i), \tilde{t}_n(i))$ where $\tilde{t}_e(i)$ is the estimated end-to-end throughput, $\tilde{t}_l(i)$ and $\tilde{t}_n(i)$ are the throughput in the radio link level and core network, respectively.
- *Delay:* $\tilde{d}_e(i) = \tilde{d}_l(i) + \tilde{d}_n(i)$ where $\tilde{d}_e(i)$ is the estimated end-to-end delay.
- *Reliability:* $\tilde{r}_e(i) = \tilde{r}_l(i) \times \tilde{r}_n(i)$ where $\tilde{r}_e(i)$ is the estimated end-to-end reliability.

Then, the performances in the transport layer are obtained as follows:

$$\tilde{t}_n(i) = f_t(\tilde{d}_e(i), \tilde{r}_e(i), \tilde{t}_e(i)), \tag{6.58}$$

$$\tilde{d}_n(i) = f_d(\tilde{d}_e(i), \tilde{r}_e(i), \tilde{t}_e(i)), \tag{6.59}$$

$$\tilde{r}_n(i) = f_r(\tilde{d}_e(i), \tilde{r}_e(i), \tilde{t}_e(i)), \tag{6.60}$$

for throughput, delay, and reliability, respectively, where, $f_t(\cdot)$, $f_d(\cdot)$, and $f_r(\cdot)$ are the extension functions in fuzzy logic.

Finally, the degree of the fitness of each network access is compared by considering the application requirements using fuzzy logic as follows: $\zeta_t(i) = \max(\tilde{t}_n(i) \cap T)$, $\zeta_d(i) = \max(\tilde{d}_n(i) \cap D)$, and $\zeta_r(i) = \max(\tilde{r}_n(i) \cap R)$, where $(A \cap B)(u) = \min(A(u), B(u))$, which is a set intersection operator in fuzzy logic. Then, the overall performance (i.e. fitness) of the network access is computed as follows: $\zeta_i = \min(\zeta_t(i), \zeta_d(i), \zeta_r(i))$. The network access with the highest fitness ζ_i will be chosen by a cognitive radio user to transmit data. Performance evaluation results showed that this fuzzy logic decision making can achieve better throughput and fairness compared with the schemes which rely on each performance measure separately.

Transmit power control

In [294], a transmit power control method based on fuzzy logic control was proposed. With this power control algorithm, an unlicensed user can coexist with a licensed user by dynamically controlling the transmit power. The rules of this power control are established from the knowledge of a group of network experts (i.e. via survey). As a result, it provides better description of the system than that of a single expert. In the system model, a base station is used to manage the transmissions of all unlicensed users. The structure of the cognitive radio with fuzzy-logic-based transmit power control is shown in Figure 6.18. The spectrum sensing and analysis component is used to sense user activity on the target spectrum. It is also used to measure the interference level and estimate the distance between unlicensed and licensed users. The fuzzy logic control

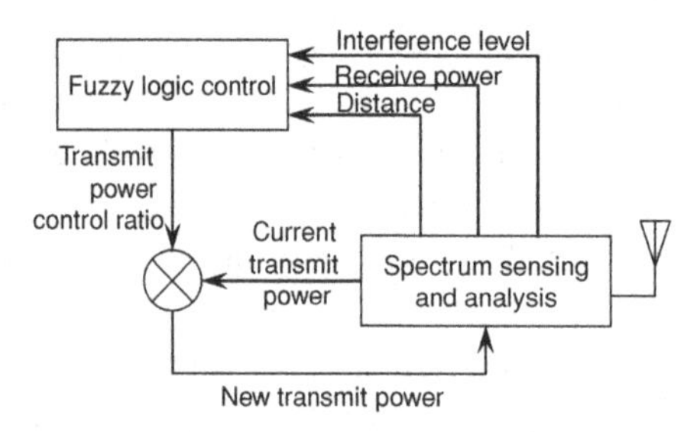

Figure 6.18 Transmit power control based on fuzzy logic.

component receives three inputs, i.e. the interference level of a licensed user, the distance, and the received power difference (i.e. the difference between actual received power and the target received power at the base station of the unlicensed user). The output is the transmit power control ratio ϕ_i, which is used to adjust the transmit power of unlicensed user i. Given the current transmit power $P_i(t)$, the new transmit power is computed from $P_i(t+1) = \phi_i P_i(t)$. The distance is estimated from the received power and the path loss. The interference is reported by a licensed user in the system. The inputs are fuzzified and then the output is computed from a set of rules. This output is defuzzified into the transmit power control ratio.

In addition, a method was presented for an unlicensed user to change the operating channel. The state (or condition) of an unlicensed user is as follows: the interference level is high, the distance from an unlicensed user to a licensed user is near or moderate or far, and the received power difference is negative. This condition indicates that a licensed user is occupying the spectrum, and the unlicensed user cannot decrease the transmit power due to the QoS constraint.

Simulation results showed that with this fuzzy-logic-based transmit power control, the transmission outage probability can be reduced. Also, compared with fixed power control, this scheme uses a lower transmit power, while the QoS performance of an unlicensed user and interference to a licensed user can be maintained at the target levels.

6.6 Summary

Intelligent algorithms such as those based on machine learning, genetic algorithms, and fuzzy logic play an important role in cognitive radio design to achieve adaptability in spectrum access. In this chapter, an introduction to these algorithms has been provided and their applications to wireless networking problems have been described. Machine learning technique based on neural networks can be used to predict network performance when the training data are available. Reinforcement learning techniques such as Q-learning can be used to solve the problem of dynamic channel assignment (DCA) and spectrum pricing. Genetic algorithms can be used to optimize the transmission parameters of a cognitive radio user. Fuzzy logic control has been used to solve the problem of channel selection and transmit power control in wireless networks.

Part III

Dynamic spectrum access and management

7 Dynamic spectrum access: models, architectures, and control

Dynamic spectrum access (DSA) models for cognitive radio can be categorized as exclusive-use, shared-use, and commons models. In the exclusive-use model, a licensed user can grant an unlicensed user the right to have exclusive access to the spectrum. In a shared-use model, an unlicensed user accesses the spectrum opportunistically without interrupting a licensed user. In a commons model, an unlicensed user can access the spectrum freely. DSA can be implemented in a centralized or a distributed cognitive network architecture. DSA can be optimized globally in a cognitive radio network if a central controller is available. On the other hand, when a central controller is not available, distributed algorithms would be required for dynamic spectrum access. Issues related to spectrum trading such as pricing will also need to be considered for dynamic spectrum access, especially with the exclusive-use model. For DSA-based cognitive radio networks, MAC protocols designed for traditional wireless networks have to be modified to include spectrum sensing and spectrum access, as well as spectrum trading between licensed and unlicensed users.

In this chapter, we describe the different spectrum access models and the system architectures for DSA. Then, two major components of dynamic spectrum access, namely, spectrum sensing and spectrum access, are presented. Spectrum sensing, which can be implemented in both physical and MAC layers, is used to detect the presence of a licensed user. In this case, an unlicensed user observes the target frequency band and searches for a signal from a licensed user. The spectrum sensing result is used by the unlicensed user to access the spectrum without interfering with the licensed user and colliding with other unlicensed users.

The major objectives of cognitive MAC protocol design are: (1) to optimize spectrum sensing and the spectrum access decision, (2) to control multiuser access in multichannel network, (3) to allocate radio spectrum and schedule traffic transmission, and (4) to support the spectrum trading function. Some of the MAC protocols that have been designed based on the above objectives are reviewed. Major research challenges and open research issues in dynamic spectrum access are also outlined.

7.1 Spectrum access models

As has been mentioned before, the traditional "command-and-control" licensing scheme, in which the radio resource is statically assigned to a particular spectrum licensee [110], results in spectrum under-utilization. This spectrum licensing scheme does not allow the flexibile usage of frequency spectrum according to the time-varying demands of the user [5]. The limitations of the "command-and-control"-based spectrum allocation scheme can be summarized as follows:

- The spectrum licensee cannot be changed. Consequently, if the radio spectrum remains under-utilized, it cannot be allocated to other service providers who may need the spectrum to accommodate new users.
- The type of wireless service in the licensed spectrum cannot be changed. Spectrum allocated to one wireless technology could be lightly utilized for a certain period of time. For example, it was observed that the 6 MHz spectrum bands allocated to analog TV services remain largely unused. However, the current spectrum licensing scheme does not allow a service provider to use the TV bands for other services.
- The spectrum license is location-invariant. Consequently, the spectrum can remain under-utilized in some locations. For example, the radio spectrum could be allocated to a cellular service in an entire region (e.g. an entire province). While the spectrum could be heavily used in urban areas, it would be under-utilized in suburban/rural areas. In fact, it may not be profitable for a service provider to extend its service to those areas. However, this spectrum could be used by other wireless systems.
- The granularity of spectrum usage is fixed. For example, the spectrum allocation for cellular service is made only in large chunks (e.g. 50 MHz in the 800 MHz band). Consequently, it is not possible to allocate the spectrum band in smaller chunks to meet traffic demands in special scenarios (e.g. in a hotspot scenario).
- The spectrum is protected from use by unlicensed users. Traditionally, a radio spectrum allocated to a licensed user cannot be used by unlicensed users. However, this spectrum may not be used by a licensed user all the time in all areas. The spectrum utilization can be improved if unlicensed users are allowed to opportunistically access the spectrum as long as their transmissions do not interfere with or interrupt the transmissions from licensed users.

To overcome the above limitations, the concept of *open spectrum* [296] was introduced. The open spectrum concept defines a set of techniques and models to support dynamic management of frequency bands for wireless communications systems. It has led to new spectrum access and licensing models which are able to improve the flexibility and efficiency of spectrum access, classified as *exclusive-use*, *shared-use of the licensed spectrum*, and *commons* models. A taxonomy of the different spectrum access models is shown in Figure 7.1.

7.1.1 Exclusive-use model

In the exclusive-use model for spectrum access, the radio spectrum is licensed to a user/service to be exclusively used under a certain rule. In this model, the licenser (e.g. the government) allocates the spectrum to a licensee. However, the licensee or spectrum owner may not fully utilize the allocated spectrum in all times and in all locations. Therefore, spectrum access rights can be granted to cognitive radio users. In

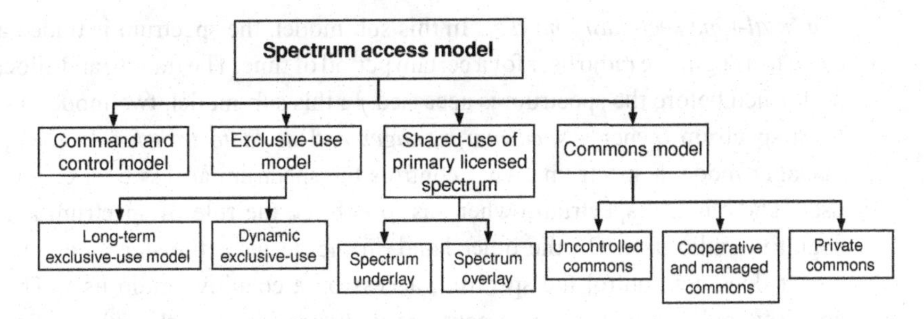

Figure 7.1 Spectrum access models.

this case, the spectrum owner is defined as the licensed user while a cognitive radio user is defined as an unlicensed user. Constrained by the rules defined by the spectrum owner, a cognitive radio user can optimize the spectrum usage to achieve the best performance. There are two variants of this exclusive-use model, namely, the long-term exclusive-use model and the dynamic exclusive-use model. In the dynamic exclusive-use model, spectrum allocation can be performed at a finer scale compared to that for a long-term exclusive-use model. For example, in a dynamic model, the spectrum can be divided into small chunks to be allocated to cognitive radio users for a relatively short time period.

Long-term exclusive-use model

In a long-term exclusive-use model, the radio spectrum is allocated exclusively to a licensed user/service provider for some period of time (e.g. a few weeks). After the spectrum is licensed, the type of wireless service using this spectrum may or may not be changed (i.e. flexible-type sub-model and fixed-type sub-model). In a fixed-type sub-model, the spectrum owner specifies the type of wireless service as well as the parameters for spectrum access by a cognitive radio user. On the other hand, in flexible-type sub-model, a cognitive radio user can change the type of wireless service.

Dynamic exclusive-use model

In a dynamic exclusive-use model, only one user can exclusively access the spectrum at any particular point in time. However, at different points in time, the users accessing the spectrum and the types of wireless service using that spectrum can be changed. In a dynamic exclusive-use model, a spectrum owner can trade its owned spectrum to a cognitive radio user, and thus can earn revenue. This spectrum can then be accessed by a cognitive radio user for a certain period of time (i.e. an on-demand time-bound spectrum lease [297]). This trading is referred to as a secondary market [298]. The three different sub-models defined based on this secondary market in a dynamic exclusive-use model are *non-real-time secondary market*, *real-time secondary market for homogeneous multi-operator sharing*, and *heterogeneous multi-operator sharing*, which are summarized below:

- *Non-real-time secondary market:* In this sub-model, the spectrum is traded and allocated to a cognitive radio user for a certain period of time. The trading and allocation are performed before the spectrum is accessed. In this sub-model, two modes are defined for a spectrum owner – *spectrum manager* and *de facto transfer*. In the spectrum-manager mode, a spectrum owner controls the spectrum access by a cognitive radio user. Therefore, a spectrum owner has to enforce the rule of spectrum access by a cognitive radio user. On the other hand, in the de facto transfer mode, a spectrum owner does not control the spectrum access by a cognitive radio user. Therefore, a cognitive radio user itself must conform its transmission to the rules defined by the licenser.
- *Real-time secondary market for homogeneous multi-operator sharing:* In this sub-model, the spectrum can be traded and allocated to a cognitive radio user in an on-demand basis. However, the wireless service type cannot be changed (i.e. homogeneous sharing). When a cognitive radio user needs radio spectrum, he/she can request to buy the spectrum from the spectrum owner. This spectrum trading is known as spectrum leasing since the agreement between a spectrum owner and a cognitive radio user is temporary.
- *Real-time secondary market for heterogeneous multi-operator sharing:* Similar to homogeneous sharing, the spectrum can be traded and allocated in an on-demand basis. However, in this heterogeneous sharing, the wireless service type over the allocated spectrum can be changed. For example, the spectrum can be traded between a TV broadcaster and a broadband wireless service provider.

Since in this exclusive-use model the spectrum is exclusively allocated by the spectrum owner to the cognitive radio user, the economic issues (e.g. pricing) become important. The spectrum owners would be able to gain revenue by providing spectrum access rights to the cognitive radio users.

7.1.2 Shared-use model

In the shared-use spectrum access model, the radio spectrum can be simultaneously shared between a primary user (i.e. a licensed user) and a secondary user (i.e. an unlicensed user). In this model, unlicensed users can opportunistically access the radio spectrum if it is not occupied or fully utilized by primary users. In other words, as long as an unlicensed user does not interrupt a primary user (e.g. the collision probability is maintained below the target level), spectrum access by an unlicensed user is allowed and it remains transparent to a primary user. In a shared-use model, the spectrum can be accessed by an unlicensed user in two different modes, namely, spectrum overlay and spectrum underlay modes (Figure 7.2).

Spectrum overlay

In the case of spectrum overlay, a primary user receives an exclusive right to spectrum access. However, at a particular time or frequency, if the spectrum is not utilized by a primary user, it can be opportunistically accessed by a secondary user. Therefore, to

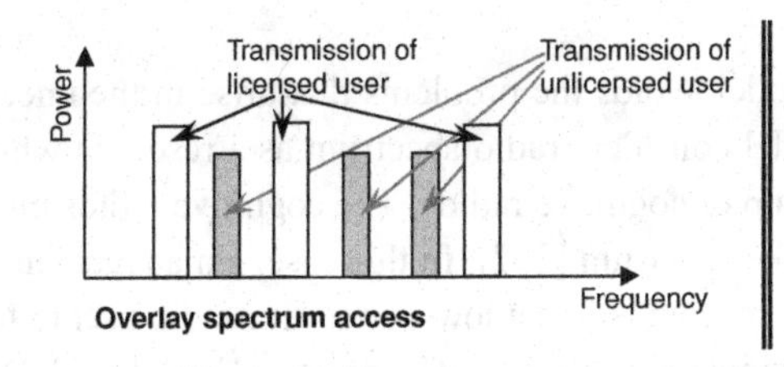

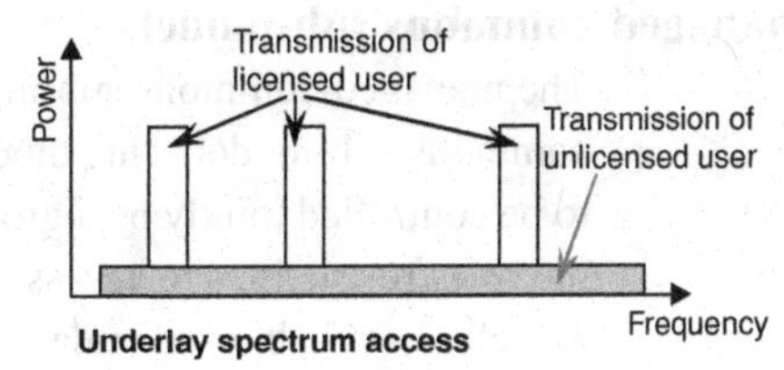

Figure 7.2 Spectrum overlay and spectrum underlay.

access a spectrum band, a secondary user has to perform spectrum sensing to detect the activity of a primary user in that band (Figure 7.2). If a spectrum hole is found, a secondary user may access the spectrum. The decision of a secondary user whether to access the spectrum or not depends on constraints such as the collision probability, which is defined as the probability that the transmission from a secondary user occurs at the same time as that from a primary user. Spectrum overlay can be used for cognitive radio in FDMA, TDMA, or OFDM wireless systems.

Spectrum underlay

In the case of spectrum underlay, a secondary user can transmit concurrently with a primary user. However, the transmit power of the secondary user should be limited so that the interference caused to the primary users remains below the interference temperature limit. Spectrum underlay can be used for cognitive radio systems using CDMA or UWB technology.

7.1.3 Spectrum commons model

In a spectrum commons model, all cognitive radio users have the same right to access the radio spectrum. There are three variants of this model, namely, uncontrolled, managed, and private-commons sub-models.

Uncontrolled-commons sub-model

This is the simplest form of spectrum access in which the spectrum is not owned by any entity. This spectrum access sub-model is already being used in the ISM (2.4 GHz) and U-NII (5 GHz) unlicensed bands. In this case, only the maximum transmit power constraint applies to a cognitive radio user. However, since there is no control on spectrum access, a cognitive radio user suffers from interference, which can be either uncontrolled interference or controlled interference (*tragedy of the commons* [299]). Uncontrolled interference results from the devices outside the network. For example, for IEEE 802.11 b/g networks operating in the 2.4 GHz band, this uncontrolled interference could be due to microwave ovens [300, 301] or Bluetooth devices. In contrast, controlled interference results from neighboring devices in the network. For example, multiple IEEE 802.11 b/g networks can interfere with each other if they decide to operate on the same channel.

Managed-commons sub-model

The managed-commons sub-model avoids the problems that arise in the uncontrolled-commons sub-model. This model considers radio spectrum as a resource which needs to be controlled jointly by a group of cognitive radios. The cognitive radios must follow the rules/restrictions to access the spectrum [302]. In this case, a management protocol as well as a reliable and scalable mechanism to allow a cognitive radio user to follow the rules are required. This management protocol should be designed based on the following objectives [124]:

- Support advanced and efficient device design, services, and business model.
- Minimize communication and coordination overheads for spectrum access.
- Provide flexibility for protocol changes in the future to support new technology.
- Promote fair spectrum access among the cognitive radio users.

The basic rules for distributed spectrum access and management are as follows [124]:

- The transmitter of all cognitive radio users must conform to the maximum transmit power limit. The limit should be defined based on aggregated transmit power from all users (e.g. the interference temperature limit).
- All cognitive radio users must be able to detect the presence of both transmitter-only and receiver-only licensed users. An example of a transmitter-only user is a paging base station, while that of a receiver-only user is a TV receiver.
- All cognitive radio users must be able to exchange information (e.g. transmission parameters) between each other to coordinate spectrum access. This information exchange can be performed through out-of-band signaling or in-band signaling.
- Resource allocation and multiple access control are required in a cognitive radio network. If multiple cognitive radio users access the spectrum at the same time, collision will occur. This collision needs to be minimized to maximize the throughput of the cognitive radio users. The resource allocation mechanism should guarantee the performance requirements for the cognitive radios admitted into the network.

An important issue in the managed-commons spectrum access sub-model is the enforcement of the spectrum access rules. In the system, there could be some cognitive radio users who violate the spectrum access agreement. The issue of protocol misbehavior in wireless networks has been studied extensively in the literature [303, 304, 305, 306] and different techniques (e.g. game theory) have been applied to avoid this problem.

To avoid misbehavior in dynamic spectrum access, proactive or reactive techniques can be used. A proactive technique includes the rule (e.g. maximum power limit) and an enforcement mechanism (e.g. power allocation). This proactive technique is applied before the cognitive radio users start misbehaving. On the other hand, a reactive requirement is applied as a punishment to a misbehaving cognitive radio.

Private-commons sub-model

In this model, a spectrum owner can specify a technology and a protocol for the cognitive radio users to access the spectrum. A cognitive radio user may receive a command from the spectrum owner. This command may contain the transmission parameters (e.g. time,

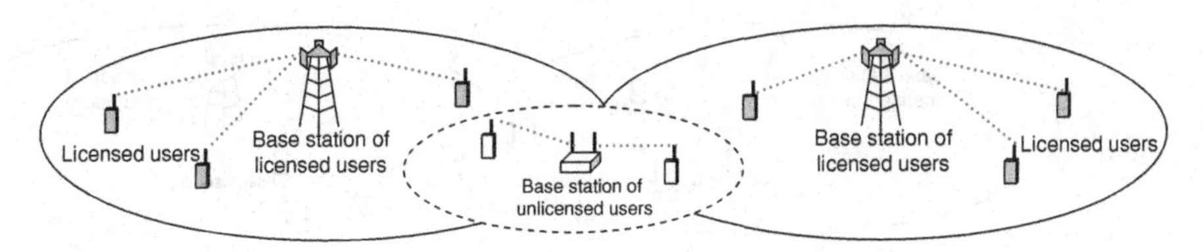

Figure 7.3 Single-hop infrastructure-based cognitive radio network.

frequency band, and transmit power) to be used by a cognitive radio user. Alternatively, a cognitive radio user may opportunistically sense and access the spectrum without interrupting a spectrum owner. Note that the major features of the private-commons sub-model and the shared-use model are as follows:

- A cognitive radio user must receive approval from the spectrum owner before opportunistically accessing the spectrum.
- A cognitive radio user can access the spectrum owned by a particular owner.
- The sensing and coordination protocols must be approved by the spectrum owner.

This private-commons sub-model is currently limited only for peer-to-peer communication between the cognitive radio users in a flat network. In this model, a spectrum owner can control the spectrum access of the cognitive radio users, while the cognitive radio users have flexibility to access the spectrum. Therefore, a secondary market can be established for spectrum trading as in an exclusive-use model.

7.2 Dynamic spectrum access architecture

The architecture of a cognitive radio network can be either infrastructure-based or infrastructureless. In the former case, the network topology rarely changes, while in the latter case the topology changes frequently. Also, a cognitive radio network may require both single-hop and multihop communications between a transmitter and a receiver. Dynamic spectrum access architecture for different types of cognitive radio networks can be either centralized or distributed. In a centralized architecture, the decision of spectrum access is made by a central controller, while in a distributed architecture, this decision is made locally by each of the unlicensed users.

7.2.1 Infrastructure-based versus infrastructureless cognitive radio network

As shown in Figure 7.3, in a single-hop infrastructure-based cognitive radio network, every unlicensed user transmits data through a central controller (e.g. a base station). Dynamic spectrum access by an unlicensed user is controlled (e.g. through the allocation of time, frequency band, and transmit power) by this central controller. This cognitive radio network can coexist/overlap with multiple licensed networks.

In a multihop infrastructure-based cognitive radio network, multiple base stations (i.e. relay nodes) can form a multihop network (Figure 7.4). Unlicensed users (e.g. unlicensed

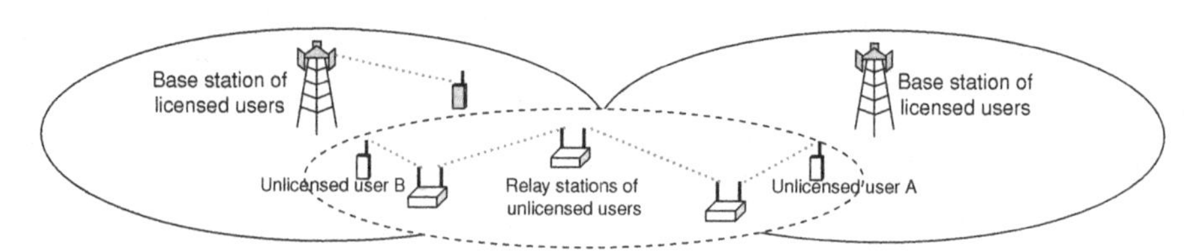

Figure 7.4 Multihop infrastructure-based cognitive radio network.

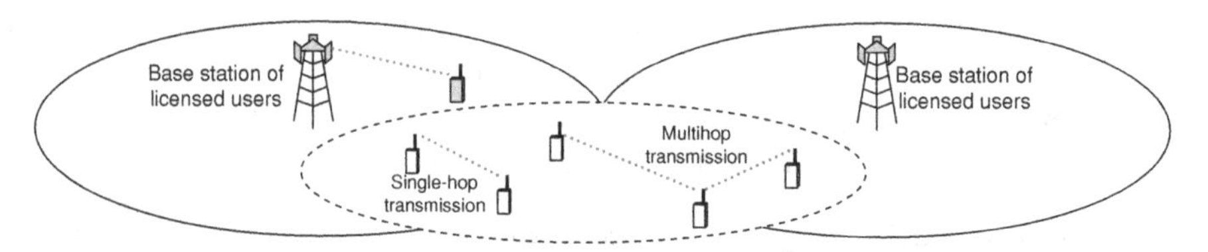

Figure 7.5 Single-hop and multihop ad hoc cognitive radio networks.

user A and B in Figure 7.4) can exchange data by multihop communications through the base stations even though they are not in the transmission range of each other.

In an infrastructureless cognitive radio network, the unlicensed users communicate with each other directly (i.e. in a peer-to-peer fashion) without requiring any base station. The communication can be either single-hop or multihop (Figure 7.5). For multihop communications, some unlicensed users can temporarily assume the role of relay stations.

7.2.2 Centralized versus distributed dynamic spectrum access

In the case of centralized dynamic spectrum access, a central controller makes the decision on spectrum access by all unlicensed users. For this, the central controller will need to collect information about the spectrum usage of the licensed users as well as information about the transmission requirements of the unlicensed users. Based on this information, an optimal solution (e.g. one which maximizes total network throughput) on dynamic spectrum access can be obtained. The decisions of the central controller are broadcast to all unlicensed users in the network. However, information collection and exchange to and from the central controller can incur a considerable overhead.

In the case of distributed dynamic spectrum access, an unlicensed user can make a decision on spectrum access independently and autonomously. Since each unlicensed user has to collect information about the ambient radio environment and make its decision locally, the cognitive radio transceiver of each unlicensed user requires greater computational resources than are required in centralized dynamic spectrum access. However, the communications overhead in this case would be smaller. Since each user has only local information, the optimal solution for spectrum access may not be achievable by all unlicensed users.

As shown in Figure 7.6, distributed dynamic spectrum access can be implemented in both infrastructure-based and infrastructureless cognitive radio networks. However,

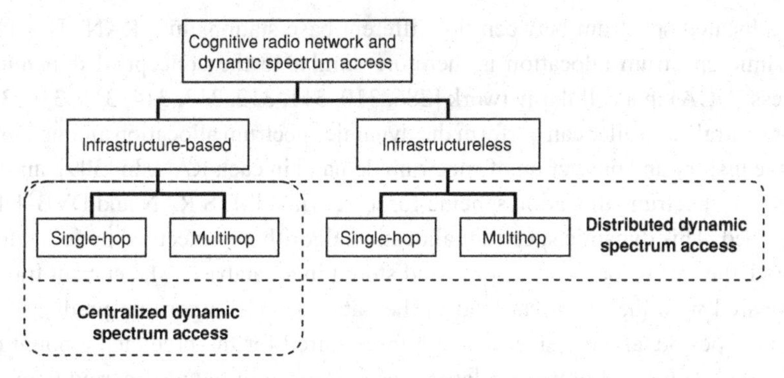

Figure 7.6 Infrastructure-based/infrastructureless cognitive radio networks and centralized/distributed dynamic spectrum access.

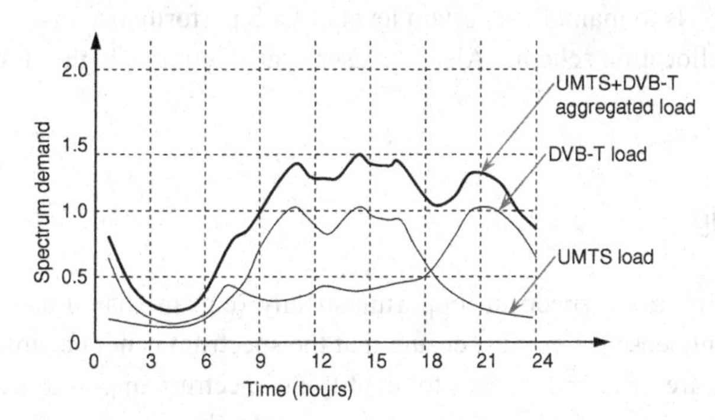

Figure 7.7 Time-varying spectrum demand.

a centralized dynamic spectrum access can only be applied in an infrastructure-based network since it requires a central controller to plan, schedule, and optimize spectrum access by the unlicensed users. In the case of a multihop infrastructure-based network, one of the relay stations can assume the responsibility of controlling dynamic spectrum access.

7.2.3 Inter- and intra-RAN dynamic spectrum allocation

In a wireless communications environment, the radio access network (RAN) provides an interface between a user's mobile and the core network. Dynamic spectrum allocation will be a major functionality of the RAN in a heterogeneous cognitive wireless access environment. The radio spectrum can be allocated to different RANs (e.g. UMTS and DVB-T (digital video broadcasting terrestrial) RAN) dynamically depending on time-varying traffic demand (e.g. as shown in Figure 7.7 [307, 308]). Dynamic spectrum allocation can be implemented using two different approaches, namely, intra-RAN and inter-RAN approaches [309].

The inter-RAN approach decides how the overall radio spectrum can be partitioned among the different RANs. The intra-RAN approach is responsible for distributing the allocated spectrum between the different base stations in a RAN. This intra-RAN dynamic spectrum allocation is therefore similar to the concept of dynamic channel access (DCA) in a cellular network [286, 310, 311, 312, 313, 314, 315, 316, 317].

A central controller can perform the dynamic spectrum allocation among RANs based on the history and prediction of spectrum demand in each RAN. In [309], an inter-RAN dynamic spectrum allocation scheme for terrestrial UMTS RAN and DVB-T RAN was proposed. This dynamic spectrum allocation algorithm is executed periodically. In each period, the traffic load is measured and stored in a database. The current traffic load is compared with the historical load in the database. The algorithm predicts the load in the next period, and the size of spectrum required for the predicted amount of load is estimated. If a RAN requires a larger spectrum size in the next period than that it has in the current period, the RAN borrows spectrum from other RANs. Simulation results showed that with dynamic spectrum allocation, the total amount of radio spectrum required for both RANs to maintain a certain level of QoS performance is less than that required in a static allocation scheme. Also, the user's satisfaction is higher for both the RANs.

7.3 Spectrum sensing

In order to access the radio spectrum opportunistically (e.g. in shared-use spectrum access model), an unlicensed user must ensure that the spectrum is not occupied by any licensed user. There are three approaches to identifying spectrum opportunities – *database registry*, *beacon signals*, and *spectrum sensing*. In the database registry approach, the information about spectrum opportunity is exchanged between the licensed and unlicensed users through a central database. This information can be synchronized using beacon signals transmitted between licensed and unlicensed users over a common channel. However, both approaches suffer from high infrastructure cost and have limited ability for dynamic spectrum access. On the other hand, the spectrum sensing approach relies only on unlicensed users and requires them to identify and track spectrum opportunities. An unlicensed user transmits when the spectrum is not occupied by the licensed user. As a result, no modification to the existing infrastructure of a licensed system is required. Therefore, dynamic spectrum access through spectrum sensing is compatible with legacy wireless communications systems.

To protect a licensed user from interference, an unlicensed user must periodically sense the spectrum, e.g. for every T_{p} units of time. If the spectrum is indicated to be idle, the unlicensed user will access the spectrum. However, during the transmission, this unlicensed user will be unaware of the licensed user until the next spectrum sensing is performed. With a single radio transceiver, the spectrum cannot be sensed and accessed simultaneously by the same unlicensed user. Therefore, if a licensed user starts transmission in between two sensing points (Figure 7.8), it will be interfered with by the transmission of the unlicensed user. Therefore, the QoS performances of both licensed and unlicensed users will depend on the sensing period T_{p}. In particular, if T_{p} is large, the duration of possible interference for a licensed user is large. For example, if $T_{\mathrm{p}} = 10$ ms,

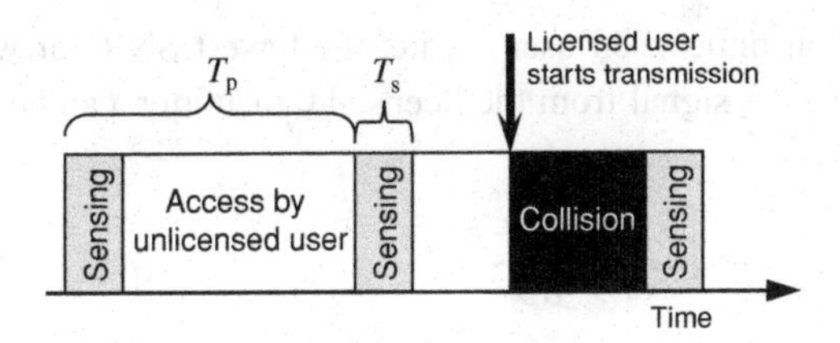

Figure 7.8 Spectrum sensing period.

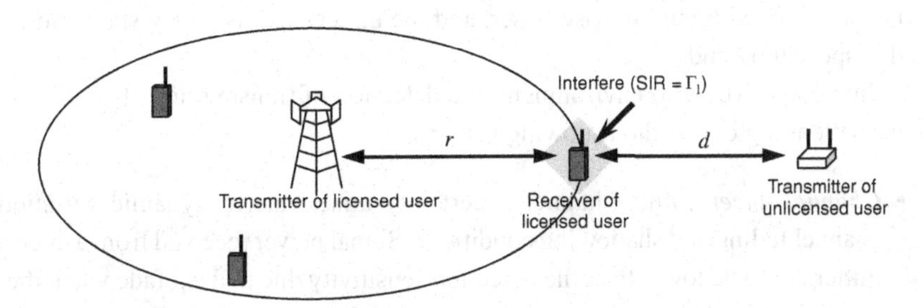

Figure 7.9 Interference range.

the longest duration that a licensed user will be interfered with by an unlicensed user will be 10 ms. However, the length of the spectrum sensing period introduces a performance tradeoff. The shorter the sensing period T_p, the shorter will be the interference duration to a licensed user, but the lower will be the throughput for an unlicensed user. For example, with the sensing duration $T_s = 1$ ms, if $T_p = 2$ ms, the total transmission time for an unlicensed user in a time interval of 10 ms is only 5 ms. However, if $T_p = 10$ ms, the transmission time becomes 9 ms. This tradeoff between the QoS performances of the licensed and unlicensed users has to be considered when determining this sensing period T_p.

If the transmitter of an unlicensed user is far from the receiver of a licensed user (i.e. primary user), depending on the interference temperature limit at the receiver of a licensed user, both the licensed and the unlicensed users (i.e. secondary users) could transmit their data simultaneously. In this case, the *interference range* is defined as the minimum distance that an unlicensed transmitter should be away from so that it does not cause "unacceptable" interference to a receiver of a licensed user. This interference range can be obtained from

$$\Gamma_l = \frac{P_l h(r)}{P_u h(d) + I_B}, \tag{7.1}$$

where P_l and P_u are the transmit power of the licensed and unlicensed users, respectively, $h(r)$ is the channel gain at a distance r from the transmitter, I_B is the background interference power at the receiver of a licensed user (Figure 7.9), Γ_l is a given target signal-to-interference ratio (SIR) at a receiver of a licensed user which is at a maximum distance r from a transmitter of a licensed user, and d is the interference range. To avoid causing an unacceptable interference to a receiver of a licensed user, an unlicensed user must be able to detect a signal from the licensed transmitter within the range of $d + r$.

The detection sensitivity of an unlicensed user γ_u, i.e. the lowest SNR for which the secondary user can still detect the signal from the licensed transmitter, can be obtained from

$$\gamma_u = \frac{P_l h(d+r)}{\sigma^2}, \tag{7.2}$$

where σ^2 is the noise power. During sensing, if the SNR of received signal from a licensed transmitter is lower than γ_u, an unlicensed user may assume that the spectrum is not occupied by the licensed user, and the unlicensed user may start transmission in this spectrum band.

In a cognitive radio environment, the detection of transmissions from licensed users is challenging due to the following reasons:

- *Channel uncertainty:* Channel uncertainty arises due to dynamic variations in the channel fading and shadowing conditions. Signal power received from a licensed transmitter could be lower than the detection sensitivity due to deep fade when the licensed receiver is in the interference range of an unlicensed user. A performance study on spectrum sensing under the effect of correlated shadowing was presented in [318].
- *Noise uncertainty:* To calculate the detection sensitivity of an unlicensed user, the noise power is required, which is usually not known. The uncertainty in noise power will affect the estimation of detection sensitivity, especially in the case of spectrum sensing through an energy detector, since this type of detection cannot distinguish between the signal from a licensed user and the noise signal. However, spectrum sensing based on feature detection is not severely impacted by this noise uncertainty.
- *Aggregated-interference uncertainty:* When there are many unlicensed users in the same cognitive radio network, they can interfere with each other. Since the number of unlicensed users and their transmission parameters may not be known, estimation of the interference due to the unlicensed users becomes very challenging. In particular, an unlicensed user may not be able to detect transmissions from a nearby licensed user due to interference caused by transmissions from other unlicensed users. Modeling of this aggregated interference will be useful to characterize the effects of network parameters (e.g. the number, location, transmit power, and propagation detail of users) on the performance of the cognitive radio network [319].

To mitigate the problem of uncertainty in spectrum sensing in a cognitive radio network, cooperative spectrum sensing can be used. Cooperative spectrum sensing provides diversity gain, which improves the detection performance of an unlicensed user [135]. In this case, multiple unlicensed users cooperatively sense the target spectrum and share the spectrum sensing results with each other [320]. One advantage of cooperative spectrum sensing is shown in Figure 7.10. In this figure, unlicensed user U_1 may not be able to detect transmission from licensed user L_1 due to channel fading. If U_1 starts transmission, it will interfere with data reception at the licensed node L_2. However, if unlicensed user U_2 senses the spectrum and reports the presence of licensed user L_1 to the controller (e.g. over a dedicated control channel), U_2 can be notified by this controller and will defer its transmission to avoid any interference to the licensed node L_2.

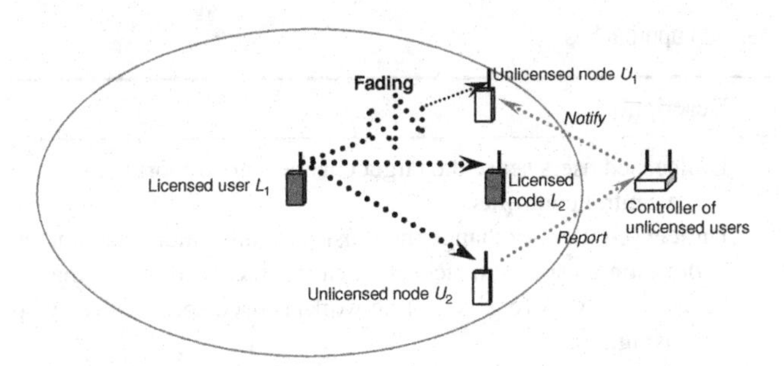

Figure 7.10 Cooperative spectrum sensing.

Alternatively, multiple relay users can forward the signal received from a licensed user to an unlicensed user to improve the performance of spectrum sensing through spatial diversity. This system model was analyzed in [321].

Clearly, cooperative spectrum sensing can be used to combat the noise and channel uncertainties, and, thereby, the probability of mis-detection and false alarm can be decreased. Note that the probability of mis-detection is defined as the probability that an occupied spectrum is sensed to be idle, while the probability of false alarm is the probability that an idle spectrum is sensed to be occupied by a licensed user. Also, cooperative spectrum sensing reduces the sensing time while improving the spectrum sensing accuracy. However, cooperative spectrum sensing incurs higher complexity and overhead (e.g. an increase in energy consumption [196]).

In the literature, different techniques were proposed for combining the spectrum sensing results from different unlicensed users and making decision in cooperative spectrum sensing. The simplest method is to use an OR operation among the received sensing results [322]. Combining techniques based on maximal ratio combining (MRC) and equal gain combining (EGC) were investigated in [323]. Based on a likelihood ratio test (LRT), an optimization problem can be formulated to obtain the optimal cooperative sensing parameters [324] for a given objective function and constraint (e.g. maximize the probability of detection while the probability of false alarm is lower than a target level). In [325], the presence of a licensed user was determined based on a false discovery rate measure, which is defined as the expected ratio of the number of channels falsely sensed to be occupied and the total number of channels sensed to be occupied.

7.3.1 Design tradeoff in spectrum sensing

Due to different choices of spectrum sensing design (Table 7.1), several performance tradeoffs exist, which can be summarized as follows:

- *Performance improvement from cooperation vs. communication overhead:* When cooperative spectrum sensing is employed, the overhead of communication among unlicensed users becomes a significant issue. In this case, a spectrum sensing manager or a central controller is required to collect and process the sensing results. The communication among unlicensed users and the sensing manager may be performed over a dedicated common channel. This requires an extra radio spectrum band. To minimize

Table 7.1 Summary of spectrum sensing approaches.

Approach	Description
Cooperative centralized spectrum sensing	Unlicensed users sense the target channels and report the sensing results to the central controller.
Cooperative distributed spectrum sensing	Unlicensed users exchange the sensing results among each other, and the decision on spectrum access by each unlicensed user is made locally.
Spectrum sensing using sensor networks	External network (e.g. sensor network) is used specifically for spectrum sensing.
Spectrum sensing and prediction	From the spectrum sensing database, unlicensed user anticipates and identifies spectrum opportunities.
Spectrum sensing standards	New wireless standards integrate spectrum sensing mechanism to improve the spectrum usages.

the communications overhead, a simple and effective protocol needs to be developed for unlicensed users. For example, each unlicensed user obtains its spectrum sensing result. Then, a censoring method can be applied to reduce the communications overhead. In this case, sensing data which is reliable enough is only used to determine whether a licensed user exists or not [195]. Alternatively, multiple unlicensed users can form a cluster. Then, the spectrum sensing results are combined and processed locally by a cluster head. Then, the cluster head of each cluster reports the result to a central controller to make a final decision. This approach can reduce the overhead due to communications with the central controller [194].

- *Reactive vs. proactive sensing:* Spectrum sensing can be either reactive or proactive. With reactive sensing, an unlicensed user performs spectrum sensing only when this unlicensed user needs to access the spectrum (i.e. on-demand basis). On the other hand, with proactive sensing, an unlicensed user continuously senses the spectrum, and the spectrum sensing results are maintained in the database. When an unlicensed user wants to access the spectrum, it uses the database to locate the spectrum opportunity to be accessed. Reactive sensing incurs a lower overhead since the sensing is performed when it is needed. However, the delay is longer since unlicensed users have to finish sensing before accessing the spectrum. In contrast, proactive sensing incurs a larger overhead but lower access delays. Selection between these spectrum sensing methods depends on application requirements. For a delay-sensitive application, proactive spectrum sensing is preferred. However, for an energy-constrained application, reactive spectrum sensing could reduce energy consumption since sensing activities, which may consume a considerable amount of energy, can be minimized.
- *Rate and reliability tradeoff due to multiband access:* Instead of transmitting on a single spectrum band, a cognitive radio can spread its transmission over multiple bands to achieve a higher rate and reliability. Transmission over multiple bands can be achieved using OFDM techniques which provides the flexibility of using non-contiguous frequency bands for wireless transmissions. The data rate can be improved when multiple streams are transmitted concurrently over multiple bands, while data reliability can be enhanced when the same data stream is transmitted redundantly

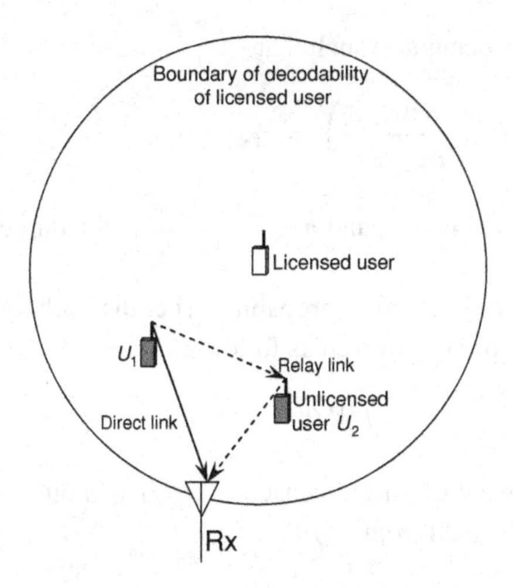

Figure 7.11 Cooperation in a cognitive radio network.

over multiple bands. However, accessing multiple bands requires an unlicensed user to sense and maintain the spectrum usage statistics for these bands. This multi-band sensing could incur considerable overhead. The more the number of bands, the larger will be the overhead but the higher will be transmission rate or reliability. An optimal decision would therefore be required for frequency band selection (e.g. how many and which bands to select for sensing).

7.3.2 Cooperative centralized spectrum sensing

In cooperative centralized spectrum sensing, the unlicensed users sense the target channels and report the sensing results to the central controller. The controller processes the spectrum sensing results and uses them to make decisions on spectrum access by the unlicensed users [197, 198, 326, 327, 328, 329, 330, 331]. A number of such schemes are described below.

A central controller was assumed in [197, 198] to coordinate spectrum sensing among multiple unlicensed users (e.g. Figure 7.11). In the system model, amplify and forward (AF) cooperative diversity transmission [332] was considered. Let us consider the case of two unlicensed users (i.e. U_1 and U_2) and a licensed user where the unlicensed users perform spectrum sensing based on energy detection [333]. The detection works as follows. User U_1 transmits the sensed information in the first time slot, while U_2 listens. In the second time slot, U_2 transmits the received data from U_1 during the first time slot, while U_1 also listens to this relayed data. In this case, the detection probability of U_1 with cooperation from U_2 is defined as follows:

$$P_{\mathrm{c}}^{(1)} = \phi(\lambda, P_1 + 1, \beta(P_2 + 1)), \tag{7.3}$$

where $\phi(t; a, b) = \int_0^\infty \mathrm{e}^{-h - \frac{t}{a+bh}} \, \mathrm{d}h$, P_i is the transmit power of unlicensed user U_i. The threshold λ is obtained from a given probability of false alarm P_{f} (i.e. the probability that an idle spectrum is sensed to be occupied) such that

$$\phi\left(\lambda; 1, \frac{Ph_{1,2}}{\tilde{P}h_{1,2}+1}\right) = P_{\mathrm{f}}, \tag{7.4}$$

where $\tilde{P}$ is the maximum transmission power, and $h_{1,2}$ is the channel gain between the users U_1 and U_2.

With a central controller, the overall detection probability (i.e. the probability that a licensed user is detected by user U_1 or U_2) is given as follows:

$$P_{\mathrm{c}}^{(1)} + P_{\mathrm{n}}^{(2)} - P_{\mathrm{c}}^{(1)} P_{\mathrm{n}}^{(2)}, \tag{7.5}$$

where $P_{\mathrm{n}}^{(2)}$ is the detection probability of an unlicensed user U_2 without cooperation from user U_1. This probability is obtained from

$$P_{\mathrm{n}}^{(2)} = (P_{\mathrm{f}})^{\frac{1}{P_2+1}}. \tag{7.6}$$

It was shown that this cooperative spectrum sensing can significantly improve the probability of detection for a given false alarm probability.

7.3.3 Cooperative distributed spectrum sensing

In cooperative distributed spectrum sensing, the unlicensed users exchange the sensing results with each other, and the decision on spectrum access by each unlicensed user is made locally. This type of spectrum sensing does not require any infrastructure (e.g. a central controller) and would therefore be suitable for a practical cognitive radio network [334, 335, 336, 337]. Several such schemes are described below.

In [334], a decentralized coordinated spectrum sensing scheme was proposed for a low-cost wireless transceiver. This scheme is scalable and robust. If one unlicensed user cannot detect a signal from a licensed user, there will be other unlicensed users which can detect and share the results among all unlicensed users. This algorithm is developed based on a discrete-time gossip protocol [338] which works as follows. At each time slot, an unlicensed user senses the target spectrum. Then, an unlicensed user randomly selects one or more neighboring users and transmits the sensing result to them. This dissemination strategy of the sensing data is referred to as uniform gossip. Also, a random walk strategy was considered in which only a subset of unlicensed users (or designated users) communicate with each other in a time slot. That is, k users are randomly elected as the designated users. Each designated user transmits its sensing result to a random neighbor, which becomes the designated user for the subsequent time slot. These steps are repeated until all unlicensed users receive the sensing results. After each unlicensed user receives the sensing data from its neighbors, this data is combined with local sensing results by using the algorithm proposed by Flajolet and Martin (FM) [322], which is based on the order and duplicate insensitive (ODI) technique. This FM algorithm is based on OR operations carried out on the sensing data from the unlicensed users.

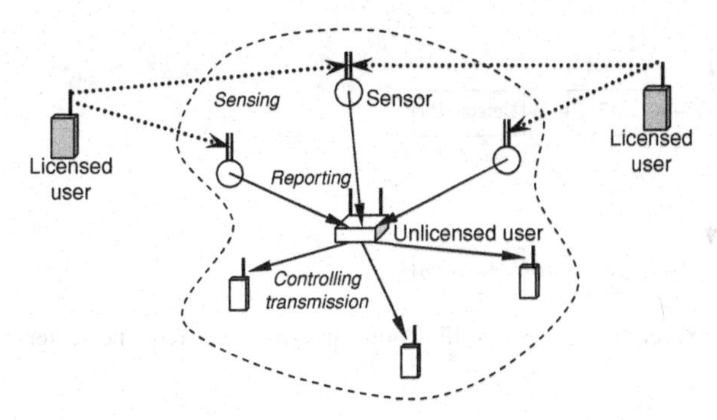

Figure 7.12 Spectrum sensing using external sensors.

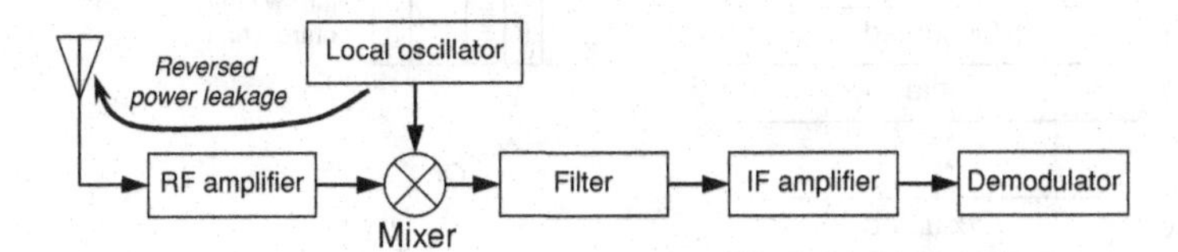

Figure 7.13 Superheterodyne receiver.

7.3.4 Spectrum sensing using sensor networks

Instead of using the resources of the unlicensed users for spectrum sensing, an external network (e.g. a sensor network) can be used specifically for spectrum sensing (Figure 7.12). The sensing results from this sensor network are sent to an unlicensed user to make a decision on spectrum access. Using this sensor network reduces the sensing overhead as well as improving the sensing performance under noise and channel uncertainties. Such a scheme was described in Chapter 3 (see Section 3.3). However, this approach requires special infrastructure in the target location which may not be available or could be costly to deploy in practice.

Sensor network to detect passive licensed users

A spectrum sensor network was proposed to detect a passive licensed user based on the local oscillator (LO) leakage power [134]. Since a passive licensed user (e.g. a TV receiver) does not transmit, an unlicensed user cannot observe the transmit signal directly. However, with a superheterodyne receiver (Figure 7.13) at the spectrum sensor, the power leakage from a local oscillator can be used as an indicator to detect the presence of a licensed user. For example, a TV receiver does not transmit any signal, but it can be interfered by transmission from an unlicensed user. In a TV receiver, the leakage oscillator frequency is about 41 MHz above the actual channel frequency. However, since this leakage oscillator power is small, direct detection of the licensed user activity (i.e. transmission from TV transmitter) may not be feasible for an unlicensed user (e.g. the unlicensed user may not be close enough to the TV receiver). Therefore, an external sensor network composed of small and cheap sensor nodes was used. The architecture of these sensor nodes is shown in Figure 7.14.

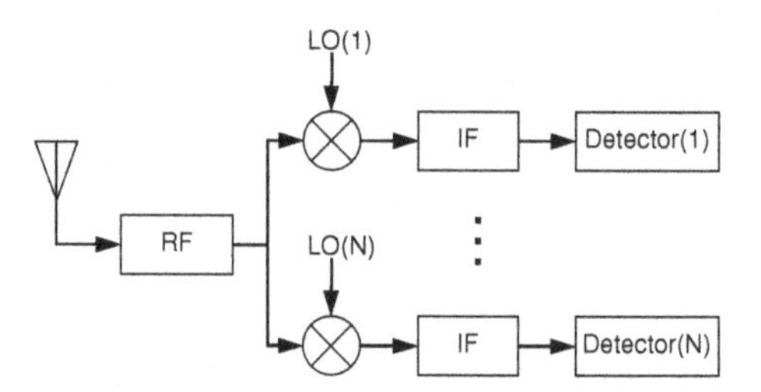

Figure 7.14 Sensor node architecture (where IF denotes intermediate frequency filters).

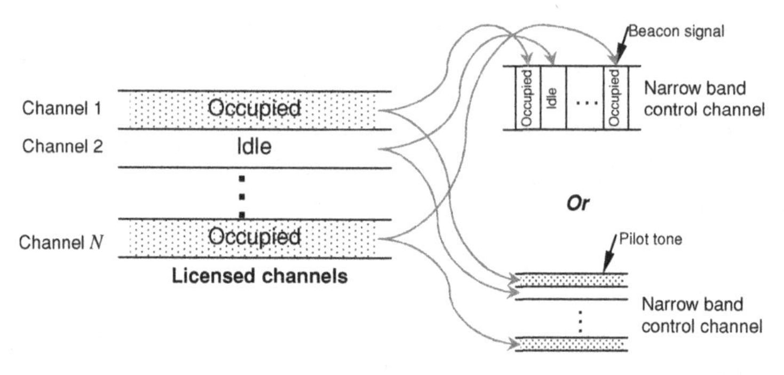

Figure 7.15 Pilot tone and beacon signal.

When a power leakage is detected, a sensor node will report it to an unlicensed user through a dedicated control channel. This can be done through a beacon message in the time slots associated with the licensed channels to indicate which of the channels are occupied by the licensed users (Figure 7.15, top right). An alternative is to transmit fixed power pilot tones over narrowband subchannels (Figure 7.15, bottom right) which correspond to the different licensed channels.

7.3.5 Spectrum sensing and prediction

To efficiently use the spectrum, an unlicensed user can maintain a database of information about spectrum usage. Based on historical information about spectrum occupancy by licensed users, estimation and prediction algorithms can be used to anticipate and identify spectrum opportunities as needed by an unlicensed user.

A study on spectrum occupancy characteristics and estimation of spectrum occupancy was presented in [339]. The activity of spectrum usage was modeled using a first-order Markov chain. The parameters of this Markov chain were obtained through sampled data. A measurement of spectrum occupancy over several years was performed in [340]. The spectrum occupancy in central Australia was measured and modeled as a Markov model in [341]. The model captures the effects of propagation condition as well as solar radiation. However, these works did not consider any prediction models, and are more useful for modeling long-term spectrum occupancy (e.g. over several months) rather than short-term spectrum occupancy which is required for dynamic spectrum access networks.

Prediction models based on a time series method for a cognitive radio network were proposed in [342, 343]. In [342], binary time series was used for spectrum prediction. The state of spectrum occupancy is indicated by a random variable $\mathcal{X}_t$ for which $\mathcal{X}_t = 1$ if the frequency band is not occupied by a licensed user at time step t (e.g. the measured power from the licensed user is below a threshold), and $\mathcal{X}_t = 0$ otherwise (e.g. the measured power is above a threshold). The objective of prediction is to obtain

$$\mu_t = \sum_{j=1}^{d} \beta_j \mathcal{X}_{t-j}, \tag{7.7}$$

where μ_t is the expected value for the spectrum status at time step t (i.e. time slot t), β_j is the prediction model parameter, and d is the maximum number of prediction model parameters. In other words, an attempt is made to predict the state of the frequency band at time step t by using the historical information from time step $t-1$ up to $t-d$. Any standard method such as the least squares method [344] can be used to obtain the set of parameters for the prediction model. In [342], a model for selecting the order (i.e. the optimal value of d in (7.7)) of the model was discussed. The optimal value of d can be chosen based on the Akaike information criteria (AIC) [344]. The performance of the prediction model was analyzed using two activity models, namely, non-deterministic and deterministic spectrum usage models.

When the CSMA/CA protocol in the IEEE 802.11 standard is used by a licensed user to access the spectrum, an unlicensed user can utilize the network allocation vector (NAV) to estimate the spectrum occupancy duration by a licensed user. This NAV information, which is embedded in the request-to-send (RTS) and clear-to-send (CTS) messages of the IEEE 802.11 MAC protocol, indicates the length of data transmissions by a licensed user winning the contention for spectrum access. In [343], an autoregressive (AR) model was used to predict the value of the NAV. In particular, based on measurement data over the 2.412 GHz frequency band, the parameters of the autoregressive model were estimated. For example, if a prediction is performed from 0 to 300 seconds, the regression model can be expressed as follows:

$$\begin{aligned}\hat{n}_{t+1} = {} & 1.1614 n_t - 1.0146 n_{t-1} + 1.0733 n_{t-2} - 0.8625 n_{t-3} + 0.8308 n_{t-4} \\ & - 0.6628 n_{t-5} + 0.5197 n_{t-6} - 0.2635 n_{t-7} + 0.1792 n_{t-8} \\ & - 0.0378 n_{t-9} + 0.0018,\end{aligned} \tag{7.8}$$

where $\hat{n}_{t+1}$ is a prediction of the value of the next NAV, and n_t is the value of the NAV at step t. It is observed that the autocorrelation function of the NAV decreases as the number of lags increases.

In the above works, the prediction results were not used for making spectrum access decisions. However, using the predicted information, the network performance can be improved by reducing the interference caused to the licensed users and the probability of collision with licensed users.

In [345], the cyclostationary behavior in the activity of a licensed user was used for predictive dynamic spectrum access. Based on the cyclostationary behavior, a simple statistic, namely, the probability that the spectrum is busy at a time step $t \bmod \tau$ (denoted by $S_0(t) = Pr(\mathcal{X}_t = 0)$ for $t \in [0, \tau]$), can be determined. Here the period of cyclostationarity τ can be obtained from the historical observations. Then, given the knowledge of $S_0(t)$, and the required transmission duration of an unlicensed user denoted by T_{tr}, the optimal transmission time t^* for dynamic spectrum access can be optimized as follows:

$$t^*(T_{tr}) = \arg \min_{t \in [0,\tau]} \int_t^{t+T_{\text{tr}}} S_0(t) \mathrm{d}t. \tag{7.9}$$

That is, the transmission time t^* minimizes the probability of collision between a licensed user and an unlicensed user. The accuracy of statistic $S_0(t)$ can be improved by considering the current state of the spectrum. This improved statistic, which is denoted by $S_1(t, t_0)$, is given by

$$S_1(t, t_0) = Pr(\mathcal{X}'(t_0, t) = 0 | \mathcal{X}_{t_0} = 1, \mathcal{X}_{t_0-1} = 0), \tag{7.10}$$

where $\mathcal{X}'(t_1, t_2) = 1 - \prod_{t=t_1}^{t_2} (\mathcal{X}_t)$. That is, $S_1(t, t_0)$ denotes the probability that the spectrum will remain unoccupied by a licensed user between time t_0 and time t. An optimization similar to that in (7.9) can be formulated to obtain the optimal transmission length for an unlicensed user.

7.3.6 Spectrum sensing standards

IEEE 802.11k

The IEEE 802.11k standard includes many channel measurements and management features [346] (e.g. channel load, noise histogram, traffic activity, and performance in a station). In this standard, an access point can collect and process channel information from each station. Then, the results can be used by an access point to control channel access by the users. For example, if it is observed that the interference level in one channel is higher than a given threshold, an access point could arrange for all stations to switch to a channel with a lower interference level. Also, if one access point is heavily loaded, a new station could be directed to connect to an access point with a lower load [347]. In this case, the information about average access delay can be used to estimate the amount of network load. IEEE 802.11k standard-based signaling can be used to determine the coverage area of an access point and its neighbors [348]. This coverage area information can be used for mobility management and load assignments in WLANs.

Bluetooth

Since Bluetooth® radio operates in the ISM band, detection of other devices (e.g. IEEE 802.11 devices and cordless phones) sharing the same spectrum, for example, through adaptive frequency hopping (AFH) [349, 350], would be useful to avoid performance degradation. Spectrum sensing is a part of this AFH to measure channel statistics and make a decision whether a target channel is occupied or not. The channel statistics

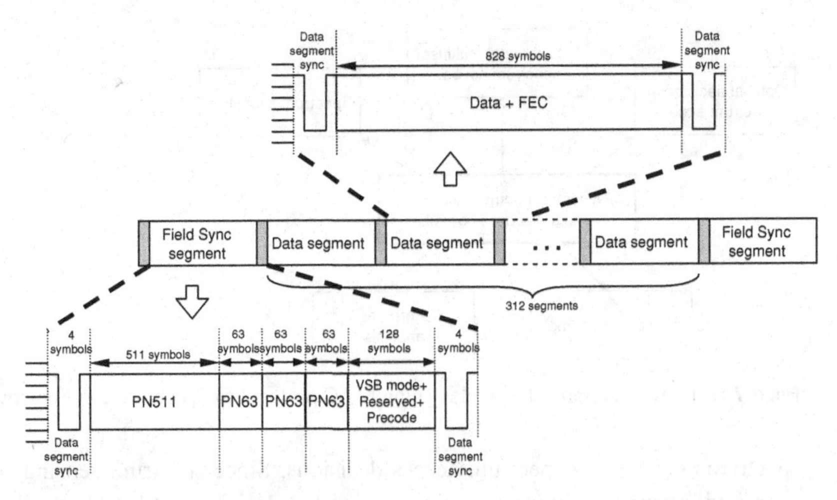

Figure 7.16 ATSC DTV frame format.

include packet-error rate, bit-error rate, received signal strength indicator (RSSI), and carrier-to-interference noise ratio (CINR).

IEEE 802.22

Spectrum sensing is a crucial part of the IEEE 802.22 standard to detect a signal from a TV transmitter. The standard will support fast and fine sensing [20]. Different techniques can be used for these fast and fine sensing. These techniques include energy detection, waveform-based sensing, cyclostationary feature detection, and matched filtering [351, 352, 353, 354, 355, 356, 357].

In [356], the signature of ATSC DTV signals was used for spectrum sensing. The DTV data are modulated by using 8-vestigial sideband (8-VSB) with 828 symbols in a data segment (Figure 7.16). The data segment sync (Segment Sync) is inserted at the beginning of each data frame. This length of Segment Sync is four symbols and each symbol can have either of two levels. Therefore, the total length of a data segment becomes 832 symbols. A total of 313 data segments are grouped together and the first data segment in a group is called the data field sync segment (Field Sync). The Field Sync occurs periodically every 24.2 ms. Due to this periodicity of the Field Sync, spectrum sensing can be performed easily to detect and identify the DTV signal (e.g. by using Field Sync-based algorithms in [356]).

7.4 Medium access control for dynamic spectrum access

To implement dynamic spectrum access for a cognitive radio network, different MAC protocols have been designed. Design of MAC protocols for cognitive radios poses significant challenges due to the requirement for the "peaceful" coexistence of unlicensed users with licensed users. Such a protocol needs to achieve the highest spectrum utilization by detecting all spectrum opportunities and access the spectrum so that collision with the other unlicensed users is minimized. The MAC protocol is responsible for

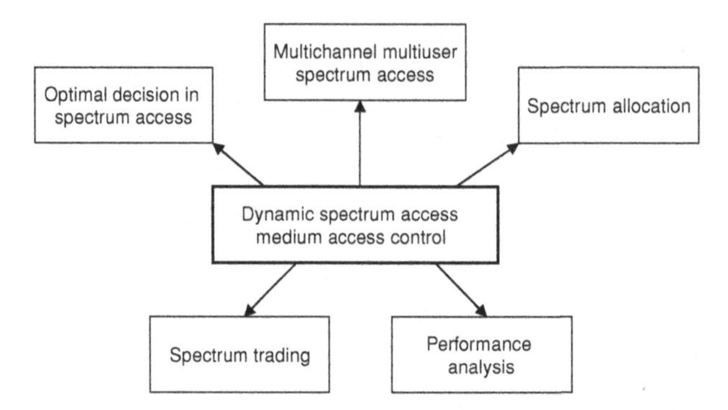

Figure 7.17 Different approaches in designing MAC protocol for a cognitive radio network.

spectrum sensing and spectrum access decisions. Since spectrum sensing is required over a wide range of frequencies (or channels), a decision needs to be made on which channels should be sensed (and in which sequence). Spectrum access decisions should consider the fact that the spectrum sensing results may not be completely accurate. Synchronization (in time, and the frequency band of transmission) between the unlicensed transmitter and the unlicensed receiver is also required for successful communication between the cognitive radios. Depending on the channel quality, transmission parameters (e.g. the modulation and coding level) can be adapted at the MAC level.

Different approaches to designing and analysis of MAC protocols for a cognitive radio network can be summarized as follows (Figure 7.17):

- *MAC protocol with optimal decision on spectrum sensing and spectrum access:* The main idea in this approach is to use some optimization model [152, 358, 359, 360] for optimizing spectrum sensing and spectrum access decision. The model can be for single or multiple channels and single or multiple users. For example, an unlicensed user has to decide which channel to be sensed and given the sensing result (i.e. observation), the unlicensed user decides further which channel to be accessed. These decisions are based on the objective (e.g. to maximize the transmission rate) and the constraints (e.g. the collision probability with a licensed user must be lower than the threshold). Usually, a stochastic optimization model is applied, since the activity of the licensed user is generally random.
- *MAC protocol for multichannel and multiuser cognitive radio systems:* The main objectives of this approach are to perform negotiations among unlicensed users for spectrum access in multichannel environments and to avoid collisions due to simultaneous transmissions [361, 362, 363, 364, 365, 366, 367, 368, 369].
- *MAC protocol based on spectrum allocation and scheduling:* In this approach, the MAC protocol is designed to optimally assign available frequency bands/time slots/transmit power to the unlicensed users [370, 371, 372, 373, 374, 375, 376].
- *MAC protocol to support spectrum trading:* A spectrum owner can sell the available spectrum opportunities to unlicensed users. MAC protocols can be developed to facilitate the spectrum trading functions (e.g. spectrum bidding and spectrum pricing) [377, 378, 379].

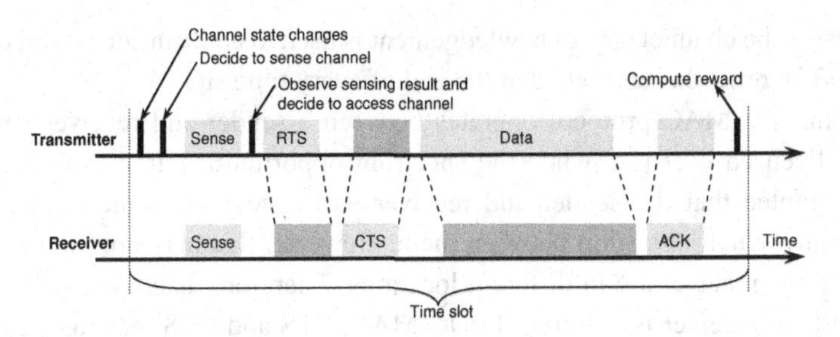

Figure 7.18 DC-MAC: sequence of operations in a time slot.

- *Performance analysis models for cognitive radio MAC protocols:* Analytical models were developed to investigate MAC and radio-link layer performances in a cognitive radio network. These analytical models would be useful for system optimization [380, 381, 382, 383, 384, 385, 386, 387].

7.4.1 Optimal decision on spectrum sensing and spectrum access

In a traditional wireless system, channel sensing involves carrier sensing in a single wireless channel (e.g. in CSMA/CA MAC). Channel access can be optimized according to the traffic load condition, QoS requirements, and wireless channel quality [388, 389, 390, 391]. However, in a cognitive radio network, channel sensing as well as channel access need to be optimized. This optimization must take the licensed users' activity behavior into account.

Optimization of joint spectrum sensing and spectrum access

To achieve the optimal performance in a distributed cognitive radio network, a MAC protocol must determine a set of channels to sense and a subset of channels to access. Assuming that the activity of a licensed user can be modeled as an on–off model represented by a two-state Markov chain, the problem of channel sensing and access in a spectrum overlay system was formulated as a partially observable Markov decision process (POMDP) [358]. For this POMDP model, the state of the system was defined in terms of channel occupancy (i.e. occupied or idle). The actions referred to sensing and accessing the subset of channels, and an observation referred to the channel sensing result. The reward was defined as the number of transmitted bits. The objective was to maximize the expected total number of transmitted bits in a certain number of time slots under the constraint that the collision probability with a licensed user should be maintained below a target level. This POMDP model can be solved using linear programming. However, when the number of channels increases, the state space of the POMDP formulation becomes large. Therefore, a suboptimal strategy based on a reduced number of states was also considered.

Based on the POMDP formulation, a decentralized cognitive MAC (DC-MAC) was proposed [358]. In DC-MAC, a few operations are performed in each time slot (Figure 7.18). First, the channel is sensed and the result of sensing is observed. The unlicensed user makes a decision on whether to access the channel or not. If an unlicensed user accesses the channel, an acknowledgement is used to confirm successful data transmissions. The reward is computed at the end of every time slot.

When the DC-MAC protocol operates between a sender and receiver neighboring the same licensed user (i.e. when the spectrum opportunities are spatially invariant), it can guarantee that the sender and receiver will access the same channel without requiring any synchronization between them. However, this may not be true when the sender and the receiver are in different locations. Therefore, handshaking between the sender and the receiver is required. In DC-MAC, RTS and CTS messages can be used for synchronization and also to mitigate the hidden and exposed terminal problems (Figure 7.18).

It was observed that the throughput obtained from the suboptimal solution of the POMDP model is close to that of the optimal solution, and both of these strategies achieve much higher throughput than that of random channel sensing and access policy. The performance of the proposed DC-MAC was also evaluated with an energy detector (i.e. a Neyman–Pearson detector) used in the physical layer. When the maximum probability of collision with licensed users increases, with the suboptimal strategy, the throughput of a secondary user approaches that of the optimal strategy. In such a case, the probability of false alarm (i.e. the probability of missing spectrum opportunities) decreases and an unlicensed user can make greater use of the spectrum opportunities. However, as the maximum probability of collision increases beyond a certain point, throughput decreases since only a small number of transmissions are successful.

However, in this work the number of channels to be sensed in one time slot is not optimized. This optimization is especially important for the cognitive transceiver with limited hardware.

Optimal sensing for hardware-constrained cognitive radio

With a constraint on radio hardware (e.g. a single radio transceiver), spectrum sensing must be optimally performed to achieve the best performance. With this in mind, a hardware-constrained multichannel cognitive MAC (HC-MAC) protocol was proposed in [359]. Channel sensing in HC-MAC was formulated as an optimal stopping problem [392], with the objective of maximizing the throughput.

Let T_s denote the time period used by an unlicensed user to sense the channel, and let T_t denote the length of the transmission period which depends on the acceptable level of interference to the licensed users. The problem of optimal channel sensing arises due to the fact that when an unlicensed user spends a longer amount of time in sensing, a shorter amount of time is available to it for channel access. An example is shown in Figure 7.19. In Figure 7.19(a), an unlicensed user senses two channels (i.e. $n-1$ and n) and then stops. In this case, channel $n-1$ is unoccupied while channel n is occupied. The unlicensed user accesses channel $n-1$. As a result, the achievable data rate is $rWT_t/(T_t+2T_s)$, where W is the channel bandwidth and r is the spectral efficiency (i.e. rate) of transmission. However, if an unlicensed user does not stop and continues sensing channel $n+1$ (Figure 7.19(b)), which is occupied by a licensed user, the data rate of the unlicensed user becomes $rWT_t/(T_t+3T_s)$. Therefore, better performance is achieved when the unlicensed user stops sensing at channel n. By using the theory of

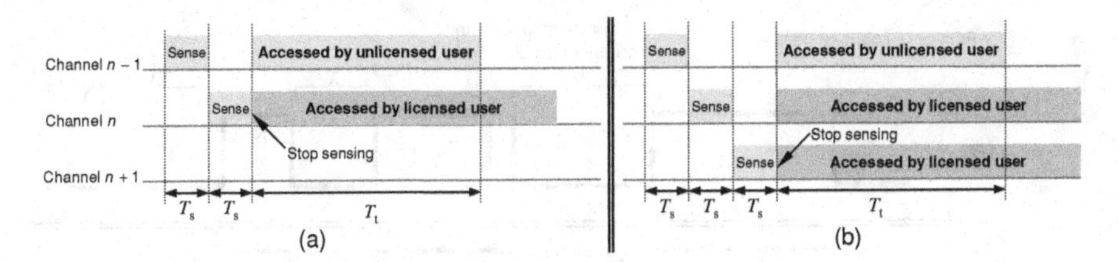

Figure 7.19 Overhead of channel sensing.

optimal stopping, the optimal number of sensing channels can be obtained so that the data rate is maximized.

The maximum number of idle channels within n' adjacent channels (where channel $1, \ldots, n'$ are sensed) can be obtained from (given activity x_n in channel n) [392]

$$\mathscr{M}_{n'}(x_1, \ldots, x_{n'}) = \max_{\substack{1 \leq n_s \leq n_e \leq n' \\ n_e - n_s + 1 \leq N_a \\ \mathscr{F}(n_s, n_e) \leq N_f}} \sum_{n=n_s}^{n_e} (x_n), \tag{7.11}$$

where $\{n_s, n_s + 1, \ldots, n_e\}$ denotes a set of adjacent channels, $\mathscr{F}(n_s, n_e)$ is a function indicating the number of fragments of idle channels, N_a denotes the maximum number of adjacent channels that an unlicensed user can simultaneously access, and N_f is the maximum number of spectrum fragments which can be aggregated and accessed by an unlicensed user. The reward function of sensing n channels can be defined as follows:

$$\mathscr{J}_n = \frac{T_t}{T_t + nT_s} \mathscr{M}_n(x_1, \ldots, x_n). \tag{7.12}$$

To obtain n, this problem can be solved using backward induction, starting from the maximum number of channels which an unlicensed user can sense.

It was shown that HC-MAC can achieve a higher throughput than a fixed channel assignment MAC. Also, the throughput of an unlicensed user increases as the number of available channels increases since there is a higher chance that an idle channel can be found.

Since the size of the time slot is assumed to be constant, the proposed schemes may not be applicable to an unslotted system, i.e. where the licensed user can start transmission at any time arbitrarily.

Optimal discovery of spectrum opportunities

When a channel is periodically sensed, the sensing period can be optimized to maximize the discovery of spectrum opportunities [152]. For example, if the sensing period is large (i.e. a channel is sensed infrequently), spectrum opportunities may not be fully discovered. However, if the sensing period is small, the sensing overhead becomes large. Therefore, the ratio $T_{s,i}/T_{p,i}$, where $T_{s,i}$ is the sensing time interval and $T_{p,i}$ is the length of duration between two consecutive channel sensing (i.e. sensing period) of channel i, can be optimized. An example is shown in Figure 7.20. In Figure 7.20(c), the

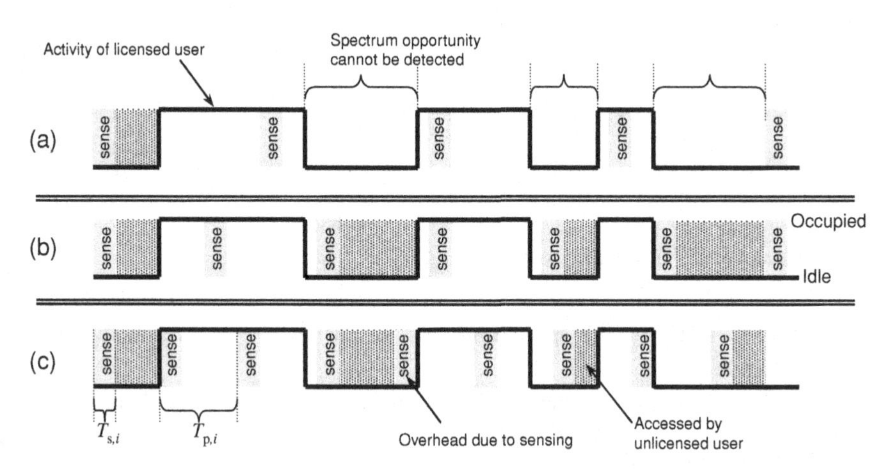

Figure 7.20 Sensing time and sensing period.

sensing period is too small, and, therefore, channel access by an unlicensed user is not maximized due to the large overhead. On the other hand, in Figure 7.20(a), the sensing period is too large, and only a small number of spectrum opportunities are found. In Figure 7.20(b), the sensing period is optimally chosen so that the maximum number of spectrum opportunities are found.

If u_i denotes the average proportion of time that channel i is occupied by the licensed users, an optimization problem can be formulated as follows:

$$\text{Maximize:} \sum_{i=1}^{N}(1 - u_i) - \text{SSOH}_i(\mathbf{T}_\text{p}) - \text{UOPP}_i(T_{\text{p},i}), \tag{7.13}$$

where N is the total number of channels, and $\mathbf{T}_\text{p}$ is a vector of sensing periods for all channels. Function $\text{SSOH}_i(\mathbf{T}_\text{p})$ is defined as the *sensing overhead* and $\text{UOPP}_i(T_{\text{p},i})$ is defined as the *unexplored opportunity*. While the sensing overhead can be simply expressed as the ratio of sensing time and sensing period, the unexplored opportunity is obtained by considering the probability distribution of the "off" period for a licensed user. From this optimization problem, the optimal vector of sensing periods can be obtained numerically.

In addition, the channel-switching latency incurred due to searching idle channels can be optimized through a channel sensing sequence. In this case, an unlicensed user senses the channels one-by-one, and stops if an idle channel is found. To obtain the optimal channel sensing sequence, the probability for a channel to become idle is calculated based on the historical channel sensing data.

The performance of the optimized spectrum sensing in the MAC layer was evaluated using achieved opportunity ratio (AOR) and channel switching latency (CSL) metrics. The achieved opportunity ratio is a measure of the ratio of the total discovered spectrum opportunities to the total available opportunities. The channel switching latency is defined as the average time for an unlicensed user to find an idle channel. The proposed spectrum sensing was compared with a static scheme in which the spectrum sensing period is fixed. In all the evaluation scenarios, the proposed spectrum sensing scheme outperformed the static scheme in terms of both achieved opportunity ratio and channel switching latency. That is, with the proposed spectrum sensing, more spectrum opportunities can be discovered and an unlicensed user can access the spectrum opportunities as soon as they are detected.

While this work can optimize the parameters of spectrum sensing, it relies on the information available from the licensed user. However, if this information is not available, the learning algorithm can be applied to estimate and gain knowledge about the environment.

Learning-based optimal channel selection

In [360], a channel selection scheme based on stochastic control theory was proposed. A system model with N channels was considered. The activity of the licensed users on each channel was modeled as an on–off process where the durations of on and off periods were geometrically distributed random variables.

To estimate the probability that a time slot of channel n is idle, a Bayesian learning algorithm was applied. The posterior distribution $f_t^{\text{idle}}(\mathcal{X})$ of the random variable $\mathcal{X} \in \{0, 1\}$ at period t indicating an idle channel was updated by the standard Bayesian rule as follows:

$$\tilde{p}_t^{\text{idle}} = \frac{Y_{\text{idle}}(t) + 1}{Y_{\text{busy}}(t) + Y_{\text{idle}}(t) + 2}, \tag{7.14}$$

where $Y_{\text{busy}}(t)$ and $Y_{\text{idle}}(t)$ are the number of times that the channel is sensed to be busy and idle, respectively. Similarly, the successful channel access probability was updated as follows:

$$\tilde{p}_t^{\text{succ}} = \frac{Y_{\text{succ}}(t) + 1}{Y_{\text{succ}}(t) + Y_{\text{fail}}(t) + 2}, \tag{7.15}$$

where $Y_{\text{succ}}(t)$ and $Y_{\text{fail}}(t)$ are the number of times the transmission of a packet is successful and unsuccessful, respectively.

An optimal channel allocation was formulated based on a multi-armed bandit problem [393]. There are two steps in this channel allocation algorithm. The first step is to learn (i.e. estimate) the probability of successful transmissions in each channel. The second step is to choose the channel with the highest probability of successful transmissions. This problem can be solved by using a dynamic allocation index for each channel, which is a function of state $x_n(t)$ of channel n. This dynamic allocation index can be defined as follows:

$$v_n(x_n) = \max_{\tau} \frac{E\left[\sum_{\tau=1}^{t_1} \beta^{\tau} \mathcal{I}_n(t) | x_n(1) = x_n\right]}{E\left[\sum_{\tau=1}^{t_1} \beta^{\tau} x_n(1) = x_n\right]}, \tag{7.16}$$

where β is the discount factor, τ is the stopping time, and $\mathcal{I}_n(t)$ is the reward function defined as follows:

$$\mathcal{I}_n(t) = \begin{cases} 0, & \text{channel is busy} \\ L, & \text{successful transmission} \\ -\alpha L, & \text{transmission failure,} \end{cases} \tag{7.17}$$

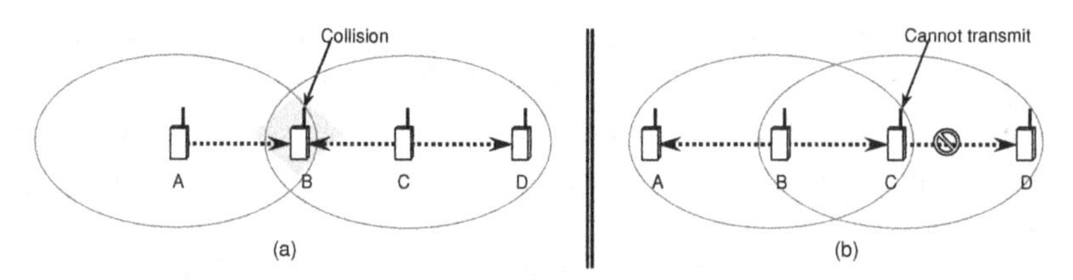

Figure 7.21 (a) Hidden and (b) exposed terminal problems.

where L is the packet size and α is the energy cost factor indicating the energy consumption of transmitting a packet of size L bits. In this case, the expected reward in period t can be obtained from

$$E\left[\mathcal{J}(t)\right] = \tilde{P}_t^{\text{idle}} \tilde{P}_t^{\text{succ}} L - \tilde{P}_t^{\text{idle}}(1 - \tilde{P}_t^{\text{succ}})\alpha L. \tag{7.18}$$

It was observed that the proposed channel allocation and selection can quickly converge to an optimal solution. Consequently, the achieved reward is higher than that for other schemes (i.e. it minimizes the channel selection failure rate).

The above works considered the problem of optimizing the channel sensing and access to achieve the optimal performance. However, they were not designed specifically for cognitive radio transmissions in a multiuser environment. In such an environment collision avoidance will be a key requirement.

7.4.2 Multichannel and multiuser MAC

A number of multichannel multiuser MAC protocols were proposed for wireless networks [394, 395, 396, 397, 42]. However, these protocols were designed based on an assumption that the channels are available all the time. This assumption does not hold in a multichannel and multiuser cognitive radio network since the channels could be occupied by licensed users. Therefore, the channels available to the different unlicensed users can be different. Also, if there are multiple unlicensed users in the same area, the hidden and exposed terminal problems arise and these need to be mitigated.

The hidden terminal problem arises when some of the unlicensed users are unaware of the transmissions of other unlicensed users, and as a consequence collision can occur. For example, in Figure 7.21(a), node C is unaware of transmissions from node A to node B since node C is out of the transmission range of node A. As a result, node C may start transmission to node D, but this transmission will collide with the reception of node B.

The exposed terminal problem arises when some of the nodes overhear the transmissions from other nodes. In this case, the overheard nodes will defer transmission, which will result in performance degradation. For example, in Figure 7.21(b), node B transmits to node A, but node C overhears the transmission. As a result, node C defers transmission to node D, although this would not interfere with the reception at node C.

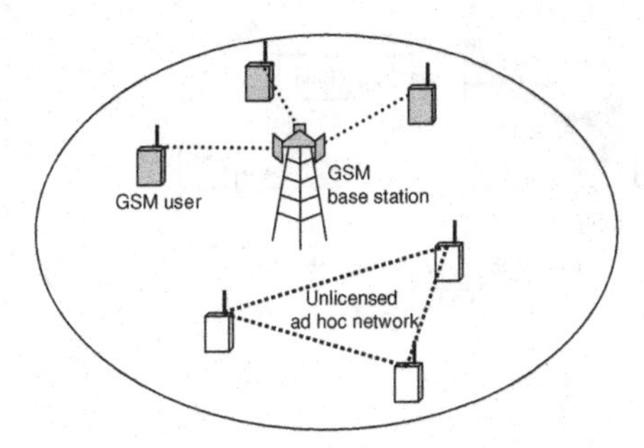

Figure 7.22 An ad hoc network of secondary users overlaid on a GSM network.

Ad hoc secondary-system MAC (AS-MAC) for spectrum sharing

The problem of spectrum sharing and management in a cellular wireless system has been studied in the literature [361, 398, 399, 400, 401, 402, 403, 404, 405, 406]. However, the concepts of opportunistic spectrum access and interference-controlled spectrum access and the issues related to the coexistence of primary and secondary users have not been addressed in these works.

An opportunistic MAC protocol, namely, ad hoc secondary system MAC (AS-MAC), was proposed for an ad hoc network overlayed on a GSM cellular network (Figure 7.22) [361]. AS-MAC provides the ad hoc nodes with the necessary functionality to observe licensed user activity on GSM channels and opportunistically access the available channels without interfering with the GSM users.

With AS-MAC, a cognitive radio detects the boundary of the time slot of a GSM transmission by decoding the frequency correction channel (FCCH) from the GSM base station. Then, the cell global identity (CGI) and cell channel description (CCD) are decoded from the synchronization channel (SCH) and the broadcast control channel (BCCH). The channels currently being used in a GSM cell can be determined from this information. After the information is obtained, the cognitive radio node observes the channel usage by sensing licensed user activity at the beginning of a time slot. With time slot size of 577 μs, the sensing duration of a cognitive radio node is less than $T_s = 15$ μs.

If a cognitive radio node in the ad hoc network has data to transmit, it senses the channel to detect an available time slot. Then, to avoid collision, this sender waits for a time period of T_{rts} and transmits an RTS message. This T_{rts} is uniformly distributed between 40 and 140 μs. The RTS message contains the number of time slots for data transmission, which is referred to as the NAV and is similar to that in the IEEE 802.11 CSMA/CA protocol. After the receiver receives the RTS message, it senses the channel for an available time slot and transmits the CTS message. This CTS message contains the ID of the receiver, the NAV, and the time slot for communication. This time slot is selected from among the free slots in a transmission frame. After the sender receives the CTS message, the data is fragmented into multiple packets and transmitted over the designated time slots. To enhance the transmission reliability, a selective-repeat automatic repeat request (ARQ) mechanism is used along with AS-MAC. After all the packets are transmitted, the sender waits for an ACK message. This ACK message

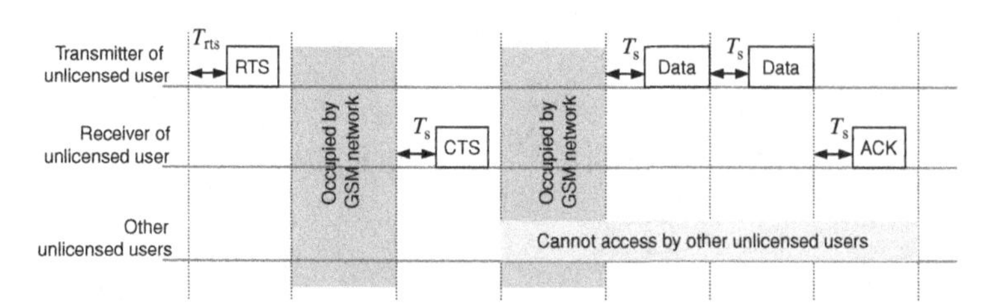

Figure 7.23 AS-MAC protocol.

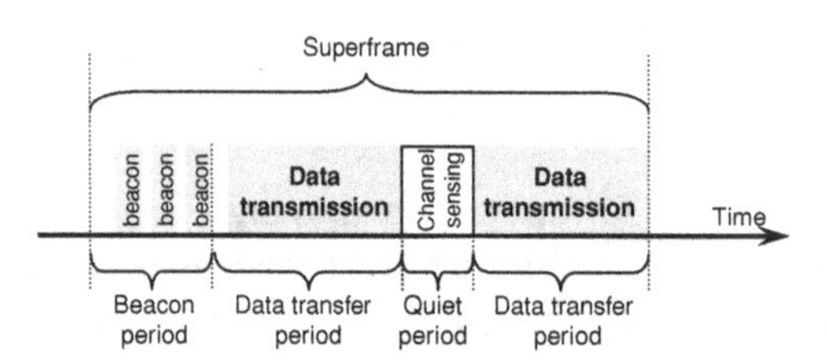

Figure 7.24 Superframe structure of C-MAC.

contains a list of successfully received packets. Based on this information, the sender re-transmits only the lost packets. The AS-MAC protocol is shown in Figure 7.23.

Note that in designing this AS-MAC, the propagation delay was taken into account. This propagation delay could result in 200 μs of overlapping duration of transmission from GSM base station to mobile and transmission between ad hoc nodes. To avoid this problem, the guard band can be lengthened, but this results in a performance reduction. Therefore, in AS-MAC, overlapping duration is allowed if interference to GSM users can be maintained below the target level.

Performance of the AS-MAC protocol was evaluated in terms of bandwidth utilization, throughput, and queueing delay for an unlicensed user. It was observed that the performances (e.g. throughput and queueing delay) of an unlicensed user depend largely on the traffic load in the GSM network. However, this work did not consider the problem of synchronization between the unlicensed users.

Slotted-beaconing period and rendezvous channel

One of the challenges in designing MAC protocols for a multihop cognitive radio network is to establish synchronization among cognitive radio transmitters and the corresponding receivers. To support this synchronization, a cognitive MAC (C-MAC) protocol was proposed in [362]. The major component of this C-MAC is a rendezvous channel (RC), which is selected dynamically by the secondary users through a distributed mechanism. The RC is used mainly for information exchange among secondary nodes.

The major functions of the rendezvous channel are: (1) to support network-wide group communication (i.e. multicast, broadcast), (2) to discover neighboring unlicensed users, and (3) to balance the channel load among the available channels. In the C-MAC protocol, each channel uses a superframe structure (Figure 7.24), and one channel is identified as an RC. A superframe is composed of a beacon period (BP) and a data

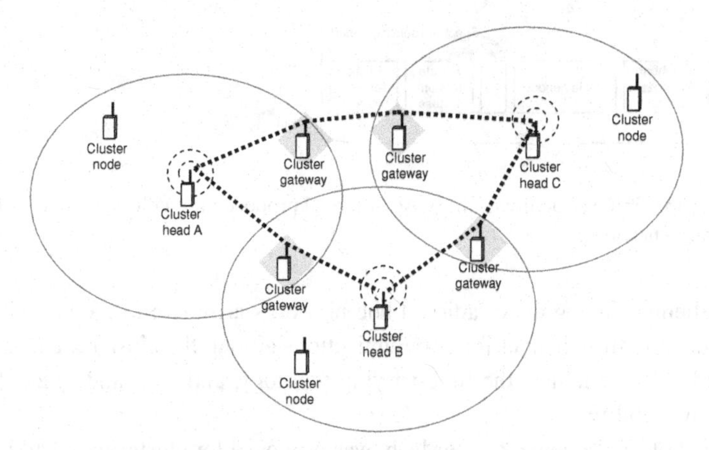

Figure 7.25 A cluster-based ad hoc cognitive radio network.

transfer period (DTP). The beacon period is divided into multiple slots. The first two slots are used by the new unlicensed users to join the network, while the rest are used by the existing unlicensed users. Here, every unlicensed user operating in different channels periodically visits the rendezvous channel to resynchronize with the network. After a new unlicensed user scans all channels, the rendezvous channel will be found, and the new node can obtain information on the channel usage and activity of other unlicensed users in order to identify any neighboring users and for use in load balancing.

During a beacon period, a cognitive radio user can observe channel activities in the ambient RF environment. This information (i.e. a new set of neighboring users and their transmission schedule) can mitigate the hidden terminal problem. For group communication, an unlicensed user with a group message visits the rendezvous channel and finds a transmission schedule. Then, the group message is transmitted during several superframes. Since all other unlicensed users will visit the RC periodically, the unlicensed users in the network are guaranteed to receive the group message. Finally, C-MAC can detect the presence of licensed users by using both in-band and out-of-band channel sensing. If a licensed user is found, an unlicensed user informs its neighbors by using the beacon signal, and the transmissions of all unlicensed users in the occupied channel are re-scheduled. Note that this work considered transmission between individual unlicensed users. However, the performance of a cognitive radio network can be improved by using the cluster-based communications, where communications in a cluster are controlled by the cluster head. With these cluster-based communications, the channel congestion can be alleviated.

MAC protocol for cluster-based cognitive radio networks and CogMesh

For a cognitive ad hoc network, cluster-based architectures are promising to reduce congestion in channel access by separating local and remote traffic [144, 407, 408, 409, 410, 411, 412, 413] (Figure 7.25). A cluster can be formed by a group of nodes, and the local traffic within a cluster can be exchanged among these nodes through the cluster head. The inter-cluster traffic can be transmitted through the gateway. The

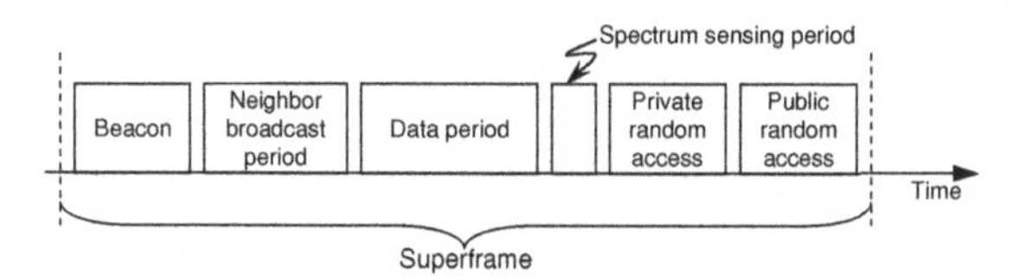

Figure 7.26 Superframe structure of the MAC protocol proposed for a cluster-based ad hoc cognitive radio network.

major challenges in cognitive radio ad hoc networks arise due to the unavailability of any common control channel for communications among the cluster members, cluster heads, and gateway nodes, the time-varying topology, and time and space-dependent spectrum availability.

In [366, 414], a framework, CogMesh, was proposed for cluster-based ad hoc cognitive radio networks. In this network, the unlicensed users/nodes form an ad hoc network by exploiting the spectrum opportunities. It is assumed that there is no common channel for the entire network. The unlicensed nodes can establish clusters on available channels. To access the same channel, nodes can join a cluster by communicating with the corresponding cluster head. A MAC protocol was proposed to provide multiple channel access to avoid collision, spectrum opportunity detection, neighbor discovery and cluster formation, inter-cluster communication, and topology management. In this MAC protocol, the transmission time is divided into multiple superframes. Each superframe is divided into five parts: beacon period, neighborhood broadcasting period (NBP), data transmission period, intra-cluster random access period (RAP), and public random access period (Figure 7.26).

In the superframe structure, a beacon period is transmitted by a cluster head for time synchronization, and resource allocation information for the corresponding cluster. The neighborhood broadcasting period is used for the nodes in a cluster to broadcast information to other nodes. This period is divided into fixed length slots, and each slot is used by a member node to transmit its identity, its list of neighboring nodes, and the corresponding channels. Also, the random access period of each member is transmitted in this slot. The slot allocation is performed by a cluster head. In the data transmission period, the unlicensed nodes in a cluster transmit data. In this case, the transmission can be concurrent if different channels are used by different members. The multiple access scheme for the same channel is based on time-division multiple access (TDMA). The intra-cluster random access period is used for the unlicensed nodes in a cluster to exchange control messages with each other. The public random access period is used by an unlicensed node to join the cluster. Also, this period is used to exchange information between neighbors. Transmission in these random access periods is based on slotted-ALOHA. In addition, to detect spectrum opportunity, a cluster head assigns a certain period after a data transmission period (as shown in Figure 7.26). In this period, all unlicensed nodes in a cluster perform spectrum sensing and standard methods for spectrum sensing (e.g. energy detection and cyclostationary feature detection) can be applied.

When an unlicensed node wants to join a network, it scans one of the channels and looks for a beacon message:

- If no beacon message is detected, the unlicensed node becomes a cluster head and uses the scanned channel to form a cluster.
- If a beacon message is detected, the unlicensed node sends a request message to join a cluster during the public random access period. A slot in the neighborhood broadcasting period will be assigned to the node if a cluster head accepts a request. However, if the request is rejected (e.g. there is no available slot), the unlicensed node scans other channels to join another cluster or becomes a cluster head itself.
- If an unlicensed node detects only messages from neighboring nodes, but no beacon messages, there is a cluster head within a distance of two hops. In this case, the unlicensed node exchanges information with the neighboring nodes during the public random access period, and scans other channels. Again, the node can join another cluster or become a cluster head depending on whether a beacon message is received in other channels or not.

For inter-cluster connections, there are two possible cases where two clusters are either overlapping or non-overlapping. In the overlapping case, there is at least one unlicensed node which can communicate with the cluster heads of both clusters in one hop. In the non-overlapping case, the members of two clusters can communicate with each other, but not to the cluster head directly. For topology management, the MAC protocol provides support for an unlicensed node to join and leave a network, to detect spectrum opportunities, to change the common channel used by a cluster head (e.g. due to the unavailability of spectrum), and for cluster merging.

The performance of the proposed protocol for a cluster-based ad hoc cognitive radio network was evaluated through simulations. When the number of unlicensed nodes in the network increases, the number of clusters increases. The proposed protocol can merge small clusters to larger clusters so that network congestion and communication overhead are reduced.

Dynamic open spectrum sharing (DOSS) protocol

The dynamic open spectrum sharing (DOSS) scheme proposed in [363] for an ad hoc cognitive radio environment allows an unlicensed user to adaptively select an arbitrary frequency band. The DOSS scheme uses a busy tone [415] to mitigate the hidden and exposed terminal problems. This busy tone is transmitted in a narrow band channel which can be mapped from a wideband data channel. Let the band corresponding to a busy tone channel and a data channel be denoted by $[f_{\mathrm{l}}, f_{\mathrm{u}}]$ and $[F_{\mathrm{l}}, F_{\mathrm{u}}]$, respectively. The mapping can be expressed as follows:

$$\mathscr{G}(z) = \frac{1}{F_{\mathrm{u}} - F_{\mathrm{l}}}((f_{\mathrm{u}} - f_{\mathrm{l}})z + f_{\mathrm{l}}F_{\mathrm{u}} - f_{\mathrm{u}}F_{\mathrm{l}}), \tag{7.19}$$

where $\mathscr{G}(z)$ is a mapping function and z is a design parameter. An illustration of this mapping is shown in Figure 7.27.

When an unlicensed user has data to transmit, the DOSS protocol works as follows. First, both the transmitter and the receiver locate an idle frequency band by listening to the busy tone channel. Then, the transmitter sends a REQ message over the control channel to indicate the center frequency and the bandwidth. From the REQ message,

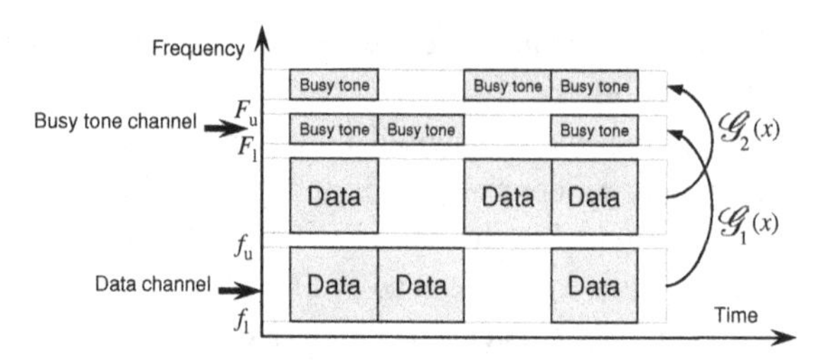

Figure 7.27 Channel mapping in DOSS.

the receiver identifies the frequency band to be used by the transmitter. A REQ_ACK message is then transmitted by the receiver over the control channel. After receiving the REQ_ACK message, the transmitter starts transmitting data. While the receiver receives data, it transmits a busy tone on the mapped narrow band channel to inform the other nodes of the transmission. If the data is successfully received, the receiver replies with DATA_ACK.

An analytical model was developed to study the performance of this DOSS protocol. This analytical model can be used to obtain the capacity of an unlicensed user operating on the spectrum allocated to a licensed user. It was shown that the DOSS protocol can achieve a high capacity gain by opportunistically accessing the spectrum. Also, the access delay due to negotiation between a cognitive radio transmitter and receiver was analyzed.

While most of the proposed MAC protocols were designed with the local sensing capability, the collaborative spectrum sensing can be used to improved the performance. However, the MAC protocols need to facilitate the information sharing for collaborative spectrum sensing.

Multichannel MAC protocol with collaborative channel sensing

In a distributed environment, the performance of a cognitive MAC protocol can be improved through collaborative sensing. In this case, each unlicensed user performs spectrum sensing on different channels, and the sensing results are exchanged among each other. For a multichannel cognitive radio environment, two collaborative channel sensing policies, namely, random sensing policy [416] and negotiation-based sensing policy, were proposed and analyzed in [417].

In the system model, there are N channels for the licensed users and the channel access activity for each user is modeled by a two-state Markov chain (i.e. on–off model). There are I unlicensed users, and each of them has two radio interfaces. One interface is used for communication/negotiation on a control channel (i.e. control interface), while the other interface uses software-defined radio to sense and access the channel (i.e. an SDR interface). Both licensed and unlicensed users use time-slot-based transmissions. On a control channel, each time slot is divided into reporting and negotiation phases (Figure 7.28). The reporting phase is divided into N mini-slots, each of which corresponds to one of the N channels. The length of a mini-slot was assumed to be 9 μs based on the IEEE 802.11a standard.

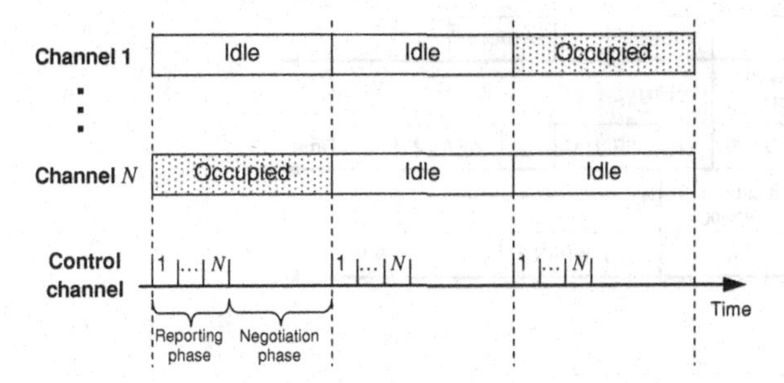

Figure 7.28 Reporting and negotiation phases.

The MAC protocol works as follows. At the beginning of a time slot, an unlicensed user performs spectrum sensing by using the SDR interface. Then, the channel status is reported by the control interface, which sends a beacon signal in the corresponding mini-slot in the control channel. Based on the information obtained in the reporting phase, the unlicensed users can use the control interface to contend for transmission by using the RTS-CTS mechanism. A successful user will transmit data on the data channel using the SDR interface.

With a random sensing policy, each unlicensed user chooses one of the N channels to sense with probability $1/N$. This random channel sensing policy can be modeled as a Markov chain. For a given number of unlicensed users I, the steady state probability that n channels will be sensed in a time slot can be obtained.

With a negotiation-based sensing policy, one unlicensed user observes the channels sensed by its neighbors. Initially, each unlicensed user selects a channel to be sensed randomly, and then the status of the sensed channel is reported by a beacon signal on the control channel. When an idle channel is found, an unlicensed user piggybacks the information on the sensed channels in the RTS/CTS messages. Observing these messages, the neighboring unlicensed users can select different channels. It was shown that this negotiation-based sensing policy converges to the desired state if the number of unlicensed users is equal to or more than the number of channels.

The performance of the proposed MAC protocol was analyzed with both random sensing and negotiation-based sensing policies. It was observed that the saturation throughput of the proposed negotiation-based sensing policy is higher than that of a random sensing policy. However, this throughput is affected by the number of unlicensed users. In particular, the throughput linearly increases as the number of unlicensed users increases until it becomes constant when the number of unlicensed users becomes equal to or larger than the number of channels. Note that the queueing delay and service delay for the non-saturation case were also analyzed in [418] using an $M/G^y/1$ queueing model. The average delay for the negotiation-based sensing policy was observed to be smaller than that for the random sensing policy.

Dynamic frequency hopping MAC for IEEE 802.22

In an IEEE 802.22-based cognitive radio network, in-band channel sensing for a single frequency band is inefficient since a base station (BS) and the customer premise

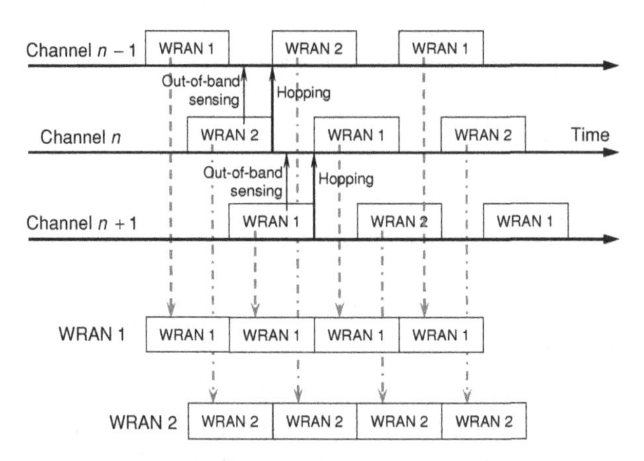

Figure 7.29 Dynamic frequency hopping (DFH).

equipments (CPEs) must stop transmission during the sensing period. To improve the performance of channel sensing in an IEEE 802.22 network, a dynamic frequency hopping (DFH) scheme was proposed in [419]. According to the standard, an IEEE 802.22 network is designed to operate in a single channel (i.e. non-hopping mode), and channel sensing must be performed every two seconds. In contrast, with DFH, the network can hop over a set of channels. While utilizing one channel, the 802.22 network may perform out-of-band sensing for other channels. The decision of hopping can be made based on the result of channel sensing. In this way, transmissions in the 802.22 network will not be interrupted due to in-band channel sensing. Also, if multiple BSs perform collaborative channel sensing, the problem of self-coexistence of BSs in the same dynamic frequency hopping community can be mitigated. An example of frequency hopping is shown in Figure 7.29. When WRAN2 accesses channel n, it can perform out-of-band sensing of channel $n-1$ simultaneously. If the incumbent signal is not detected on channel $n-1$, WRAN2 hops to channel $n-1$.

A community of IEEE 802.22 networks supporting DFH operation was defined. In this community, there is a leader to manage the member networks, collect channel occupancy/availability information from the networks in the same community, calculate the hopping pattern, and broadcast this information back to the other member networks. Based on this mechanism, all IEEE 802.22 networks must be able to communicate with the leader through the coexistence beacon protocol (CBP), which was designed for wireless inter-network communication. To form a community, the leader with the smallest MAC address was first selected from among the BSs. Then, the leader chooses the hopping channels and broadcasts the information to other networks. The BSs which receive this message may join the community.

After a community is formed, the leader decides the hopping pattern for a certain period of time. All member networks perform channel sensing and send the sensing results to the leader. A new hopping pattern is decided and broadcast back to the community if an incumbent service is detected. The protocol also supports community merging and splitting.

The throughput of this hopping MAC protocol was observed to be higher than that of a traditional MAC since the network can transmit data continuously without interruption due to spectrum sensing. Also, it was shown that splitting networks into different communities can achieve advantages in terms of the smaller number of channels required for hopping.

7.4.3 Spectrum allocation and scheduling

Channel allocation and traffic scheduling problems for wireless transmission in both single-hop and multihop networks were extensively studied in the literature [420, 421, 422, 423, 424, 425, 426, 427]. However, these mechanisms were not required to consider the problem of interference caused to the licensed users, as in a cognitive radio environment.

Cognitive radio MAC with rate and power adaptation

The major challenges in a multichannel cognitive radio network are: (1) an unlicensed user must choose an optimal set of channels to access, (2) an unlicensed user must choose the transmission parameters (i.e. rate and power) to satisfy the transmission requirements (e.g. the interference to a licensed user), and (3) collision due to the transmission of multiple secondary users must be avoided. In [370], these challenges were considered jointly.

First, the channel allocation was solved by enumerating and searching for an optimal channel selection. The algorithm starts from the minimum number of channels. Then, a rate and power allocation algorithm was performed. If there is a feasible optimal solution for rate and power, the algorithm terminates and the unlicensed users access these channels. Otherwise, the algorithm continues searching until the constraint on the maximum number of channels is reached. In this case, if the maximum number of channels accessed by the unlicensed users is M and the total number of available channels is N, where $N > M$, the total number of combinations to be searched is $\sum_{m=1}^{M} N!/m!(N-m)!$.

For rate and power allocation, an optimization problem can be formulated as follows:

$$\text{minimize:} \quad \sum_{n=1}^{m} (e^{r_n} - 1)\frac{\sigma^2}{h} \tag{7.20}$$

$$\text{subject to:} \quad \sum_{n=1}^{m} r_n W_n = R_{\text{req}} \tag{7.21}$$

$$0 \leq r_n \leq \ln\left(1 + \frac{P_{\max} h}{\sigma^2}\right), \tag{7.22}$$

where r_n is the transmission rate on channel n, m is the number of channels to be accessed by secondary users, σ^2 is the noise power, h is the channel gain, W_n is the spectrum size corresponding to channel n, R_{req} is the transmission rate requirement of secondary users, and $P_{\max}$ is the maximum power. The objective of this optimization problem is to minimize the transmit power, which is a function of the transmission rate. The constraint in (7.21) indicates that the total transmission rate must meet the requirement of the unlicensed users, while the constraint in (7.22) limits the transmit power of an unlicensed user. With these constraints, an algorithm was proposed to obtain the solution [370]. The rate and power allocation problem for a cognitive radio system has also been studied in [428, 429, 430, 431, 432, 433].

For channel contention among multiple unlicensed users, an extension of the CSMA/CA protocol (used in the IEEE 802.11 standard) was also proposed in [370]. In this case, a free-channel table (FCT) was used to maintain the list of unoccupied channels by other unlicensed users, and a dedicated control channel was used to exchange information among unlicensed users. The protocol works as follows:

1. An unlicensed user listens to the control channel and updates FCT if a transmission is detected.
2. If an unlicensed user wants to transmit data, it transmits an RTS message on the control channel. This RTS message contains the FCT and the duration of the flow (DOF). However, if the FCT at a source is empty (i.e. none of the channels is available), the unlicensed user backs off and retransmits the RTS message.
3. Upon receiving the RTS message, the destination unlicensed user obtains a set of available channels by comparing the FCT in the RTS message with the local FCT. Then, the rate and power allocation algorithm is performed on a set of available channels. If there is a solution of this algorithm, the set of channels and the transmission parameters are included in the CTS message. Otherwise, an empty flag is used to indicate to the source unlicensed user that there is no feasible solution.
4. After the source unlicensed user receives the CTS message, it transmits an echo-CTS (ECTS) message over the control channel to inform other unlicensed users of the reception of the CTS message. Then, the source unlicensed user starts transmitting data over the set of available channels by using the optimal rate and power control strategy.

The performance of this MAC protocol was compared with the receiver-based channel selection (RBCS) scheme in [434]. The RBCS scheme only selects a channel with the best quality for data transmission. Simulation results showed that the throughput of the proposed MAC protocol is much higher than that of the RBCS scheme, while at the same time the connection blocking probability is lower for the proposed scheme.

Statistical channel allocation MAC (SCA-MAC) for cognitive radio

The statistical channel allocation MAC (SCA-MAC) with channel aggregation feature was proposed in [371]. In this protocol, a control channel is used for exchanging CRTS and CCTS messages (Figure 7.30). To avoid the saturation problem, this control channel is accessed by an unlicensed user using the CSMA/CA mechanism [435, 436]. It is assumed that m consecutive channels can be aggregated for transmission at the radio interface. The problem of selection of the range of channels was formulated as an optimization problem as follows:

$$\{n^*, m^*\} = \arg\max_{n,m} P_{\text{succ}}([n, n+m-1], L), \tag{7.23}$$

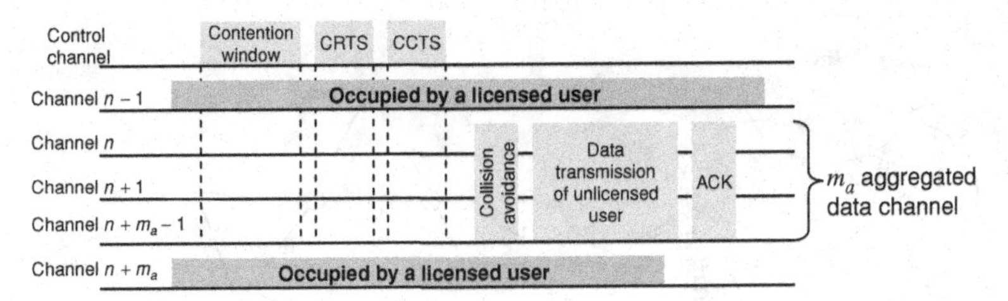

Figure 7.30 The SCA-MAC protocol.

where $P_{\text{succ}}(\cdot, \cdot)$ is the successful rate of transmitting a packet with length L on an aggregated channel $[n, n+m-1]$ (i.e. from channel n to $n+m-1$). This successful rate is computed from

$$P_{\text{succ}}([n, n+m-1], L) = P_{\text{succ,c}}([n, n+m-1])P_{\text{suc},L}, \tag{7.24}$$

where $P_{\text{succ,c}}(q)$ is the probability of successful channel allocation for a given range of channels q. This probability can be obtained from

$$P_{\text{succ,c}}(q) = \left(1 - \frac{\overline{\rho}_{\text{c}} l \overline{m}}{(1-\rho)q}\right)^m, \tag{7.25}$$

where ρ is the utilization of the licensed users, $\overline{\rho}_{\text{c}}$ is the average utilization of neighboring cognitive radio nodes, l is the number of neighboring nodes, and $\overline{m}$ is the average number of aggregated channels of neighboring nodes. Also, $P_{\text{succ},L}$ denotes the probability that a packet with length L can be transmitted in an idle aggregated channel $[n, n+m-1]$ for a duration τ, and is given by

$$P_{\text{succ},L} = \prod_{j=l}^{n+m-1} \Pr\left(n'_j : \tau \geq t_{0,k} + L/M | \tau \geq t_{0,k}\right), \tag{7.26}$$

where $\Pr\left(n'_j : \tau \geq t_{0,k} + L/M | \tau \geq t_{0,k}\right)$ is the probability that channel n'_j is idle during $[t_{0,k}, \tau]$. Based on this rate estimation, solutions for variables n and m (i.e. the first channel index and the total number of channels to be aggregated) can be obtained numerically to achieve the best performance.

The proposed SCA-MAC was benchmarked with a random scheme where an idle channel is selected randomly. It was observed that, with a larger range of operating frequencies, an unlicensed user has more flexibility in choosing the best frequency band, which increases its chances of finding an idle band. Also, the proposed SC-MAC was observed to outperform the random scheme in terms of the throughput achieved by an unlicensed user.

Enhancements to the IEEE 802.22 air interface and MAC protocol

The problems of self coexistence and hidden incumbent service in IEEE 802.22 WRAN were considered in [372]. In particular, a channel allocation algorithm based on the graph coloring model [437] was presented for self coexistence. Also, an enhanced MAC protocol was proposed for detection of hidden incumbent service. The utility of an IEEE 802.22 network was defined in terms of the bandwidth of the operating channel.

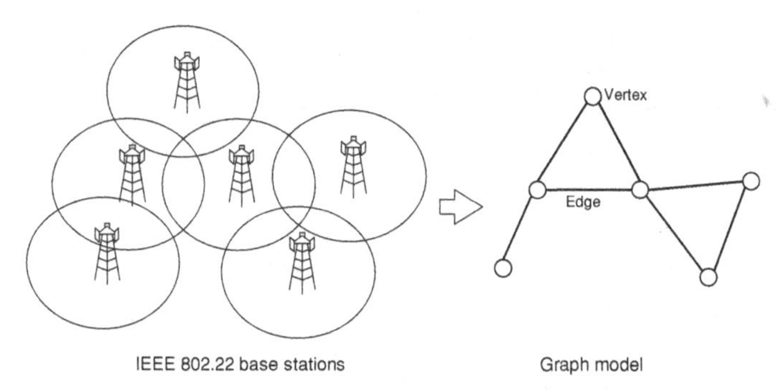

Figure 7.31 Graph coloring model.

Transmissions are constrained by the fact that neighboring BSs cannot use the same or overlapping channel. Based on the utility definition and the constraint, an undirected graph $\mathscr{G} = \{\mathbb{V}, \mathbb{E}, \mathbb{N}\}$ was defined, where $\mathbb{V}$ is the set of vertices (i.e. base stations), $\mathbb{E}$ is the set of all undirected links (i.e. a link between two vertices indicates that transmissions from the corresponding BSs interfere), and $\mathbb{N}$ is the set of available channels unused by an incumbent service. An example of this graph model is shown in Figure 7.31.

To obtain the solution of this graph coloring model, it was assumed that the topology information is available to all BSs. Three different objectives were defined for this graph coloring model. The first objective was to maximize the total utility of all BSs. The second objective was to provide proportional fair utility. That is, the BS which interferes with the least number of other BSs is prioritized to maximize the utility. The third objective was to provide complete fairness utility regardless of the interference.

To obtain the optimal solution, a two-phase algorithm, namely, the utility graph coloring (UGC) algorithm, was proposed. In the first phase, the graph was colored by a standard graph coloring algorithm to obtain the minimum number of colors to meet the constraint. In the second phase, a progressive algorithm was used to collect the number of occurrences of the different colors in the graph. This information was used together with the number of edges corresponding to the vertices (i.e. the number of other BSs that one BS interferes with) to assign channels to the BSs. The assignment of channels was performed to achieve the above three objectives. For example, with an objective to maximizing utility, the color with the highest number of occurrences was assigned with the largest bandwidth channel. For the other colors, the minimum bandwidth channel is allocated.

To mitigate interference to an incumbent service, the MAC protocol was enhanced by using dynamic multiple out-of-band broadcasting. In the standard IEEE 802.22 MAC protocol, messages from a BS are transmitted using one channel only. On the other hand, in the scheme proposed in [372], messages from a BS are periodically broadcast in different channels and can be changed dynamically depending on feedback from the CPEs. As a result, if a CPE detects an incumbent service through in-band channel sensing, the CPE will still be able to report this information to the BS. Another enhancement was the spectrum usage report, which is broadcast by the BS to the CPEs. Using this report, the CPEs can change the uplink channels to transmit to the BS.

The performance of this collaborative graph coloring-based channel allocation algorithm was compared with the greedy non-collaborative algorithm (i.e. each BS acquires as many available channels as possible). It was observed that the utility from the collaborative graph coloring algorithm is much higher than that from the greedy non-collaborative algorithm. This is due to the fact that the greedy algorithm incurs wider and larger interference in a network. Also, the performance in terms of average startup delay was evaluated. This startup delay is affected by the contention among IEEE 802.22 users and incumbent services. In this case, the startup delay increases as the number of IEEE 802.22 users and the number of incumbent services increase. However, this average startup delay for the proposed MAC protocol was observed to be significantly smaller than that for a traditional MAC protocol.

MAC-layer scheduling in cognitive radio networks

A scheduling algorithm for a multihop cognitive radio network was proposed in [373]. This scheduling algorithm determines the time slot and channel for transmission by unlicensed users in the network. However, each unlicensed user in the network has a different set of available channels. Let the network be presented as a graph containing nodes and links. Let $x_{l,n,k}$ be a binary variable defined as follows:

$$x_{l,n,k} = \begin{cases} 1, & \text{if link } l \text{ accesses time slot } k \text{ in channel } n \\ 0, & \text{otherwise.} \end{cases} \tag{7.27}$$

Let $\mathbb{L}_i$ denote the set of links used at node i, $\mathbb{N}_l^{(l)}$ denote the set of available channels for link l, and $\mathbb{N}_i^{(c)}$ denote the set of available channels at node i. The first constraint for the scheduling policy is defined as follows:

$$\sum_{l \in \mathbb{L}_i} \sum_{n \in \mathbb{N}_l^{(l)}} x_{l,n,k} \leq 1, \quad \forall i, k. \tag{7.28}$$

This constraint indicates that the same time slot cannot be allocated to more than one link at the same node. The second constraint is

$$\sum_{\forall k} \sum_{n \in \mathbb{N}_l^{(l)}} x_{l,n,k} = 1, \tag{7.29}$$

which ensures that each link in the network is allocated with a time slot and a channel for transmission. The third constraint is

$$x_{l,n,k} + x_{l',n,k} \leq 1 \tag{7.30}$$

for all links l' at nodes which are neighbors of node with link l. This constraint is used to ensure that none of the neighboring nodes transmits in the same time slot and in the same channel. Based on these constraints, an integer linear program can be formulated with the objective of minimizing the number of time slots allocated for transmission in all links given a set of available channels in the network.

Since solving the above integer linear programming formulation incurs a large computational complexity, a heuristic-based distributed algorithm was proposed. This algorithm consists of two phases. In the first phase, each node chooses the time slot and

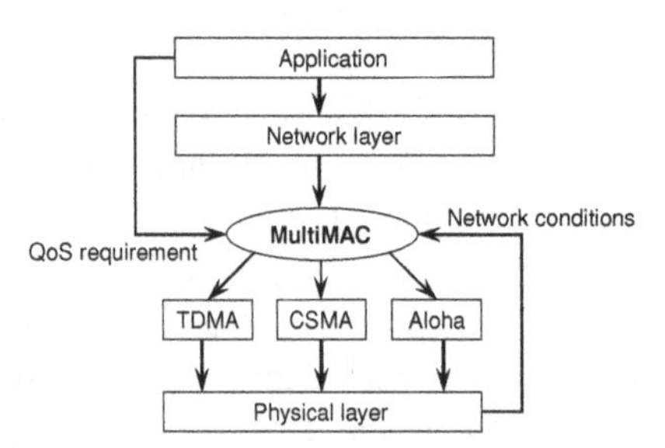

Figure 7.32 Structure of MultiMAC for MAC protocol adaptation.

the channel to access for all the outgoing links. In the second phase, this information is propagated throughout the network and each node adjusts the time slot and the channel allocation accordingly.

The performances in terms of schedule length (i.e. the number of time slots allocated to each node for transmission) obtained from a heuristic-based distributed algorithm were observed to be close to those obtained from the integer linear programming approach. When the number of nodes in a network increases, each node has a higher chance to find an available neighboring node. As a result, the schedule length increases.

A cognitive radio transceiver may use several different MAC protocols instead of using just one. It should be able to switch between these different protocols adaptively according to the network condition.

Adaptive MAC

Different MAC protocols can be used to achieve different objectives under different situations. For example, a fixed assignment TDMA MAC protocol can achieve better performance under heavy load conditions, while a CSMA/CA provides better performance under light load situations. Therefore, based on the different network conditions and characteristics, a cognitive radio can select an appropriate MAC protocol dynamically. A framework for MAC protocol adaptation, namely, MultiMAC was proposed in [376]. The structure of MultiMAC is shown in Figure 7.32. In MultiMAC, two simple policies for MAC protocol selection were considered. The first policy was based on the load level in the network. The second policy was based on the type of traffic. For best-effort traffic, a low complexity protocol such as ALOHA can be used. However, for QoS-sensitive traffic, a fixed assignment TDMA scheme would be more appropriate.

7.4.4 Spectrum trading

Rental protocol for dynamic channel allocation

When a licensed radio spectrum is not being used, the spectrum owner can sell the spectrum opportunities to the unlicensed users. In such a case, a protocol is required to support negotiations (e.g. on price and spectrum share) between the licensed and unlicensed users. A credit token-based rental protocol for dynamic spectrum access was proposed

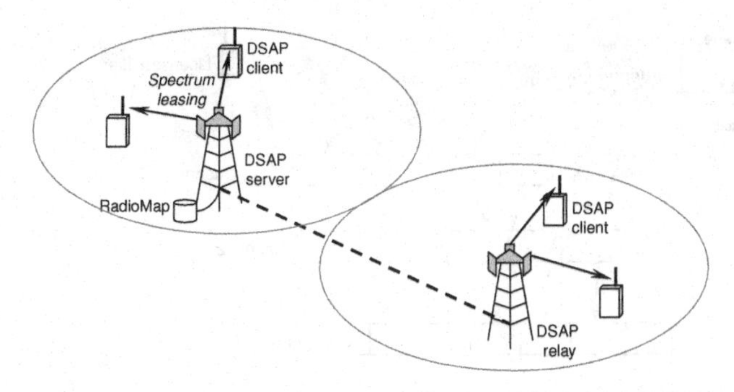

Figure 7.33 Spectrum leasing in DSAP.

in [377]. In the system model, a licensed user wants to sell spectrum opportunities to multiple unlicensed users. These unlicensed users join an auction [237] by bidding for the spectrum offered by a licensed user (or a licensed service provider). A credit token is used as the medium for negotiations between the licensed and unlicensed users.

The proposed protocol works as follows. First, a licensed user advertises the information on spectrum opportunities (i.e. channel and time period). Then, the interested unlicensed users contact that licensed user to register their interests in spectrum bidding. The licensed user replies with the auction period (i.e. time period that spectrum will be allocated to the unlicensed users). In order to access the spectrum, the unlicensed users provide information on the amount of bidding credit tokens (or price), the size of the spectrum, and the starting and ending times. Through a licensed network controller, the licensed user chooses the unlicensed users which are going to share the spectrum, the clearing price to charge for credit token, and the time period allocated to each unlicensed user. Note that this rental protocol can be applied to an IEEE standard (e.g. IEEE 802.16h [438]) to support the self-coexistence of multiple IEEE 802.16 networks.

Centralized cognitive MAC protocol for coordinated spectrum access

Spectrum assignment to the unlicensed users can be performed by a central controller. A protocol to support this spectrum leasing, namely, the dynamic spectrum access protocol (DSAP), was developed in [378]. The system model for spectrum leasing considered in DSAP is shown in Figure 7.33. A *DSAP client* is an unlicensed node to lease the spectrum from a *DSAP server*. This DSAP server is a licensed user or a licensed service provider. A *DSAP relay* is used for spectrum leasing to DSAP clients which are not in the coverage of a DSAP server. At the DSAP server, a *RadioMap* (i.e. a database), is used to maintain information on spectrum allocation and leasing between the licensed and unlicensed users.

With DSAP, spectrum leasing works as follows. The DSAP client sends a request for spectrum usage (i.e. a *ChannelDiscover* message). The DSAP server searches for available spectrum in the RadioMap and responds by sending a *ChannelOffer* message to the DSAP client. This message contains information on available spectrum that can be leased to the DSAP clients. Based on this information, the DSAP client chooses the

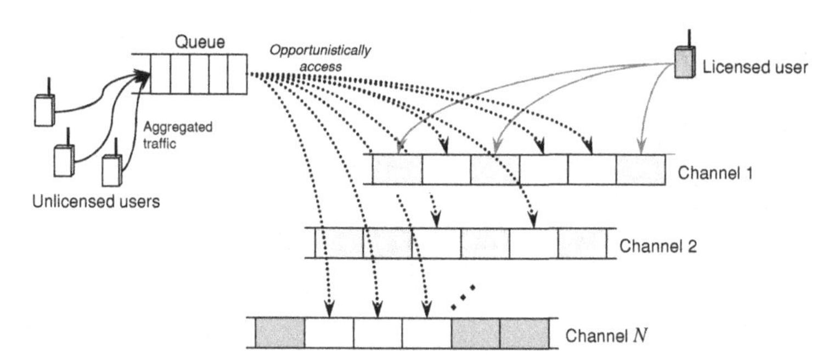

Figure 7.34 Queueing model for cognitive radio.

spectrum access parameters (e.g. the frequency and transmit power). Then, a *Channel-Request* message is sent back to the DSAP server. If DSAP agrees to lease the spectrum, the RadioMap is updated and a *ChannelACK* message is sent to the DSAP client. In addition, a *ChannelReclaim* message is used by the DSAP server to reallocate or terminate the lease of the spectrum to the DSAP client.

An experiment was conducted based on the 2.4 GHz and 5.2 GHz frequency bands used by IEEE 802.11g and 802.11a. It was observed that a DSAP server can successfully balance the interference generated by all nodes in the network. In particular, the nodes close to the 802.11a access point use the 5.2 GHz band, while the nodes far from an access point use the 2.4 GHz band, which has better propagation characteristics.

7.4.5 Performance analysis of cognitive MAC protocols

An analytical model can be used not only to investigate the system behavior quantitatively, but also to optimize the parameter settings of a cognitive radio MAC protocol to achieve optimal performance [439, 440, 441]. Several analytical models were proposed in the literature to evaluate the performance of cognitive radio MAC protocols.

Queueing analysis

To investigate the QoS performance of cognitive radio transmissions in a dynamic spectrum access environment, a queueing model was proposed in [381] considering both physical and MAC parameters. In the system model, when a licensed user does not access the spectrum, spectrum opportunities are generated. These spectrum opportunities are aggregated and the unlicensed users use this aggregated bandwidth as a single channel for transmission. This aggregation of spectrum opportunities can be achieved, for example, through the OFDM technique in the physical layer. In the queueing model, it was assumed that there are I unlicensed users sharing the spectrum with N licensed users. The packets arriving at all of the unlicensed users were assumed to be aggregated in a single queue (Figure 7.34).

Packet arrivals at each unlicensed user were assumed to follow a Poisson process. For a licensed user n, the spectrum is occupied for period $T_{\mathrm{oc}}^{(n)}$ and available for period $T_{\mathrm{av}}^{(n)}$.

Therefore, the probability of availability of k channels can be obtained from

$$P_k^{\text{av}} = \sum_{c=1}^{\frac{N!}{k!(N-k)!}} \left(\prod_{n \in \mathbb{N}_c^{(k)}} (1 - \rho_n) \left(\prod_{n' \in \{1,2,\ldots,N\}/\mathbb{N}_c^{(k)}} \rho_{n'} \right) \right), \tag{7.31}$$

where $\rho_n = T_{\text{oc}}^{(n)}/(T_{\text{oc}}^{(n)} + T_{\text{av}}^{(n)})$ is the probability that a licensed user n occupies the channel, and $\mathbb{N}_c^{(k)}$ is a set of the licensed users for which k channels are occupied. For the unlicensed users, the capacity (i.e. the transmission rate or service rate of a queue) can be expressed as

$$C = \frac{T - T_s}{T} \times P_k^{\text{av}} \times W \times \log_2 \left(1 + \frac{P}{\sigma^2 \times P_k^{\text{av}} \times W} \right), \tag{7.32}$$

where T is the duration of a time slot, T_s is the time used to sense the channel, W is the bandwidth of the channel, P is the transmit power, and σ^2 is the noise power. An $M/G/1$ queueing model was used to analyze the transmission performance for a set of unlicensed users, where the service rate was assumed to be equal to the capacity of the unlicensed users.

This queueing model can be extended by considering the correlation in spectrum access by licensed users. For this, a multi-dimension Markov chain can be established to obtain the performance measures [380]. Adaptive modulation and coding used at the cognitive radio transceiver can be modeled by using a finite-state Markov channel (FSMC). The evolution of the number of packets in the queue of an unlicensed user can be modeled as a quasi-birth-and-death (QBD) process [442] and the corresponding probability transition matrix can be expressed as follows:

$$\mathbf{P} = \begin{bmatrix} \mathbf{B} & \mathbf{E} & & & \\ \mathbf{C} & \mathbf{D} & \mathbf{H}_0 & & \\ & \mathbf{H}_2 & \mathbf{H}_1 & \mathbf{H}_0 & \\ & & \mathbf{H}_2 & \mathbf{H}_1 & \mathbf{H}_0 \\ & & & \ddots & \ddots & \ddots \end{bmatrix}, \tag{7.33}$$

where matrices $\mathbf{E}$ and $\mathbf{H}_0$ correspond to the case that the number of packets in the queue increases, matrices $\mathbf{C}$ and $\mathbf{H}_2$ correspond to the case that the number of packets in the queue decreases, and matrices $\mathbf{B}$, $\mathbf{D}$, and $\mathbf{H}_1$ correspond to the case that the number of packets in the queue remains the same. From this queueing model, various performance measures (e.g. queue length distribution and access delay distribution) can be obtained for an unlicensed user.

The queueing model considers a contentionless MAC protocol. However, contention-based MAC protocols will be required for distributed multiuser cognitive radio networks. The contention for channel access in such a network will incur latency in data transmission, and its impact on the performance of cognitive radio users has to be analyzed.

Latency analysis

In [382], analytical models were developed for two types of dynamic spectrum access MAC protocols, i.e. with dedicated control channel and with embedded dedicated control

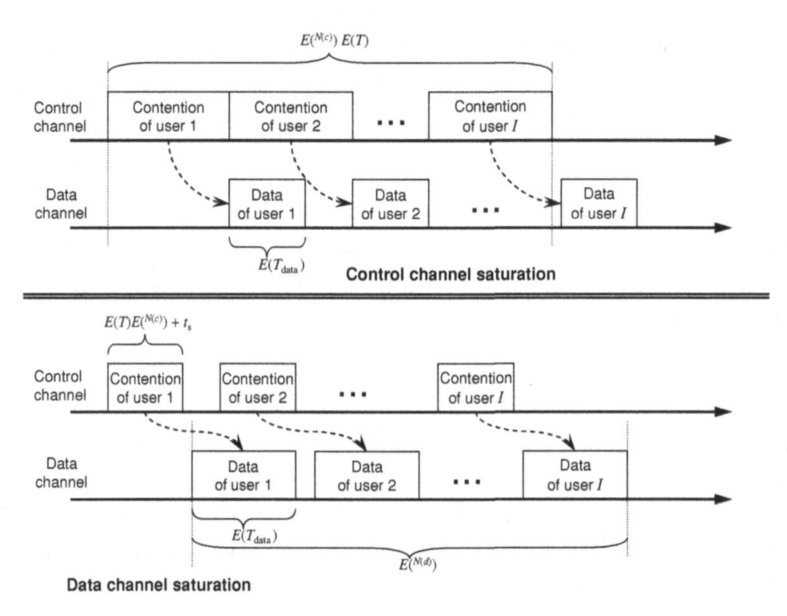

Figure 7.35 Saturation of control channel and data channel.

channels. The models assume that each unlicensed user is equipped with two radio interfaces, i.e. one for receiving and another for transmitting.

The protocol with a dedicated control channel works as follows. When an unlicensed user has data to transmit, it contends for a channel by exchanging RTS and CTS messages over a control channel as well as negotiates for a data channel. After negotiation and contention, data is transmitted over an assigned data channel. Note that this MAC protocol can introduce a control channel saturation problem. This problem occurs when $E(T_{\text{data}})/E(T_{\text{c}}) < 1$, where $E(T_{\text{data}})$ is the expected data transmission duration and $E(T_{\text{c}})$ is the sum of the contention duration and the exchange duration of the control messages. To solve this control channel saturation problem, the bandwidth of a control channel needs to be increased. However, this will decrease bandwidth for the data channel. The transmission rate of an unlicensed user could be degraded if a control channel has too large a bandwidth (i.e. data channel saturation occurs) or $E(T_{\text{data}})/E(T_{\text{c}}) > 1$. These phenomena on control channel and data channel saturation are shown in Figure 7.35.

The access delay was analyzed based on the concept of virtual slots [443]. The access delay of a MAC protocol with dedicated control channel can be expressed as follows: $E(D_{\text{de}}) = P_{\text{cs}}E(D|\text{control channel saturation}) + P_{\text{ds}}E(D|\text{data control saturation})$, where P_{cs} and P_{ds} denote the probability of control channel and data channel saturation, respectively. When control channel saturation occurs, the average access delay is given by $E(D|\text{control channel saturation}) = E(T)E(N^{(c)}) + E(T_{\text{data}})$ (Figure 7.35). Here, $E(T)$ is the expected duration of a virtual slot, $E(N^{(c)})$ is the expected number of virtual slots an unlicensed user has to wait before RTS transmission. This $E(T)$ can be obtained based on the calculation of the probability of successful transmission, and probability of collision as in [283], while $E(N^{(c)})$ can be obtained as in [443]. When data channel saturation occurs, the average access delay is denoted by $E(D|\text{data control saturation}) = \left(E(T)E(N^{(c)})\right) + E(T_{\text{data}})E(N^{(d)})$, where $E(N^{(d)})$ is the expected number of virtual slots an unlicensed user has to wait until the data transmission of other users ends.

A protocol with embedded control channel works as follows. When an unlicensed user has data to transmit, it contends for a data channel by transmitting RTS and CTS messages over that data channel. In this case, the contention matches if both a transmitter and a receiver choose the same data channel; otherwise, the contention mismatches. The probability of matched contention is defined by $P_{\text{match}} = 1/N$ and the probability of mismatched contention is $P_{\text{mismatch}} = (N-1)/N$, where N is the total number of channels. The average access delay of a MAC protocol with embedded control channel can be obtained from $E(D_{\text{em}}) = E(T)E(\tilde{N})$, where $E(T)$ is the expected virtual slot duration and $E(\tilde{N})$ is the expected number of virtual slots before an RTS transmission. The expected virtual slot duration $E(T)$ and the expected number of virtual slots $E(\tilde{N})$ can be obtained for both the cases, i.e. when the contention is matched and the contention is mismatched.

It was observed that as the control channel bandwidth increases, the control channel saturation probability decreases while the data channel saturation probability increases. As a result, the average access delay first decreases since the overhead due to the channel contention decreases. However, at a particular value of control channel bandwidth, the average access delay increases since the data channel has small bandwidth. Note that the average access delay of a MAC protocol with an embedded control channel is not affected by the bandwidth of the control channel. In addition, it was observed that, in some cases, the performance of a MAC protocol with a dedicated control channel can be optimized to achieve smaller average delays than that of a MAC protocol with an embedded control channel.

While the above analysis considers packet level performance only, connection-level performance will also need to be analyzed. Connection-level performance is typically analyzed based on the connection blocking effect.

Blocking analysis

With dynamic spectrum sharing, unlicensed users can access the radio spectrum when it is not occupied by licensed users. However, if both licensed and unlicensed users request to access the spectrum simultaneously, the requests from unlicensed users will be blocked. Similarly, if the spectrum is currently occupied by the unlicensed user, an incoming request from the licensed users will be blocked. To analyze this blocking performance in a cognitive radio environment, an analytical model was developed based on the concept of admission control used in traditional wireless networks [63, 444, 445, 446, 65, 447]. As in [384, 385], a cognitive radio network has N channels shared between two groups of users (i.e. licensed and unlicensed users). The arrivals of licensed users follow a Poisson process and a licensed user occupies N_l channels (i.e. $N_l < N$). The duration of channel occupancy is exponentially distributed with mean $1/\mu_l$. The arrival of unlicensed users can follow either a Poisson process [385] or a general renewal process [384] (e.g. the inter-arrival time is Gamma distributed). The requests from

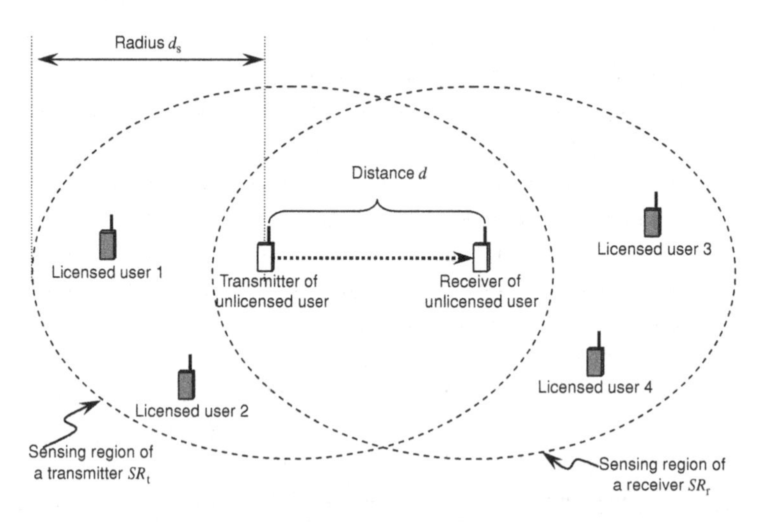

Figure 7.36 Sensing regions of a transmitter and receiver of an unlicensed user.

licensed or unlicensed users are not buffered. In this case, if the number of idle channels is less than N_l, an incoming request from a licensed user will be blocked. A request from an unlicensed user is blocked if all N channels are occupied. Based on this system model, the state space is defined as (n_l, n_u), where n_l and n_u are the number of occupied channels by a licensed user and an unlicensed user, respectively. This model is a variant of a $G/M/N$ queueing model [448] for which a standard method in renewal theory can be applied to obtain the steady state probability of the system. The blocking probability of both licensed and unlicensed users can be derived from the steady state probabilities.

Note that a similar performance analysis model for narrowband licensed users and wideband unlicensed users was discussed in [387].

Throughput analysis

When the transmitter and receiver of an unlicensed user are in different locations, the spectrum availability of the transmitter and receiver can be different, since the set of licensed users occupying the spectrum can be different. In such an environment, the *two-switch* interweave model can be applied to analyze the throughput of unlicensed users [449, 450, 451]. An example is shown in Figure 7.36 in which an unlicensed user can perfectly detect the presence of a licensed user. In this case, the set of licensed users (i.e. licensed users 1 and 2) in the sensing region of an unlicensed transmitter is different from that of an unlicensed receiver (i.e. licensed users 3 and 4). Therefore, the spectrum opportunities detected by the transmitter and receiver are not identical. This cognitive radio network can be abstracted by a two-switch mathematical model shown in Figure 7.37. In this case, the transmitter of an unlicensed user transmits data to a receiver if the switches $s_t, s_r \in \{0, 1\}$ are closed (i.e. $s_t = s_r = 1$). That is, both transmitter and receiver of an unlicensed user detect the activities of the licensed users in both sensing regions SR_t and SR_r. These states s_t and s_r are random, and there is a correlation between

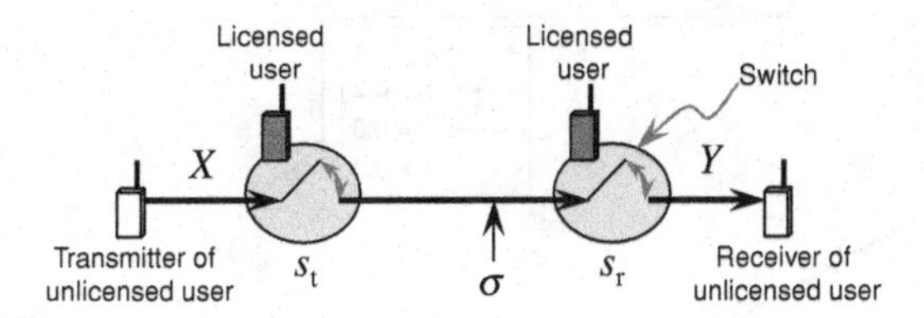

Figure 7.37 Two-switch model.

states s_t and s_r. In particular, if the distance d between transmitter and receiver is small, the correlation between these states becomes larger.

Based on the two-switch model, the throughput of an unlicensed user can be derived. In this case, the received signal Y at the receiver of an unlicensed user can be expressed as follows:

$$Y = s_r(s_t X + \sigma), \tag{7.34}$$

where X is the transmitted signal, and σ is the additive white Gaussian noise at the receiver of the unlicensed user. With received power P, the upper bound on capacity (i.e. rate) can be obtained from

$$r = \Pr(s_t = s_r = 1) \log\left(1 + \frac{P}{\Pr(s_t = 1)}\right), \tag{7.35}$$

where $\Pr(s_t = s_r = 1)$ is the probability that both the transmitter and receiver of the unlicensed user detect no activity from the licensed users, and $\Pr(s_t = 1)$ is the probability that a transmitter detects no activity.

To analyze the throughput of the unlicensed user in the cognitive radio network shown in Figure 7.36, it is assumed that the spatial distribution of the licensed users follows a Poisson distribution with a density of λ licensed users per unit area. In this case, the probability of having n licensed users in area A is obtained from

$$\Pr(n \text{ users in } A) = \Pr(\mathscr{F}(A) = n) = \frac{e^{-\lambda A}(\lambda A)^n}{n!}. \tag{7.36}$$

The sensing region of an unlicensed user is a circle with radius d_s. The probability that an unlicensed user detects no activity from a licensed user is obtained from

$$\begin{aligned} \Pr(s_t = 1) &= \Pr(\mathscr{F}(\mathrm{SR}_t) = 0) = \Pr(\mathscr{F}(\pi(d_s)^2) = 0) \\ &= e^{-\lambda\pi(d_s)^2}, \end{aligned} \tag{7.37}$$

$$\begin{aligned} \Pr(s_t = s_r = 1) &= \Pr(\mathscr{F}(\mathrm{SR}_t \cup \mathrm{SR}_r) = 0) \\ &= e^{-\lambda\left(2(d_s)^2\left(\pi - \cos^{-1}\left(\frac{d}{2d_s}\right)\right) + dd_s\sqrt{1 - \frac{d^2}{4(d_s)^2}}\right)}. \end{aligned} \tag{7.38}$$

Therefore, the capacity of an unlicensed user is given by

$$r = e^{-\lambda\left(2(d_s)^2\left(\pi - \cos^{-1}\left(\frac{d}{2d_s}\right)\right) + dd_s\sqrt{1 - \frac{d^2}{4(d_s)^2}}\right)} \log\left(1 + Pe^{\lambda\pi(d_s)^2}\right). \tag{7.39}$$

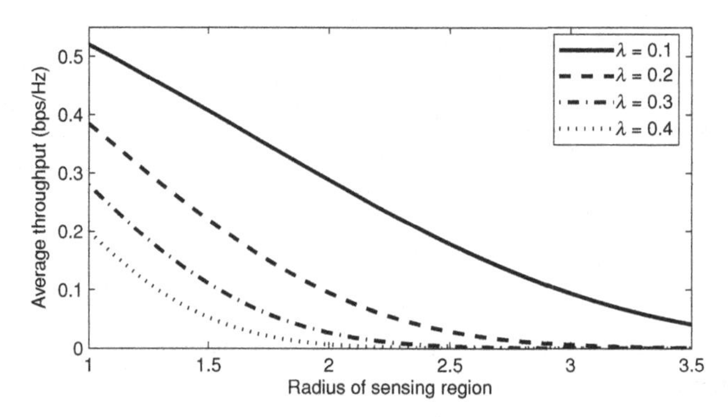

Figure 7.38 Throughput under different spatial densities of licensed users.

Figure 7.38 shows the throughput under different licensed user densities and different sensing region radii. As the radius of the sensing region or the density of licensed users increases, there is a higher chance that a licensed user will occupy the spectrum. Consequently, the throughput of an unlicensed user decreases.

7.5 Open issues in dynamic spectrum access

The open research issues in dynamic spectrum access can be summarized as follows:

- *MAC protocol for dynamic spectrum access:* Designing efficient MAC protocols for the emerging wireless standards (e.g. IEEE 802.16h and IEEE 802.22) is an open problem. The MAC protocols must be designed to support the physical layer specifications. Also, the QoS requirements of the spectrum owner (e.g. interference temperature, collision rate) and those of the cognitive radio user (e.g. throughput, delay, and loss rate) have to be taken into account. In addition, a customized MAC protocol for dynamic spectrum access can be designed for a specific application (e.g. wireless intelligent transportation systems or healthcare applications). Specific application requirements (e.g. mobility management in vehicular network and interference avoidance in passive medical devices in healthcare services) have to be integrated when designing such a MAC protocol.
- *Centralized dynamic spectrum access:* In an infrastructure-based cognitive radio network, dynamic spectrum access by the unlicensed users can be facilitated by a central controller. This central controller can be used to collect spectrum usage information on the licensed users. This information can be used to plan and schedule spectrum access by the unlicensed users. The decisions on spectrum access are then distributed to the unlicensed users in the network. Since there is a central controller, dynamic spectrum access can be optimized globally to achieve the system objectives given the constraints of the licensed users. One of the challenges in this centralized dynamic spectrum access is to minimize the overhead due to communications among the unlicensed users and the overhead due to computation of the optimal solution for dynamic spectrum access.

- *Distributed dynamic spectrum access:* In an infrastructureless cognitive radio network there is no central controller available and the unlicensed users have to make their decisions on spectrum access locally and independently. Since each of the unlicensed users makes its decision independently, cooperation and competition between the cognitive users become important issues. If the unlicensed users cooperate, a communication protocol will be required to exchange information with each other to achieve the desired objective. When the unlicensed users compete with each other, spectrum access decisions should be made such that all unlicensed users are satisfied. Efficient distributed solutions (e.g. equilibrium solutions) will be required for such competitive spectrum access.
- *Dynamic spectrum access from economic perspective:* In a cognitive radio environment, the radio spectrum can be traded between the licensed and unlicensed users. The economic issues related to spectrum trading, namely, spectrum pricing, modeling of spectrum demand, and utility from spectrum access, are important for practical system design. Microeconomic theory and models (e.g. game theory, market equilibrium, and auction) can be used to address these spectrum trading issues.

7.6 Summary

The basic concepts of dynamic spectrum access in a cognitive radio network have been discussed. Different spectrum access models for cognitive radio, namely, the exclusive-use, shared-use, and commons models, have been described. In the exclusive-use model, an unlicensed user can obtain the right to use the spectrum access for exclusive usage from a licensed user. In the shared-use model, an unlicensed user can access the spectrum opportunistically if it is not utilized by a licensed user. In the commons model, all unlicensed users can freely access the spectrum. These spectrum access models provide improved flexibility in spectrum access and higher spectrum utilization. Spectrum sensing and spectrum access are the two major functionalities of a cognitive MAC protocol. A comprehensive survey on the cognitive MAC protocols has been provided. To this end, major research issues related to dynamic spectrum access and management have been briefly described. Dynamic spectrum access poses different sets of challenges in centralized and distributed cognitive radio networks. Besides the technical aspects of dynamic spectrum access, economic issues such as those related to spectrum pricing should be also considered for practical system design. The design approaches for both centralized and distributed dynamic spectrum access as well as the economic aspects of dynamic spectrum access will be dealt with in detail in the following chapters.

8 Centralized dynamic spectrum access

8.1 Introduction

In a centralized dynamic spectrum access architecture, a central controller is deployed to gather and process information about the wireless environment. With a central controller, the decision of cognitive radio users to access the spectrum can be made such that the desired system-wide objectives are achieved.

In this chapter, we review centralized dynamic spectrum access schemes. A summary of these schemes is provided in Table 8.1. In a centralized scheme, every cognitive radio user communicates with a central controller to inform their states and objectives/requirements. The central controller then makes the decisions in terms of the action for each cognitive radio user to access the spectrum so that their requirements are satisfied under given system constraints. To implement centralized dynamic spectrum access, two approaches, namely, optimization approach and auction-based approach, can be used. With an optimization-based approach, different types of optimization problems can be formulated (e.g. convex optimization, assignment problem, linear programming, and graph theory). Standard methods in optimization theory can then be applied to obtain the optimal solution for dynamic spectrum access. Alternatively, centralized dynamic spectrum access can be designed based on auction theory which is well developed in the field of economics. In this approach, cognitive radio users submit their bids to the spectrum owner. The winning cognitive radio user is determined from the bids, and the spectrum is allocated accordingly.

8.2 Optimization-based approach

8.2.1 Quality of service (QoS)-constrained dynamic spectrum access

With spectrum underlay access (i.e. the shared-use model), an optimization problem was formulated by considering QoS differentiation for different unlicensed users and also interference temperature constraints [452]. The objective here is to maximize the total transmission rate by adjusting the transmit power of the unlicensed users under an interference temperature constraint. With this constraint, at the measurement point, the interference temperature due to the transmissions of the unlicensed users must be maintained below the target limit (Figure 8.1).

Table 8.1 Summary of centralized dynamic spectrum access schemes.

Article	Brief description	Access model	Approach
Xing *et al.* [452]	Dynamic Spectrum Access Under QoS and Interference Constraints	Shared-use	Optimization
Wang *et al.* [453]	Primary-Prioritized Dynamic Spectrum Access	Shared-use	Optimization
Yu *et al.* [454]	Dynamic Control of Open Spectrum Management	Exclusive-use	Optimization
Zhang *et al.* [455]	Joint Admission Control and Power Allocation	Shared-use	Optimization
Gao *et al.* [456]	Power and Rate Control for Dynamic Spectrum Access	Shared-use	Optimization
Kamakaris *et al.* [406]	Coordinated Dynamic Spectrum Access in Cellular Networks	Exclusive-use	Optimization
Kovács and Vidács [457]	Spatio-Temporal Dynamic Spectrum Allocation	Exclusive-use	Optimization
Kim *et al.* [458]	Dynamic Spectrum Allocation Among Network Service Providers	Exclusive-use	Optimization
Subramanian *et al.* [459]	Coordinated Dynamic Spectrum Access	Exclusive-use	Optimization
Pan *et al.* [460]	Cooperative Game for Dynamic Spectrum Access	Exclusive-use	Optimization
Raman *et al.* [461]	Transmission Scheduling via Spectrum Server	Shared-use	Optimization
Hoang and Liang [462]	Spectrum Sensing and Access Scheduling	Shared-use	Optimization
Gandhi *et al.* [463]	General Framework of Spectrum Auction	Exclusive-use	Auction
Kovács and Vidács [464]	Multibid Auction for Dynamic Spectrum Allocation	Exclusive-use	Auction
Sengupta and Chatterjee [465]	Dynamic Spectrum Allocator Knapsack Auction	Exclusive-use	Auction
Feng and Zhen [466]	Weighted Proportional Fair Spectrum Allocation	Exclusive-use	Auction
Grandblaise *et al.* [467]	Bilateral Bargain in Spectrum Access	Exclusive-use	Auction

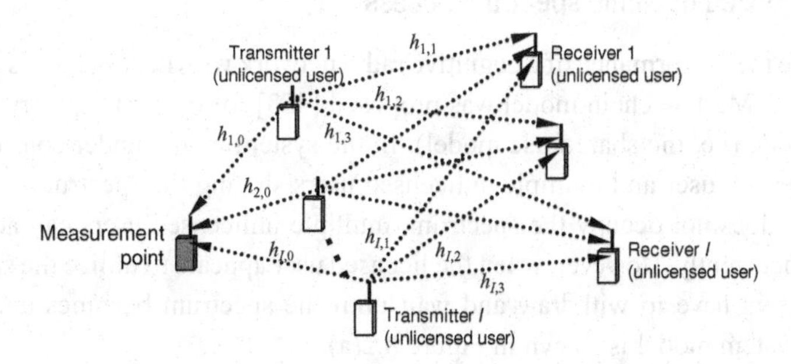

Figure 8.1 Interference generated at a measurement point due to transmissions from unlicensed users.

With I licensed users, for spread spectrum transmission, the received SIR can be expressed as follows:

$$\gamma_i = \frac{L P_i h_{i,i}}{\sum_{j \neq i} P_j h_{i,j} + \sigma^2}, \tag{8.1}$$

where P_i is the transmit power of user i, $h_{i,j}$ is the channel gain between transmitter i and receiver j, and σ^2 is the background noise power. The interference temperature constraint is denoted by

$$\sum_{i=1}^{I} P_i h_{i,0} \leq I_L, \tag{8.2}$$

where $h_{i,0}$ is the channel gain from the transmitter of user i to the measurement point, and I_L is the interference temperature limit.

To maximize the utility, an optimization problem is formulated as follows:

$$\text{maximize:} \quad \sum_{i=1}^{I} \log(\gamma_i), \tag{8.3}$$

$$\text{subject to:} \quad \eta_i \geq \gamma_i^t, \tag{8.4}$$

$$\sum_i h_{i,0} P_i \leq I_L, \tag{8.5}$$

$$0 \leq P_i \leq P_i^{\max}, \tag{8.6}$$

where η_i is the signal-to-interference ratio (SIR) at the receiver of the unlicensed user, γ_i^t is the SIR requirement, and $P_i^{\max}$ is the maximum transmit power of user i. Since $\sum_{i=1}^{I} \log(\gamma_i) = \log\left(\prod_{i=1}^{I} \gamma_i\right)$, the objective function is a posynomial, and the constraints can be represented in posynomial and monomial forms. Therefore, this is a convex optimization problem in the form of a geometric program which can be solved centrally to obtain the globally optimal solution.

8.2.2 Primary-prioritized dynamic spectrum access

To investigate the performance of a cognitive radio network with licensed users and unlicensed users, a Markov chain model was proposed [453] for dynamic spectrum access in overlay mode (i.e. the shared-use model). In the system model under consideration, there is a licensed user and multiple unlicensed users sharing the spectrum. When the licensed user does not occupy the spectrum, multiple unlicensed users can access the spectrum concurrently. However, when the licensed user appears to utilize the spectrum, unlicensed users have to withdraw and wait until the spectrum becomes unoccupied again. This system model is shown in Figure 8.2(a).

In this system model, traffic arrivals at both licensed and unlicensed users are assumed to follow independent Poisson processes. The average traffic arrival rate at user i is denoted by λ_i (e.g. $i \in \{P, A, B\}$ for the licensed user, unlicensed user A, and unlicensed user B, respectively). The average service time is denoted by μ_i. The state transition

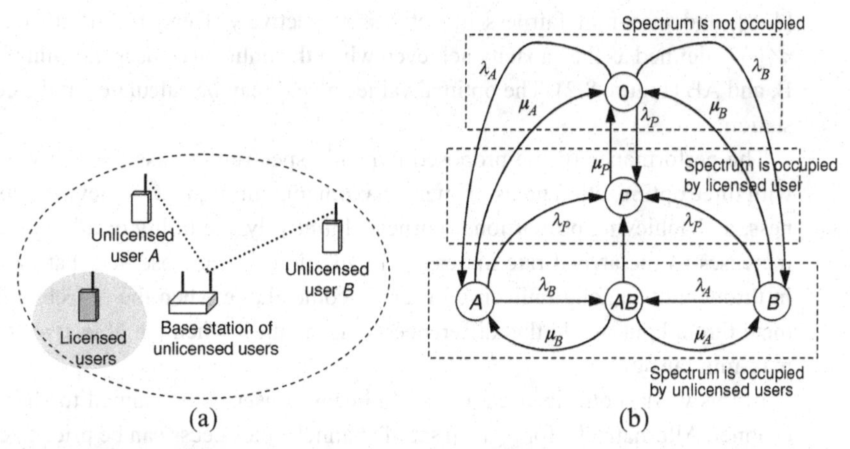

Figure 8.2 (a) System model for licensed-prioritized dynamic spectrum access and (b) the corresponding Markov chain.

diagram for the case of two unlicensed users is shown in Figure 8.2(b) where state 0, P, A, B indicate the events that no user occupies the spectrum, licensed user P occupies the spectrum, and unlicensed users A and B occupy the spectrum, respectively. State AB denotes the case that both of the unlicensed users occupy the spectrum. In this case, the capacities of both unlicensed users are computed from an interference model (e.g. as in (8.1)), and there is no collision in the medium access control (MAC) layer. Note that in this system model, an unlicensed user is not allowed to access the spectrum if it is occupied by a licensed user. However, this model was extended to consider the case that traffic from unlicensed users can be buffered when a licensed user occupies the spectrum. For this, waiting states are added in the Markov model. Based on this Markov chain model, the utilization of the spectrum and the throughput of each user can be obtained analytically.

To prioritize the spectrum access of licensed users over that of unlicensed users, an admission probability $P_i^{(a)}$ is used when unlicensed user i wants to access the spectrum. That is, an unlicensed user can successfully access the spectrum with probability $P_i^{(a)}$. As a result, the arrival rate of the traffic from an unlicensed user becomes $P_i^{(a)}\lambda_i$. This admission probability is applied to lower the spectrum occupancy from an unlicensed user, so that the licensed user can access the spectrum when required. Assuming that there is a base station to control the admission of traffic from unlicensed users (Figure 8.2(a)), an optimization problem can be formulated to maximize the utility of unlicensed users as follows:

$$\text{maximize: } \mathscr{U}, \tag{8.7}$$

$$\text{subject to: } \mathscr{U}_i(P_i^{(a)}) \geq 0, \quad 0 \leq P_i^{(a)} \leq 1, \tag{8.8}$$

where i is the index of the unlicensed user. For function $\mathscr{U}$, three different definitions were provided, i.e. $\mathscr{U} = \sum_i \mathscr{U}_i(P_i^{(a)})$, $\mathscr{U} = \prod_{i=1}^{I} \mathscr{U}_i(P_i^{(a)})$, and $\mathscr{U} = \min_i \mathscr{U}_i(P_i^{(a)})$, which correspond to maximum sum-throughput objective, proportional fairness

[468], and max-min fairness objectives, respectively. Here, the utility of each user $\mathcal{U}_i(\cdot)$ is defined as the capacity achieved when the unlicensed user transmits in states A, B, and AB (Figure 8.2). The optimal values of $P_i^{(a)}$ can be calculated at the central base station.

The performance of the proposed dynamic spectrum access scheme was evaluated with three optimization goals, namely, maximizing throughput, achieving max-min fairness, and achieving proportional fairness. Intuitively, the utility of all unlicensed users increases as the arrival rate increases. In addition, it was observed that the difference in throughputs among unlicensed users becomes large when the objective is to maximize throughput, while this difference is the smallest when the objective is to achieve max-min fairness.

In this work, both licensed and all unlicensed users are assumed to share the same channel. Alternatively, for a given set of channels, the access can be prioritized between licensed and unlicensed users in the connection level using admission control.

8.2.3 Dynamic control of open spectrum management

In a cognitive radio network, spectrum management can be controlled by a spectrum broker. One of the challenges for a spectrum broker is to handle the non-stationary nature of the spectrum usage behavior of licensed users and the spectrum demand from unlicensed users. A method for dynamic control of open spectrum management at the spectrum broker, namely, optimal stochastically controlled dynamic guard bandwidth (OSC-DGB), was proposed in [454]. Different from other dynamic spectrum access schemes, this OSC-DGB was designed to achieve the best connection-level performance (i.e. minimum connection blocking probability). In the system model, there are low- and high-priority unlicensed systems which share the radio channels not occupied by a licensed user. The prioritization between the two unlicensed systems (i.e. low- and high-priority systems) was implemented through the guard band concept [469]. In this guard band scheme, given N number of available channels from a licensed system during a time period, new unlicensed users from either the low-priority unlicensed system or the high-priority unlicensed system will be accepted if the number of currently occupied channels is less than the threshold L. Otherwise, only users from the high-priority unlicensed system are accepted. The allocation of the available channels from the licensed system to the unlicensed systems is performed by a spectrum broker.

To obtain the optimal policy for this dynamic guard band spectrum sharing policy, a finite-time horizon Markov decision process (MDP) was formulated for non-stationary traffic of unlicensed systems. However, in the unlicensed systems, the prediction of the future spectrum demands (i.e. the connection arrival rates of the unlicensed systems) could be inaccurate. In this formulation, each period with a fixed number of available channels consists of multiple decision epochs (Figure 8.3). At the end of epoch j, a decision on the threshold L is made to minimize the connection blocking probability of the unlicensed system. The state of the system observed at decision epoch j is the average number of blocked connections of the high-priority unlicensed system. To reduce the number of states (which affects the complexity of the Markov decision process model),

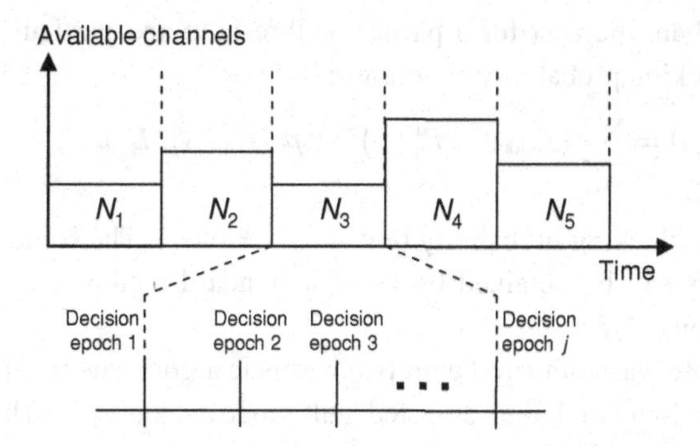

Figure 8.3 Decision making framework.

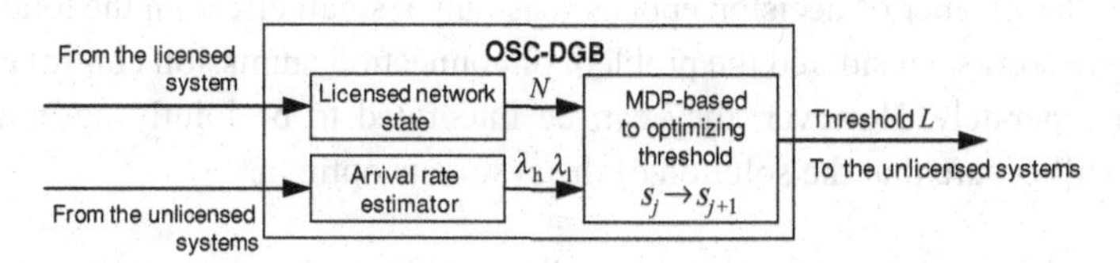

Figure 8.4 Block diagram of OSC-DGB.

this average number of blocked connections is truncated to the nearest integer. The average number of blocked connections can be obtained from

$$P_{\text{block}} = \frac{1}{N!}\left(\frac{\lambda_{\text{h}}}{\mu}\right)^{N-L}\left(\frac{\lambda_{\text{h}}+\lambda_{\text{l}}}{\mu}\right)^{L}$$

$$\times\left[1+\sum_{i=1}^{L}\frac{1}{i!}\left(\frac{\lambda_{\text{h}}+\lambda_{\text{l}}}{\mu}\right)^{i}+\sum_{i=L+1}^{N}\frac{1}{i!}\left(\frac{\lambda_{\text{h}}}{\mu}\right)^{i-L}\right]^{-1}. \tag{8.9}$$

Then, the state is defined as

$$s_{j+1} = \left\lfloor \frac{j\times s_j + \lambda_{\text{h}} \times P_{\text{block}}}{j+1} \right\rfloor, \tag{8.10}$$

where $\lfloor\cdot\rfloor$ denotes the floor function, λ_{h} and λ_{l} are the mean connection arrival rates for high- and low-priority unlicensed systems, respectively, and μ is the mean connection holding time during decision epoch j. The decision of the Markov decision process is defined as the threshold L. The block diagram of OSC-DGB is shown in Figure 8.4. Due to the inaccurate prediction, these connection arrival rates can be expressed as $\lambda_{\text{h}} = \tilde{\lambda}_{\text{h}} + \epsilon_{\text{h}}$ and $\lambda_{\text{l}} = \tilde{\lambda}_{\text{l}} + \epsilon_{\text{l}}$, where ϵ_{h} and ϵ_{l} are random variables. Therefore, the state transition probability $p_j(\cdot)$ of the Markov decision process from state s_j to state s_{j+1} when action L_j is taken, can be defined as follows:

$$p_j(s_{j+1}|s_j, L_j) = \int\!\!\int f_{\epsilon_{\text{h}}}(e_{\text{h}}) \times f_{\epsilon_{\text{l}}}(e_{\text{l}})\mathrm{d}e_{\text{h}}\mathrm{d}e_{\text{l}}, \tag{8.11}$$

where $f_{\epsilon_{\text{h}}}(e_{\text{h}})$ and $f_{\epsilon_{\text{l}}}(e_{\text{l}})$ are the probability distribution functions of the prediction errors (i.e. ϵ_{h} and ϵ_{l}) for the connection arrival rates for high- and low-priority unlicensed systems, respectively. Then, the cost for a particular threshold and particular state is defined based on the blocking probability as follows:

$$\mathscr{C}_j(s_j, L_j) = \sum_{s_{j+1}} \left(P_{\text{block}} - P_{\text{block}}^{\text{tar}}\right)^2 \times p_j(s_{j+1}|s_j, L_j), \tag{8.12}$$

where $P_{\text{block}}^{\text{tar}}$ is the target blocking probability (e.g. $P_{\text{block}}^{\text{tar}} = 0.01$). The solution of this Markov decision process can be obtained by using a standard method (i.e. dynamic programming (see Section 4.6)).

This OSC-DGB scheme was compared with two heuristic algorithms (i.e. predictive dynamic guard adaptation [470] and fairly adjusted multimode dynamic guard bandwidth (FMA-DGB) admission control [471]), and the static scheme. It was observed that when the prediction error increases, the expected total cost increases. Also, the length of the period (i.e. the number of decision epochs) has only a small effect on the total cost.

The above works considered the problem of connection admission control and power allocation separately. However, they can be integrated to be jointly optimized. This approach will ensure that the solution of the system is optimal.

8.2.4 Joint admission control and power allocation for dynamic spectrum access

For dynamic spectrum access in a spectrum underlay scenario, power control is a key mechanism to satisfy the interference temperature constraint for the licensed users while maximizing the performance of unlicensed users. Although the problem of power control for cellular networks (e.g. CDMA) was extensively investigated in the literature [472, 473, 474, 475, 476, 477, 478, 479, 480, 481, 482, 483], these works may not be directly applicable to a cognitive radio network where the constraint on the interference temperature limit at the licensed receiver is crucial. In [455], the problem of admission control and power allocation in a cognitive radio environment was considered. While the admission control mechanism is used to select a subset of unlicensed users to access the spectrum, the power allocation mechanism is used to determine the transmit power of these unlicensed users so that the interference temperature limit is not violated. In this case, a candidate group of unlicensed users is selected, and then power allocation is performed based on a gradient descent based algorithm.

For admission control, an optimization problem can be formulated as follows:

$$\text{maximize:} \quad \sum_{i=1}^{I} w_i x_i, \tag{8.13}$$

$$\text{subject to:} \quad \sum_{i=1}^{I} h_{i,0} P_i \le I_L, \quad \eta_i \le \gamma^t x_i, \tag{8.14}$$

where w_i is the weight indicating the revenue obtained by admitting unlicensed user i, x_i is the binary variable which indicates whether user i is allowed to transmit or not, $h_{i,0}$ is the channel gain from the transmitter of unlicensed user i to the measurement point (Figure 8.1), P_i is the transmit power, η_i is the SIR requirement of user i, and γ^t is the SIR requirement of unlicensed users.

However, this optimization is an integer programming (see Section 4.5) which is NP-hard. This problem is transformed into an equivalent optimization formulation as follows:

$$\text{maximize:} \left(\sum_{i=1}^{I} w_i \mathscr{F}_s(\gamma_i) + w_o \mathscr{F}_s(\gamma_o) \right), \tag{8.15}$$

where

$$\gamma_o = I_L - \sum_{i=1}^{I} h_{i,0} P_i, \quad \gamma_i = \frac{h_{i,i} P_i}{\sum_{j \neq i} h_{i,j} P_j + \sigma_i^2}, \tag{8.16}$$

and w_o is the revenue weight of the licensed user. The function $\mathscr{F}_s(\gamma)$ is defined as $\mathscr{F}_s(\gamma) = 1/(1 + \mathrm{e}^{a(\gamma+b)})$, where a and b are the parameters of a logistic function which is applied due to differentiability property, and $h_{i,j}$ and σ_i^2 denote the channel gain and the noise power, respectively.

To solve this optimization formulation, the problem is decoupled into two subproblems which are solved iteratively. First, a group of unlicensed users is considered. Then, the power allocation algorithm is applied. If the solution does not converge, some unlicensed users will be removed and the power allocation performed again. This procedure is repeated until the algorithm converges.

This joint admission control and power allocation algorithm was compared with the algorithm based on single and multiple accumulative removals technique (SMART) [484], SRA algorithm [485], and game theory based algorithm (GTBA) [486]. It was shown that the proposed joint admission control and power allocation algorithm can maximize the number of admitted unlicensed users.

8.2.5 Power and rate control for dynamic spectrum access

In [456], the power control problem for dynamic spectrum access was formulated as a dynamic programming problem taking the energy constraint into account. In the system model, activity of a licensed user is modeled as a two-state Markov chain. However, an unlicensed user can observe only the state of the licensed user in the previous time slot, but not in the current time slot. Two power levels, i.e. P_{L} and P_{H} where $P_{\mathrm{H}} > P_{\mathrm{L}}$ for low and high levels, respectively, are defined for the spectrum access of an unlicensed user according to the state of a licensed user in the previous time slot. That is, P_{H} and P_{L} are applied if the state of the licensed user is idle and busy, respectively. However, since the state of a licensed user in the current time slot can be different from that in the previous time slot, a high power level P_{H} used by an unlicensed user may cause interference to the licensed user. The probability of disturbance or probability of collision P_{col} is defined based on the event that the unlicensed user transmits with high power when the spectrum is occupied by a licensed user in the current time slot, i.e.

$$P_{\mathrm{col}} = \frac{pq}{p+q}, \tag{8.17}$$

given the transition probabilities p and q corresponding to the two-state Markov model of a licensed user. This model is defined by the probability transition matrix $\mathbf{C} = \begin{bmatrix} 1-p & p \\ q & 1-q \end{bmatrix}$ for which the first and second rows correspond to the states in which the spectrum is occupied and not occupied by the licensed user, respectively. For an unlicensed user, this disturbance probability must be maintained below the threshold P_{th}. It is assumed that the unlicensed user operates by battery power where the remaining capacity of battery at time k is denoted by B_k. The battery capacity evolves over time slots, i.e. $B_{k+1} = B_k - p_k\tau$, where τ is the length of a time slot, and p_k is the level of transmit power at time slot k.

For a duration of K time slots, an optimization problem can be formulated as follows:

$$\begin{aligned} \text{maximize:} \quad & E\left(\frac{1}{K}\sum_{k=0}^{K-1} r_k\right), \\ \text{subject to:} \quad & p_k \le \frac{B_k}{\tau}, \quad k = 0, 1, \ldots, K-1, \\ & p_k \le P_k^{\text{mask}}, \quad k = 0, 1, \ldots, K-1, \\ & P_{\text{col}} \le P_{\text{th}}, \end{aligned} \tag{8.18}$$

where P_k^{mask} is the power mask used to limit the interference to the licensed user. This power mask is defined by utilizing the correlation of the licensed user's activity model, i.e.

$$P_k^{\text{mask}} = \begin{cases} P_{\text{L}}, & \text{if spectrum is occupied in time slot } k-1 \\ P_{\text{H}}, & \text{if spectrum is idle in time slot } k-1. \end{cases} \tag{8.19}$$

The transmission rate r_k in time slot k is obtained from $r_k = \log_2\left(1 + p_k h_k/\sigma^2\right)$ where h_k is the channel gain which is assumed to be an independently and identically distributed exponential random variable, and σ^2 is the variance of Gaussian noise.

To solve the optimization problem defined in (8.18), dynamic programming was applied. The state of the dynamic programming is the state of the spectrum occupancy in the previous time slot (i.e. X_{k-1} where $X_{k-1} = 0$ if the spectrum is occupied by a licensed user and $X_{k-1} = 1$ if the spectrum is idle in time slot $k-1$), and the remaining power in the battery. The action here corresponds to selection of the transmit power. The dynamic programming is solved by considering two cases, i.e. $P_{\text{col}} \le P_{\text{th}}$ and $P_{\text{col}} > P_{\text{th}}$. In the case of $P_{\text{col}} \le P_{\text{th}}$, only the first two constraints in (8.18) are considered, and the reward function $\mathcal{J}_k(B_k, X_{k-1})$ can be defined as

$$\begin{aligned} \mathcal{J}_k(B_k, X_{k-1} = 0) &= \max_{0 \le p_k \le \min(B_k, P_{\text{L}})} \left(E_{h_k}(r_k(p_k)) + \overline{\mathcal{J}}_{k+1}(B_k - p_k\tau)\right) \\ &= \max_{0 \le p_k \le \min(B_k, P_{\text{L}})} \left[\exp\left(\frac{\sigma^2}{p_k}\right) \text{Ei}\left(\frac{\sigma^2}{p_k}\right)\right. \\ &\quad + (1-p)\mathcal{J}_{k+1}(B_k - p_k\tau, X_k = 0) \\ &\quad \left. + p\,\mathcal{J}_{k+1}(B_k - p_k\tau, X_k = 1)\right], \end{aligned} \tag{8.20}$$

$$\begin{aligned} \mathcal{J}_k(B_k, X_{k-1} = 1) &= \max_{0 \le p_k \le \min(B_k, P_{\text{H}})} \left(E_{h_k}(r_k(p_k)) + \overline{\mathcal{J}}_{k+1}(B_k - p_k\tau)\right) \\ &= \max_{0 \le p_k \le \min(B_k, P_{\text{H}})} \left[\exp\left(\frac{\sigma^2}{p_k}\right) \text{Ei}\left(\frac{\sigma^2}{p_k}\right)\right. \\ &\quad + q\,\mathcal{J}_{k+1}(B_k - p_k\tau, X_k = 0) \\ &\quad \left. + (1-q)\mathcal{J}_{k+1}(B_k - p_k\tau, X_k = 1)\right], \end{aligned} \tag{8.21}$$

where p and q denote the probabilities of changing from occupied state to idle state and from idle state to occupied state in the two-state Markov model for activity of a licensed user. $\mathrm{Ei}(\cdot)$ is defined as follows: $\mathrm{Ei}(x) = \int_x^{\infty} (\exp(-t))/t dt$.

However, in the case of $P_{\mathrm{col}} > P_{\mathrm{th}}$, the power cannot be bounded by only P_{L} and P_{H}, since the decision can violate the disturbance probability, for example, when an unlicensed user follows the rule in (8.19) strictly to maximize the transmission rate. Therefore, an unlicensed user transmits with probability P_0 at time slot k if the spectrum was idle in the previous time slot. This probability can be obtained from $P_0 = P_{\mathrm{th}}/P_{\mathrm{col}}$. Again, the reward function can be defined as

$$\begin{aligned}
\mathcal{J}_k(B_k, X_{k-1} = 1) &= \max_{0 \le p_k \le \min(B_k, P_H)} \Big[P_0 E_{h_k}(r_k(p_k)) + P_0 \overline{\mathcal{J}}_{k+1}(B_k - p_k\tau) \\
&\quad + (1 - P_0)\overline{\mathcal{J}}_{k+1}(B_k) \Big] \\
&= \max_{0 \le p_k \le \min(B_k, P_H)} \Big[P_0 \exp\left(\frac{\sigma^2}{p_k}\right) \mathrm{Ei}\left(\frac{\sigma^2}{p_k}\right) \\
&\quad + P_0 q \, \mathcal{J}_{k+1}(B_k - p_k\tau, X_k = 0) \\
&\quad + P_0(1 - q)\mathcal{J}_{k+1}(B_k - p_k\tau, X_k = 1) \\
&\quad + (1 - P_0)(1 - q)\mathcal{J}_{k+1}(B_k, X_k = 1) \Big]. \qquad (8.22)
\end{aligned}$$

The solution of dynamic programming for power control was observed to achieve a high expected sum rate for unlicensed users. It was also observed that in the case of $P_{\mathrm{col}} > P_{\mathrm{th}}$, an unlicensed user uses a small amount of transmit power at the beginning. However, the transmit power increases as the time approaches K (i.e. the optimization period). Since there is an uncertainty in the state of the licensed user, the unlicensed user initially tries to maintain low energy consumption and waits for better opportunity in the future. However, when the end of the period approaches, the unlicensed user has to use all of its remaining battery power to maximize the transmission rate. Note that this work did not consider the coordination of the dynamic spectrum access among unlicensed users. This coordination is important to ensure that the spectrum utilization is maximized.

8.2.6 Coordinated dynamic spectrum access in cellular networks

With dynamic spectrum access in the exclusive-use model, the spectrum allocation among multiple service providers can be coordinated through a spectrum broker which is connected to the radio access networks (RANs) (Figure 8.6). In [406], a coordinated dynamic spectrum access at the spectrum broker was proposed based on the concept of coordinated access band (CAB) and statistically multiplexed access to the spectrum (SMA).

CAB is a spectrum band owned by the spectrum broker (e.g. TV band or under-utilized public safety bands). This CAB can be allocated to a particular service provider (e.g. cellular or broadband service provider) for a fixed time duration. In this case, base stations and mobile devices have to be able to operate on different ranges of the

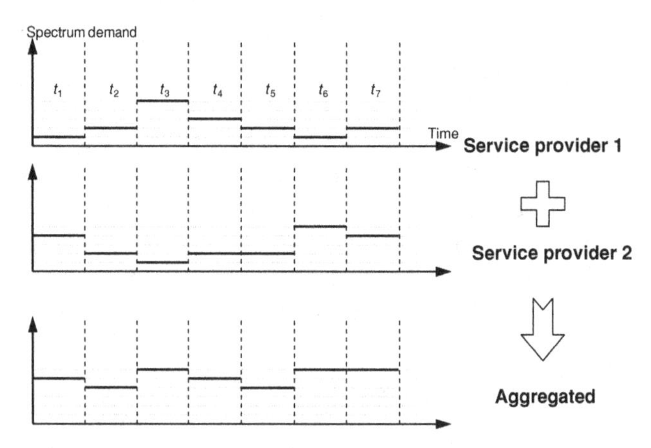

Figure 8.5 Example of multiplexing spectrum demand.

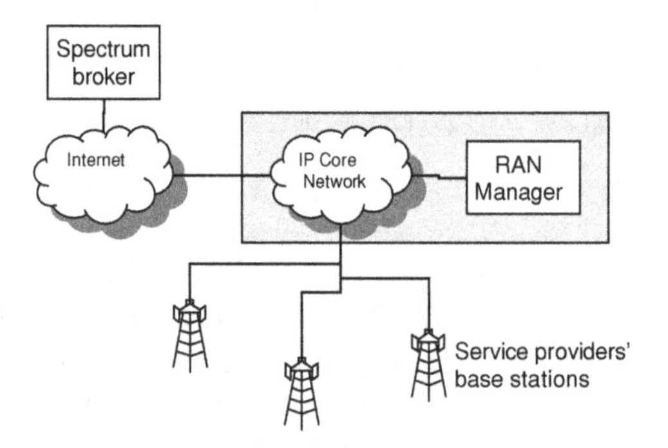

Figure 8.6 Cellular architecture with coordinated DSA.

frequency spectrum. Namely, re-configurable software defined radio is required. When the spectrum is allocated, the service provider must meet the usage constraint of this CAB, e.g. maximum transmit power.

Due to the time-varying nature of spectrum demand of each service provider, the utilization of the spectrum can be improved by using the concept of SMA. For example, in Figure 8.5, there are two service providers with a time-varying spectrum demand. If static spectrum allocation is used, five channels will be required for both service providers (10 in total). However, if the spectrum demands are aggregated, and CAB is dynamically allocated, the total number of required channels becomes 6. This example shows that if the spectrum demand can be multiplexed, the required amount of radio spectrum decreases and the utilization increases.

To implement CAB and SMA, an architecture of coordinated DSA designed for cellular network was presented in [406] (Figure 8.6). In this architecture, the spectrum broker manages the CAB spectrum in the region, while the RAN manager negotiates with the spectrum broker to obtain the CAB. After the spectrum broker allocates the CAB, base stations use this spectrum to communicate with the mobile.

Experiments were performed on a weekday in Hoboken, New Jersey for a period of 27 hours. The utilization of the CDMA code channels and GSM time slots was measured which showed the traffic load fluctuations during the day. These measurement results indicated that spectrum multiplexing can be used to improve the utilization of the radio spectrum. However, this work did not consider the spectrum sharing among multiple service providers. This issue is important especially for the exclusive-use spectrum access model.

8.2.7 Collaboration and fairness in spectrum allocation

Graph theory (e.g. the graph-coloring model) was applied to the problem of centralized spectrum allocation [374, 487, 488, 154]. In [374], collaboration among unlicensed users is used to achieve fair allocation in opportunistic spectrum access. In this case, the allocation is performed by a central controller which performs a graph-coloring-based algorithm for spectrum allocation. In the system model, there are I unlicensed users accessing N channels. The availability of a channel to an unlicensed user is denoted by the variable $v_{i,n}$, for which $v_{i,n} = 1$ if the channel n is available to unlicensed user i. The throughput (i.e. rate) of unlicensed user i due to the transmission using channel n is denoted by $r_{i,n}$. The variable $c_{i,i',n}$ indicates the interference between unlicensed user i and i' on channel n (i.e. $c_{i,i',n} = 1$, if interference occurs).

The spectrum allocation is formulated as an optimization problem with three different objectives:

- *Max-sum-throughput:* The objective is to maximize the total system throughput, i.e.

$$\max_{a_{i,n}} \sum_{i=1}^{I} \sum_{n=1}^{N} a_{i,n} r_{i,n}. \tag{8.23}$$

- *Max-min-throughput:* The objective is to maximize the minimum throughput of an unlicensed user, i.e.

$$\max_{a_{i,n}} \min_{i} \sum_{n=1}^{N} a_{i,n} r_{i,n}. \tag{8.24}$$

- *Max-proportional-fair:* The objective is to achieve proportional fairness, i.e.

$$\max_{a_{i,n}} \sum_{i=1}^{I} \log \left(\sum_{n=1}^{N} a_{i,n} r_{i,n} \right). \tag{8.25}$$

In (8.23)–(8.25) $a_{i,n}(= 1)$ is a decision variable indicating the assignment of channel n to unlicensed user i.

To solve the optimization problem with the above objective functions, a color-sensitive group coloring model was proposed based on progressive minimum neighbor first algorithm [489]. In this case, a vertex represents an unlicensed user, an edge represents the interference constraint between unlicensed users, and a color represents a channel. The proposed color-sensitive group coloring model was designed to maximize the channel utility according to the different objectives. This objective is different from that of a general graph coloring model which is to minimize the number of assigned colors.

The general algorithm to solve the color-sensitive group coloring model is as follows. The algorithm labels all the unlicensed users with a non-empty channel list according to the labeling rule which will be presented later. Each label is associated with a channel. Then, the algorithm chooses an unlicensed user with the highest label and assigns the channel associated with the label. This channel is removed from the list of that unlicensed user, and also from the lists of neighboring unlicensed users. The labels of the unlicensed users who are assigned with the channel as well as their neighboring users are revised according to the resulting graph. The algorithm repeats until all users' lists of channels are empty. The labeling rules in the collaborative cases which require a central controller are as follows:

- *Collaborative-max-sum-throughput:* The rule will assign channel n to unlicensed user i, and the total throughput is obtained from $r_{i,n}/I_{i,n}^{\text{neigh}}$ where $I_{i,n}^{\text{neigh}}$ denotes the number of neighboring users that cannot use channel n if this channel is assigned to unlicensed user i. The label and channel assignment works as follows:

$$\text{label}_i = \max_{n \in \mathbb{N}_i} r_{i,n}/(I_{i,n}^{\text{neigh}} + 1), \quad \text{channel}_i = \arg\max_{n \in \mathbb{N}_i} r_{i,n}/(I_{i,n}^{\text{neigh}} + 1) \tag{8.26}$$

  where $\mathbb{N}_i$ is the list of channels available to user i. This list is updated in each iteration of the algorithm.
- *Collaborative-max-min-throughput:* The rule will assign the channels so that all users have the same number of channels. The label and channel assignment works as follows:

$$\text{label}_i = -\sum_{n=1}^{I-1} a_{i,n} r_{i,n}, \quad \text{channel}_i = \arg\max_{n \in \mathbb{N}_i} r_{i,n}/(I_{i,n}^{\text{neigh}} + 1). \tag{8.27}$$

- *Collaborative-max-proportional-fair:* The label and channel assignment works as follows:

$$\text{label}_i = \frac{\max_{n \in \mathbb{N}_n} r_{i,n}/(I_{i,n}^{\text{neigh}} + 1)}{\sum_{n=1}^{N} a_{i,n} r_{i,n}}, \quad \text{channel}_i = \arg\max_{n \in \mathbb{N}_i} r_{i,n}/(I_{i,n}^{\text{neigh}} + 1). \tag{8.28}$$

Also, an algorithm for the non-collaborative case was presented. In this algorithm, an unlicensed user chooses a channel which maximizes its own objective. However, it does not consider the impact of its decision on other users in the network. Simulation results showed that although the non-collaborative algorithm can operate without a central controller, its performance is, however, worse than the performance of the collaborative algorithms. Note that the spectrum demand was not used in the optimization formulation in this work.

8.2.8 Spatio-temporal dynamic spectrum allocation

Spectrum opportunity can be identified in the dimensions of space and time. Accordingly, a spatio-temporal dynamic spectrum access model was proposed in [457]. In this work, a dynamic spectrum access scheme for exclusive-use model was considered in which a

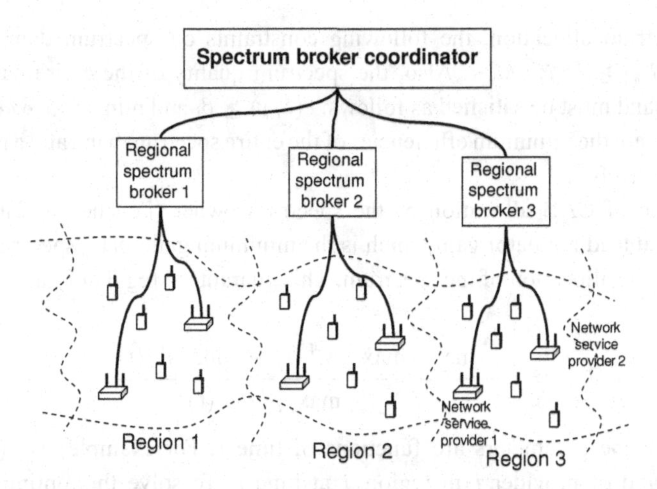

Figure 8.7 Spatio-temporal dynamic spectrum access model.

spectrum owner allocates frequency bands to I network service providers. In the system model, the service area is divided into multiple regions. In region k, network service provider i provides wireless services to users, and the spectrum demand for this service provider is denoted by $D_{i,k}$. This spectrum demand of each network service provider in terms of coordinated access band (CAB) [406] in the same region is collected and maintained by a regional spectrum broker. This broker is connected to a spectrum broker coordinator operated by the spectrum owner to optimize CAB allocation for all regions and all network service providers (Figure 8.7).

The challenges of CAB allocation arise due to the interference which affects the efficiency of the system. In this case, interference can be characterized by the geographic location and the radio access technique used by each network service provider. The parameters $\epsilon_{l,k}^{(i)}$ denote the normalized noise level caused by provider i operating in region k to the region l (i.e. geographic coupling parameter), and $\eta_{i,i'}$ denotes the normalized interference of provider i to provider i' (i.e. radio technology coupling parameter). Let the spectrum block $S_{i,k}$, which is allocated to provider i in area k, be defined by upper and lower frequencies $s_{i,k}^{\text{up}}$ and $s_{i,k}^{\text{lo}}$, respectively. The size of this spectrum block is denoted by $|S_{i,k}| = s_{i,k}^{\text{up}} - s_{i,k}^{\text{lo}}$. The efficiency of this contiguous spectrum block can be expressed as follows:

$$e(S_{i,k}) = \frac{1}{|S_{i,k}|} \int_{s_{i,k}^{\text{lo}}}^{s_{i,k}^{\text{up}}} f_{i,k}(\lambda) \mathrm{d}\lambda, \tag{8.29}$$

where $f_{i,k}(\lambda)$ is the efficiency of frequency λ, defined by

$$f_{i,k}(\lambda) = \prod_{i'=1}^{I} \prod_{j=1}^{K} \left(1 - \epsilon_{j,k}^{(i)} \eta_{i,i'} x_{\lambda, S_{i',j}}\right), \tag{8.30}$$

in which $x_{\lambda, S_{i,j}}$ is a binary variable indicating whether frequency λ is allocated to provider i in region j or not, and K is the total number of regions.

During spectrum allocation, the following constraints on spectrum demand must be satisfied: $|S_{i,k}| \geq D_{i,k}, \quad \forall i, k$. Also, the spectrum quality of the entire band and a particular subband must be satisfied as follows: $e(S_{i,k}) \geq \beta_i$ and $\min_{s_{i,k}^{\text{lo}} \leq \lambda \leq s_{i,k}^{\text{up}}} e(\lambda) \geq \alpha_i$, where β_i and α_i are the minimum efficiencies of the entire spectrum band and a particular subband, respectively.

The objective of CAB allocation of the spectrum owner (i.e. the regulator) is to maximize guaranteed regulator gain which is the minimum gain that a spectrum owner receives due to the allocation of the spectrum. This guaranteed regulator gain is defined as follows:

$$\text{RG}_{\text{min}} = 1 - \frac{\max_t \left(\max_{i,j} s_{i,j}^{\text{up}}(t) - \min_{i,j} s_{i,j}^{\text{lo}}(t)\right)}{\sum_{i=1}^{M} \max_{j,t} \left(D_{i,j}(t)\right)}. \tag{8.31}$$

In this case, all the parameters are functions of time t. For example, $D_{i,j}(t)$ is the spectrum demand of provider i in region j at time t. To solve this optimization, a heuristic method based on simulated annealing (see Section 4.4.3) was applied. Note that a variant of the model which considers the gain of a network service provider can be found in [490].

8.2.9 Dynamic spectrum allocation among network service providers

In [458], the problem of dynamic spectrum allocation among network service providers with three different types of spectrum bands was considered. These spectrum types are dedicated band, shared band, and auction band. The sizes of these bands are denoted by B_{d}, B_{s}, and B_{a}, respectively. The spectrum broker allocates a dedicated band to service provider i regardless of the required bandwidth B_i^{req} by the users served by service provider i. On the other hand, the shared band can be accessed by two service providers. The service providers can request to utilize this shared band according to their required bandwidth. Similarly, the auction band can be shared between two service providers. However, a negotiation among the service providers would be required.

The allocation scheme works as follows:

1. Service providers contact the spectrum broker to exclusively allocate a dedicated band and provide information on the occupancy ratio among the service providers in the shared bands.
2. The dedicated band allocated to each service provider is fixed and cannot be changed during the allocation period.
3. The occupancy ratio in a shared band can be adjusted based on the requirement of each service provider. If the total required bandwidth is less than or equal to the available spectrum, i.e. $\sum_{i=1}^{I} B_i^{\text{req}} \leq (B_{\text{s}} + B_{\text{a}})$, where I is the total number of service providers, the spectrum broker allocates the spectrum as the service providers request. Otherwise, the shared band is proportionally allocated among the service providers as follows:

$$B_k = W_k \frac{\sum_j \alpha_{k,j} w_{k,j}}{\sum_i \sum_j \alpha_{i,j} w_{i,j} B_{\text{s}}} \tag{8.32}$$

where B_k is the bandwidth allocated to service provider k, W_k is a control parameter, $\alpha_{i,j}$ and $w_{i,j}$ are the parameters indicating the priority and the required bandwidth, respectively, for service class j of service provider i.

4. If a service provider is not satisfied by the dedicated and shared bands allocated by the spectrum broker, this service provider can contact other service providers to rent bandwidth. In this case, compensation will be paid by the renting service provider to the leasing service provider. However, if the negotiation between service providers fail, the negotiation among service providers switches to competition mode. In this mode, available spectrum is sold to the requesting service provider by auction.

In this work, however, no method to obtain the payment for the service provider was specified. Also, the optimal strategy for competing for the auction band was not presented.

The performance of the proposed dynamic spectrum allocation method was evaluated in terms of satisfaction ratio, which is defined as the allocated bandwidth divided by the required bandwidth for a given service class. This satisfaction ratio decreases as the maximum required bandwidth increases.

8.2.10 Coordinated dynamic spectrum access

Similar to [457], a dynamic spectrum access model for cellular networks was considered [459] in which graph theory was applied to obtain coordinated dynamic spectrum allocation. In a region, there could be multiple base stations operated by different service providers. These base stations contact the regional spectrum broker to obtain the radio spectrum (i.e. dynamic spectrum access is based on the exclusive-use model). The spectrum allocation algorithm is executed by the broker. In particular, two optimization problems were formulated for this algorithm. In the first optimization problem, the objective is to maximize the overall spectrum demand while the interference constraint is satisfied. The objective of the second optimization problem is to minimize overall interference in the network while the spectrum demands of all base stations are satisfied. To obtain the solution, an interference graph is constructed to model the interference among base stations in the regions.

In the system model, it was assumed that the base station requires between $d_{\min}$ and $d_{\max}$ channels to support end users. The spectrum is allocated for a certain number of time windows. The propagation characteristics of the radio spectrum in each region are assumed to be known by the spectrum broker.

The interference graph is denoted by $G(\mathbb{V}, \mathbb{E})$, where $\mathbb{V}$ and $\mathbb{E}$ are the sets of nodes and edges, respectively. While a node represents a base station, an edge indicates interference between the corresponding base stations (i.e. $(i, j) \in \mathbb{E}$ if the transmissions of base stations i and j interfere with each other). An example of an interference graph is shown in Figure 8.8. In this case, base station i interferes with base station j, but not with base station k, since base stations i and k are far from each other.

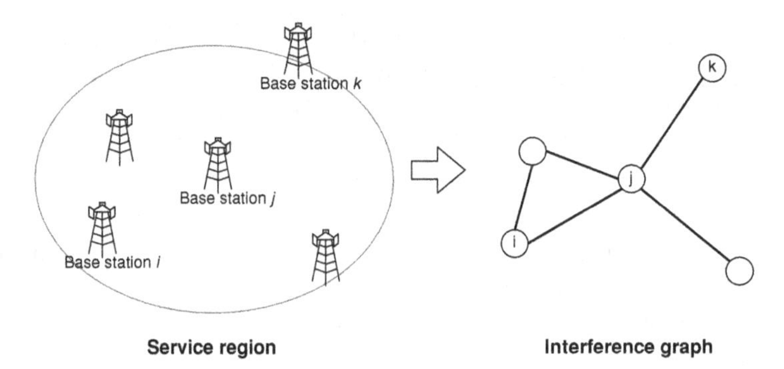

Figure 8.8 Example of interference graph.

Based on this graph model, an optimization problem can be formulated as follows:

$$\text{maximize:} \quad \sum_{i \in \mathbb{V}} \left(|\mathbb{N}_i| - d_{\min}(i) \right), \tag{8.33}$$

$$\text{subject to:} \quad d_{\min}(i) \leq |\mathbb{N}_i| \leq d_{\max}(i), \tag{8.34}$$

$$\forall (i, j) \in \mathbb{E}, \quad \mathbb{N}_i \cap \mathbb{N}_j, \tag{8.35}$$

where $\mathbb{N}_i$ is a set of channels allocated to base station i, and $|.|$ denotes the number of allocated channels.

An optimization problem with the objective of minimizing interference can then be formulated as follows:

$$\text{minimize:} \quad \sum_{(i,j) \in \mathbb{E}} c_{i,j} \left| \mathbb{N}_i \cap \mathbb{N}_j \right|, \tag{8.36}$$

$$\text{subject to:} \quad d_{\min}(i) \, |\mathbb{N}_i| \, , \tag{8.37}$$

where $c_{i,j}$ is the penalty due to interference between base stations i and j.

To solve these optimization problems, two algorithms, namely, max-demand DSA and min-interference DSA, were proposed. Max-demand DSA has two major phases. In phase one, the condition whether the minimum demand of all nodes in the network can be served by using N available channels is checked. If the minimum demand can be satisfied, then in phase two the algorithm maximizes the demand which can be satisfied beyond the minimum demand of each node. In min-interference DSA, first the random heuristic algorithm was proposed. Then, the Tabu search is applied to improve the solution obtained from the heuristic algorithm.

The performance of max-demand DSA was evaluated for different cases. It was shown that when the number of available channels is small or the node density increases, there could be a high chance that the minimum demand of the user may not be satisfied. For min-interference DSA, the remaining interference obtained from the Tabu search is much better than that of the heuristic algorithm.

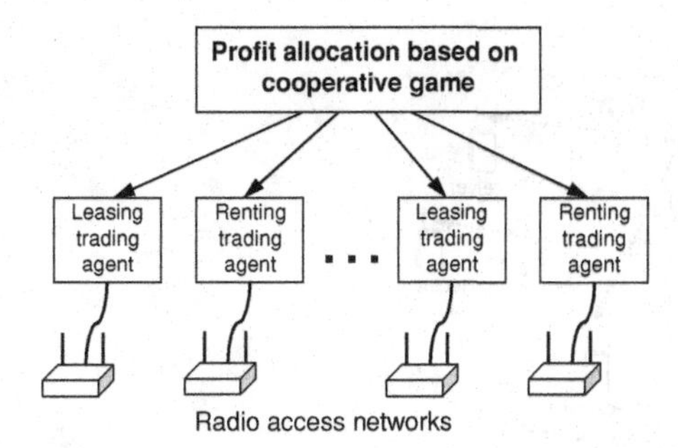

Figure 8.9 System model for integrated dynamic spectrum access and radio resource management.

8.2.11 Cooperative game for dynamic spectrum access

To improve spectrum utilization, radio resources can be leased and rented by different radio access networks (RANs) (i.e. exclusive-use model). With this approach, an integrated dynamic spectrum access and radio resource management framework was proposed [460]. For the dynamic spectrum access method used in this framework, the radio resource can be traded with other RANs in a particular period of time (i.e. trading period). Then, given the allocated spectrum, radio resource management is performed by a RAN to serve the users in the corresponding service area. A trading agent handles the information exchange and negotiations among different RANs. This trading agent performs tasks similar to those of the RAN manager in Figure 8.6. To distribute the profit earned through spectrum leasing and spectrum renting, a cooperative game theory model is used among the RANs. The system model of this framework is shown in Figure 8.9.

In this integrated framework, the trading agent performs four steps in each trading period, i.e. profit settlement, radio resource management tuning for current period, dynamic spectrum access trading for the next period, and profit sharing. In the profit settlement step, the profit due to spectrum leasing and renting in the previous period is obtained for each RAN. Trading information repository (TIR) is used to maintain profit sharing information which will be used to compute the profit and to adapt the trading behavior of the trading agent in the future. In the management tuning step, the trading agent performs radio resource allocation according to the available spectrum. The traffic load information is used to determine whether a RAN has to rent or lease the radio spectrum. In the dynamic spectrum access trading step, the spectrum is traded among RANs through the trading agent. In the last step, i.e. profit sharing, the profit of each RAN is computed and shared among each other in a fair manner by using the Shapley value method in cooperative game theory (see Section 5.5.2).

The performance of this joint dynamic spectrum access and radio resource management framework was evaluated by simulations. The system is composed of three overlaid RANs, which includes UMTS, GMS, and DVB-T. The scheme was compared with bargaining-based dynamic spectrum management (BDSM) [491] and adaptive threshold load balancing (ATLB) [492] schemes and was observed to provide the highest profit.

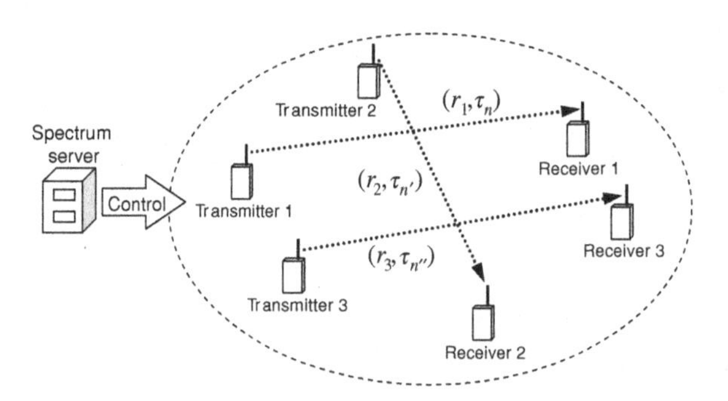

Figure 8.10 Transmission scheduling for cognitive radio users through a spectrum server.

8.2.12 Transmission scheduling via spectrum server

With the use of a spectrum server (i.e. spectrum broker), transmissions from the cognitive radio users can be optimally scheduled [461] if perfect knowledge about the transmission links (e.g. channel gains in the transmission links) is available to the spectrum server. For example, in a particular area, there are multiple cognitive radio users for which the rate r_i of user i and the time share τ_n of the transmission on channel n are controlled by the spectrum server (Figure 8.10). Three different scheduling schemes for this shared-use spectrum access model, namely, maximum sum rate, max-min fair rate, and proportional fair scheduling were proposed in [461]. In the system model under consideration, the transmission of user i can be scheduled over channel n (N is total number of channels) where the SNR of link i is obtained from $\gamma_{i,n} = \frac{x_{i,n} h_{i,i} P_i}{\sum_{j \neq i} x_{j,n} h_{i,j} P_j + \sigma_i^2}$, where $h_{i,j}$ is the channel gain between the transmitter of user i to the receiver of user j, and σ_i^2 is the background noise power. Here, $x_{i,n}$ is a binary vector indicating the activity of link i and it is defined as follows:

$$x_{i,n} = \begin{cases} 1, & \text{link } i \text{ is active using channel } n \\ 0, & \text{otherwise.} \end{cases} \tag{8.38}$$

The transmission rate of user i using channel n is obtained from $c_{i,n} = \log(1 + \gamma_{i,n})$, and the average transmission rate is obtained from

$$r_i = \sum_{n=1}^{N} c_{i,n} \tau_n. \tag{8.39}$$

Let us define matrix $\mathbf{C}$ as follows:

$$\mathbf{C} = \begin{bmatrix} c_{1,1} & & \cdots & & c_{1,N} \\ & \ddots & & \iddots & \\ \vdots & & c_{i,n} & & \vdots \\ & \iddots & & \ddots & \\ c_{I,1} & & \cdots & & c_{I,N} \end{bmatrix}. \tag{8.40}$$

Equation (8.39) can be expressed in a vector form as $\mathbf{r} = \mathbf{C}\mathbf{x}$, where $\mathbf{r}$ and $\mathbf{x}$ denote the vectors of average rate of user $i = \{1, \ldots, I\}$ and the time share for channel $n = \{1, \ldots, N\}$, respectively.

For the maximum sum rate scheduling scheme, an optimization problem can be formulated based on linear programming with the minimum rate constraint for each link as follows:

$$\text{maximize: } \mathbf{1}^{\mathrm{T}}\mathbf{C}\mathbf{x}, \tag{8.41}$$

$$\text{subject to: } \mathbf{C}\mathbf{x} \geq \mathbf{r}_{\min}, \tag{8.42}$$

$$\mathbf{1}^{\mathrm{T}}\mathbf{x} \leq 1, \quad \mathbf{x} \geq \mathbf{0}, \tag{8.43}$$

where $\mathbf{r}_{\min}$ is a vector of minimum transmission rate requirement of each link, $\mathbf{0}$ and $\mathbf{1}$ are the vectors of zeros and ones, respectively.

For max-min fair rate scheduling and proportional fair scheduling, optimization problems are formulated based on max-min [493] and proportional fairness [468] criteria, respectively.

The max-min fairness is achieved when for each i, r_i cannot be increased without decreasing $r_{i'}$ for some link i' where $r_{i'} \leq r_i$. In other words, the transmission rate is max-min fair when the minimum transmission that a link receives is maximized. To achieve max-min fairness, an optimization problem based on linear programming can be formulated as follows:

$$\text{maximize: } r_{\min}, \tag{8.44}$$

$$\text{subject to: } \mathbf{r} = \mathbf{C}\mathbf{x}, \tag{8.45}$$

$$\mathbf{r} \geq r_{\min}\mathbf{1}, \tag{8.46}$$

$$\mathbf{1}^{\mathrm{T}}\mathbf{x} \leq 1, \quad \mathbf{x} \geq \mathbf{0}. \tag{8.47}$$

Proportional fairness is achieved for a transmission rate vector $\mathbf{r}$ if for all other transmission rate vectors $\mathbf{r}'$, and the aggregate of proportional change is negative, i.e. $\sum_{i=1}^{I} (r_i' - r_i)/r_i \leq 0$. To achieve a proportional fair transmission rate vector, an optimization problem similar to that in [468] is formulated as follows:

$$\text{maximize: } \sum_{i=1}^{I} \log r_i, \tag{8.48}$$

$$\text{subject to: } \mathbf{r} = \mathbf{C}\mathbf{x}, \tag{8.49}$$

$$\mathbf{1}^{\mathrm{T}}\mathbf{x} \leq 1, \quad \mathbf{x} \geq \mathbf{0}. \tag{8.50}$$

This is a convex optimization problem, and therefore the solution can be obtained by using a gradient search algorithm.

With max-min fairness, it was observed that the average transmission rate decreases as the minimum rate requirement increases. Since the scheduling must achieve the minimum rate requirement for all users, the available spectrum cannot be used to maximize the transmission rate of all users. The solutions for max-min fairness, proportional fairness, and maximum total transmission rate were compared. It was observed that the

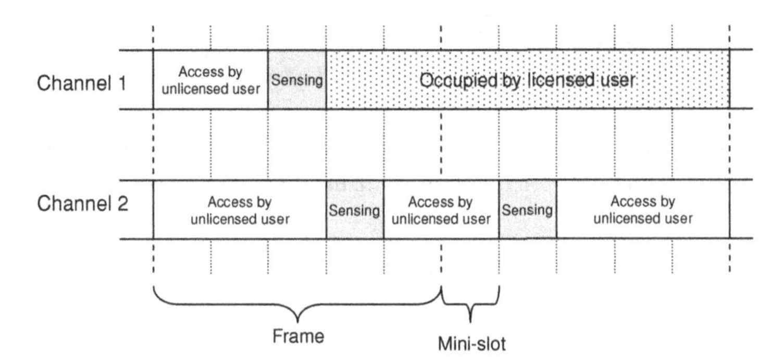

Figure 8.11 Frame and mini-slot structure for scheduling of spectrum sensing.

difference among transmission rates of all users is the smallest with max-min fairness. With proportional fairness, this difference is smaller than that with maximum total transmission rate formulation. Note, however, that the optimization of spectrum sensing was not considered here.

8.2.13 Spectrum sensing and access scheduling

For spectrum sharing based on spectrum overlay, spectrum sensing and access have to be scheduled to achieve the desired system objective. In [462], a centralized scheduling algorithm for dynamic spectrum access was proposed. This scheduling problem is divided into two cases, namely to maximize the transmission rate and to minimize packet loss due to delay constraint violation. Time is divided into frames (Figure 8.11), where a frame corresponds to the maximum tolerable duration of interference caused to a licensed user by an unlicensed user. For example, if a licensed user can tolerate interference for at most 200 ms, an unlicensed user must sense the channel at least every 200 ms. Each frame is divided into T mini-slots and each mini-slot corresponds to data transmission or channel sensing by the unlicensed users. In the system model, there are N channels and I unlicensed users. In each frame, unlicensed users must sense the channels to ensure that they are idle and can be accessed without interfering with the licensed users. Transmissions from unlicensed users utilize adaptive modulation so that the transmission rate can be adjusted according to the channel quality. This channel state can be modeled by a finite-state Markov chain (FSMC) [10].

To maximize the transmission rate, a channel state information (CSI)-based sensing and scheduling algorithm can be used for scheduling channel sensing activity in a frame duration. Intuitively, channel sensing should not be performed when channel quality is good (i.e. transmission rate is high). This scheduling algorithm can be formulated as a finite-horizon dynamic programming [232]. The control horizon is T mini-slots or one frame. The state of the system is denoted by $(s_n, m_1^n, \ldots, m_I^n)$, where s_n is a binary variable (i.e. 1 if the sensing is performed and 0 otherwise), and m_i^n is the state of channel n used by unlicensed user i. The action, i.e. to sense or to access the channel, is made every mini-slot and the reward of these actions is obtained

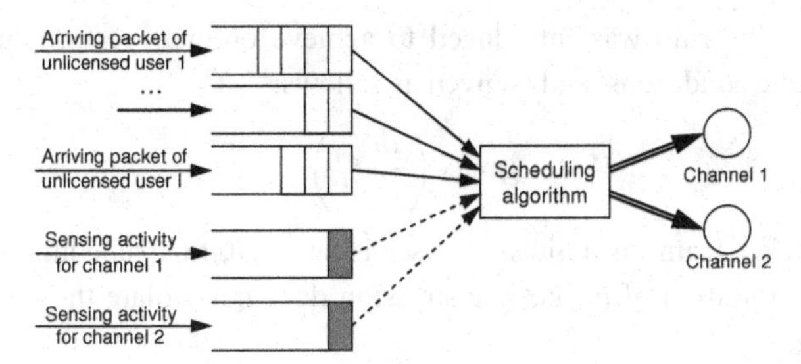

Figure 8.12 Queueing system of packet arrival and spectrum sensing activity for QCI-based scheduling algorithm.

from

$$R = \begin{cases} \max_i(r_i^n), & \text{channel is accessed} \\ 0, & \text{channel is sensed,} \end{cases} \tag{8.51}$$

where r_i^n is the transmission rate of user i on channel n. Note that at the end decision horizon, if variable $s_n = 0$ (i.e. channel n is never sensed), then the only available action is to sense the channel. In order to obtain an optimal policy, a standard method in dynamic programming with a finite-horizon time can be applied.

To minimize packet loss, a QCI-based scheduling algorithm was designed under stochastic packet arrival with delay constraint (i.e. a packet is dropped if the corresponding delay constraint D_i is violated). This scheduling is formulated as a queue selection problem and the sensing activity is modeled as a virtual packet (Figure 8.12). The scheduling algorithm in each mini-slot selects the actual/virtual packet from the queue and transmits/senses the channel based on a cost function. Let d_i denote the waiting time (i.e. delay) of the actual/virtual packet in queue i, where $i \in \{1, \ldots, I\}$ denotes the queue for the actual packet and $i = \{I+1, \ldots, I+N\}$ denotes the queue for the virtual packet for the sensing activity. Note that the delay constraint for a queue with a virtual packet is denoted by T, which is the latest mini-slot required for spectrum sensing. Based on this queueing model, three scheduling rules can be applied for queue selection:

- *Cost-based rule:* This rule was introduced in [494] for which the cost function is defined as $C(d_i) = (d_i/D_i)^\alpha$, where $\alpha > 1$ is the constant. The queue selection rule is as follows:

$$i^* \in \arg\max_i(C_i'(d_i)r_i^n), \tag{8.52}$$

  where $C'(\cdot)$ is the first derivative of the cost function.
- *Modified largest weighted delay first (M-LWDF) rule:* This rule was proposed in [495] and is given as follows:

$$i^* \in \arg\max_i(c_i(d_i)^\beta r_i^n), \tag{8.53}$$

  where c_i is a weighting factor and β is a constant.

- *QCI-based rule:* This rule was introduced to achieve optimal performance under different traffic load conditions, and is given as follows:

$$i^* \in \arg\max_i \left(\frac{d_i}{D_i} r_i^n \right). \tag{8.54}$$

Note that if a packet from an unlicensed user is selected, this scheduling rule has to ensure that the duration of packet transmission does not violate the deadline for spectrum sensing.

Performance evaluation results showed that the packet loss rate increases as the packet arrival rate increases. It is also observed that the packet loss rate of QCI-based sensing is much lower than that of static sensing. The packet loss rate decreases as the number of mini-slots per frame increases. Since more mini-slots can provide increased flexibility to the algorithm to optimize the scheduling of spectrum sensing, the performance of the system is better when there are more mini-slots.

The above model considers only single-hop communication among unlicensed users. However, multihop communications will also be common in cognitive radio networks. For multihop cognitive radio networks the routing issue needs to be taken into account when designing dynamic spectrum access schemes.

8.2.14 Joint spectrum allocation and routing

In a multihop cognitive radio network, different sets of channels will be available at the different nodes. In such a network, the issues of channel allocation and traffic routing can be jointly considered [496, 497, 498]. In [496], for joint spectrum sharing and fair routing an optimization problem was formulated as a mixed integer linear programming (MILP). In the system model, there are I unlicensed nodes and N orthogonal channels. A node i can perform spectrum sensing to identify a set of available channels. The link between nodes i and j is denoted by $l = (i, j)$. $\mathbb{L}_i$ denotes a set of links associated with node i, and $\mathbb{L}'_l$ denotes a set of links which interfere with link l. The set of channels available for link l is denoted by $\mathbb{N}_l$. Let $x_{n,l} = 1$ indicate the assignment of channel n to link l, and $x_{n,l} = 0$ otherwise. The constraint on interference can be expressed as follows:

$$x_{n,l} + x_{n,l'} \leq 1, \quad \text{for } n \in \mathbb{N}_l \cap \mathbb{N}_{l'}, l \in \bigcup_{i=1}^{I} \mathbb{L}_i, l' \in \mathbb{L}'_l. \tag{8.55}$$

A node i has m_i interfaces, and hence m_i channels can be accessed by this node concurrently. The constraint on concurrent transmission can be defined as follows:

$$\sum_{l \in \mathbb{L}_i} \sum_{n \in \mathbb{N}_l} x_{n,l} \leq m_i. \tag{8.56}$$

For routing, assume that there are S sessions in this network. Let $n_{\text{sou}}(s)$, $n_{\text{des}}(s)$, and $d(s)$ denote the source node, the destination node, and the traffic demand of session s, respectively. For each link, $f_{i,j}(s)$ denotes the flow from node i to j corresponding to the session s. The constraints on traffic routing are as follows:

- $f_{i,j}(s) \geq 0$ indicates that the flow rate is non-negative.
- $\sum_{(k,i)\in\mathbb{L}_i, k\neq n_{\text{des}}(s)} f_{k,i}(s) = \sum_{(i,j)\in\mathbb{L}_i, j\neq n_{\text{sou}}(s)} f_{i,j}(s)$ for $i \neq n_{\text{des}}(s), i \neq n_{\text{sou}}(s)$ indicates that at each node except the source and destination nodes, the total rate of incoming flows from all nodes k is equal to that of outgoing flows to all nodes j.
- $\sum_{(i,j)\in\mathbb{L}_i} f_{i,j}(s) \geq \lambda d(s)$ for $i = n_{\text{sou}}(s)$ indicates that the total rate of outgoing flows from source node i is equal to or larger than λ percent of traffic demand of session s.
- $\sum_{(j,i)\in\mathbb{L}_i} f_{j,i}(s) \geq \lambda d(s)$ for $i = n_{\text{des}}(s)$ indicates that the total rate of incoming flows to destination node i is equal to or larger than λ percent of traffic demand of session s.
- $\sum_s \left(f_{i,j}(s) + f_{j,i}(s)\right) \leq \sum_{n\in\mathbb{N}_l} x_{n,l} r_l$, where r_l is the capacity (i.e. rate) of link l, indicates that the total rate of the flows between nodes i and j has to be equal or smaller than the total capacity between these nodes. This capacity can be obtained from $r_l = W \log_2\left(1 + P d_{i,j}^{-\alpha}/\sigma^2 W\right)$, where W is the bandwidth of the channel, P is the transmit power, $d_{i,j}$ is the distance between nodes i and j, α is a path loss exponent, and σ^2 is the noise power.

An optimization problem can be formulated with the following objective:

$$\text{maximize: } \lambda, \tag{8.57}$$

which is equivalent to maximizing the traffic rate of all sessions in a network. In this case, $x_{n,l}$, which is a binary integer (i.e. $x_{n,l} \in \{0, 1\}$) for channel allocation, and $f_{i,j}(s)$, which is a real number for traffic routing, are the decision variables. This optimization problem is in the form of mixed integer linear programming and can be solved by using standard tools. However, to obtain the optimal solution, a central controller is required.

Performance evaluation results showed that the solution of this optimization problem can provide fair resource allocation and routing for all sessions in the network. Since λ is used as a weight for traffic demand, the achievable rates of all sessions are maximized in proportion to this weight.

8.2.15 DSA based on the water-filling algorithm

The water-filling algorithm [499] was extensively studied for power and rate allocation in multichannel wireless transmission scenarios (e.g. OFDMA) [241, 500, 501, 502, 503, 504, 505, 506, 507, 508, 509]. The term water-filling describes the calculation of rate or power allocated to a user or a channel. In water-filling, more power is allocated to users/channels with better quality in order to maximize sum data rate for all users/channels. This water-filling algorithm was applied to the dynamic spectrum access problem in a cognitive radio network with multiple channels [510, 511, 512, 513, 514, 515, 516, 517].

8.3 Auction-based approach

Auction is a process of selling and buying a commodity (or service) whose price is undetermined. In an auction process, the bidders submit their bids (e.g. in terms of

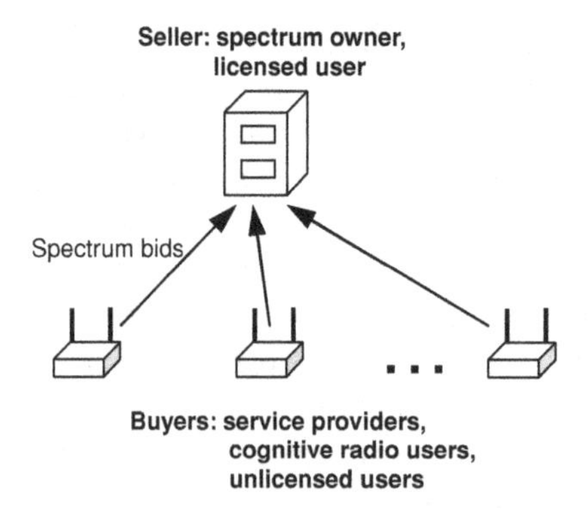

Figure 8.13 Spectrum auction.

bidding price and quantity) to the auctioneer. The auctioneer determines the winning bidder. Then, the commodity is sold at the trading price. Auction theory can be applied to a cognitive radio network, allowing a licensed user or a spectrum owner to sell its licensed spectrum to unlicensed or cognitive radio users. In this case, the price of the spectrum is undetermined and could vary with the availability of spectrum opportunities as well as spectrum demand. The main objectives of spectrum auction are: (1) to determine the clearing price of the spectrum to be sold and (2) to determine the parameter of spectrum access (e.g. frequency band, time slot, and transmit power) by a cognitive radio user. In this case, the first objective is related to the economic aspect while the second objective is related to the optimization aspect of dynamic spectrum access. In this chapter, we emphasize the optimization aspect of dynamic spectrum access. The economic aspects of spectrum trading will be addressed in Chapters 11 and 12.

8.3.1 General framework of spectrum auction

To support spectrum allocation and pricing, a real-time spectrum auction framework was proposed in [463]. In the considered system model, the seller sells the available spectrum to the buyers (Figure 8.13). The seller could be a spectrum owner or a licensed user with spectrum opportunities to be sold. The buyers could be service providers, cognitive radio users, or unlicensed users.

This framework was designed to support a large number of cognitive radio users under interference constraints. A multi-unit auction scheme was used to support diverse spectrum demand due to both short- and long-term traffic loads. The objective is to maximize the revenue of the spectrum owner and to maximize spectrum utilization. To obtain the solution, fast auction clearing algorithms were used.

In order to communicate the bidding information between spectrum seller and buyer, a bidding language, namely, piecewise linear price-quality (PLPQ), was used. The buyers express their demand in terms of the requested spectrum size and per-unit price. This bid is submitted to the seller. In PLPQ, spectrum demand is expressed as a concave piecewise linear demand function. For example, the spectrum demand can be expressed as

$$p_i(f_i) = -a_i f_i + b_i, \tag{8.58}$$

where p_i is the spectrum price, f_i is the spectrum size, a_i and b_i are parameters of the linear function. In this case, the size of the spectrum demand can be expressed as a function of price as follows: $f_i(p_i) = (b_i - p_i)/a_i$, and the revenue can be expressed as $\mathscr{R}_i(p_i) = f_i(p_i) = (b_i p_i - p_i^2)/a_i$.

The spectrum seller collects the spectrum bids (i.e. spectrum demand) from all buyers. Then, the auction clearing algorithm is used to obtain the optimal clearing price. Two pricing models were used, namely, uniform and discriminatory pricing. In the case of uniform pricing, the clearing price p is fixed for all users and the spectrum will be sold to the buyer if $b_i \geq p$, where b_i is a bidding parameter from the buyer (in (8.58)). An optimization problem was formulated to obtain price p such that the revenue is maximized as follows:

$$\text{maximize:} \quad \sum_{\{i|b_i>p\}} \frac{b_i p - p^2}{a_i}, \tag{8.59}$$

$$\text{subject to:} \quad f_i = \frac{b_i - p}{a_i}, \tag{8.60}$$

$$f_i + \sum_{j \in \mathbb{I}_L(i)} f_j \leq 1. \tag{8.61}$$

The constraint in (8.61) specifies that the sum of the spectrum share f_i and f_j must be less than or equal to one. $\mathbb{I}_L(i)$ is the set of nodes lying to the left of node i, and is used (instead of the set of all neighboring nodes) to reduce the complexity of optimization.

In discriminatory pricing, the clearing price could be different for each buyer. In this case, an optimization problem can be formulated as follows:

$$\text{maximize:} \quad \sum_i (-a_i f_i^2 + b_i f_i), \tag{8.62}$$

$$\text{subject to:} \quad -a_i f_i + b_i \geq 0, \tag{8.63}$$

$$f_i \geq 0, \tag{8.64}$$

$$f_i + \sum_{j \in \mathbb{I}_L(i)} f_j \leq 1, \tag{8.65}$$

where f_i is a function of p_i, which is the optimal discriminatory price.

To obtain the optimal solution, an approximation algorithm based on linear programming was proposed in [518]. For uniform pricing, the feasible region of the clearing price p is first obtained. Then, the algorithm searches for an optimal clearing price which maximizes the revenue. For discriminatory pricing, the algorithm uses separable programming [519] to solve a special class of non-linear programs using linear programming. The details of the derivation can be found in [520].

It was observed that the discriminatory pricing model can achieve higher revenue and higher utilization than those of a uniformed pricing model. Under the discriminatory pricing model, the spectrum broker charges cognitive radio users based on their requirements with an objective to achieving higher revenue and higher spectrum utilization. These two different pricing models were also evaluated under different user behaviors, namely, aggressive, normal, and conservative behaviors. For these behaviors, the unit price functions were defined as follows: $p(f) = -f + 1$, $p(f) = 1/2(-f + 1)$, and $p(f) = 2(-f + 1)$, respectively. In the case of uniform pricing model, the aggressive user obtains all the spectrum. This is due to the fact that the clearing price is high, so only the aggressive user can afford it. In contrast, with the discriminatory pricing model, although the aggressive user obtains the largest portion of the spectrum, normal and conservative users also obtain parts of the spectrum. For an approximation algorithm, it was observed that the performance is lower than the optimal solution by only 10 percent. However, the optimal solution takes much longer time to compute.

8.3.2 Multibid auction for dynamic spectrum allocation

In [464], a one-shot multibid auction framework for dynamic spectrum allocation was proposed. This framework explicitly considers different wireless services in the same or different regions which could interfere with each other. Similar to that in [463], the framework determines the solution in terms of spectrum size and the price for service providers who bid for the spectrum from the spectrum owner (Figure 8.13). However, in this framework, a service provider submits the bid as a set of two-dimensional bids instead of a piecewise linear function as in [463]. Consequently, the solution space becomes smaller, which results in lower computational complexity.

Let the set of bids from service provider i be defined as follows:

$$\mathbb{B}_i = \{b_{i,1}, \ldots, b_{i,D_i}\}, \tag{8.66}$$

for $i = \{1, \ldots, I\}$ where D_i is the total number of bids from the service provider i, and $b_{i,d}$ is defined as $b_{i,d} = (q_{i,d}, \Theta_i(q_{i,d}))$. Here $q_{i,d}$ denotes the requested spectrum size, and $\Theta_i(\cdot)$ denotes the total value that the provider is willing to pay for the given requested spectrum size.

The profile of the bids from all service providers is denoted by $\mathbb{B} = \{\mathbb{B}_1, \ldots, \mathbb{B}_I\}$. The optimal feasible allocation of spectrum is defined as $\mathbb{A}^*(\mathbb{B}) = \{a_1, \ldots, a_I\}$, where $a_i = q_{i,d_i^*}$, and

$$\left(d_1^*, \ldots, d_i^*, \ldots, d_I^*\right) = \arg \max_{(d_1, \ldots, d_I)} \sum_{i=1}^{I} \Theta_i(q_{i,d_i}), \tag{8.67}$$

and q_{i,d_i} is an element from the set of feasible allocation $\mathbb{B}$. This solution was obtained using the simulated annealing algorithm.

The price charged to the service provider is based on an exclusion–compensation principle (i.e. second-price auction). That is, service provider i considers price c_i which covers the social opportunity cost. This cost quantifies the loss of utility due to the presence of service provider i. In this case, the set of all bids except that of service provider i is defined as $\mathbb{B}_{-i} = \{\ldots, \mathbb{B}_{i-1}, \mathbb{B}_{i+1}, \ldots\}$. The optimal feasible spectrum allocation is defined as

$$\mathbb{A}^*_{-i} = \{a_1^{-i*}, \ldots, a_{i-1}^{-i*}, a_{i+1}^{-i*}, \ldots a_I^{-i*}\}.$$

The price for service provider i is then obtained from

$$c_i(\mathbb{B}) = \sum_{j \neq i} \Theta_j(a_j^{-i*}) - \Theta_j(a_j). \tag{8.68}$$

The proposed scheme was evaluated by considering two different cognitive radio service providers and one DVB-T service provider. The DVB-T provider pays the highest price since it serves a larger coverage region and its interference limit is low.

8.3.3 Dynamic spectrum allocator knapsack auction

A model for spectrum auction was proposed in [465] in which the solution of the auction problem was formulated as a knapsack problem (see Section 4.5.2). In the system model, I service providers bid for N frequency bands from the spectrum broker. Service provider i submits bid b_i, which is defined as $b_i = \{x_i, v_i\}$, where x_i and v_i are the number of requested bands and the bidding price, respectively. Given the bids from all service providers, the spectrum broker solves the knapsack problem, which is defined as follows:

$$\begin{aligned} \text{maximize:} & \sum_{i \in \mathbb{W}} v_i, \\ \text{subject to:} & \sum_{i \in \mathbb{W}} x_i \leq N, \end{aligned} \tag{8.69}$$

where $\mathbb{W}$ is the set of service providers winning the auction.

For an asynchronous auction, a service provider submits a bid to the spectrum broker whenever a spectrum band is required (e.g. due to instantaneous load). If the spectrum broker has enough spectrum bands, these bands are immediately allocated to the service provider. However, if the number of requested bands is larger than the number of available spectrum bands at the spectrum broker, the knapsack spectrum auction in (8.69) is solved to obtain the optimal solution for the spectrum broker. For the winning service providers, the spectrum bands will be allocated according to the duration requested in the bid. A similar auction process is used in a synchronous auction. However, the spectrum band will be allocated over equal durations for all service providers.

It was proven in [465] that the revenue generated by service providers in an asynchronous auction cannot be better than that in a synchronous auction for a given set of bids. This is due to the fact that in a synchronous auction, the spectrum band is allocated for a fixed and equal duration for all service providers. Therefore, revenue can be optimized in every period. Let us consider an example where the number of bands is $N = 2$. At time slot t, only service provider 1 submits the bid $b_1 = \{1, 1\}$ (i.e. 1 band for price 1) for two time slots. At time slot $t + 1$, service provider 2 submits the bid $b_2 = \{2, 5\}$ (i.e. 2 bands for price 5 for one time slot). In the case of an asynchronous auction, one band is allocated to service provider 1 in both time slot t and $t + 1$, while another band is allocated to service provider 2 only in time slot $t + 1$. In this case, both spectrum bands cannot be allocated in time slot $t + 1$ to service provider 2 whose bidding price is higher than that of service provider 1. In contrast, in a synchronous auction, one band is allocated to service provider 1, while two bands are allocated to service provider 2 at time slot $t + 1$. Therefore, the total revenue earned by the spectrum broker in a synchronous auction is $(1 \times 1) + (2 \times 5) = 11$ which is clearly higher than $(1 \times 1) + (1 \times 1 + 1 \times 5) = 7$ in an asynchronous auction. Since the solution of the auction is obtained at every time slot, the spectrum band allocation can be adjusted according to the bids, and the revenue of the spectrum broker is higher for a synchronous auction.

8.3.4 Weighted proportional fair spectrum allocation

A spectrum allocation algorithm based on weighted proportional fairness for OFDMA-based cognitive radio was proposed in [466]. This algorithm is based on the asymmetric-Nash-bargaining-solution (ANBS) utility function, which captures spectrum sensing contribution of the unlicensed users. These unlicensed users are grouped to cooperate in spectrum sensing and sharing. The user who contributes more in spectrum sensing will receive better allocation while the fairness and efficiency for all users can be achieved. The coordination among unlicensed users in the same group is performed by a central base station.

In the system model, for subcarrier n, the throughput (i.e. rate) of the unlicensed user i is denoted by $r_{i,n}$ and its allocation is denoted by variable $x_{i,n} = 1$ ($x_{i,n} = 0$ if subcarrier n is not allocated to user i). This throughput is computed from the channel quality of the unlicensed user on that subcarrier where adaptive modulation is used for transmission. If ω_i denotes the spectrum sensing contribution of unlicensed user i, an optimization problem can be defined as follows:

$$\text{maximize:} \quad \mathscr{U} = \prod_{i=1}^{I} \left(\sum_n x_{i,n} r_{i,n} - R_{\min}^{(i)} \right)^{\omega_i}, \tag{8.70}$$

$$\text{subject to:} \quad \sum_n x_{i,n} r_{i,n} \geq R_{\min}^{(i)}, \quad \sum_{i=1}^{I} = 1, \tag{8.71}$$

$$\sum_n P_{i,n} \leq P_{\max}, \quad \sum_{i=1}^{I} \omega_i = 1, \tag{8.72}$$

where $\mathscr{U}$ is the utility, I is the total number of unlicensed users, $R_{\min}^{(i)}$ is the minimum throughput (i.e. rate) requirement of unlicensed user i, $P_{i,n}$ is the transmit power of user i on subcarrier n, and $P_{\max}$ is the power budget. An algorithm to obtain the solution was presented and the convergence of the algorithm was proved.

For performance evaluation, an OFDMA system with adaptive modulation was considered. As the spectrum sensing contribution of one unlicensed user increases, from the optimization formulation, its throughput increases while that of other users decreases. In addition, there is an optimal point such that the aggregated throughput is maximized. This proposed scheme was compared with the maximum total throughput and max-min fairness formulations. Intuitively, the maximum total throughput formulation provides the highest throughput, but it is not fair for the user with bad channel quality. On the other hand, the max-min fairness formulation achieves fair solutions (i.e. individual throughputs of the users are almost the same), but the total system throughput is not maximized. In this case, the proposed scheme which is based on the asymmetric Nash bargain solution compromises both formulations. That is, the user with better channel quality will receive higher throughput, while the throughput of the user with bad channel quality is not significantly different. Also, the user that contributes more in spectrum sensing is rewarded by gaining higher throughput. This work considered only the auction from the buyer side (i.e. the unlicensed user side). However, the sellers (i.e. spectrum owners) can adapt their price strategies to achieve the highest profit.

8.3.5 Bilateral bargain in spectrum access

Within a microeconomic framework, radio resource allocation can be modeled as a bargaining game [491, 440, 521, 254, 522, 523, 524, 525]. In a cognitive radio network, the dynamic spectrum allocation can be optimized using the concepts of auction and bargaining [467]. In particular, two algorithms based on Anglo-Dutch split award auction and bilateral bargaining models were used in [467] for short-term and long-term spectrum trading, respectively, among multiple radio access networks (RANs). In the first algorithm, which is based on auction model, spectrum trading is divided into four stages. In the first stage, service providers owning RANs submit the bids for the auction. These service providers are allocated with the minimum pre-specified amount of available spectrum. In addition, two service providers with the highest bid prices are selected to proceed to the second stage. In the second stage, these two service providers submit additional bids to obtain the spectrum remaining from the first stage. In the third stage, the spectrum owner allocates the spectrum to maximize the total bids. Finally, in the fourth stage, two service providers compete with each other to sell their services to the market using the obtained spectrum.

In the second algorithm, which is based on the bilateral bargaining model, there are leasing and renting RANs negotiating for the spectrum to maximize their utilities. In this case, the leasing RAN submits the asking price while the renting RAN submits the bidding price. Let c and v denote the cost and the value of the spectrum for leasing and renting RANs, respectively. Based on the asking price p_{ask} and the bidding price p_{bid}, the trading price is determined from $p_{\text{tr}} = p_{\text{bid}} + p_{\text{ask}}/2$, and the profits of these RANs can be expressed as follows:

$$\mathscr{P}_{\text{leasing}} = \frac{p_{\text{ask}} + p_{\text{bid}}}{2} - c, \tag{8.73}$$

$$\mathscr{P}_{\text{renting}} = v - \frac{p_{\text{ask}} + p_{\text{bid}}}{2}, \tag{8.74}$$

for $p_{\text{ask}} \leq p_{\text{bid}}$. Both leasing and renting RANs search for the optimal asking and bidding prices to maximize their profits. However, they do not have the information of each other. Namely, the leasing RAN does not have perfect information of value v of the renting

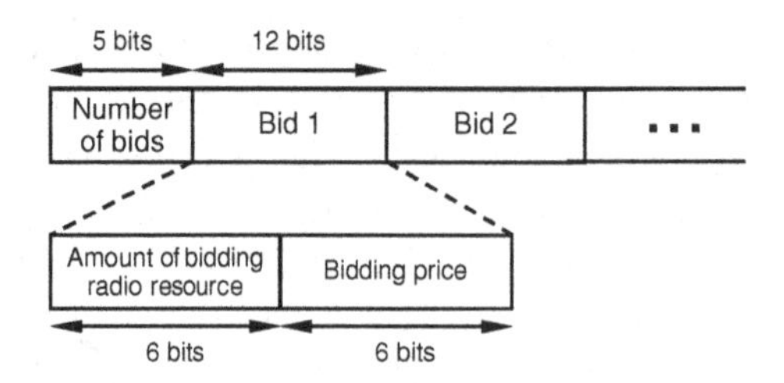

Figure 8.14 Bidding message.

RAN, and the renting RAN does not have perfect information of cost c of the leasing RAN. Therefore, an optimization problem can be formulated for the leasing RAN as follows:

$$\text{maximize:} \quad \mathscr{P}_{\text{leasing}} = \frac{1}{2}\left(p_{\text{ask}} + E\left(\tilde{p}_{\text{bid}}^{\text{lease}}(v)|\tilde{p}_{\text{bid}}^{\text{lease}}(v) \geq p_{\text{ask}}\right) - c\right) \times \Pr\left(\tilde{p}_{\text{bid}}^{\text{lease}}(v) \geq p_{\text{ask}}\right), \tag{8.75}$$

where $\tilde{p}_{\text{bid}}^{\text{lease}}(v)$ is the estimated value of the bidding price from the renting RAN, which is used by the leasing RAN, $E\left(\tilde{p}_{\text{bid}}^{\text{lease}}(v)|\tilde{p}_{\text{bid}}^{\text{lease}}(v) \geq p_{\text{ask}}\right)$ is the expected value of the estimated bidding price given that this estimated bidding price is larger than or equal to the asking price, and $\Pr\left(\tilde{p}_{\text{bid}}^{\text{lease}}(v) \geq p_{\text{ask}}\right)$ is the probability that the estimated bidding price is larger than or equal to the asking price. Similarly, an optimization problem for the renting RAN can be formulated as follows:

$$\text{maximize:} \quad \mathscr{P}_{\text{renting}} = v - \frac{1}{2}\left(p_{\text{bid}} + E\left(\tilde{p}_{\text{ask}}^{\text{rent}}(c)|p_{\text{bid}} \geq \tilde{p}_{\text{ask}}^{\text{rent}}(c)\right)\right) \times \Pr\left(p_{\text{bid}} \geq \tilde{p}_{\text{ask}}^{\text{rent}}\right), \tag{8.76}$$

where $\tilde{p}_{\text{ask}}^{\text{rent}}(c)$ is the estimated value of the asking price from the leasing RAN, which is used by the renting RAN. An optimal Bayesian equilibrium, which is considered to be the solution of this bilateral bargaining model, is defined as follows:

$$\frac{d\mathscr{P}_{\text{leasing}}}{dp_{\text{ask}}} = 0, \quad \frac{d^2\mathscr{P}_{\text{leasing}}}{dp_{\text{ask}}^2} < 0, \tag{8.77}$$

$$\frac{d\mathscr{P}_{\text{renting}}}{dp_{\text{bid}}} = 0, \quad \frac{d^2\mathscr{P}_{\text{renting}}}{dp_{\text{bid}}^2} < 0. \tag{8.78}$$

After a RAN obtains the radio spectrum, it sells this radio resource to the users through an auction (i.e. between UMTS RAN and UMTS users). For a UMTS system, the bidding radio resources are considered as the CDMA code channels which are shared over a time frame. A time frame is divided into 15 slots and each of these slots can be allocated to a user. Each user has to bid for the radio resource in both uplink and downlink. The bid consists of the amount of radio resources and the bidding price. The bid from each user is submitted to the service provider by using the bidding message (Figure 8.14). Then, the service provider allocates radio resource according to the users' bids to maximize its revenue. This allocation is based on discriminatory auction.

Simulation results showed that the solution of the auction among service providers can be obtained with smaller overhead compared to a case when no estimation is used. This result is similar to that in [526]. For the auction between UMTS service providers and the users, the discriminatory pricing solution of the auction was compared with the uniform pricing model. Similar to the results in [463], the discriminatory pricing model was observed to achieve higher revenue compared to that from a uniform pricing model.

8.4 Summary

In this chapter, different models for a centralized dynamic spectrum access in a cognitive radio network have been presented. In centralized dynamic spectrum access, a central controller collects information about spectrum usage by the licensed users and also collects information about the requirements of the unlicensed users. The decision and scheduling of spectrum access by the unlicensed users can be determined by the central controller so that system performance is optimized. The action of unlicensed users then can be controlled so that the objective of the system can be achieved while all the constraints are satisfied.

These centralized dynamic spectrum access schemes were applied to two spectrum sharing models, i.e. the exclusive-use and shared-use models. In the exclusive-use model, the most common system model is composed of a spectrum broker and service providers. While the spectrum broker owns the spectrum band, a service provider owns the wireless infrastructure to provide wireless access services to the cognitive radio users. A service provider has to obtain the spectrum band from the broker and use it to provide a wireless access service to the users. In the shared-use model, the common system model is composed of licensed and unlicensed users. While the spectrum band is not accessed by the licensed users, the central controller of the unlicensed users can command the unlicensed users to access the spectrum.

The centralized dynamic spectrum access schemes were designed following two approaches, i.e. based on optimization and based on auction theory. In an optimization-based approach, the spectrum access by unlicensed users is formulated as an optimization problem which can be solved by a variety of techniques. In such a formulation, the common objective is to maximize the transmission rate while the common constraint is to maintain the interference temperature at a licensed receiver below the tolerable limit. Alternatively, centralized dynamic spectrum access schemes can be designed from the economic point of view. In this case, auction theory can be applied considering the spectrum bands as the commodity. The auction theory can be used to determine not only the spectrum access and allocation strategies, but also to determine the price, cost, and utility of spectrum access by the unlicensed users. In short, an optimization-based approach emphasizes the technical aspect, while an auction-based approach emphasizes the economic aspect of dynamic spectrum access.

9 Distributed dynamic spectrum access: cooperative and non-cooperative approaches

In many scenarios such as in ad hoc cognitive radio networks, deploying a central controller may not be feasible. Therefore, distributed dynamic spectrum access would be required in such cognitive radio networks. Due to the absence of any central controller, each unlicensed user has to gather, exchange, and process the information about the wireless environment independently. Also, an unlicensed user has to make decisions autonomously based on the available information to access the spectrum, so that the unlicensed user can achieve its performance objective under interference constraints. The common behaviors of an unlicensed user in a cognitive radio network without a central controller are as follows:

- *Cooperative or non-cooperative behavior:* Since a central controller which controls a decision of spectrum sharing is not available, an unlicensed user can adopt either cooperative or non-cooperative behavior. An unlicensed user with cooperative behavior will make a decision on spectrum access to achieve a network-wide objective (i.e. a group objective), even though this decision may not result in the highest individual benefit for each unlicensed user. In other words, an unlicensed user is concerned more about the overall performance of the network than its individual performance. On the other hand, an unlicensed user with non-cooperative behavior will make a decision only to maximize its own benefits (i.e. an individual objective). In this case, the unlicensed user will not be aware of the effect on the overall network performance.

  In a cognitive radio network with cooperative unlicensed users, optimization or cooperative game theories can be applied to obtain an optimal and fair solution for distributed dynamic spectrum access. Alternatively, a non-cooperative game can be used to analyze and obtain the solution when an unlicensed user adopts non-cooperative behavior. Note that the choice of adopting either cooperative or non-cooperative behavior depends on both the type of the cognitive radio network and the users. For example, if the cognitive radio users are in the same group with the same objective, these users typically have cooperative behavior. In contrast, if the unlicensed users are independent and have different objectives, non-cooperative behavior would be more common.
- *Collaborative or non-collaborative behavior:* Again, since a central controller which can coordinate the gathering and broadcasting of network information (e.g. spectrum usage by a licensed user) is not available, the unlicensed users have to collect

network information themselves. In this case, an unlicensed user can be collaborative or non-collaborative to exchange network information for distributed dynamic spectrum access. With non-collaborative behavior, all network information is gathered and processed locally by each unlicensed user. In this case, there is no interaction among the unlicensed users. Conversely, with collaborative behavior, the unlicensed users can exchange network information with each other. Again, the choice to become either collaborative or non-collaborative depends on the type of the cognitive radio network and the users. Also, the collaboration of the unlicensed users relates to cooperative and non-cooperative behavior.

Typically, collaboration among unlicensed users to exchange network information is required to achieve cooperative behavior. Also, if the unlicensed users are non-collaborative, they are also typically non-cooperative. However, if the unlicensed users are collaborative, they could be either cooperative or non-cooperative. For example, the unlicensed users may agree to reveal some information (e.g. the chosen spectrum access action), but they make a decision to achieve their own objectives (i.e. non-cooperative), rather than a group objective. Therefore, an unlicensed user can be classified into one of the following three categories: collaborative–cooperative, collaborative–non-cooperative, and non-collaborative–non-cooperative. In the case of collaboration among unlicensed users, a protocol will be required for exchanging network information. However, in the case of non-collaborative unlicensed users, the network information has to be observed and learned locally. Therefore, learning ability will be crucial for a cognitive radio.

- *Learning ability:* The ability to learn and make intelligent decisions is important for distributed dynamic spectrum access. In particular, the unlicensed users have to observe and learn the system state (e.g. the occupancy of the radio spectrum at a particular time). The output of this learning process is the knowledge about the RF environment and the system, which would be useful for an unlicensed user to make a decision on spectrum access. Again, the learning process can be either non-collaborative or collaborative. In the case of non-collaborative learning, the knowledge about the system is produced by each individual unlicensed user without interaction with other users. On the other hand, the unlicensed users can collaborate not only to exchange network information, but also to process and produce the system knowledge. Then, based on this knowledge, an unlicensed user can make the decision whether to achieve the group objective or its individual objective.

The different types of behavior of unlicensed users in distributed dynamic spectrum access environments are shown in Figure 9.1.

We discuss the related work on distributed dynamic spectrum access in this chapter and also in Chapter 10. Specifically, in this chapter, distributed dynamic spectrum access schemes based on the cooperative and non-cooperative approaches are discussed. In Chapter 10, distributed dynamic spectrum access schemes based on learning algorithms will be described. Also, distributed protocols required to support collaboration among unlicensed users for dynamic spectrum access will be reviewed.

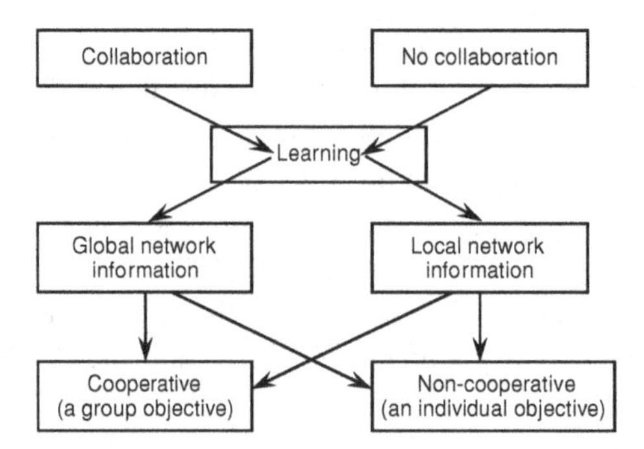

Figure 9.1 Characteristics of distributed dynamic spectrum access.

A summary of the distributed dynamic spectrum access schemes is provided in Table 9.1.[1] The related work in the literature is categorized into four classes based on the above behaviors, i.e. cooperative and optimization-based approach, non-cooperative approach, learning-based approach, and distributed network protocol. In cooperative and optimization-based approaches, distributed dynamic spectrum access methods are designed using the theories from cooperative game and optimization, respectively. In a learning-based approach, learning and artificial intelligent algorithms are applied to design distributed schemes.

9.1 Cooperative and optimization-based distributed dynamic spectrum access

9.1.1 Power and admission control of cognitive radio with antenna arrays

In a wireless network, while a power control mechanism is used to obtain optimal transmit power to maximize the throughput of the system, an admission control mechanism is used to limit the number of users so that interference in the system is maintained below a target level [553, 554, 555]. Along with these mechanisms, the issue of coexistence between the licensed and unlicensed users needs to be taken into account [527, 510]. In [527], adaptive power and admission control mechanisms were designed specifically for a cognitive radio network with antenna arrays. In the system model, there is a licensed user (i.e. transmitter and receiver). There are I unlicensed users in the same area sharing the same spectrum with a licensed user (Figure 9.2). These unlicensed users communicate with a base station in the uplink direction. While a base station uses an M-element antenna, the transmitter of each unlicensed and licensed user will use a single antenna.

The channel gain between an M-antenna base station and a transmitter of an unlicensed user is denoted by $\mathbf{h}_i^{(u)} = \alpha_i \mathbf{a}_i(\theta_i)$, where α_i accounts for the fading, path loss, and pulse shaping parameters. $\mathbf{a}_i(\theta_i)$ is the M-component array response vector for the

[1] The cooperative behavior implies that the unlicensed users are collaborative.

Table 9.1 Summary of distributed dynamic spectrum access schemes.

Article	Access model	Behavior
Power and admission control of cognitive radio with antenna arrays [527]	Shared-use (underlay)	Cooperative
Genetic algorithm to optimize distributed dynamic spectrum access [528]	Shared-use (overlay)	Cooperative
Distributed spectrum sharing based on cooperative game [254]	Shared-use (underlay)	Cooperative
Optimal channel sensing and allocation in cognitive radio mesh network [529]	Shared-use (underlay)	Cooperative
Cooperative spectrum sharing using repeated game [530]	Commons-use	Cooperative
Cooperative strategy for distributed spectrum sharing [531]	Commons-use	Cooperative and non-cooperative, collaborative
Decentralized channel selection for QoS-sensitive cognitive radio network [532]	Shared-use (overlay)	Non-cooperative, collaborative
Joint distributed power control and channel selection in CDMA cognitive radio network [533]	Shared-use (underlay)	Non-cooperative, non-collaborative
Random access and interference channel in cognitive radio network [534]	Shared-use (underlay)	Non-cooperative, collaborative and non-collaborative
Dynamic channel selection in multihop networks [535]	Shared-use (underlay)	Non-cooperative, collaborative
Competitive power allocation based on genetic algorithm in spectrum access [536]	Shared-use (underlay)	Non-cooperative, non-collaborative
Cognitive medium access and coexistence with WLAN [537, 538]	Commons-use	Non-cooperative, collaborative and non-collaborative
Time domain spectrum allocation [539]	Exclusive-use	Non-cooperative, collborative
Distributed resource management in multihop cognitive radio network [540]	Shared-use (overlay)	Cooperative
Opportunistic channel selection based on fuzzy learning algorithm [541]	Commons-use	Non-cooperative, non-collaborative
Distributed dynamic spectrum access and no-regret learning [542]	Shared-use (overlay)	Non-cooperative, collaborative and non-collaborative
Agent-based dynamic spectrum access [543]	Exclusive-use, Shared-use (overlay), Commons-use	Non-cooperative, non-collaborative
Biologically-inspired spectrum sharing algorithm (BIOSS) [544]	Shared-use (underlay)	Non-cooperative, non-collaborative
Distributed rule-regulated spectrum sharing [545, 546]	Exclusive-use	Non-cooperative, collaborative and non-collaborative
Game theory based spectrum etiquette [547]	Shared-use (underlay)	Cooperative, collaborative and non-cooperative, non-collaborative
Opportunistic channel selection in cooperative mobility network [548]	Exclusive-use	Cooperative
Channel evacuation protocol [549]	Shared-use (overlay)	Non-cooperative, collaborative
Distributed channel management in uncoordinated wireless environment [550]	Commons-use	Non-cooperative, non-collaborative
Spectrum aware on-demand routing protocol [551, 552]	Shared-use (overlay)	Cooperative

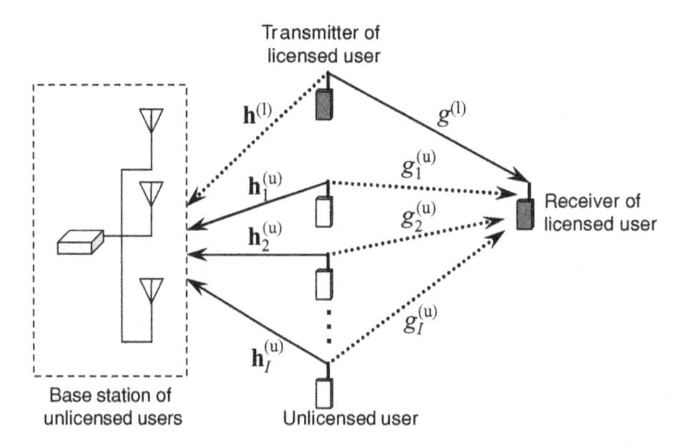

Figure 9.2 System model of a cognitive radio network with antenna arrays.

direction of signal arrival θ_i. The channel gain between a transmitter of an unlicensed user and a receiver of a licensed user is denoted by $g_i^{(u)}$. Similarly, the channel gains from a transmitter of a licensed user to a base station of unlicensed users and to a receiver of a licensed user are denoted by $\mathbf{h}^{(l)}$ and $g^{(l)}$, respectively. The SINR for a licensed user $\gamma^{(l)}$ and an unlicensed user $\gamma_i^{(u)}$ can be obtained from

$$\gamma^{(l)} = \frac{g^{(l)} P^{(l)}}{\sum_{i=1}^{I} g_i^{(u)} P_i^{(u)} + \sigma^2}, \tag{9.1}$$

$$\gamma_i^{(u)} = \frac{|\mathbf{w}_i^\dagger \mathbf{h}_i^{(u)}|^2 P_i^{(u)}}{\sum_{j \neq i} |\mathbf{w}_i^\dagger \mathbf{h}_j^{(u)}|^2 P_j^{(u)} + |\mathbf{w}_i^\dagger \mathbf{h}^{(l)}|^2 P^{(l)} + ||\mathbf{w}_i||^2 \sigma^2}, \tag{9.2}$$

where $\mathbf{w}_i$ is the unity-norm beamforming receiver weight for unlicensed user i, σ^2 is the noise power, $P^{(l)}$ and $P_i^{(u)}$ are the transmit power of licensed user and unlicensed user i, respectively, and $\mathbf{w}_i^\dagger$ denotes the Hermitian transpose of matrix $\mathbf{w}_i$.

It is assumed that a licensed user will increase its transmit power until the target SINR is achieved. Also, as long as the target SINR of a licensed user is maintained, the unlicensed users can increase transmit power to gain better performance. In this case, the maximum transmit power constraint is enforced for the unlicensed users. An optimization problem can be formulated as follows:

$$\text{minimize:} \sum_{i=1}^{I} P_i^{(u)}, \tag{9.3}$$

$$\text{subject to:} \quad \gamma_i^{(u)} \geq \eta^{(u)}, \tag{9.4}$$

$$P_i^{(u)} \leq P_{\max}, \tag{9.5}$$

$$\gamma^{(l)} \geq \eta^{(l)}, \tag{9.6}$$

where $\eta^{(l)}$ and $\eta^{(u)}$ are the target SINR for licensed and unlicensed users, respectively, and $P_{\max}$ is the maximum power of an unlicensed user. Let $\mathbf{p}$ denote the vector of the transmit power of a licensed user and the unlicensed users (i.e. $\mathbf{p} = \begin{bmatrix} P^{(l)} & P_1^{(u)} & \cdots & P_I^{(u)} \end{bmatrix}$). $\mathbf{I}_N$ is an identity matrix with size $N \times N$. The optimal power allocation under SINR constraints can be obtained from the following relations:

$$\mathbf{u} = (\mathbf{I}_{I+1} - \mathbf{D}\boldsymbol{\Phi})\mathbf{p}, \tag{9.7}$$

$$\mathbf{u} = \begin{bmatrix} \frac{\eta^{(l)}\sigma^2}{g^{(l)}} & \frac{\eta^{(u)}\|\mathbf{w}_1\|^2\sigma^2}{|\mathbf{w}_1^\dagger \mathbf{h}_1|^2} & \cdots & \frac{\eta^{(u)}\|\mathbf{w}_I\|^2\sigma^2}{|\mathbf{w}_I^\dagger \mathbf{h}_I|^2} \end{bmatrix}, \tag{9.8}$$

$$\mathbf{D} = \begin{bmatrix} \frac{\eta^{(l)}}{g^{(l)}} & & & \\ & \frac{\eta^{(u)}}{|\mathbf{w}_1^\dagger \mathbf{h}_1^{(u)}|^2} & & \\ & & \ddots & \\ & & & \frac{\eta^{(u)}}{|\mathbf{w}_I^\dagger \mathbf{h}_I^{(u)}|^2} \end{bmatrix}, \tag{9.9}$$

$$[\boldsymbol{\Phi}]_{j,k} = \begin{cases} 0, & j = k,\, j, k \in \{1, \ldots, I+1\} \\ |\mathbf{w}_{j-1}^\dagger \mathbf{h}^{(l)}, & k = 1,\, j \in \{2, \ldots, I+1\} \\ g_{k-1}^{(u)}, & j = 1,\, k \in \{2, \ldots, I+1\} \\ |\mathbf{w}_{j-1}^\dagger \mathbf{h}_{k-1}|^2, & \text{otherwise}, \end{cases} \tag{9.10}$$

where $[\boldsymbol{\Phi}]_{j,k}$ denotes the element at row j and column k of a matrix $\boldsymbol{\Phi}$. Basically, the optimal transmit power is obtained from $\mathbf{p}^* = (\mathbf{I}_{I+1} - \mathbf{D}\boldsymbol{\Phi})^{-1}\,\mathbf{u}$. This solution can be obtained if all the channel gains are available. However, this may not be the case for a cognitive radio network where the licensed and the unlicensed users are non-cooperative and/or distributed. Therefore, a distributed algorithm was proposed.

Let $P^{(l)}(t)$ and $P_i^{(u)}(t)$ denote the transmit power of licensed and unlicensed users at iteration t, respectively. Then, the update of the transmit power can be expressed as follows:

$$P^{(l)}(t) = \begin{cases} \frac{\delta\eta^{(l)}}{\gamma^{(l)}(t)} P^{(l)}(t), & \text{if the link is active} \\ \delta P^{(l)}(t), & \text{if the link is inactive,} \end{cases} \tag{9.11}$$

$$P_i^{(u)}(t) = \begin{cases} \min(P_{\max}, \frac{\delta\eta^{(u)} P_i^{(u)}(t)}{\gamma_i^{(u)}(t)}), & \text{if the link is active} \\ \min(P_{\max}, \delta P_i^{(u)}(t)), & \text{if the link is inactive,} \end{cases} \tag{9.12}$$

where $\delta > 1$ is a parameter to control the update size of the transmit power. Here, the link is defined to be active if $P^{(l)} \geq \eta^{(l)}$ and $P_i^{(u)} \geq \eta^{(u)}$ for the licensed and unlicensed users, respectively.

Although the transmit power of an unlicensed user can be adjusted, the constraints on SINR may not be satisfied (especially when the number of unlicensed users is large). To ensure that the SINR requirements of all active users are met, an admission control method is applied. This admission control method works as follows:

1. Each inactive unlicensed user increases its transmit power by a factor δ. This increment must not violate the constraint on the maximum power.

2. If the maximum power is reached while its SINR is still lower than $\eta^{(u)}$ (i.e. inactive), the unlicensed user drops its link with a probability $P^{(u)}_{i,\text{drop}} = \rho\big(1 - \frac{\gamma_i^{(u)}}{\eta^{(u)}}\big)$, where ρ is a constant which determines the speed of an unlicensed user leaving the system.

This algorithm is performed when the transmit power is updated.

Simulation results showed that the transmit power converges to the solution within 20 iterations. Without admission control, there could be a situation where all unlicensed users transmit at the maximum power, but their SINR requirements cannot be satisfied. In this case, an admission control mechanism will randomly drop some unlicensed users to reduce the interference to a licensed user and other unlicensed users in the network.

9.1.2 Genetic algorithm to optimize distributed dynamic spectrum access

Genetic algorithms can be applied to solve the channel allocation problem in a wireless network (e.g. a single-hop cellular network) [556, 557, 558]. In a cognitive radio environment where channel availability is dynamic, an optimization problem can be formulated for dynamic spectrum access and a distributed solution can be obtained by using a genetic algorithm [528] (Section 6.4). In the system model, a wireless platform similar to that in [559] is assumed and used to implement distributed learning algorithms. There are I unlicensed users/nodes and L one-hop communication links. Each unlicensed node can access multiple channels simultaneously. A set of all links is denoted by $\mathbb{L}$ and the link between node i and j is denoted by $l_{i,j}$. In this case, if the unlicensed nodes are defined to interfere with each other, a channel can be accessed by one of these unlicensed nodes. To access a channel, an unlicensed node senses the target channel. Each unlicensed node can identify a set of available channels. The optimization problem can be described as follows:

$$\max_{\mathbf{n}} \left(\mathscr{F}(\mathbf{n}) = \sum_{l_{i,j} \in \mathbb{L}} \frac{r_{i,j}}{1 + |\mathbb{L}_{i,j}|} \right), \tag{9.13}$$

where $\mathbf{n} = \begin{bmatrix} \cdots & n_{i,j} & \cdots \end{bmatrix}$, and $n_{i,j}$ is a variable representing the assignment of a channel to a link between unlicensed nodes i and j. $r_{i,j}$ is the capacity (i.e. rate) of the channel corresponding to $n_{i,j}$, and $|\mathbb{L}_{i,j}|$ is the number of elements in set $\mathbb{L}_{i,j}$. The links in $\mathbb{L}_{i,j}$ cannot be activated at the same time as $l_{i,j}$. $\mathbb{L}_{i,j}$ can be constructed through channel sensing to estimate the interference due to transmissions from unlicensed nodes. The objective function $\mathscr{F}(\mathbf{n})$ is the total capacity of an unlicensed user discounted by the number of conflicting links. This optimization problem to obtain the optimal solution in terms of channel assignment $n_{i,j}$ is NP-hard. Also, the number of elements in $\mathbf{n}$ increases exponentially as the number of links L increases.

To solve the optimization problem in (9.13) in a distributed network, a genetic algorithm can be applied. In this case, $\mathbf{n}$ is an individual and $\mathscr{F}(\mathbf{n})$ is defined as a fitness function. Given the initial population which is randomly generated (i.e. a set of random $\mathbf{n}$), the algorithm works as follows:

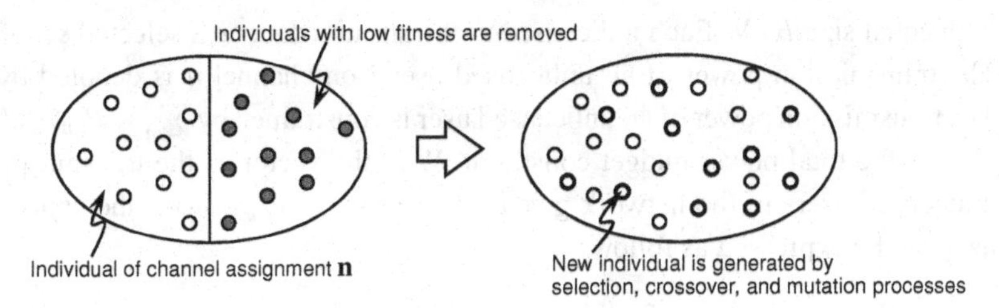

Figure 9.3 Genetic algorithm to generate channel assignment population.

1. The fitness of all individuals is computed (i.e. $\mathscr{F}(\mathbf{n})$).
2. Half of the population with the lowest fitness are removed from the population. The new population is defined as a parent population (Figure 9.3).
3. The new individuals are generated from the parent population by tournament selection and crossover processes. Specifically, a crossover point is determined from the random integer number which is generated from the set $\{2, \ldots, L-1\}$. A tournament selection process is used to select a channel assignment $\mathbf{n}$ for crossover. In this process, three individuals are randomly selected from a population. Then, these individuals are crossovered with the highest fitness.
4. A mutation process of each individual, except the individual with the highest fitness, is randomly performed with a certain probability.

In addition, to avoid the local maxima problem, this genetic algorithm is modified to keep track of the number of iterations during which the maximum fitness of a population does not change. If the number of iterations reaches a certain value (i.e. the solution is locally optimal but may not be globally optimal), the algorithm will re-initialize the population to search for other alternative solutions.

Simulation results showed that, for 25 unlicensed nodes, the proposed algorithm converges to the final solution in 47.5 iterations on average. This final solution achieves the network throughput close to that for optimal assignment. However, the number of iterations to reach a final solution depends on the number of unlicensed nodes in the network. In particular, if the number of unlicensed nodes increases, the number of iterations increases.

9.1.3 Distributed spectrum sharing based on cooperative game

In an underlay spectrum access scenario, the unlicensed users can cooperate, for example, by controlling the transmit power. The theory of bargaining game can be applied to obtain an efficient and fair solution of dynamic spectrum sharing. However, due to the distributed structure of a cognitive radio network, the standard methods used to obtain the solution of a bargaining game may not be applicable. Therefore, an algorithm utilizing only local information was proposed in [254], where a bargaining game with a Nash bargaining solution was considered. The game is formulated as follows. The **players** are the unlicensed users, the **strategy** for each player is the transmit power, and the **payoff** is the capacity. In the system model, the available bandwidth B is divided into N channels with equal size B/N. Each unlicensed user transmits data on a selected set of channels. The transmission power of an unlicensed user i on channel n is denoted by $P_{i,n} \geq 0$. The transmission power of an unlicensed user is constrained by $\sum_{n=1}^{N} P_{i,n} \leq P_{\max}$ where $P_{\max}$ is the total power budget constraint. With the vector of the transmit power of all unlicensed users in the network given by $\mathbf{p} = \begin{bmatrix} \cdots & P_{i,n} & \cdots \end{bmatrix}$, the capacity of each user can be expressed as follows:

$$\mathscr{C}_i(\mathbf{p}) = \frac{B}{N} \sum_{n=1}^{N} \log \left(1 + \frac{h_{i,i,n} P_{i,n}}{\sum_{j \neq i} h_{i,j,n} P_{j,n} + \sigma^2/N} \right), \tag{9.14}$$

where $h_{i,j,n}$ is the channel gain between the transmitter of an unlicensed user i and the receiver of an unlicensed user j on channel n, and σ^2 is the noise power for the entire bandwidth B.

Based on the strategies of all unlicensed users, it was shown that the payoff space is non-convex if $N = 1$, which corresponds to the case of high interference, since all unlicensed users transmit on the same channel. However, as N increases, the payoff space becomes convex, and also more efficient solutions (i.e. higher capacity) can be achieved by the unlicensed users.

In order to achieve the solution of the bargaining game in a distributed network, an algorithm is required which operates by using only local information without a central controller. In this case, the unlicensed users can be considered as a group (i.e. a set of unlicensed users in the interference zone with a given interference radius) whose members' transmissions interfere with each other. The Nash bargaining solution can be obtained for this group. The requirements for the distributed algorithm are as follows. First, the unlicensed users can exchange information with their one-hop neighbors. This period of information exchange and algorithm execution is shorter than the duration of one iteration in the algorithm. Second, each unlicensed user executes the algorithm randomly so that the probability of having two unlicensed users making the decision simultaneously is very small. The capacity of an unlicensed user i can be expressed as

$$\mathscr{C}'_i(\mathbf{p}) = \frac{B}{N} \sum_{n=1}^{N} \log \left(1 + \frac{h_{i,i,n} P_{i,n}}{\sum_{j \in \mathbb{J}_i} h_{i,j,n} P_{j,n} + \sigma^2/N} \right), \tag{9.15}$$

where $\mathbb{J}_i$ is a set of interferers for unlicensed user i. The Nash bargaining solution is defined as

$$\mathbf{p}^* = \arg\max_{\mathbf{p}} \prod_{i=1}^{I} \mathscr{C}'_i(\mathbf{p}), \tag{9.16}$$

where I is the total number of unlicensed users. To obtain this solution, the distributed Nash bargaining solution algorithm of each unlicensed user works as follows:

1. Obtain $\mathbb{J}_i$, which is the set of other unlicensed users in the interference zone.
2. $C_{\text{old}} = \prod_{j \in \mathbb{J}_i} \mathscr{C}_i(\mathbf{p})$.
3. Obtain $\mathbf{p}^*$ from (9.16).

4. $C_{\text{new}} = \prod_{j \in \mathbb{J}_i} \mathscr{C}_i(\mathbf{p}^*)$.
5. If $C_{\text{new}} > (1 + \beta)C_{\text{old}}$, then update $P_{i,n}$.

This algorithm will update the solution **p** only if the payoff of the new solution is larger than that of the old solution weighted by $1 + \beta$ (i.e. Step 5 in the above algorithm). The convergence of this algorithm is analytically proven to ensure that the solution close to the Nash bargaining solution is achieved.

Simulation results showed that the Nash bargaining solution of this bargaining game formulation is efficient and fair. The proposed formulation was compared with the max-min and max-sum schemes. In the max-min scheme, the minimum capacity of all unlicensed users is maximized, while in the max-sum scheme the summation of the capacities of all unlicensed users is maximized. This Nash bargaining solution can achieve larger capacity compared with the max-min scheme. Also, the minimum capacity of an unlicensed user obtained from the Nash bargaining solution is larger than that of the max-sum scheme. It is observed that as N increases, the total capacity of all unlicensed users increases. However, at a certain point, this total capacity cannot be improved when it reaches the capacity limit. Note that this work considered single-hop communication between the transmitter and receiver of an unlicensed user. The model can be extended for multihop communications as well.

9.1.4 Optimal channel sensing and allocation in cognitive radio mesh network

In a wireless mesh network, a mesh client connects to a mesh router. The user traffic is transmitted over multiple mesh routers to the destination. The concept of cognitive radio can be applied to this wireless mesh network in which the mesh clients and routers can opportunistically access the channels which are licensed to a licensed user. In this cognitive wireless mesh network, dynamic spectrum access is required for the mesh clients and routers to identify and access the channels which are not occupied by the licensed users. The cognitive mesh network (COMNET) framework was presented for such a network in [529]. This framework uses an optimal spectrum sensing algorithm. An analytical model was developed for the mesh routers to estimate interference in the network. This analytical model was used to optimize channel assignments for mesh clients and routers.

There are multiple mesh clients connecting to a mesh router (Figure 9.4). This can be referred to as a cluster where a mesh client is a cluster member, while a mesh router is a cluster head. The mesh routers connect to each other to relay traffic to the destination (i.e. the gateway). Communications between the mesh routers is on different channels to avoid contention with the mesh client. The mesh clients and routers communicate with each other using the IEEE 802.11b standard-based radio transceivers. However, this mesh network is in the same area as a licensed service (i.e. TV services). Therefore, these transceivers are assumed to have an agility and ability to operate in both unlicensed bands (e.g. ISM) and TV bands (i.e. 16 channels with 5 MHz spacing between 712 and 187 MHz). In particular, a TV band can be used if it is detected to be unoccupied by a

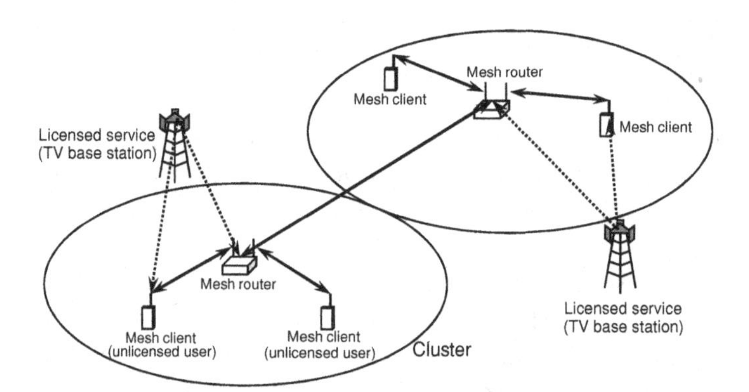

Figure 9.4 Cognitive Mesh NETwork (COMNET) framework.

licensed system. It is assumed that a mesh client knows the distance from the TV stations which can be achieved by localization techniques (e.g. triangulation).

The mesh clients sense the TV bands periodically by measuring the total received power. The result of this sensing in terms of the aggregated received power is sent by a mesh client to a mesh router in the same cluster. This mesh router is assumed to have high computational power to compute different environmental parameters (e.g. interference in the cluster). Based on these parameters, the mesh router will make a decision in a distributed manner whether to switch to operate on the TV band or remain in the unlicensed band.

The frequency of the TV band which is not occupied by a licensed service has to be identified through spectrum sensing. With the free space path loss model, the average normalized power received at mesh client i due to the transmission of TV transmitter 1 on frequency band f_1 can be obtained from

$$P_{1,i} = \sigma_{1,i}\alpha_1 D_{1,i}^{-\beta}, \tag{9.17}$$

where $\sigma_{1,i}$ is the spectral overlap factor between the frequency of a transmitter of a TV station and the frequency of a receiver of a mesh client, $\alpha_1 = G_t G_r c^2/(4\pi f_1)^2$ where G_t and G_r are the transmit and receive antenna gains, c is the speed of light. β is the path loss exponent, and $D_{1,i}$ is the distance between the transmitter and receiver. Since the receiving power at a mesh client is the superposition of the power from all TV transmitters, this receiving power can be expressed as follows:

$$\begin{aligned} P_i &= \sum_{n=1}^{N} \sigma_{n,i}\alpha_n D_{n,i}^{-\beta} \\ &= \sigma_{1,i}\alpha_1 D_{1,i}^{-\beta} + \cdots + \sigma_{n,i}\alpha_n D_{n,i}^{-\beta} + \cdots + \sigma_{N,i}\alpha_N D_{N,i}^{-\beta}, \end{aligned} \tag{9.18}$$

where N is the total number of TV transmitters, and P_i is the aggregated received power. Given that at least N mesh clients perform spectrum sensing, there will be N equations to be solved for N unknown variables of $\sigma_{n,i}\alpha_n$. These unknown variables can be solved, namely ϕ_n in this case. If $G_t = G_r = 1$, the channel frequency can be estimated from

$$f_n = \arg\min_{n'} \left(\beta_{n',1} \frac{c^2}{(4\pi f_{n'})^2} - \phi_n \right), \tag{9.19}$$

$$\beta_{n,1} = \Omega\left(\frac{|f_n - f_1|}{\Delta f} \right), \tag{9.20}$$

where $\Omega(\cdot)$ is a function of spectral overlap factor, and Δf is the inter-channel spacing. Also, the load of the computing channel frequency can be shared among mesh clients. In this case, the Gauss elimination technique for solving linear equations [560] can be applied.

Then, the band and channel switching algorithm is developed to allow some clusters to operate in the TV bands so that the congestion in unlicensed bands can be alleviated. Let $\mathbb{N}$ denote the set of the TV transmitters, $\mathbb{I}_{\text{ub}\rightarrow\text{tv}}$ denote the set of clusters switching to operate on the TV bands, and $\mathbb{I}_{\text{ub}\rightarrow\text{ub}}$ is the set of clusters remaining in the unlicensed band. An optimization problem was formulated to obtain $\mathbb{I}_{\text{ub}\rightarrow\text{tv}}$ and f_j, which is the frequency for the clusters switching to the TV bands. The constraints are as follows:

- $\sum_{j\in\mathbb{N}, f_j\neq f_i} P_{i,j}\sigma_{i,j}(f_i, f_j) + \sum_{k\in\mathbb{I}_{\text{ub}\rightarrow\text{tv}}} P_{i,k}\sigma_{i,k}(f_i, f_k) < \text{Th}$, where $\sigma_{i,j}(f_i, f_j)$ is the spectral overlap at the location i between frequency f_i and the frequency of transmitter f_j. This constraint indicates that the total interference power from the TV transmitters (i.e. the first summation) and the switching clusters (i.e. the second summation) is less than the threshold Th.
- $\sum_{k\in\mathbb{I}_{\text{ub}\rightarrow\text{tv}}} P_{i,k} - \sum_{j\in\mathbb{I}_{\text{ub}\rightarrow\text{ub}}} P_{i,j} < P_{\text{th}}$ indicates that the difference between interference from the clusters in the unlicensed band and the clusters in TV bands is maintained close to each other (i.e. less than the threshold P_{th}).
- $||\mathbb{I}_{\text{ub}\rightarrow\text{tv}}| - |\mathbb{I}_{\text{ub}\rightarrow\text{ub}}|| < Z_{\text{th}}$, where $|\mathbb{I}_{\text{ub}\rightarrow\text{tv}}|$ determines the number of elements in the set $\mathbb{I}_{\text{ub}\rightarrow\text{tv}}$. This constraint indicates that the number of clusters in the two bands are balanced (i.e. the difference is less than the threshold Z_{th}).

Simulation results showed that the proposed spectrum sensing scheme can accurately estimate the frequency used by the licensed transmitters. However, the accuracy of the proposed spectrum sensing decreases as the noise level increases. However, this inaccuracy can be removed, if the mean noise power is known so that it can be subtracted from the aggregated receiving power. Also, the accuracy of spectrum sensing can be significantly improved when many sets of measurements are used. However, the computational overhead to solve the set of linear equations is high. The overhead of centralized and distributed spectrum sensing is also evaluated among mesh clients. The distributed spectrum sensing takes more time to compute the solution since the measurement data has to be distributed. However, this distributed sensing can reduce the load at the mesh router, which may not always have enough computational resources. In addition, it was shown that with the band and channel switching algorithm, the interference in the unlicensed and TV bands can be balanced. Some of the clusters are switched to the TV bands when these bands are not occupied by the TV transmitters. As a result, the interference in unlicensed bands can be reduced.

If the unlicensed users interact among each other to cooperate, a repeated game can be applied to analyze the solution (i.e. whether the unlicensed user is willing to cooperate or not).

9.1.5 Cooperative spectrum sharing using repeated game

In a cognitive radio network with independent unlicensed users, an incentive is the most important factor for an unlicensed user to make a decision on spectrum access. In [530], the spectrum sharing (i.e. power allocation) algorithm was devised so that the unlicensed users have an incentive to use the fair and efficient solution. This algorithm is self-enforcing and hence can be used without a central controller. In the system model, there are I unlicensed users accessing the spectrum of size W in a CDMA-based system. The objective of the spectrum sharing algorithm is to find the transmission power strategy which is fair and efficient given the constraint on total transmit power.

First, an optimal power allocation is considered to represent the case that all unlicensed users are cooperative. The Pareto optimality (see Section 5.7.2) of the rate can be defined as follows:

$$\mathbb{R}^* = \big\{(r_1, \ldots, r_I) \in \mathbb{R} : r_i \geq r_i' \;\; \forall (r_1, \ldots, r_{i-1}, r_i', r_{i+1}, \ldots, r_I) \in \mathbb{R}, \\ \text{for } i = 1, \ldots, I\big\}, \tag{9.21}$$

where $\mathbb{R}$ is a feasible set of rates, defined as follows:

$$\mathbb{R} = \left\{ r_i = \int_0^W \log\left(1 + \frac{h_{i,i} P_i(f)}{\sum_{j\neq i} h_{i,j} P_j(f) + \sigma^2}\right) df \right. \\ \left. \text{and } \int_0^W P_i(f) df \leq P_{\max,i},\ P_i(f) \geq 0, \text{ for } i = 1, \ldots, I \right\}, \tag{9.22}$$

where $h_{i,j}$ is the gain, $P_i(f)$ is the transmit power at the frequency f, and σ^2 is the noise power. The utility of a network can be defined as $\mathcal{U} = \sum_{i=1}^{I} r_i$ or $\mathcal{U} = \sum_{i=1}^{I} \log(r_i)$, which denote the sum rate and proportional fair utility functions, respectively.

Then, this power allocation is formulated as a one-shot non-cooperative game which represents the case that the unlicensed users are non-cooperative. The game is defined as follows. The **players** are the unlicensed users. The **strategies** correspond to the set of power allocations. The payoff is assumed to be a utility of the rate, i.e. $\mathcal{U}(r_i)$. The solution is given by the Nash equilibrium. In the case of a Gaussian interference game [561], it was shown that a set of frequency-flat allocations, i.e. $P_i(f) = P_{\max,i}/W$ for $i = 1, \ldots, I$ gives the Nash equilibrium. In addition, it was proven that if $\sum_{j=1, j\neq i}^{I}(h_{i,j}/h_{i,i}) < 1$ for $i = 1, \ldots, I$, then the above Nash equilibrium power allocation is unique. However, in many cases, this Nash equilibrium is not Pareto optimal (i.e. not in $\mathbb{R}^*$). Therefore, the solution of the one-shot non-cooperative game is not efficient, and there are other power allocation strategies that the unlicensed users can gain better rates from. To solve this problem, the long-term interactions of the unlicensed users are considered and a repeated game (see Section 5.3) is used to model this situation.

An infinite horizon repeated game was assumed in which all unlicensed users interact forever. The utility of each player is defined as

$$\mathscr{U}_i = (1 - \delta) \sum_{t=0}^{\infty} \delta^t r_i(t), \tag{9.23}$$

where $\delta \in [0, 1]$ is a discount factor, and $r_i(t)$ is the rate of an unlicensed user i at time t. In this repeated game, each unlicensed user can observe the outcome of the game, and the decision in the future can be made accordingly. Since the game is repeated, using the Nash equilibrium power allocation from the one-shot game every time is not a single subgame perfect Nash equilibrium, which is a desirable solution. Using any other power allocation, which achieves higher rates, can be a subgame perfect Nash equilibrium. When a trigger strategy is applied, an unlicensed user adopts a transmit power which yields a higher rate than that of the Nash equilibrium (e.g. a solution which is Pareto optimal, or a globally optimal power allocation) in the one-shot game as long as other users do so. Otherwise, an unlicensed user uses the Nash equilibrium in the one-shot game. In other words, the unlicensed users cooperate to achieve higher rates. The deviating unlicensed user will be punished forever. It was shown that if the discount factor δ is sufficiently close to one, the unlicensed users will maintain their cooperation [562, 563]. As a result, the achievable rates of the unlicensed users are higher than those using the Nash equilibrium in the one-shot game.

However, in a non-cooperative environment, each unlicensed user can send false information (e.g. channel gain information). Protocols are proposed for measurement and exchange of system parameters. In this case, to measure the channel gain, each transmitter of an unlicensed user sends pilot signals using predefined power levels. With this pilot signal, the target receiver and also other unlicensed users can estimate channel gains from the transmitter. However, some unlicensed users can falsify other users by transmitting the pilot signal using a power different from the predefined level or an unlicensed user can report false channel measurement results. The following approaches are used to detect the falsification of an unlicensed user:

- *Test message:* To verify that the information about the channel gain is correct, an unlicensed user can use the reported channel gain to transmit a test message with rate $W \log(1 + h_{i,j} P_{\max,i} / \sigma^2 / W)$. If $h_{i,j}$ is the true value, an unlicensed user j should receive the test message with negligible error probability. Otherwise, if $h_{i,j}$ is smaller than the reported value, the test message might contain errors. With this approach, an unlicensed user can detect whether the reported channel gain is larger than the true value or not.
- *Multiple pilots:* In this approach, all transmitters $j \neq i$ transmit a pilot with random power $P_{\max,j}$. The unlicensed user j also broadcasts the value of the total interference power received at a receiver. Other unlicensed users can exchange the random power values and check whether the reported interference power from an unlicensed user j matches $\sum_{j \neq i} h_{i,j} P_{\max,j}$ or not.
- *Triangulation:* This approach is used to determine the locations of the transmitter and the receiver from the channel gain measurement. Since the value of channel gain

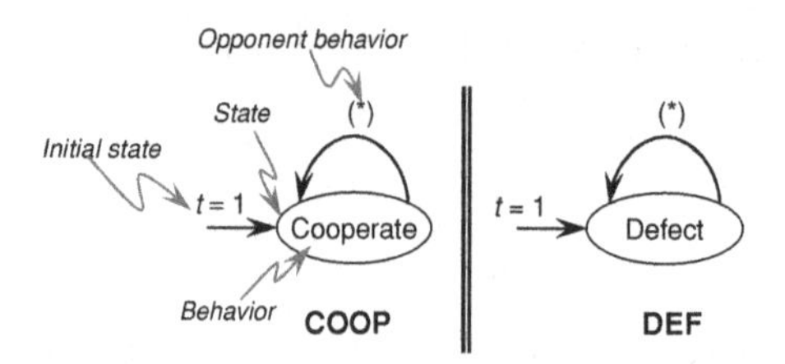

Figure 9.5 Static strategy of cooperation and defection.

typically corresponds to the location, the false reported information about channel gain can be detected by comparing with the estimated path loss, which is a function of the location of the transmitter.

- *Rate detection:* An unlicensed user j decodes the message to determine the rate of an unlicensed user i. This rate measurement is then exchanged with other unlicensed users to detect whether an unlicensed user i uses the correct rate or not.

9.1.6 Cooperative strategy for distributed spectrum sharing

Coexistence is an important issue especially for the unlicensed users sharing the spectrum in a commons-use model. The independent unlicensed users usually competitively access the radio spectrum. However, the performance of a cognitive radio network can be improved if the unlicensed users cooperate with each other to achieve optimal spectrum access. The cooperation and defection behaviors can be modeled using repeated games [531, 564, 565, 566, 567]. In [531], a system model based on IEEE 802.11 networks was analyzed for two unlicensed users. Each user can observe the action of the opponent user. Here, a user can choose an action in terms of demanded throughput and access period. The utility of a user is determined from the QoS performances (i.e. achievable throughput and delay) as in [568, 569]. In this case, each user may adopt a cooperation or defection behavior in choosing its action:

- *Defection:* If a user defects, that user will attempt to maximize its own payoff. This defection can be achieved by using aggressive MAC parameter (e.g. probability of accessing a channel is large). Although a user can gain higher payoff with this defection, the payoff of an opponent user will be degraded. Therefore, the defection can be used as a punishment to the opponent.
- *Cooperation:* If a user cooperates, that user will attempt to maximize the payoff of all users. This cooperation can be achieved by using less aggressive MAC parameters (e.g. a lower channel access probability, which results in smaller collisions). However, this cooperation is susceptible to the defection of an opponent user. That is, the payoff of a user with cooperation may be degraded if an opponent user defects.

The strategy in a one-shot game (i.e. the static strategy) can be represented as in Figure 9.5, where COOP and DEF denote the cooperation and defection strategies, respectively. The strategy is defined by four parameters, namely, initial state, state, behavior, and opponent behavior. Note that the opponent behavior $(*)$ indicates any

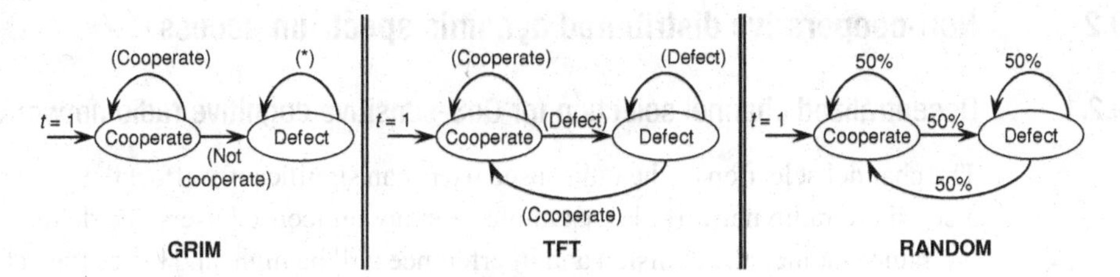

Figure 9.6 Dynamic strategy of cooperation and defection.

behavior. In this one-shot game, a user chooses a strategy given the strategy chosen by the opponent user. The past strategy is not used in the current decision making (i.e. memoryless behavior).

In a multistage game, where a one-shot game is played repeatedly, the strategy chosen by a user depends not only on the current strategy of the opponent user, but also on the strategy in the past (i.e. in previous stages). Also, the decision of a user will be made considering the reaction from the opponent user. Therefore, the payoff of a user in this multistage game is weighted in which the payoff in the current stage is higher than that in the future stage. In particular, the payoff of a user is discounted as follows:

$$\mathcal{U}_i = \sum_{t=0}^{\infty} (\delta_i)^t \mathcal{U}_i(t) = \frac{1}{1-\delta_i} \mathcal{U}_i(t) \tag{9.24}$$

for the case that the multistage game is played forever (i.e. an infinite time horizon game). $\mathcal{U}_i(t)$ denotes the payoff of user i which is the utility of achievable throughput and delay at stage t of user i. $0 < \delta < 1$ is the discount factor. The strategies considered in this multistage game are shown in Figure 9.6 [531]. These strategies are based on the trigger strategy [570] (see Section 5.3.2) in which the transition from one state to another state is event-driven. Two strategies are GRIM and TFT (TitForTat). In a GRIM strategy, a user will be punished if this user deviates by using the defection behavior. The punishment is performed forever. In contrast, with a TFT strategy, a user cooperates as long as the opponent user cooperates. However, if one user deviates by adopting the defection behavior, this user will be punished. The user will cooperate again if the opponent user cooperates. Also, a RANDOM strategy was considered in which the decision of a user is random with probability 0.5 given the current state (i.e. either cooperation or defection).

If the game is played once where the users do not have a memory of the past strategy, simulation results showed that a user adopting the DEF strategy yields a better performance than a user adopting the COOP strategy. Therefore, all users will prefer the DEF strategy. However, this preference will degrade network performance (e.g. high collision probability), since the channel is accessed by all users aggressively. However, in a multistage game where the user has a memory of past strategies and is aware of the future (i.e. if it uses DEF strategy, it will be punished). Therefore, there is an incentive for all users to use the COOP strategy, since if a user is punished, its utility is lower than when a user cooperates. Therefore, TFT and GRIM strategies are stable.

9.2 Non-cooperative distributed dynamic spectrum access

9.2.1 Decentralized channel selection for QoS-sensitive cognitive radio networks

The channel selection of the unlicensed users can significantly affect the performance of a cognitive radio network. For example, if many unlicensed users decide to transmit on the same channel, the collision and interference will be high, and hence the network performance will degrade. This channel selection becomes more challenging for a cognitive radio network with QoS-sensitive users/traffic. A decentralized channel selection algorithm based on game theory for QoS-sensitive cognitive radio network was presented in [532]. In the system model, there are N channels and I unlicensed users. Two QoS performance measures considered for the unlicensed users are delay and packet loss rate. The channel access by the unlicensed users is based on time slots. It is assumed that R packets can be transmitted per unit of time. The traffic arrival rate for an unlicensed user is denoted by λ_i. The unlicensed users transmit packets in an uncontrolled manner. That is, an unlicensed user randomly chooses a time slot to transmit a packet. If more than one user chooses to transmit in the same time slot, a collision occurs.

With an additive white Gaussian noise channel, the packet loss rate for an unlicensed user i can be expressed as follows:

$$P_{\text{loss},i} = 1 - (1 - P_{\text{e}}(\gamma_{n_i})) \prod_{j \neq i, n_j \neq n_i} \left(1 - \frac{r_j}{R}\right), \tag{9.25}$$

where n_i is the channel selected by unlicensed user i, γ_{n_i} is the SNR of the channel n_i, and P_{e} is the packet error rate. The ARQ error control is used by an unlicensed user. Therefore, with the maximum number of retransmissions q_i by unlicensed user i, the effective packet loss rate is $P_{\text{efloss},i} = (P_{\text{loss},i})^{q_i+1}$. Then, the average transmission duration can be obtained from

$$D_{\text{avg},i} = (q_i + 1)\,(P_{\text{loss},i})^{q_i+1}\, T_{\text{b}} + \sum_{k=0}^{q_i} (P_{\text{loss},i})^k (1 - P_{\text{loss},i})(kT_{\text{b}} + T_{\text{g}}), \tag{9.26}$$

where T_{g} and T_{b} denote the time duration due to successful and unsuccessful transmissions, respectively. The average transmission delay is $D_i = \lambda_i D_{\text{avg},i}$.

The game formulation for channel selection can be described as follows. The **players** of this game are the unlicensed users. The **strategies** correspond to the selection of channels. The **payoff** is the utility, which is expressed as follows:

$$\mathcal{U}_i = 1 - (P_{\text{loss},i})^{q_i+1}, \quad \text{given } D_i < D_{\max}, \tag{9.27}$$

where $D_{\max}$ is the maximum transmission delay. This utility function represents the throughput of an unlicensed user. The solution of this game is the Nash equilibrium. In this channel selection game, it is assumed that all unlicensed users perfectly know the information of the channel (e.g. packet error rate). To reach this Nash equilibrium, each unlicensed user is scheduled to adapt its selected channel according to the following schemes:

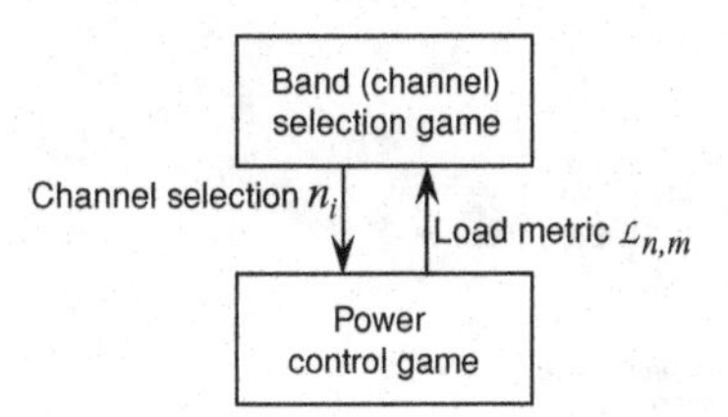

Figure 9.7 Interaction between power control game and band (channel) selection game.

- Round-robin, where each unlicensed user selects a channel sequentially.
- Ordered round-robin, where an unlicensed user with a larger arrival rate selects a channel before other users with smaller arrival rate.
- Higher packet loss priority, where an unlicensed user with high packet loss rate selects a channel before other users.
- Random, where each unlicensed user selects a channel randomly with identical probability to other users.

Simulation results showed that most of the above channel selection schemes can converge to the Nash equilibrium within 30 iterations. When a licensed user starts transmission on a particular channel, the unlicensed user has to stop transmitting on that channel. Consequently, the number of unlicensed users with unsatisfied delay requirements increases.

9.2.2 Joint distributed power control and channel selection in CDMA cognitive radio networks

In a CDMA cognitive radio network, the transmit power of the unlicensed users has to be controlled so that the highest SINR and hence the highest data rate can be achieved. However, in a multichannel environment, an unlicensed user can select a channel to operate so that its benefit is maximized. This power control and channel selection was considered jointly for a cognitive radio network in [533, 571]. In [533], a non-cooperative game was formulated to obtain both the transmit power and the operating channel for an unlicensed user. In the system model, there are N channels to be accessed by I unlicensed users. There are M base stations to provide service to the unlicensed users. An unlicensed user i first chooses the operating channel n_i. Then, in a certain access period, an unlicensed user optimizes its transmit power P_i to maximize its utility. Also, an unlicensed user observes its received utility, and adjusts the channel selection in the next period accordingly. Therefore, for an unlicensed user, the power control game and the band (i.e. channel) selection game can be formulated together and their interaction is shown in Figure 9.7.

The SINR of an unlicensed user i associated with base station m on channel n given the transmit power P_i is denoted by

$$\gamma_i = \frac{R_c}{R_t} \frac{h_{m,i} P_i}{\sum_{j \neq i, n_j = n} h_{m,j} P_j + \sigma^2}, \tag{9.28}$$

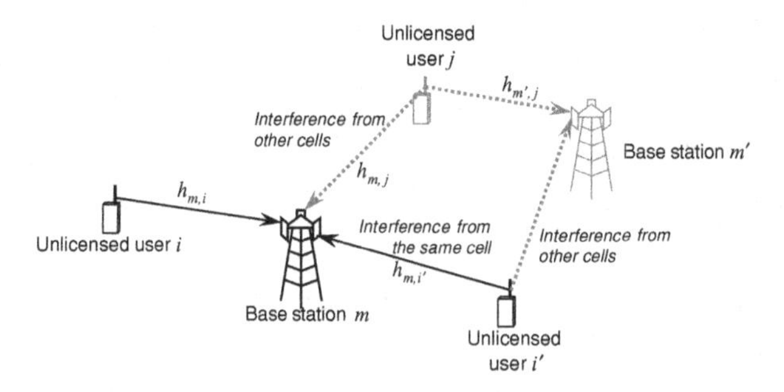

Figure 9.8 Interference from other cells.

where R_c is the chip rate, R_t is the transmission rate, $h_{m,i}$ is the channel gain between an unlicensed user i and a base station m, and σ^2 is the noise power. The utility (i.e. the payoff) of an unlicensed user given the SNR is defined as follows:

$$\mathscr{U}_i(\mathbf{p}, \mathbf{c}) = \frac{L R_t}{L_h P_i}(1 - \exp(-0.5\gamma_i))^{L_h}, \tag{9.29}$$

where $\mathbf{p}$ and $\mathbf{c}$ denote the vectors of transmit power and selected channels of all unlicensed users, respectively. L_h and L denote the length of a frame with and without headers, respectively. Note that the term $(1 - \exp(-0.5\gamma_i))^{L_h}$ indicates the probability of successful frame reception (for asynchronous FSK modulation). Therefore, this utility function represents the throughput of an unlicensed user. Similar to the models in [238, 572], the Nash equilibrium for the transmit power can be obtained as follows:

$$\hat{P}_i^* = \left(\frac{R_c}{R_t \gamma_i^*}\right)^{-1} \frac{\tilde{\sigma}_{n,m}^2}{1 - \rho_{n,m}}, \tag{9.30}$$

where $\hat{P}_i^* = h_{m,i} P_i^*$, and $\tilde{\sigma}_{n,m}^2$ indicates the interference to an unlicensed user i from other cells at a base station m on channel n. Figure 9.8 shows the interference from the same cell and from other cells to an unlicensed user i. This interference can be obtained from

$$\tilde{\sigma}_{n,m}^2 = \sum_{j \neq i, n_j = n, m_j \neq m} h_{m,j} P_j + \sigma^2. \tag{9.31}$$

In (9.30), $\rho_{n,m}$ (< 1) is a load factor which can be obtained from

$$\rho_{n,m} = \sum_{j=1, n_j = n, m_j = m}^{I} \left(1 + \frac{R_c}{R_t \sigma_i^*}\right)^{-1}. \tag{9.32}$$

In this case, as more of the unlicensed users select base stations m and channel n, the load factor $\rho_{n,m}$ increases, and hence the transmit power and interference to all unlicensed users choosing this channel will increase.

It is logical for an unlicensed user to switch to other channels if the load or interference in the current channel increases. The load metric, which is used by an unlicensed user to make decisions on channel selection, is given as

$$\mathscr{L}_{n,m} = \tilde{\sigma}^2_{n,m} \rho_{n,m}. \tag{9.33}$$

For an unlicensed user i, the equilibrium of channel selection n_i is achieved if and only if unlicensed user i associates with base station m such that $\mathscr{L}_{n_i,m} \geq \mathscr{L}_{n'_i,m}$ for $n'_i \neq n_i$. Note that this band selection game is similar to the *dispersion game* [573]. To achieve the solution of the band selection game, a distributed probabilistic strategy for band selection was proposed. In this case, an unlicensed user choosing any channel in the current period observes the outcome of the power control game (i.e. $\tilde{\sigma}^2_{n,m}$ and $\rho_{n,m}$). Then, in the next period, an unlicensed user adapts its decision as follows. If $\mathscr{L}_{n,m} \leq F_m/N$, then an unlicensed user selects channel n with probability 1. Here, F_m/N is the average load metric of all channels, where $F_m = \sum_{n=1}^{N} \mathscr{L}_{n,m}$. Otherwise, an unlicensed user remains in channel n with probability $1 - P_{\text{same}} = (F_m/N)/\mathscr{L}_{n,m}$. With probability $1 - P_{\text{same}}$, this unlicensed user switches to channel n' with $\mathscr{L}'_{n',m} < \mathscr{L}_{n,m}$ and $\rho'_{n,m} < 1$, where $\mathscr{L}'_{n,m} = \tilde{\sigma}^2_{n,m} \rho'_{n,m}$ and $\rho'_{n,m} = \rho_{n,m} + \left(1 + R_c/R_t\gamma_i^*\right)^{-1}$. In this algorithm, if an unlicensed user observes that the load of its operating channel exceeds the average load metric F_m/N, this user will switch to another channel with a probability which is proportional to the average load metric of all channels and the load of the current channel.

This distributed probabilistic strategy for band selection was compared with *freeze* [573] and *elementary step system (ESS)* [574] schemes. In the *freeze* scheme, an unlicensed user selects a channel randomly. If the load metric is smaller than the average load, an unlicensed user remains in the same channel. However, in the *ESS* scheme, an unlicensed user selects the channel based on the best response (e.g. lowest load). This algorithm requires a central controller to collect and broadcast load information of all channels. It was shown that although the proposed algorithm converges to the solution slower than the *freeze* scheme, it is faster than the *ESS* scheme. Also, the proposed algorithm achieves the Nash equilibrium solution in most of the periods, while *freeze* does not.

9.2.3 Random access and interference channel in cognitive radio networks

With an underlay spectrum access model, multiple unlicensed users can transmit data simultaneously. However, concurrent transmissions can interfere with each other, and each unlicensed user has to make an optimal decision whether to access the spectrum or not independently. This situation can be modeled using game theory with the consideration of packet arrivals at an unlicensed user [534]. The solution of the game formulation can be obtained in terms of random spectrum access probability of an unlicensed user.

In the system model, two unlicensed users access a common spectrum [575, 576]. Each unlicensed user has a buffer of size of one packet. The transmission is based on time slots and one packet is transmitted in one time slot. For an unlicensed user i, packet arrivals follow a Bernoulli process with probability λ_i. If there is a packet in the buffer, the newly arrived packet will be dropped. An unlicensed user with a packet in the buffer will transmit the packet with a certain probability which can be obtained using a game formulation. Under this system model, both the cases of perfect and partial state information were considered. In a perfect information case, it is assumed that an unlicensed user knows the buffer state of another unlicensed user. In this case, the channel access probability is denoted by $P_i^{(1)}$ and $P_i^{(2)}$ if the buffer of the other unlicensed user is full or empty, respectively. On the other hand, in a partial information case, an unlicensed user cannot observe the buffer state of the other user, and the channel access probability is denoted by P_i. An ARQ-based error recovery mechanism is used. In this case, if a transmitted packet is in error (e.g. indicated by acknowledgement message from the receiver), this packet remains in the transmitter buffer to be transmitted in the next slot.

The action of an unlicensed user is to transmit or wait in a certain slot (i.e. $A_i \in \{T, W\}$ where $\mathbf{a} = (A_1, A_2)$). The state space of the entire system is denoted by $\mathbf{s} = \{S_1, S_2, S_3, S_4\} = \{(0,0), (1,0), (0,1), (1,1)\}$ where the first and second elements indicate the buffer states of unlicensed users 1 and 2, respectively. With a fading channel, the transmission outage probability given that the other unlicensed user waits in the current slot is given by

$$P_{\text{out}}^{(1)} = \Pr(\text{SINR}_i < \gamma) = 1 - \exp\left(-\frac{\gamma}{\text{SNR}}\right), \tag{9.34}$$

where γ is a threshold for the signal-to-interference ratio, and SNR is determined from the transmit power and noise power. Similarly, the outage probability given that the other unlicensed user also transmits in the same slot is given by

$$P_{\text{out}}^{(2)} = \Pr(\text{SINR}_i < \gamma) = 1 - \frac{\exp\left(-\frac{\gamma}{\text{SNR}}\right)}{1+\gamma}. \tag{9.35}$$

In general, $P_{\text{out}}^{(2)} \geq P_{\text{out}}^{(1)}$. To avoid a trivial solution, the following rules are enforced:

- If $P_i^{(1)} = 0$, then $P_i^{(2)} > 0$. This rule is used to avoid zero spectrum access probability for an unlicensed user i.
- If $P_i^{(2)} = 0$, then $P_i^{(1)} > 0$. Similar to the above rule, this rule is used to avoid zero spectrum access probability.
- If $P_i^{(2)} = 0$, then $P_j^{(2)} > 0$, where $i \neq j$. This rule is used to avoid leaving the spectrum unoccupied (i.e. if one user does not access the spectrum, the other user should access it).

Similar to the formulation in [577, 578], if an unlicensed user decides to transmit, a fixed cost c will be incurred (e.g. due to energy consumption). The reward that can be obtained for the case that only one unlicensed user can transmit a packet is

$$\rho^{(1)} = -cP_{\text{out}}^{(1)} + (1-c)(1-P_{\text{out}}^{(1)}) = 1 - c - P_{\text{out}}^{(1)}, \tag{9.36}$$

and similarly for the case that two unlicensed users transmit in the same slot, $\rho^{(2)} = 1 - c - P_{\text{out}}^{(2)}$. The payoff matrices corresponding to the state space can be expressed as follows:

	$A_2 = T$	$A_2 = W$
$A_1 = T$	$\times$	$\times$
$A_2 = W$	$\times$	$(0, 0)$

for $S_1 = (0, 0)$

	$A_2 = T$	$A_2 = W$
$A_1 = T$	$\times$	$(\rho^{(1)}, 0)$
$A_2 = W$	$\times$	$(0, 0)$

for $S_2 = (1, 0)$

	$A_2 = T$	$A_2 = W$
$A_1 = T$	$\times$	$\times$
$A_2 = W$	$(0, \rho^{(1)})$	$(0, 0)$

for $S_3 = (0, 1)$

	$A_2 = T$	$A_2 = W$
$A_1 = T$	$(\rho^{(2)}, \rho^{(2)})$	$(\rho^{(1)}, 0)$
$A_2 = W$	$(0, \rho^{(1)})$	$(0, 0)$

for $S_4 = (1, 1)$

where the elements denoted by $\times$ are infeasible.

Two game formulations were presented. These formulations have the same players (i.e. two unlicensed users) and strategies (i.e. the spectrum access probability). However, the differences are the payoff and the solution. The first formulation considers the steady-state behavior. Therefore, the payoff of a player can be computed based on the steady-state probability of a Markov chain of the state space. This formulation is basically a non-cooperative game for which Nash equilibrium is a solution. In this case, the transition matrices can be derived under different actions taken by the unlicensed users. For example, if both unlicensed users decide not to transmit in a certain slot (i.e. $A_1 = A_2 = W$), the transition matrix can be expressed as follows:

$$\mathbf{P}([W, W]) = \begin{bmatrix} (1-\lambda_1)(1-\lambda_2) & 0 & 0 & 0 \\ \lambda_1(1-\lambda_2) & 1-\lambda_2 & 0 & 0 \\ \lambda_2(1-\lambda_1) & 0 & 1-\lambda_1 & 0 \\ \lambda_1\lambda_2 & \lambda_2 & \lambda_1 & 1 \end{bmatrix} \qquad (9.37)$$

The rows in (9.37) correspond to the state space S_1, S_2, S_3, and S_4. In this case, the payoff function of an unlicensed user can be computed from

$$R_i = \sum_{k=1}^{4} \pi_k E(\mathscr{R}_i(S, \mathbf{a}) | S = S_k), \qquad (9.38)$$

where π_k is the steady-state probability. The reward function is defined as follows $\mathscr{R}_1(S_k, [T, W]) = \rho^{(1)}$ for $k = 2, 4$, $\mathscr{R}_2(S_k, [W, T]) = \rho^{(1)}$ for $k = 3, 4$, and $\mathscr{R}_1(S_4, [T, T]) = \mathscr{R}_2(S_4, [T, T]) = \rho^{(2)}$. The Nash equilibrium can be obtained numerically as a point where the best responses of both unlicensed users intersect. The numerical results show that with this non-cooperative game formulation, the Nash equilibrium does not lie on the Pareto optimality.

The second formulation considers the dynamics of the state and action. Therefore, this formulation is a stochastic (Markov) game whose solution is a Markov perfect equilibrium. The payoff of an unlicensed user is the average discounted payoff [579, 580], which can be defined as follows:

$$\overline{R}_i = \lim_{\tau \to \infty} E\left(\sum_{t=0}^{\tau} \delta^t \mathscr{R}_i(S(t), \mathbf{a}(t)) \middle| S(0) = S_0 \right), \qquad (9.39)$$

where δ is a discount factor, $S(t)$ is a state, $\mathbf{a}(t)$ is the vector of actions at time t, and $\mathbf{s}_0$ is the initial state. To obtain the solution, the Jacobi iterative algorithm is applied [581]. Numerical results show that this algorithm can converge to the final solution within 20 iterations. However, the convergence condition depends on the parameters of the system, and, in some cases, the algorithm may not converge to the desired solution.

9.2.4 Dynamic channel selection in multihop networks

Channel selection is a challenging issue for multihop cognitive mesh networks, since it has to ensure that each mesh router can utilize the assigned channel without being affected by interference. In [535], a channel selection algorithm was proposed considering the interference temperature model. The objective of this algorithm is to allocate channels (in both fixed and adaptive transmit power mode) to the mesh nodes so that the interference temperature limit requirements are met. In the system model, a static multihop multichannel wireless mesh network was considered. The network is composed of multiple mesh nodes which can measure the interference temperature of each channel locally. These mesh nodes can exchange the measurement results among each other in their interference range so that the channel selection algorithm can be performed. Based on the definition of interference temperature defined in [582], a channel n with central frequency f_n and bandwidth B_n is available for transmission by mesh node i if the transmission of this node does not violate the interference temperature at any other node j. This constraint can be defined as follows:

$$\frac{P}{KB_n} + \frac{h_{i,j,n} P_i(f_n, B_n)}{KB_n} < I_L(n), \tag{9.40}$$

where P is the average interference power, $h_{i,j,n}$ is the path loss of the transmission from node i to node j on channel n, $P_i(f_n, B_n)$ is the transmit power of node i, $I_L(n)$ is the interference temperature limit for channel n, and K is the Boltzmann's constant (i.e. $K = 1.38 \times 10^{-23}$ joules per kelvin degree). In this case, the path loss is estimated from the distance between the transmitter and the receiver. An estimation of the path loss is performed when each mesh node initializes the channel selection algorithm.

Channel selection algorithms were proposed for both the cases that the transmit power is fixed and can be adjusted. For the case of fixed transmit power, the algorithm works as follows. Every mesh node measures the interference temperature corresponding to all channels. This measurement is sent to all nodes within its interference range periodically through a dedicated control channel (e.g. unlicensed band using IEEE 802.11 interface). Each node then estimates the interference temperature to other nodes for all channels. Only the channel for which the condition in (9.40) is satisfied is selected as an available channel. However, if the transmit power can be adjusted by a mesh node, that node communicates with its receiver to determine the transmit power. This transmit power is obtained so that both the constraints of interference temperature to other mesh nodes and also the signal-to-interference radio (SIR) requirement at its receiver are met. With this transmit power, a set of available channels is obtained in the same way as that in the case of fixed transmit power.

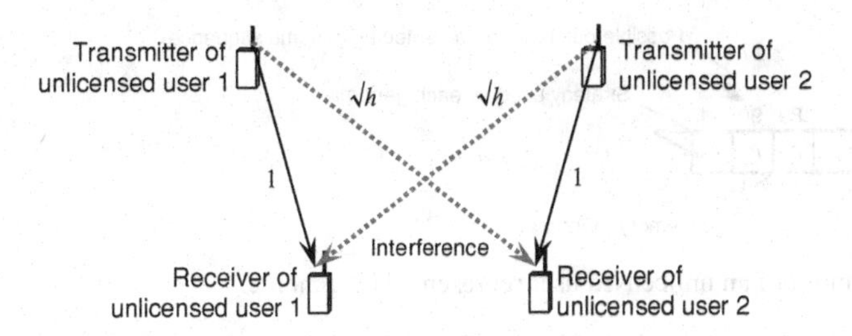

Figure 9.9 System model of two unlicensed users sharing two channels.

Once the nodes obtain a set of available channels, a specific channel is chosen for communication between two nodes. This channel has to be available at both nodes. If there is more than one available channel, two nodes will choose the channel according to the required transmission time, switching cost [583], and channel stability factor in the case of fixed transmission power. The link cost of channel n can be defined as follows:

$$\mathscr{LC}_n = c_1 \mathrm{ExTT}_n + c_2 \mathrm{SC}_n + c_3(1/\mathrm{SF}_n), \tag{9.41}$$

where c_1, c_2, and c_3 are weights, ExTT_n is the expected transmission time, SC_n is the channel switching cost, and SF_n is the channel stability factor which is given by the average channel availability time. In the case of adaptive transmit power, the amount of transmit power is also considered for channel selection. The link cost is defined as follows:

$$\mathscr{LC}_n = c_1 \mathrm{ExTT}_n + c_2 \mathrm{SC}_n + c_3(1/\mathrm{SF}_n) + c_4 P_i(f_n, B_n), \tag{9.42}$$

where $P_i(\cdot)$ is the transmit power. For routing in a mesh network, the end-to-end routing metric of route r with H hops can be defined as follows:

$$\mathscr{RM}_r = (1-\alpha)\sum_{i=1}^{H} \mathscr{LC}_{n,i} + \alpha \max_{n\in\{1,\dots,N\}} X_n, \tag{9.43}$$

where α is a weight, $\mathscr{LC}_{n,i}$ is the link cost of assigning channel n to node i, N is the total number of channels, and X_n is the total number of times that channel n is used in route r. With this routing metric, the nodes can search for the channels and the routes to transfer data in a mesh network.

9.2.5 Genetic algorithm-based competitive power allocation for dynamic spectrum access

To obtain a more efficient solution than a non-cooperative game model, in [536], a competitive power allocation method based on genetic algorithm was proposed for dynamic spectrum access by two unlicensed users. A system model with two channels and two unlicensed users were considered (Figure 9.9). For transmission by an unlicensed user, the power received at the corresponding receiver is normalized to be 1, while the interference caused to the receiver by the other unlicensed transmitter is assumed to be $\sqrt{h}$. The total transmit power is assumed to be 1. Each unlicensed user has to choose a

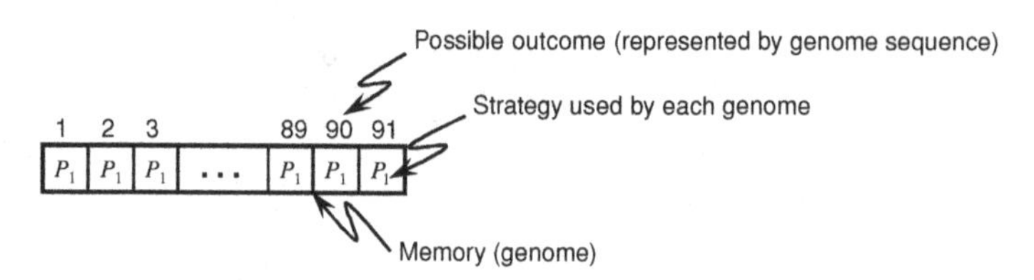

Figure 9.10 Memory of an unlicensed user represented by genome.

fraction of the total transmit power for the transmission on channels 1 and 2 which are denoted by $P_{i,1}$ and $P_{i,2}$, respectively (i.e. $P_{i,1} = P_i$ and $P_{i,2} = 1 - P_i$). With Gaussian noise, the capacity (i.e. the rate) of each unlicensed user can be expressed as

$$\mathscr{R}_i(P_i) = \frac{1}{2} \log \left(1 + \frac{P_i}{hP_j + \sigma^2}\right) + \frac{1}{2} \log \left(1 + \frac{1 - P_i}{h(1 - P_j) + \sigma^2}\right), \tag{9.44}$$

where σ^2 is noise power and $i \neq j$.

Based on the above model, a non-cooperative game can be formulated as follows. The **players** of this game are the unlicensed users. The **strategy** corresponds to the transmit power allocated to each channel. It is assumed that the possible strategy set is $(P_i, 1 - P_i) \in \{(1.0, 0.0), (0.5, 0.5), (0.0, 1.0)\}$. The **payoff** of each user is the capacity. Given the channel gain h, which is optimally computed from

$$h > \left(\frac{1}{\sqrt{2\rho + 1} - 1} - \frac{1}{\rho}\right), \tag{9.45}$$

where $\rho = 1/2\sigma^2$ and $\sigma^2 = 0.001$, the payoff matrix of both unlicensed users is given as follows [536]:

(9.46)

$(P_1, 1 - P_1)$ \ $(P_2, 1 - P_2)$	(1.0, 0.0)	(0.5, 0.5)	(0.0, 1.0)
(1.0, 0.0)	(1.46, 1.46)	(2, 7.2)	(6.9, 6.9)
(0.5, 0.5)	(7, 2.2)	(2.6, 2.6)	(7, 2.2)
(0.0, 1.0)	(6.9, 6.9)	(2, 7.2)	(1.46, 1.46)

where each element in the matrix indicates $(\mathscr{R}_1(P_1), \mathscr{R}_2(P_2))$.

The model was extended to capture the memory of an unlicensed user (Figure 9.10). Assume that an unlicensed user can recall the strategy adopted in the two previous steps. This memory is modeled by the genome in a genetic algorithm. With the two-step memory and the initial condition, the total number of possible outcomes of the game is $1 + 9 + 9 \times 9 = 91$. This number determines the length of genome, and each element of genome is indicated by the color/code for P_i used by unlicensed user i. Note that an unlicensed user can use random strategies to prevent locking up if one strategy is used more frequently.

The genetic algorithm is applied by allowing the genome to evolve according to the payoff of each unlicensed user. This algorithm is based on a method to determine robust strategies for the game [584] and tournament technique [585]. The algorithm works as follows:

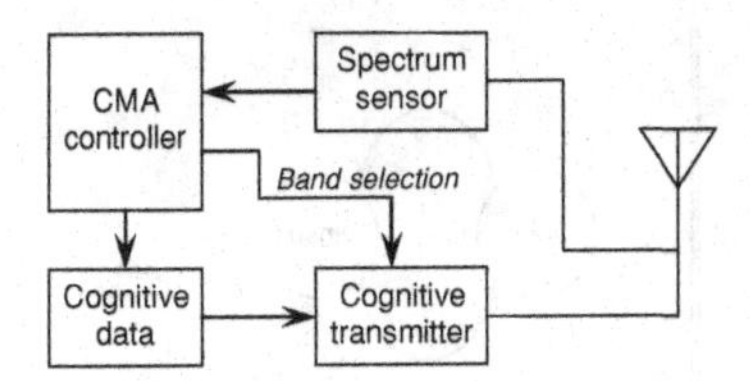

Figure 9.11 Block diagram of cognitive medium access (CMA).

1. The evaluator set $\mathbb{E}$ is created. This is a collection of strategies which is initialized to have both a random genome and a pre-defined genome (e.g. $(P_i = 1.0, 1 - P_i = 0)$).
2. A population S of strategies is generated randomly. This population will evolve in a genetic algorithm.
3. Each unlicensed user uses the strategy according to the genome. In each match, an unlicensed user plays the game with each member in the evaluator set $\mathbb{E}$. The tournament is defined when all members in the evaluator set have played the game against a particular unlicensed user.
4. After a tournament is finished, an unlicensed user measures the average payoff. The payoff in each play can be obtained from the payoff matrix defined in (9.46).
5. The genome in S is ordered according to the received payoff (i.e. fitness). Then, the next generation population is generated. In particular, two genomes with the highest fitness are copied into the new population. However, in this algorithm, the mutation process is not used.
6. Steps 3–5 are repeated for a fixed number of generations (e.g. 80). In this case, the final population will include an optimized set of strategies which returns high fitness.
7. The member with the highest fitness from the final population is added into the evaluator set $\mathbb{E}$.
8. Steps 2–7 are repeated for a fixed number of iterations.

Simulation results showed that the strategy of an unlicensed user evolves for a certain period (about 50 iterations) and then the final solution is reached. This final solution can achieve the payoff close to the highest capacity (i.e. 6.9 for both users) as indicated in the payoff matrix of (9.46).

9.2.6 Cognitive medium access and coexistence with WLAN

The problem of coexistence of unlicensed users with licensed WLAN users was considered in [537, 538]. There are N channels allocated to WLANs. Two types of users, namely, the licensed WLAN users and unlicensed users, share these channels. A WLAN user accesses the channel using the CSMA/CA protocol. When the channels are not occupied by WLAN users, unlicensed users can opportunistically access the channels. The channel access by an unlicensed user is based on frequency hopping (similar to that in Bluetooth in the ISM band).

The block diagram of the proposed cognitive medium access (CMA) scheme for an unlicensed user is shown in Figure 9.11. At the beginning of a time slot, the channel is

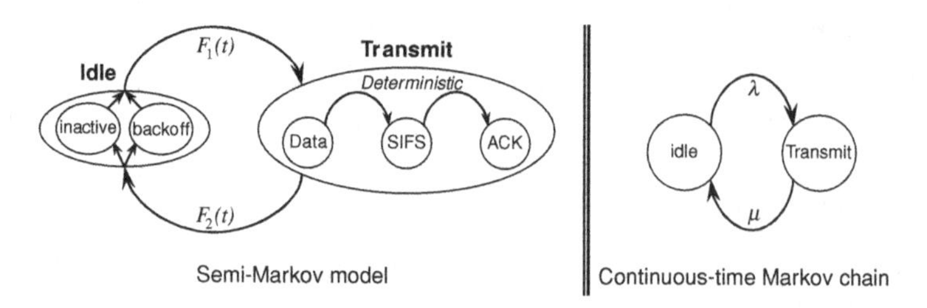

Figure 9.12 Transition diagram for the semi-Markov model and the corresponding continuous-time Markov chain.

sensed (e.g. based on energy detection [586] or feature detection). The sensing result is then used by the CMA controller to determine whether the channel can be accessed or not. In this case, the CMA controller will also ensure that the transmission will not collide with that of the WLAN users.

Experiments were conducted to investigate the impact of channel access by an unlicensed user on the licensed WLAN user. The transmission of an unlicensed user is based on static frequency hopping with the same transmission parameters as used in Bluetooth [587]. Using the experimental setup as in [588], it was observed that the transmission of an unlicensed user does not affect the backoff process in the CSMA/CA protocol. However, in the presence of transmission by an unlicensed user, the packet error rate of a WLAN user degrades. This performance degradation depends on the channel index which is used by an unlicensed user. For example, if the frequency bands used by a WLAN user and an unlicensed user are mostly overlapping, the error probability becomes very high. Therefore, an unlicensed user must make a decision on channel access carefully to avoid collision with licensed WLAN users. In this case, spectrum access by an unlicensed user will be constrained by the packet error rate of WLAN users.

Based on the experimental results and observations, a cognitive medium access protocol was developed using a prediction model [589]. Besides, to optimize the channel access, a semi-Markov model and its approximation in the form of continuous-time Markov chain were proposed. The state transition diagrams are shown in Figure 9.12. Two super-states are "transmit" and "idle." The channel is in "transmit" state if the WLAN user transmits a signal, and in "idle" state if the WLAN stops its transmission. If the channel is in "transmit" state, it could be due to the transmission of data, SIFS, and ACK packets by a WLAN user. With the assumption that the packet size is deterministic, it was observed that the transitions among data transmission, SIFS, and ACK states (Figure 9.12) of the CSMA/CA protocol are deterministic. In particular, the distribution of sojourn time in "transmit" state $F_2(t)$ is deterministic. In the "idle" state, the WLAN user could be either decreasing the backoff counter or could be inactive (e.g. due to empty queue). If the channel is in the "backoff" sub-state, it is observed that the duration in this sub-state is uniformly distributed. If none of the WLAN users transmits, the duration in the "inactive" sub-state can be approximated as a generalized Pareto distribution. Therefore, the distribution of sojourn time in "idle" state can be obtained from

$$F_1(t) = P_{\text{cw}} F_{\text{u}}(t) + (1 - P_{\text{cw}}) F_{\text{gp}}(t), \tag{9.47}$$

where P_{cw} is the probability that an idle duration of a channel is due to the backoff, $F_{\text{u}}(t)$ is a uniform random variable between [0, 0.7] ms, and

$$F_{\text{gp}}(t) = 1 - \left(1 + \frac{Kt}{S}\right)^{-1/K}, \tag{9.48}$$

in which K and S denote the shape and scale parameters, respectively. Since the semi-Markov model cannot be used in a standard optimization formulation, a continuous-time Markov chain approximation of this semi-Markov model was proposed (Figure 9.12). The parameters of this Markov chain are the transition rates λ and μ. Therefore, $F_1(t) = 1 - \exp(-\lambda t)$ and $F_2(t) = 1 - \exp(-\mu t)$. With this continuous-time Markov chain, an optimization problem can be formulated as a constrained Markov decision process [590], and a standard method [591] can be used to obtain the solution.

First, the fully observable case of the Markov decision process was considered. In this case, it is assumed that all N channels can be observed simultaneously by an unlicensed user. An optimal channel access is formulated as follows. The state is defined as the status of all N channels (i.e. "transmit" and "idle"). Note that the state of channel n is denoted by $\mathcal{X}_n$ where $\mathcal{X}_n = 1$ if the channel n is in the "transmit" state, and $\mathcal{X}_n = 0$ if the channel is in the "idle" state. For channel n, the rate transition matrix $\mathbf{Q}_n$ can be defined as follows:

$$\mathbf{Q}_n = \begin{bmatrix} -\lambda_n & \lambda_n \\ \mu_n & -\mu_n \end{bmatrix}, \tag{9.49}$$

where the steady state probability $\boldsymbol{\eta}_n$ of this chain can be obtained from

$$\boldsymbol{\eta}_n = \begin{bmatrix} \frac{\mu_n}{\lambda_n+\mu_n} & \frac{\lambda_n}{\lambda_n+\mu_n} \end{bmatrix}. \tag{9.50}$$

The corresponding probability transition matrix in discrete time is defined as follows:

$$\mathbf{P}_n = \frac{1}{\lambda_n + \mu_n} \begin{bmatrix} \mu_n + \lambda_n e^{-(\lambda_n+\mu_n)t} & \lambda_n - \lambda_n e^{-(\lambda_n+\mu_n)t} \\ \mu_n - \mu_n e^{-(\lambda_n+\mu_n)t} & \lambda_n + \mu_n e^{-(\lambda_n+\mu_n)t} \end{bmatrix}. \tag{9.51}$$

Since there are N channels, the probability transition matrix of all N channels can be obtained as follows:

$$\mathbf{P} = \bigotimes_{n=1}^{N} \mathbf{P}_n, \tag{9.52}$$

where $\otimes$ denotes the Kroneker product operator. The steady state probability is obtained from $\boldsymbol{\eta} = \bigotimes_{n=1}^{N} \boldsymbol{\eta}_n$. The action a of this Markov decision process is defined as the channel to be accessed or $a \in \{0, 1, \ldots, N\}$, where $a = 0$ if the decision is not to access any channel. The immediate reward to an action a is defined as follows:

$$\mathcal{R}(\mathbf{x}, a) = \begin{cases} 1(\mathcal{X}_a = 0)e^{-\lambda_a \tau_s}, & a \geq 1 \\ 0, & a = 0, \end{cases} \tag{9.53}$$

where $\mathbf{x}$ is a vector of channel state (i.e. $\mathbf{x} = \begin{bmatrix} \mathcal{X}_1 & \cdots & \mathcal{X}_n & \cdots & \mathcal{X}_N \end{bmatrix}$), τ_s is the length of the access slot, and 1(Cond) is an indicator function which returns one if the Cond is true.

Two alternative constraints are considered, i.e. the cumulative interference constraint (CIC) and the WLAN packet error rate constraint (PERC). The cumulative interference constraint is defined as the number of slots per unit of time during which the collision with WLAN user occurs. The expected immediate cost of cumulative interference constraint can be obtained as follows:

$$\mathscr{D}_{\text{CIC}}(\mathbf{x}, a) = \begin{cases} 1 - e^{-\lambda_a \tau_s}, & \mathcal{X}_a = 0, a \geq 1 \\ 1, & \mathcal{X}_a = 1, a \geq 1 \\ 0, & a = 0. \end{cases} \tag{9.54}$$

The packet error rate constraint is defined as the average number of collisions per transmitted WLAN packet. The expected immediate cost of packet error rate constraint can be obtained from

$$\mathscr{D}_{\text{PERC}}(\mathbf{x}, a) = \begin{cases} \frac{(\lambda_a + \mu_a)(1 - e^{-\lambda_a \tau_s})}{\mu_a \lambda_a \tau_s}, & \mathcal{X}_a = 0, a \geq 1 \\ 1, & \mathcal{X}_a = 1, a \geq 1 \\ 0, & a = 0. \end{cases} \tag{9.55}$$

Note that while the cumulative interference constraint is defined from an unlicensed user's perspective, the packet error rate constraint is defined based on the density of the WLAN traffic.

The constrained Markov decision process formulation is as follows:

$$\text{maximize:} \quad \lim_{T \to \infty} \frac{1}{T} \sum_{t=1}^{T} E\left(\mathscr{R}(\mathbf{x}_t, a_t)\right), \tag{9.56}$$

$$\text{subject to:} \quad \lim_{T \to \infty} \frac{1}{T} \sum_{t=1}^{T} E\left(\mathscr{D}_{\text{CIC}}(\mathbf{x}_t, a_t)\right) \leq \alpha \quad \text{or} \tag{9.57}$$

$$\lim_{T \to \infty} \frac{1}{T} \sum_{t=1}^{T} E\left(1(a_t = n)\mathscr{D}_{\text{PERC}}(\mathbf{x}_t, a_t)\right) \leq \alpha_n, \tag{9.58}$$

where α is the maximum acceptable interference, and α_n is the maximum acceptable packet error rate of the WLAN on channel n. To obtain the solution, in terms of Markovian randomized policies [592], a linear programming was formulated and solved. It was shown that this constrained Markov decision process formulation provides structural results. With a cumulative interference constraint, the optimal policy is to access the channels with small values of λ_n. The number of channels to be accessed by an unlicensed user can be determined from the threshold which is a function of the maximum acceptable interference α. With a packet error rate constraint, the highest reward can be achieved if all the constraints (i.e. for all channels) are tight (i.e. all channels are accessed as much as possible as long as the constraint of each channel is not violated). It was shown that if all constraints are made tight, the optimal policy can be obtained by considering each channel individually. For both cases, algorithms were designed to obtain the solution based on the structural results. In addition, the partially observable case was considered for which only a subset of channels can be observed at a time.

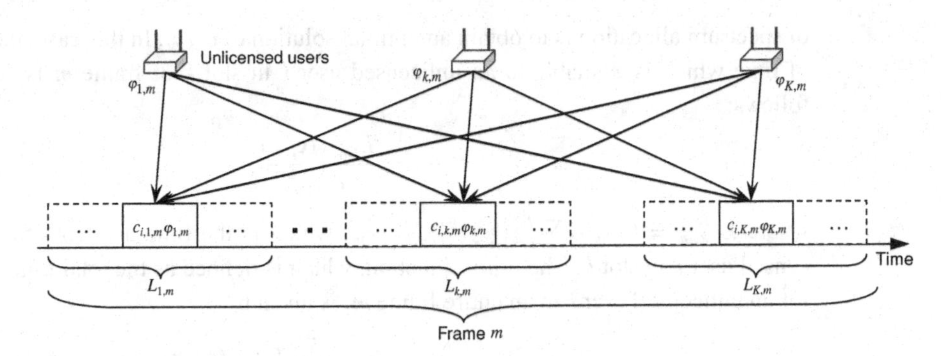

Figure 9.13 Slot and frame for spectrum allocation in time domain.

The performance of this cognitive medium access control scheme was evaluated by simulation. The performance obtained from the approximated continuous-time Markov chain was observed to be close to that of the semi-Markov chain. It was also observed that with a cumulative interference constraint, the throughput of an unlicensed user is independent of the number of channels. Since the constraint is defined for all channels, the throughput is limited by the constraint. However, with a packet error rate constraint, the throughput increases as the number of channels increases. Since this constraint is defined for each channel, if the packet error rate of one channel increases, an unlicensed user can find other channels to increase the throughput without violating the packet error rate constraint. In addition, the proposed cognitive medium access protocol was compared with the blind hopping scheme in which an unlicensed user accesses the channels without sensing. The throughput from the blind hopping scheme can be higher than that of the cognitive medium access scheme in some cases. However, with the blind hopping scheme, the interference and packet error rate for WLAN traffic cannot be controlled, which is undesirable from a QoS provisioning point of view.

9.2.7 Time domain spectrum allocation

Dynamic spectrum access can be performed in different dimensions (e.g. time, frequency, power, and space). In [539], the spectrum allocation among unlicensed users in the time domain was formulated as a non-cooperative game. In the system model, there are I unlicensed users sharing one channel. This channel is divided into frames each of which is composed of K slots. The allocation of the slots to the unlicensed users is performed over M frames. One slot can be allocated to multiple users. The structure of frames and slots is shown in Figure 9.13. The fraction of time an unlicensed user i accesses slot k in frame m is denoted by $c_{i,k,m}$, and a vector $\mathbf{c}_{i,m}$ is defined as $\mathbf{c}_{i,m} = [\, c_{i,1,m} \;\; \cdots \;\; c_{i,k,m} \;\; \cdots \;\; c_{i,K,m}\,]$. $\Phi_m = [\,\phi_{1,m} \;\; \cdots \;\; \phi_{i,m} \;\; \cdots \;\; \phi_{I,m}\,]$ denotes a vector of requested time of all I unlicensed users and

$$\Phi = [\, \Phi_1 \;\; \cdots \;\; \Phi_m \;\; \cdots \;\; \Phi_M \,]. \tag{9.59}$$

Let $L_{k,m}$ denote the length of slot and $L_{i,k,m}$ denote the length of time interval within a slot allocated to an unlicensed user i to transmit data in slot k of frame m. The objective of spectrum allocation is to obtain an optimal solution of $c_{i,k,m}$. In this case, the portion of time which is available to an unlicensed user i in slot k of frame m is defined as follows:

$$\phi_{i,k,m} = \frac{L_{i,k,m}(\mathbf{c}_{i,m})}{c_{i,k,m}}, \tag{9.60}$$

where $L_{i,k,m} = L_{k,m} - \sum_{i'=1,i'\neq i}^{I} c_{i',k,m}\phi_{i',m}$, which is the time available for an unlicensed user i in slot k. The utility function, which is defined as the total time available for an unlicensed user i in an entire frame m, is given by

$$\begin{aligned}\mathcal{U}_{i,m}(\mathbf{c}_{i,m}) &= \sum_{k=1}^{K} \phi_{i,k,m} = \sum_{k=1}^{K} \frac{L_{i,k,m}(\mathbf{c}_{i,m})}{c_{i,k,m}} \\ &= \sum_{k=1}^{K} \frac{L_{k,m} - \sum_{i'=1,i'\neq i}^{I} c_{i',k,m}\phi_{i',m}}{c_{i,k,m}}.\end{aligned} \tag{9.61}$$

Based on this utility function, a non-cooperative game was formulated. The **players** of this game are the unlicensed users. The **strategy** of each player is the requested time $c_{i,k,m}$, and the **payoff** is the utility. Given the requirement of an unlicensed user $\phi_{i,m}$, the max-min strategy is chosen as the solution of this game. The max-min strategy of player i is defined as a mixed strategy that maximizes the minimum expected payoff of that player in which the minimum is considered over all strategies of the other players. In this case, an optimization problem can be formulated as follows:

$$\max_{\mathbf{c}_{i,m}} \min \mathcal{U}_{i,m}(\mathbf{c}_{i,m}), \tag{9.62}$$

$$\text{subject to:} \quad c_{i,k,m} \geq 0, \tag{9.63}$$

$$\sum_{k=1}^{K} c_{i,k,m} = 1, \tag{9.64}$$

$$\sum_{i=1}^{I} c_{i,k,m}\phi_{i,m} < L_{k,m}. \tag{9.65}$$

The optimal solution is obtained as follows:

$$\mathcal{U}_i^*(\mathbf{c}_{i,m}) = \sum_{i=1}^{I} \phi_{i,m}^* = \frac{K^2}{\sum_{k=1}^{K}\left(\frac{c_{i,k,m}}{L_{k,m} - \sum_{i'=1,i'\neq i}^{I} c_{i',k,m}\phi_{i',m}}\right)}, \tag{9.66}$$

where

$$c_{i,k,m} = \begin{cases} \frac{1}{\phi_{i,m}}\left(L_{i,k,m} - \sqrt{L_{i,k,m}}\frac{\sum_{k=1}^{k^*} L_{i,k,m} - \phi_{i,m}}{\sum_{k=1}^{k^*}\sqrt{L_{i,k,m}}}\right), & 1 \leq k \leq k^* \\ 0, & k^* < k \leq K, \end{cases} \tag{9.67}$$

where k^* is the minimum index that satisfies the inequality

$$\sqrt{L_{i,k^*,m}} \leq \frac{\sum_{k=1}^{k^*} L_{i,k,m} - \phi_{i,m}}{\sum_{k=1}^{k^*}\sqrt{L_{i,k,m}}}. \tag{9.68}$$

The algorithm to obtain the above solution was provided in [539], which is similar to the distributed load balancing algorithm in [593]. The numerical results showed that the algorithm converges to the final solution within a limited number of iterations.

9.3 Summary

In distributed dynamic spectrum access, there is no central controller to collect network information (e.g. the spectrum usage by the licensed users and other unlicensed users) and control the spectrum access by the unlicensed users. Therefore, the unlicensed users have to observe the environment and make decisions on spectrum access themselves. In this distributed environment, the unlicensed users can have different behaviors, i.e. non-cooperative/cooperative or non-collaborative/collaborative. If unlicensed users are non-cooperative, each unlicensed user will make a decision to achieve its own individual objective. With non-cooperative behavior among the cognitive radio users, the theory of non-cooperative games can be used for distributed spectrum access. The solution of such a game (e.g. Nash or correlated equilibrium) ensures that none of the unlicensed users will want to deviate from the solution. On the other hand, the unlicensed users will make a decision to achieve an objective of a group if they are cooperative. With cooperative behavior among cognitive radios, both optimization and cooperative game theories can be applied when designing distributed dynamic spectrum access mechanisms. The objective of such a distributed scheme is to achieve optimal or fair performance in a cognitive radio network.

The unlicensed users can be non-collaborative or collaborative in exchanging network information. The choice of an unlicensed user to be non-collaborative or collaborative depends not only on its cooperative or non-cooperative behavior, but also on the benefits which it will receive. In particular, exchanging network information will consume network resources. Also, transmission performance can be degraded due to collaboration. As a result, an unlicensed user may not collaborate with other users if the benefits to be gained from collaboration are less than those without collaboration. If an unlicensed user is non-collaborative, a learning algorithm becomes crucial in estimating network information. Note that designing distributed dynamic spectrum access in a mixed environment (i.e. where some unlicensed users are cooperative while others are non-collaborative) is still open. The traditional solution (e.g. the optimal or equilibrium solution) may have to be customized for this mixed environment.

In the next chapter, distributed dynamic spectrum access methods based on learning algorithms will be discussed. Distributed protocols to support information exchange and cooperation among entities in a cognitive radio network will also be reviewed.

10 Distributed dynamic spectrum access: learning algorithms and protocols

Learning algorithms are used to build knowledge about a cognitive radio network so that the cognitive radio users can dynamically adapt their decisions on spectrum access. Learning algorithms are useful for cognitive radio networks with either collaborative or non-collaborative behavior among the network entities. In a non-collaborative scenario, there is no exchange of information in the network and a cognitive radio user has to learn from the local observations only. This chapter deals with learning-based schemes for distributed dynamic spectrum access. Different protocols to support distributed dynamic spectrum access are also discussed. These protocols can be used to facilitate spectrum handoff, exchange the information between cognitive radio users, and synchronize the transmission between cognitive radio transmitter and receiver.

10.1 Learning-based distributed dynamic spectrum access

10.1.1 Distributed resource management in multihop cognitive radio networks

In a multihop cognitive radio network, the channel selection algorithm plays an important role in optimizing the transmission performance and avoiding interference. To achieve an optimal channel selection, information of the channel availability is required at each node. Although an information exchange mechanism can be developed for this channel selection algorithm, this can incur a significant cost in the system. This information exchange not only reduces the available resources for data transmission, but also increases the delay of data transmission. Therefore, with the constraint of available information in a multihop cognitive radio network, a distributed minimum delay routing and channel selection algorithm based on learning was proposed for delay-sensitive traffic (e.g. multimedia) [540].

In the system model, there are multiple licensed users occupying the channels. In this multihop cognitive radio network, the licensed users require an interference-free environment. In particular, transmission by an unlicensed user on the same channel at the same time as a licensed user is not allowed. For an unlicensed user i, App_i denotes a delay-sensitive application, and this application sends the packet with priority $K_i = k$ in class C_k. To simplify the notation, it is assumed that C_1 is the traffic of a licensed user. For $k > 1$ (i.e. traffic from an unlicensed user), the parameters of the service class are as follows:

- λ_k, the impact factor of traffic in class k. It is assumed that for $k' > k$, $\lambda_k \geq \lambda_{k'}$;
- D_k, the delay deadline of the packets in class C_k;
- L_k, the packet length in class C_k.

The corresponding delay due to the transmission of a packet in class C_k using channel n in link l is computed from the effective transmission time (ETT) [594] as follows:

$$\mathrm{ETT}_{i,k} = \frac{L_k}{T_i(l,n)(1 - P_{\mathrm{e},i}(l,n))}, \tag{10.1}$$

where $T_i(l,n)$ and $P_{\mathrm{e},i}(l,n)$ are the transmission rate and the packet error rate of node i using channel n for link l. This transmission rate and packet error rate can be estimated from the parameters in the MAC and physical layers [595]. An action to relay and transmit data is denoted by $A_i(l,n)$, which denotes an assignment of channel n to link l of node i. This action is feasible if channel n is available to link l. In particular, channel n is available to node i, and the assignment does not interfere with a licensed user and other unlicensed nodes. The action vector can be defined as $\mathbf{A}_i = [A_{i'} | i' \in \boldsymbol{\sigma}_i]$, where $\boldsymbol{\sigma}_i$ denotes the route of packets from the source node i through intermediate nodes i' to the destination. This routing is defined as follows $\boldsymbol{\sigma}_i = \{\sigma_{i',j}\}$, where $\sigma_{i',j} = (l,n)$ is the route of packet j from a node i' through link l using channel n. In this case, the route is defined as a function of action vector $\sigma_{i,j}(\mathbf{A}_i)$ and the end-to-end delay of class C_k can be obtained from

$$d_{i,j}(\sigma_{i,j}(\mathbf{A})) = \sum_{i' \in \sigma_{i,j}} \mathrm{ETT}_{i',k}(\mathbf{A}_i). \tag{10.2}$$

If the global network information is available (e.g. at the central controller), an optimization problem can be formulated to obtain the optimal action vector as follows:

$$\mathbf{A}_i^* = \arg\max \mathscr{U}_i(\mathbf{A}_i), \tag{10.3}$$

where the utility function gives the satisfaction on delay requirement as follows:

$$\mathscr{U}_i(\mathbf{A}_i) = \sum \lambda_k \Pr(d_{i,j}(\sigma_{i,j}(\mathbf{A}_i)) \leq D_k). \tag{10.4}$$

Here $\Pr(d_{i,j}(\sigma_{i,j}(\mathbf{A}_i)) \leq D_k)$ is the probability that the delay requirement can be satisfied. However, due to the delay requirement, the solution of this optimization problem may not always be feasible. Therefore, a sub-optimal greedy algorithm was proposed. This greedy algorithm solves an optimization problem sequentially from the node with the highest priority to the node with the lowest priority. Let $\mathbf{A}_{i,k}$ denote the action vector of the node i in class C_k. Then, the optimization problem for each node can be expressed as follows:

$$\mathbf{A}_{i,k}^* = \arg\min \sum_{j \in C_k} d_{i,j}(\sigma_{i,j}(\mathbf{A}_{i,k})), \tag{10.5}$$

$$\text{subject to:} \quad d_{i,j}(\sigma_{i,j}(\mathbf{A}_{i,k})) \leq D_k. \tag{10.6}$$

However, when the network information is not available, a distributed algorithm which uses only local information would be required. Therefore, a distributed algorithm was proposed.

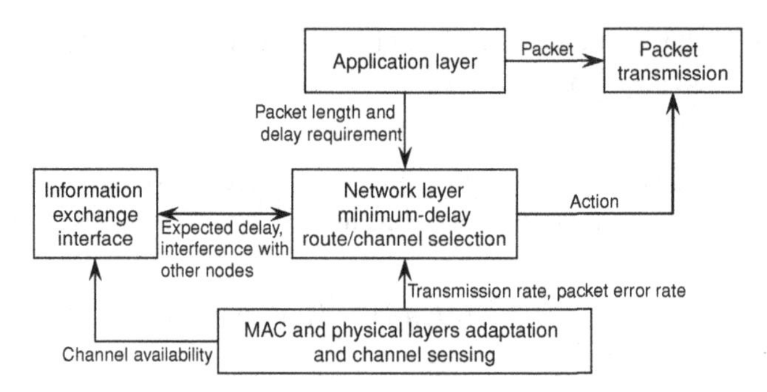

Figure 10.1 Distributed resource management in a cognitive radio network.

In the proposed algorithm, each node chooses its action to minimize the expected delay which can be expressed as follows:

$$d_{i,j}(\sigma_{i,j}) = \hat{d}_{i'}(\sigma_{i,j}) + E\left(\tilde{d}_{i'}(k, \sigma_{i,j})\right), \tag{10.7}$$

where $\hat{d}_{i'}(\sigma_{i,j})$ is the past delay that a packet j will experience before reaching node i', and $E\left(\tilde{d}_{i'}(k, \sigma_{i,j})\right)$ is the expected delay of packet j in class C_k from node i' to the destination. Given the impact factor λ_k and delay $\hat{d}_{i'}(\sigma_{i,j})$, the optimization problem for node i' can be formulated as follows:

$$A^*_{i'} = \arg\min E\left(\tilde{d}_{i'}(k, \sigma_{i,j}(A_{i'}))\right), \tag{10.8}$$

$$\text{subject to:} \quad E\left(\tilde{d}_{i'}(k, \sigma_{i,j}(A_{i'}))\right) \leq D_k - \hat{d}_{i'}(\sigma_{i,j}) - \rho, \tag{10.9}$$

where ρ is a guard interval.

The components and their interactions in the proposed distributed resource management algorithm are shown in Figure 10.1. To transmit a packet from an application layer, the information required for resource allocation (i.e. packet length, delay requirement, and class of packet) is sent to the minimum delay routing and channel selection algorithm. This algorithm interacts with the MAC and the physical layers to obtain information about the channel (i.e. transmission rate and packet error rate) and also interacts with the information exchange interface to obtain information about channel allocation to other nodes. Then, this routing and channel selection algorithm makes decisions on packet transmission.

The information exchange interface works as follows:

1. Each node performs channel sensing and identifies an available channel which is not occupied by a licensed user.
2. Each node learns the behavior of neighboring nodes by periodically observing the actions of neighboring nodes.
3. The node identifies an available channel by combining local sensing information with learned information from neighboring nodes.
4. The node updates information about interference and delays according to the action taken.
5. The node broadcasts information to neighboring nodes.

For Step 2, an adaptive fictitious play (AFP) [596] was applied to learn the feasible action. This algorithm assumes that a node can disclose the information about its current action. Let $\rho_A(i', t)$ denote the probability that node i' will choose action A. In particular, this probability can be estimated from the frequency of taking action A at time t as follows:

$$\rho_A(i', t) = \frac{q_A(i', t)}{\sum_A q_A(i', t)}, \tag{10.10}$$

where $q_A(i', t)$ is the propensity [597] of node i' for taking action A at time t. This propensity can be obtained from

$$q_A(i', t) = \alpha q_A(i', t-1) + 1(A_{i'}(t) = A), \tag{10.11}$$

where $\alpha < 1$ is a discount factor, and $1(A_{i'}(t) = A) = 1$ if the action of node i' at time t is A.

Then, the routing and channel selection algorithm works as follows:

1. The packet to be transmitted is determined from its impact factor and delay requirement, and the class of that packet.
2. Given the available channel, a node determines the feasible action set.
3. The node estimates the channel condition in terms of transmission rate and packet error rate.
4. Expected delay is computed for each traffic class.
5. The node verifies the delay requirement. If the delay requirement of a packet cannot be satisfied, this packet will be dropped.
6. The action to yield the smallest delay is chosen.
7. The node accesses the channel according to the chosen action. In this case, RTS and CTS messages are used to inform the receiver.
8. After the action is performed, the node observes and updates the delay and current action.

As the fraction of time that licensed users access the channels increases, the packet loss rate of an unlicensed user increases. It was observed that the network performance improves when the nodes have more information about the network (i.e. the channel allocation of neighboring nodes). Also, as the speed of the nodes increases, the performance degrades, since the accuracy of the information decreases. The proposed routing and channel selection algorithm was compared with AODV with load balancing (AODV/LB) [598] and dynamic least interference channel selection (DCS) [599] algorithms. With video traffic, it was shown that the proposed algorithm outperforms both AODV/LB and DCS (i.e. the Y-PSNR of the proposed algorithm is higher) due to the learning ability of the nodes.

10.1.2 Opportunistic channel selection based on the fuzzy learning algorithm

An efficient channel selection algorithm is required for WLAN hotspots to achieve high spectrum utilization and also QoS support [600, 601, 602, 603]. In [541], a learning

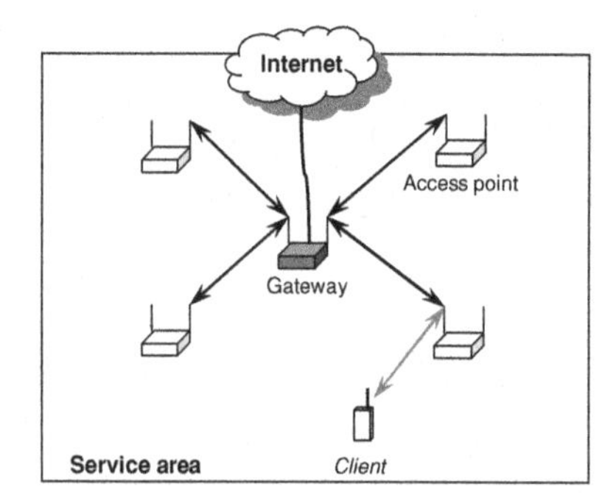

Figure 10.2 Service area of an IEEE 802.11-based wireless mesh network.

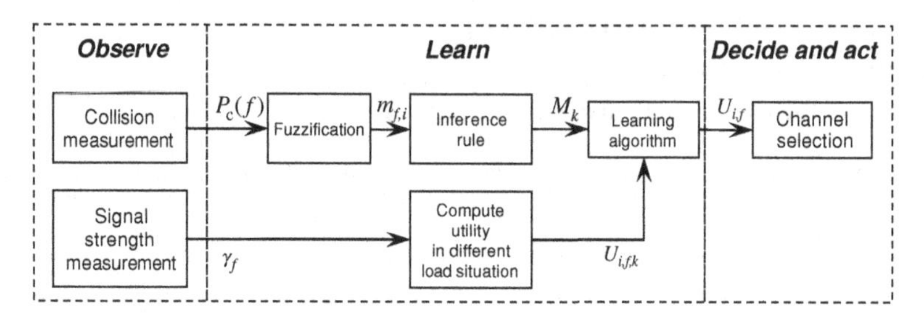

Figure 10.3 Structure of the dynamic channel selection.

algorithm was used for channel selection in IEEE 802.11-based wireless mesh networks. Access points, which provide wireless connectivity in a particular service area, are connected to a wireless gateway (Figure 10.2). A dual radio interface is used in each access point – one interface is used for communicating with the clients and the other interface is used for communicating with the gateway. Also, different access points use different channels to communicate with the clients and gateway. Since there is no central controller, each client has to select the transmission channel in a distributed manner. A cognitive radio-based dynamic channel selection scheme was developed for this channel selection in which the client observes and learns the network status in both the physical layer (i.e. signal strength) and the MAC layer (i.e. collision probability). Then, the channel is selected accordingly. In this case, a decision of each client is made independently.

The structure of the dynamic channel selection scheme used in each client is shown in Figure 10.3. The first part is the network measurement for which the signal strength and collision probability are observed in the physical and MAC layers, respectively. Then, the observation is used in the fuzzy-logic-based learning algorithm. Here, a fuzzy logic controller is used (i.e. a fuzzification process and inference rules) since the estimated collision probability $\tilde{P}_c(f)$ is often very imprecise and cannot be computed very accurately (see Section 6.5). The fuzzification process converts the imprecise collision probability in each channel into a fuzzy set which is then used by the inference rules to estimate the amount of traffic load in each channel. Similarly, the estimated signal strength γ_f is used to compute the corresponding client utility. The algorithm combines the knowledge about the traffic load with the client utility to determine the potential gain of choosing a particular channel. The output of the learning algorithm is then used for the channel selection. Note that the learning algorithm is similar to that proposed for the market selection problem in [604].

For a client, the utility of choosing a particular channel is defined as a function of measured collision probability $P_c(n)$ and received signal strength γ_n in channel n. The utility can be computed from

$$\mathscr{U}(P_c(n), \gamma_n) = \mathscr{T}(P_c(n)) \frac{\gamma_n}{\gamma_{\text{ref}}}, \tag{10.12}$$

where $\mathscr{T}(P_c(n))$ is a throughput function given the collision probability. The received signal strength is a function of distance between the access point and the client and the channel condition. Both collision probability and received signal strength impact the throughput and error performances experienced by a client.

The traffic load at an access point can be approximated from the collision probability on the corresponding channel associated with that access point. In the fuzzy logic controller, the collision probability can be either "high" or "low" (i.e. in fuzzy set c_x). Then, based on the fuzzified collision probability, the inference rules are used to gain information on the traffic load condition in a channel. Let $\tilde{P}_c(n)$ denote the estimated collision probability in channel n. Then, the inference rules can be expressed as follows:

$$\text{Rule } R_k : \text{IF } \left(\tilde{P}_c(1) \text{ in } c_x\right) \text{ AND} \ldots \text{AND } \left(\tilde{P}_c(N) \text{ in } c_y\right) \text{ THEN } U_{i,n} \text{ is } U_{i,n,k},$$

where N is the total number of available channels, $U_{i,n,k}$ is the utility corresponding to rule k (i.e. $U_{i,n,k} = \mathscr{U}(P_c(n), \gamma_n)$), and $U_{i,n}$ is the resultant utility for client i using channel n. Let $m_{n,i}$ denote the membership function for channel n obtained from fuzzification. This $m_{n,i}$ can be obtained using a standard fuzzification method. Then, the fitness of rule k to the traffic load condition can be obtained from $M_k = \prod_{n=1}^{N} m_{n,i}$. The estimated utility can then be calculated as follows:

$$U_{i,n} = \frac{\sum_{k=1}^{K} M_k \times U_{i,n,k}}{\sum_{k=1}^{K} M_k}. \tag{10.13}$$

In this case, the normalized fitness is given by

$$M'_k = \frac{M_k}{\sum_{l=1}^{K} M_l}. \tag{10.14}$$

Next, the learning algorithm is applied to obtain the utility $U_{i,f,k}$ perceived by each client corresponding to the different traffic load conditions in the service area. Inference rules of the fuzzy logic controller are used to update the utility by using estimated collision probability $\tilde{P}_c(n)$ and received signal strength γ_n for channel n as follows:

$$U_{i,n,k}^{\text{new}} = \left(1 - \alpha M'_k\right) U_{i,n,k}^{\text{old}} + \alpha M'_k \mathscr{U}(\tilde{P}_c(n), \gamma_n), \tag{10.15}$$

where $U_{i,n,k}^{\text{old}}$ denotes the utility of the previous learning iteration and α is the learning rate. In this learning algorithm, utility of a wireless node is computed based on the history and current information (i.e. $U_{i,n,k}^{\text{old}}$ and $\mathscr{U}(P_c(n), \gamma_n)$, respectively). Finally, the decision on channel selection is made based on $U_{i,n}$. In particular, a client i chooses a

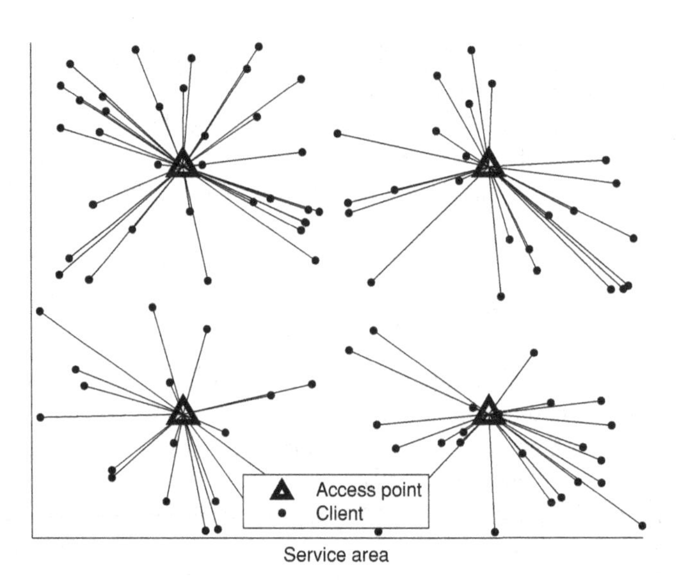

Figure 10.4 Association of clients to the access points.

channel n which provides the highest $U_{i,n}$. This channel selection scheme is executed periodically.

The results of channel selection by the clients are shown in Figure 10.4. With the learning algorithm, a client can seek for the channel which provides the highest utility. In this case, convergence to the final solution (i.e. the load from the clients is balanced among all channels) is relatively fast. For example, with 100 clients, the algorithm converges within 10 iterations.

10.1.3 Distributed dynamic spectrum access based on no-regret learning

Distributed dynamic spectrum access can be implemented based on non-cooperative game formulation of the unlicensed users sharing the spectrum. While the common solution of the non-cooperative game is the Nash equilibrium, it was shown that Nash equilibrium is inefficient in many cases to achieve the highest performance. In [542], correlated equilibrium [272] (see Section 5.7.4) was studied in distributed competitive dynamic spectrum access. To achieve this correlated equilibrium, algorithms based on linear programming and no-regret learning were proposed for the centralized and distributed implementations, respectively. A regret-based learning approach to achieve correlated equilibrium for a game model can also be found in [605].

In the system model, there are N channels which become occasionally available to the unlicensed users due to the time-varying activities of the licensed users. The total number of unlicensed users is I. The channel availability is denoted by the variable $v_{i,n}$; $v_{i,n} = 1$ if channel n is available for unlicensed user i, and $v_{i,n} = 0$ otherwise. An unlicensed user can access multiple channels simultaneously and transmit using adaptive modulation. Different modulation modes provide different transmission rates. Based on this system model, a non-cooperative game can be formulated as follows. The **players** of this game are the unlicensed users. The **strategy** of player i is the selected transmission rate $r_{i,n}$ (i.e. the modulation mode) on channel n. The transmission rates are assumed to be discrete. The **payoff** is the utility defined as follows:

$$\mathcal{U}_i = \sum_{n=1}^{N} v_{i,n} \mathcal{R}_i(r_{i,n}, \mathbf{r}_{-i,n}), \tag{10.16}$$

where $\mathcal{R}_i(\cdot)$ is the achievable rate with the transmissions of all unlicensed users. $\mathbf{r}_{-i,n}$ is a vector of selected transmission rates on channel n used by all unlicensed users except unlicensed user i. If the unlicensed users use the un-slotted 1-persistent CSMA-based MAC protocol, the achievable rate can be obtained from [493]:

$$\mathcal{R}_i(r_{i,n}, \mathbf{r}_{-i,n}) = \begin{cases} \frac{r_{i,n} S_n}{\sum_{i=1}^{I} r_{i,n}}, & \text{network load} \leq G_0 \\ 0, & \text{otherwise}, \end{cases} \tag{10.17}$$

where S_n is a channel access parameter which can be obtained as in [493]. In this case, G_0 is the maximum network load such that the delay due to the transmission is acceptable by the user. Note that if G_0 becomes larger, the user has more tolerance to the performance degradation.

With a correlated equilibrium, the players in the game will play mixed strategies with probability distribution π. The probability distribution π^* is a correlated equilibrium of the game, if and only if,

$$\sum_{\mathbf{r}_{-i}} \pi^*(\mathbf{r}_i, \mathbf{r}_{-i}) \left(\mathcal{U}_i(\mathbf{r}'_i, \mathbf{r}_{-i}) - \mathcal{U}_i(\mathbf{r}_i, \mathbf{r}_{-i}) \right) \leq 0, \tag{10.18}$$

where $\mathbf{r}_i$ is a vector of strategies applied by an unlicensed user i for all channels, and similarly, $\mathbf{r}_{-i}$ is a vector of strategies used by all unlicensed users except user i for all channels. In particular, at a correlated equilibrium, an unlicensed user cannot improve the payoff by changing the strategy to any other strategy $\mathbf{r}'_i$. In every finite game, a set of correlated equilibria is nonempty closed and convex. In general, the Nash equilibrium in a game is a special case of the correlated equilibria where the strategies adopted by different players are independent [273, 274].

To obtain the correlated equilibrium of this dynamic spectrum access game, both centralized and distributed approaches were proposed. Also, refinements of a correlated equilibrium solution were considered. In particular, the correlated equilibrium solutions are sought which (1) maximize the sum of utilities of the unlicensed users, or (2) maximize the minimum utility. In a centralized approach, an optimization problem based on linear programming can be formulated as follows:

$$\max_{\pi} \sum_{i=1}^{I} E_\pi(\mathcal{U}_i) \text{ or } \max_{\pi} \min_{i} E_\pi(\mathcal{U}_i), \tag{10.19}$$

$$\text{subject to: } \pi(\mathbf{r}_i, \mathbf{r}_{-i}) \left(\mathcal{U}_i(\mathbf{r}'_i, \mathbf{r}_{-i}) - \mathcal{U}_i(\mathbf{r}_i, \mathbf{r}_{-i}) \right) \leq 0, \tag{10.20}$$

where $E_\pi(\cdot)$ is the expectation over probability distribution π. However, to solve this optimization problem, it requires a central controller with all information in the system.

Therefore, a distributed approach based on the regret-matching algorithm [274] was also introduced. The solution of the algorithm (i.e. probability distribution of the strategies) indicates the regrets of the players to any strategy. For example, if the strategy yields high payoff, the player will highly regret it if this strategy is not used, and hence the probability to use this strategy will be high. The regret of the player who does not use strategy $\mathbf{r}_i'$ during period τ can be defined as follows:

$$\mathscr{RG}_{i,\tau}(\mathbf{r}_i, \mathbf{r}_i') = \max\left(\mathscr{G}_{i,\tau}(\mathbf{r}_i, \mathbf{r}_i'), 0\right), \tag{10.21}$$

where

$$\mathscr{G}_{i,\tau}(\mathbf{r}_i, \mathbf{r}_i') = \frac{1}{\tau}\sum_{t\leq\tau}\left(\mathscr{U}_{i,t}(\mathbf{r}_i', \mathbf{r}_{-i}) - \mathscr{U}_{i,t}(\mathbf{r}_i, \mathbf{r}_{-i})\right). \tag{10.22}$$

The learning algorithm to obtain the solution is as follows:

1. Obtain $\mathscr{G}_{i,\tau}(\mathbf{r}_i, \mathbf{r}_i')$, which is the average payoff that player i should gain, if strategy $\mathbf{r}_i'$ is used during past period τ.
2. Obtain $\mathscr{RG}_{i,\tau}(\mathbf{r}_i, \mathbf{r}_i')$, which is the average regret.
3. Update the probability distribution of a set of strategies $\mathbf{r}_i'$ for iteration $t+1$ as follows: $\pi_{i,t+1}(\mathbf{r}_i') = \frac{1}{\mu}\mathscr{RG}_{i,\tau}(\mathbf{r}_i, \mathbf{r}_i')$.
4. Update the probability distribution of a set of strategies $\mathbf{r}_i$ for iteration $t+1$ as follows: $\pi_{i,t+1}(\mathbf{r}_i') = 1 - \sum_{\mathbf{r}_i'}\pi_{i,t+1}(\mathbf{r}')$.

Here t denotes the index of current iteration and μ is a constant of the algorithm. It was shown that this algorithm converges almost exactly to the set of correlated equilibria.

Simulation results showed that the distributed algorithm based on regret matching learning can converge to the solution within 100 iterations. As expected, the utility of an unlicensed user obtained from the distributed algorithm is slightly lower than that from the centralized algorithm which is based on a linear programming formulation. Also, when the number of unlicensed users in the system increases, it is observed that the utility first increases, since the channels become highly utilized. However, at a certain point, where too many unlicensed users are sharing a limited number of channels, due to the increased collision/interference, the utility of the unlicensed users decreases.

10.1.4 Agent-based dynamic spectrum access

The concept of learning agent has been applied to many applications to support intelligent decision making [606, 607, 608, 609]. This learning agent concept can be also applied to dynamic spectrum access [543]. In the system model considered in [543], the learning agents are introduced for both unlicensed and licensed users (Figure 10.5). An unlicensed user can choose to access the spectrum for which the access model ranges from commons-use to exclusive-use. For this, the degree of control parameter $d_c \in [0, 1]$ is used (Figure 10.6). For the commons-use access model (i.e. $d_c = 0$), an unlicensed user can access the spectrum without any cost. However, interference can become severe if a number of unlicensed users access this spectrum. For the exclusive-use model (i.e.

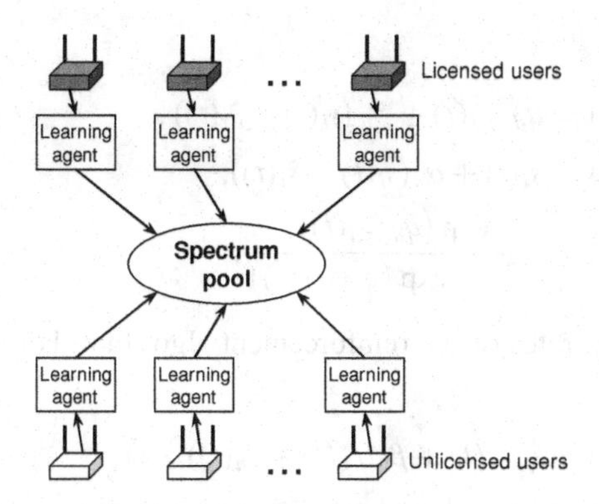

Figure 10.5 Learning agent for licensed and unlicensed users in a dynamic spectrum access environment.

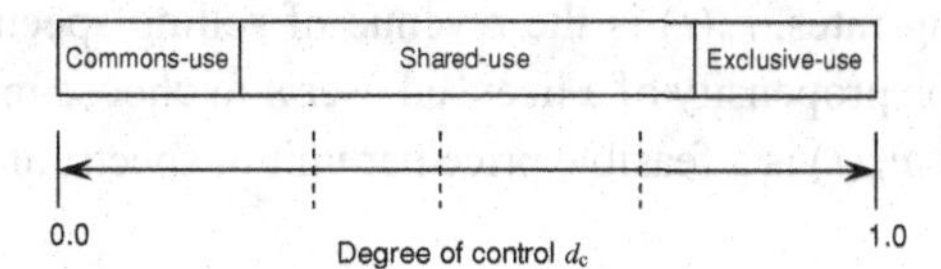

Figure 10.6 Degree of control in dynamic spectrum access.

$d_c = 1$), the spectrum is allocated exclusively to an unlicensed user, and there is no interference. However, an unlicensed user has to pay a price to the corresponding licensed user of that spectrum. For the shared-use model (i.e. $0 < d_c < 1$), an unlicensed user can pay a lower price to a licensed user, but interference can occur. In this case, if the price is high, the interference will be low which can be indicated by the value of d_c close to one. The learning agents are developed for the licensed and unlicensed users to make decisions on spectrum sharing [543]. In the system model, an unlicensed user has to choose the spectrum to access, while a licensed user can decide to offer spectrum to multiple unlicensed users to gain higher revenue. By learning from the history, the agents of the unlicensed and licensed users can adapt their actions accordingly.

The surplus $s_i(t)$ (i.e. the net utility) of accessing the spectrum by an unlicensed user i at time t is denoted by

$$s_i(t) = u_{1,i}\mathcal{U}_i(t) - u_{2,i}C_i(t), \qquad (10.23)$$

where $u_{1,i}$ and $u_{2,i}$ are the weights corresponding to the utility $\mathcal{U}_i(t)$ and $C_i(t)$ is the cost of accessing the spectrum due to the price paid to a licensed user.

The agents at both the unlicensed and licensed users use a reinforcement learning algorithm [610] (see Section 6.2). In this case, the concept of propensity for the user to take any action is used for which the user will be reinforced to take the action which yields higher surplus. For an unlicensed user, the updates to the propensity $q_{d_c,l,i}(t)$ and the probability $\rho_{d_c,l,i}(t)$ to select the spectrum with degree of control d_c for duration τ

are as follows:

$$q_{d_c,\tau,i}(t+1) = q_{d_c,\tau,i}(t) + \beta_u \left(s_i(t) - \tilde{s}_i(t)\right), \tag{10.24}$$

$$\tilde{s}_i(t+1) = \tilde{s}_i(t) + \alpha_u (s_i(t) - \tilde{s}_i(t)), \tag{10.25}$$

$$\rho_{d_c,\tau,i}(t) = \frac{\exp\left(q_{d_c,\tau,i}(t)\right)}{\sum_{d'_c,\tau'} \exp\left(q_{d'_c,\tau',i}(t)\right)}, \tag{10.26}$$

where β_u and α_u are the learning rates of the reinforcement algorithm. For a licensed user, the updates are as follows:

$$q_{m,n}(t+1) = q_{m,n}(t) + \beta_l \left(r_n(t) - \tilde{r}_n(t)\right), \tag{10.27}$$

$$\tilde{r}_n(t+1) = \tilde{r}_n(t) + \alpha_l \left(r_n(t) - \tilde{r}_n(t)\right), \tag{10.28}$$

$$\rho_{m,n}(t) = \frac{\exp\left(q_{m,n}(t)\right)}{\sum_{m'} \exp\left(q_{m',n}(t)\right)}, \tag{10.29}$$

where α_l and β_l are the learning rates, $r_n(t)$ is the revenue of selling spectrum to an unlicensed user, and $q_{m,n}(t)$ is the propensity of a licensed user n to choose market m at time t. Here, the markup choice $m_n(t)$ is a feasible price per unit of spectrum and it can be defined as follows:

$$m_n(t) = \frac{p_n(t) - c_n}{p_n(t)}, \tag{10.30}$$

where $p_n(t)$ is the price offered by licensed user n at time t, and c_n is the fixed cost to share the spectrum with an unlicensed user.

Note that a licensed user has an opportunistic behavior. This opportunistic behavior will influence a licensed user to sell the spectrum to many unlicensed users. Although this opportunistic behavior will result in interference among multiple unlicensed users choosing to buy spectrum opportunities from this licensed user, it will generate more revenue to a licensed user. Different licensed users may have different levels of opportunistic behavior. In particular, the higher the level of opportunistic behavior of a licensed user is, the more interference the unlicensed users will encounter. Therefore, on the unlicensed users' side, an unlicensed user has to develop and maintain trust information on all the licensed users. Then, this trust will be used by an unlicensed user to choose the spectrum. This trust is defined based on Klos's agent-based modeling of trust [611]. In particular, trust is the confidence that an unlicensed user has in a licensed user from which it is buying the spectrum opportunities, which allows an unlicensed user to access the spectrum opportunities exclusively without experiencing any harmful interference. The trust can be determined from

$$\mathcal{TR} = b + (1-b)\left(1 - \frac{1}{cx + 1 - c}\right), \tag{10.31}$$

where b is the base-level of trust, x is the number of consecutive allocations without interference, and c is a constant. This trust has a value between 0 and 1, where 0 indicates that an unlicensed user has no trust and 1 indicates that it has full trust in a particular licensed user. As shown in Figure 10.7, the trust increases as the number of consecutive allocations without interference increases. The rule to update the trust for an unlicensed

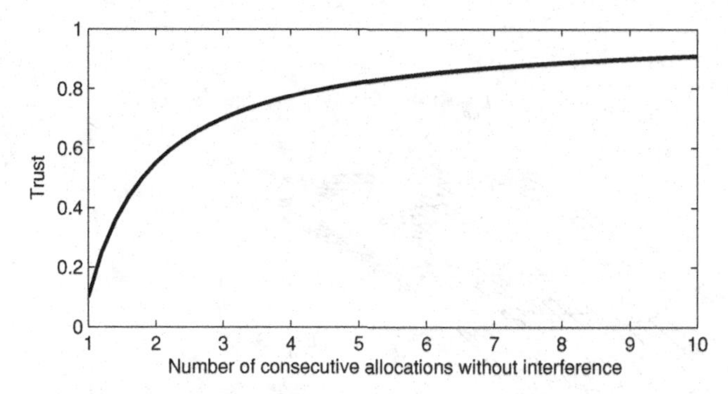

Figure 10.7 Trust versus the number of consecutive allocations without interference.

user is as follows. If an unlicensed user i experiences interference during spectrum access, the trust in the corresponding licensed user n is reduced by half. If the trust falls below 0.3, an unlicensed user i will buy the spectrum from a licensed user n with a probability of 0.5. However, if the unlicensed user does not experience any interference, the trust is updated according to (10.31).

Simulation results showed that when the number of unlicensed users is small, they will access the spectrum using the commons-use model since its cost is the lowest (i.e. no price to pay). However, when the number of unlicensed users increases, the level of interference increases, and some unlicensed users will start to access the spectrum using the exclusive-use model. Although these unlicensed users have to pay a higher cost, the utility from the spectrum access is higher due to lower interference. A similar effect is observed when the transmission range and QoS requirements of unlicensed users increase. When the transmission range increases, there is a higher probability that the transmissions of unlicensed users will interfere with each other. Also, when the level of trust between unlicensed users and licensed users increases in the network, an unlicensed user will have more choices to access spectrum opportunities from other licensed users with higher trust. Therefore, the licensed users will have to compete with each other by offering higher quality of spectrum access (i.e. low interference) to the unlicensed users.

10.1.5 Biologically inspired dynamic spectrum access

To solve the problem of distributed dynamic spectrum access, a biologically inspired spectrum sharing algorithm (BIOSS) was proposed in [544]. This algorithm is based on the adaptive task allocation mechanism in an insect colony model. The objective of the model is to perform channel selection for unlicensed users. The model is constrained by the interference and maximum allowable transmit power (i.e. the underlay spectrum access model). An unlicensed user is considered to be an insect and a cognitive radio network is considered to be an insect colony. The task of an individual insect is to select a channel to be accessed.

The problem of channel selection for an unlicensed user is similar to the task allocation as stated in [544] as follows: "in insect colonies, individuals sense the environment to

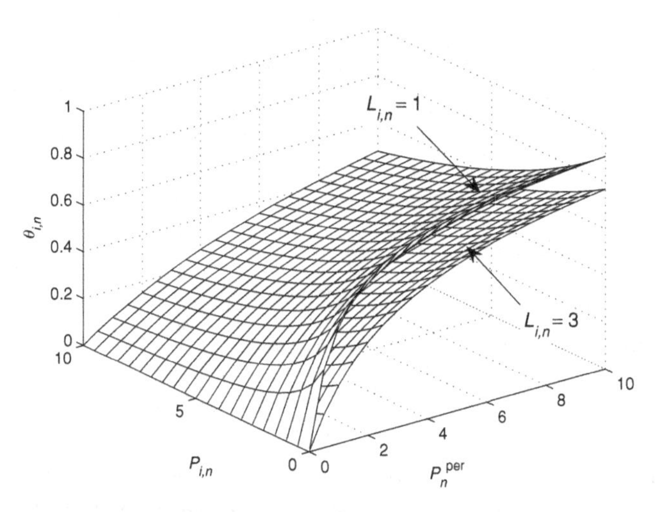

Figure 10.8 Channel selection probability in BIOSS.

detect the tasks and then the detected tasks are performed simultaneously by individuals which are better equipped for the task." In this case, the fitness of an individual to the task is determined by the response threshold [612]. This response threshold is used when a stimulus is observed by an individual. When a stimulus is higher than this threshold, a particular task is performed. This stimulus is defined as the intensity of an activator corresponding to a particular task. For example, a stimulus can be a chemical concentration, pressure, or sound sensed by an individual. In a cognitive radio network, this stimulus is the estimated permissible power P_n^{per} for channel n. The response threshold is defined as the required transmit power $P_{i,n}$ for an unlicensed user i on channel n. The channel selection probability (i.e. the probability of performing a task by an individual in an insect colony) can be expressed as follows:

$$\theta_{i,n} = \frac{\left(P_n^{\text{per}}\right)^{\gamma}}{\left(P_n^{\text{per}}\right)^{\gamma} + \alpha(P_{i,j})^{\gamma} + \beta\,(L_{i,n})^{\gamma}}, \tag{10.32}$$

where $\gamma > 1$ defines the steepness of the channel selection probability $\theta_{i,n}$, α and β are the positive constants, and $L_{i,n}$ is a learning factor. This learning factor is updated according to the QoS performance perceived by an unlicensed user, i.e.

$$L_{i,n} = \begin{cases} L_{i,n} - \zeta_0, & \text{QoS requirement is satisfied} \\ L_{i,n} + \zeta_1, & \text{otherwise,} \end{cases} \tag{10.33}$$

where ζ_0 and ζ_1 are the remember and forget rates, respectively. An example of this channel selection probability is shown in Figure 10.8 for $\alpha = \beta = \gamma = 1$.

The distributed BIOSS algorithm performed by each unlicensed user works as follows:

1. An unlicensed user determines a set of accessible channels.
2. An unlicensed user estimates the permissible power P_n^{per} for all channels n.

3. For an unlicensed user, the learning factors corresponding to all channels are initialized to the same value.
4. An unlicensed user computes the channel selection probability $\theta_{i,n}$ for all channels.
5. An unlicensed user selects the channel with the highest $\theta_{i,n}$. If an unlicensed user has a multi-interface capability, a set of channels with the highest channel selection probability is selected for data transmission. An unlicensed user observes the QoS performance and updates the learning factor according to (10.33).
6. An unlicensed user periodically observes the permissible power and repeats the above steps.

It is observed that if the positive difference between estimated permissible power P_n^{per} and the required transmit power $P_{i,n}$ increases, the channel selection probability $\theta_{i,n}$ increases, and an unlicensed user i is likely to access this channel n. If the estimated permissible power is small, this channel is likely to be allocated to an unlicensed user with small required transmission power (i.e. small transmission demand). However, if the estimated permissible power is large, this channel is likely to be allocated to an unlicensed user with either small or large required transmit power. Simulation results showed that high channel utilization can be achieved which is independent of the number of available channels. This is due to the fact that the distributed BIOSS algorithm can balance the channel access of multiple unlicensed users according to the permissible and required transmission power.

10.1.6 Secondary spectrum access

In a cellular network, a primary service provider (i.e. seller) can sell a license of service or equivalently sell the spectrum to be used by another service provider (i.e. buyer) in a particular region. In this case, the secondary pricing of the spectrum can be formulated as an optimization problem. This optimization can be solved to obtain an optimal price which maximizes the profit [613] of a seller.

Let $\mathbb{A}$ denote the set of all service areas covered by a seller. A seller can sell the spectrum opportunities to a buyer which will be used in a set of areas denoted by $\mathbb{A}_b$ (Figure 10.9). The service region of the seller and the buyer (i.e. $\mathbb{A}_s$ and $\mathbb{A}_b$) are non-overlapping (i.e. $\mathbb{A}_s \cup \mathbb{A}_b = \mathbb{A}$ and $\mathbb{A}_s \cap \mathbb{A}_b = \emptyset$). The interference between area i and j is denoted by a weight $w_{i,j}$. For CDMA-based spectrum access, it has to be ensured that

$$\sum_{i \in \mathbb{A}} u_i w_{i,j} \leq \gamma_j, \tag{10.34}$$

where u_i is the number of active users in area i. In particular, the interference from all users in areas to area j must be less than the threshold γ_j. If this interference constraint cannot be met, a new user will be blocked.

Let $\hat{\lambda}_i$ denote the arrival rate of new users in area i and $P_{\text{bl},i}$ denote the blocking probability. The long-term average rate of revenue generated by a service provider per

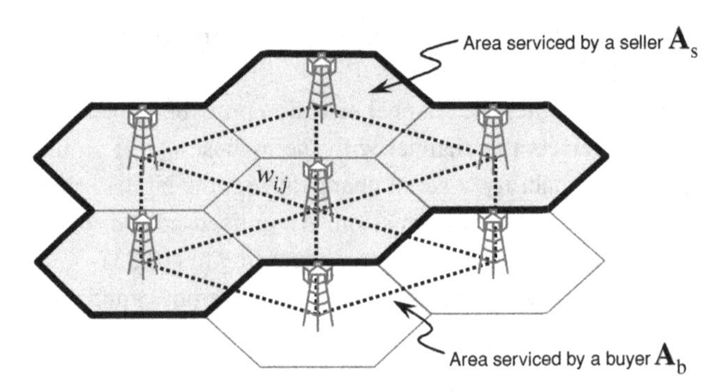

Figure 10.9 Network model for secondary spectrum access.

unit time can be expressed as follows:

$$\mathscr{R}(\hat{\boldsymbol{\lambda}}) = \sum_{i \in \mathbb{A}} (1 - P_{\text{bl},i}(\hat{\boldsymbol{\lambda}}))\hat{\lambda}_i, \tag{10.35}$$

where $\hat{\boldsymbol{\lambda}}$ is a vector of arrival rate $\hat{\lambda}_i$.

A seller sells the spectrum with price p to a buyer to be used in a region $\mathbb{A}_b$. This price will determine the service demand in terms of user arrival rate of area i. This demand is denoted by $\alpha_i(p)$. For a seller, the overall network demand after selling the spectrum can be expressed as follows:

$$\lambda_i(p) = \begin{cases} \alpha_i(p), & i \in \mathbb{A}_b \\ \hat{\lambda}_i, & i \in \mathbb{A}_s. \end{cases} \tag{10.36}$$

Let $\boldsymbol{\lambda}(p)$ denote a vector (with elements $\lambda_i(p)$) which represents network demand. There are three possible pricing schemes for a seller, namely, flat price, price per demand, and price per accepted demand. The expected rate of revenue of the lease with price p can be determined according to these pricing schemes:

- *Flat price:* The flat price is applied if the price is lower than the value of the service. This value can be represented by a random variable V. The expected rate of generating revenue of a seller given a flat price p is $\mathscr{F}(p) = p \times \Pr(V > p)$, where $\Pr(V > p)$ is the probability that the value is larger than the flat price.
- *Price per demand:* In this scheme, a seller sets a price per unit of demand (i.e. new user). The revenue rate of a seller is $\mathscr{F}(p) = p \sum_{i \in \mathbb{A}_b} \alpha_i(p)$.
- *Price per accepted demand:* In this scheme, a seller charges a price according to the interference from the accepted new user. The revenue rate of a seller is $\mathscr{F}(p) = p \sum_{i \in \mathbb{A}_b} (1 - P_{\text{bl},i}(\boldsymbol{\lambda}(p)))\alpha_i(p)$.

If a seller sells the spectrum to a buyer, the remaining rate of revenue generation in region $\mathbb{A}_s$ (Figure 10.9) is

$$\hat{\mathscr{R}}(\boldsymbol{\lambda}(p)) = \sum_{i \in \mathbb{A}_s} (1 - P_{\text{bl},i}(\boldsymbol{\lambda}(p)))\lambda_i(p). \tag{10.37}$$

Therefore, the cost of a seller in selling the spectrum becomes

$$\mathscr{C}(p) = \mathscr{R}(\hat{\boldsymbol{\lambda}}) - \hat{\mathscr{R}}(\boldsymbol{\lambda}(p)). \tag{10.38}$$

In this case, the optimal price can be obtained from

$$\text{maximize:} \quad \mathscr{P}(p) = \mathscr{F}(p) - \mathscr{C}(p). \tag{10.39}$$

To optimize the profit using a standard optimization technique, a differentiable function of blocking probability can be used as follows [614]:

$$P_{\text{bl},i}(\boldsymbol{\lambda}) = 1 - \prod_{j \in \mathbb{A}} (1 - b_j(\boldsymbol{\lambda}))^{w_{i,j}}, \tag{10.40}$$

$$b_j(\boldsymbol{\lambda}) = \text{Erl}\left((1 - b_j(\boldsymbol{\lambda}))^{-1} \sum_{i \in \mathbb{A}} w_{i,j} \lambda_i \prod_{k \in \mathbb{A}} (1 - b_k(\boldsymbol{\lambda}))^{w_{k,i}}, \gamma_j \right), \tag{10.41}$$

where $\text{Erl}(\cdot, \cdot)$ is the Erlang blocking formula.

For a flat price scheme, the optimal price can be expressed as follows:

$$p^* = \sum_{i \in \mathbb{A}_b} (1 - P_{\text{bl},i}(\boldsymbol{\lambda}(p^*)))\, \alpha_i(p^*) \beta_i(p^*), \tag{10.42}$$

$$\beta_i(p^*) = \frac{\epsilon_i(p^*)}{\mathscr{F}'(p^*)} \sum_{j \in \mathbb{A}} w_{i,j} \frac{d\hat{\mathscr{R}}(\boldsymbol{\lambda}(p^*))}{d\gamma_j}, \tag{10.43}$$

where $\mathscr{F}'(p^*)$ is the derivative of the expected revenue rate function, and $\epsilon_i(p^*) = p^* \alpha_i'(p^*)/\alpha_i(p^*)$ is the price elasticity of the demand in area i. Note that the term $w_{i,j}(d\hat{\mathscr{R}}(\boldsymbol{\lambda}(p^*))/d\gamma_j)$ determines the decrease in revenue if a new user in area i is accepted into service.

For a price per demand scheme, the optimal price can be expressed as follows:

$$p^* = (1 - P_{\text{bl},i}(\boldsymbol{\lambda}(p^*))) \left(1 + \epsilon_i^{-1}(p^*)\right)^{-1} \sum_{j \in \mathbb{A}} w_{i,j} \frac{d\hat{\mathscr{R}}(\boldsymbol{\lambda}(p^*))}{d\gamma_j}. \tag{10.44}$$

In this case, the optimal per-demand price is proportional to the marginal cost of a seller due to the accepted new user.

For a price per accepted demand, the optimal price can be obtained from

$$p^* = \left(1 + \epsilon_i^{-1}(p^*)\right)^{-1} \sum_{j \in \mathbb{A}} w_{i,j} \frac{\mathscr{Q}(\boldsymbol{\lambda}(p^*))}{d\gamma_j}, \tag{10.45}$$

$$\mathscr{Q}(p) = \sum_{i \in \mathbb{A}} (1 - P_{\text{bl},i}(\boldsymbol{\lambda}(p^*)))\, \lambda_i(p) Y_i(p), \tag{10.46}$$

$$Y_i(p) = \begin{cases} p, & i \in \mathbb{A}_b \\ 1, & i \in \mathbb{A}_s. \end{cases} \tag{10.47}$$

Numerical studies revealed that the proposed optimal pricing can achieve higher profit for a seller than that of a simple pricing scheme with an objective to maximize only the revenue. The difference of the profits from the proposed optimal pricing and simple pricing becomes larger as the demand increases. This is due to the fact that the proposed optimal pricing scheme considers the impact of the interference.

10.1.7 Distributed rule-regulated spectrum sharing

The distributed rule-based spectrum access method was proposed in [545] to maximize the channel utilization and fairness among unlicensed users in a cognitive radio network. While in [545], the algorithm requires explicit collaboration through message exchange and negotiation among the licensed users, in [546], the algorithm was improved by observing the channel activity. In this case, since an unlicensed user can make a decision on channel access locally and independently, the communications overhead can be reduced significantly. To solve the problem of channel allocation (which is an NP-hard problem [253]), a heuristic rule-based algorithm was proposed with the objective of achieving a fast convergence speed in a practical system with real-time resource allocation.

An exclusive-use spectrum access model was considered. There are I unlicensed users sharing N channels in this cognitive radio network. The transmit power is fixed and if two unlicensed users transmit on the same channel at the same time, interference will result in collision and packet loss. It is assumed that each unlicensed user can identify conflicting peers using interference detection techniques (e.g. as in [615, 550]). Let $v_{i,n}$ denote the channel availability. In particular, $v_{i,n} = 1$ if channel n is available to an unlicensed user i, and $v_{i,n} = 0$ otherwise. Also, let $x_{n,i}$ denote the channel allocation variable for which $x_{n,i} = 1$ if channel n is allocated to an unlicensed user i. The optimization problem for this spectrum sharing model can be defined as follows:

$$\text{maximize:} \sum_{i=1}^{I} \log \left(\sum_{n=1}^{N} x_{n,i} \times v_{i,n} \right). \tag{10.48}$$

Since the optimization problem defined in (10.48) is NP-hard, a heuristic algorithm was proposed. This algorithm is composed of five rules defined for an unlicensed user to access the spectrum. The procedure of the rule-based spectrum sharing algorithm is shown in Figure 10.10. First, the unlicensed user performs channel sensing to identify a set of available channels (i.e. $v_{i,n}$). Then, the unlicensed user checks the rules according to the available information (e.g. channel availability). If none of the rules are satisfied, the unlicensed user selects a new channel. The rules will determine how many and which channels are to be accessed by each unlicensed user in order to achieve the fairness and utilization criteria.

The rules are defined based on the measure of the *poverty line* of an unlicensed user, which can be expressed as follows:

$$\mathscr{P}\mathscr{L}(i) = \left\lfloor \frac{N_i}{d_i + 1} \right\rfloor, \tag{10.49}$$

where N_i is the number of channels available to unlicensed user i, and d_i is the number of conflicting neighboring unlicensed users. This poverty line is an estimation of the number of channels that each unlicensed user should obtain to maximize the fairness.

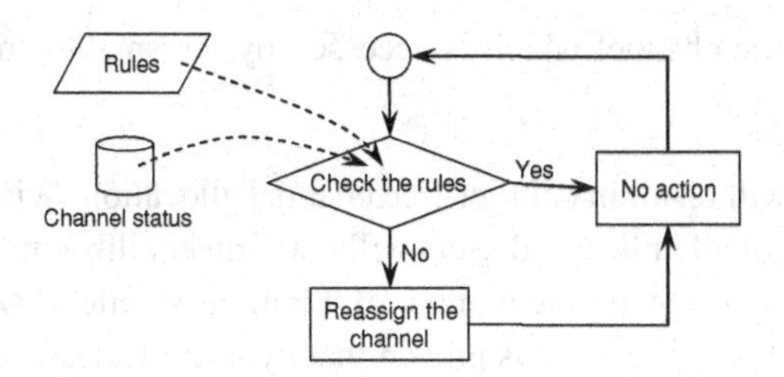

Figure 10.10 Rule-based spectrum sharing.

The rules to implement the scheduling for unlicensed users are as follows:

- *Rule A (Uniform idle preference)*: The rule states that each unlicensed user attempts to obtain the number of channels to be $\tilde{N} = \min_i \mathscr{PL}(i)$. In particular, this value $\tilde{N}$ is optimized for the entire network so that it is a lower bound for all unlicensed users (i.e. the global poverty line). Generally, this $\tilde{N}$ is determined by the unlicensed users with severe interference (i.e. small N_i) or unlicensed users in a crowded area (i.e. large d_i).
- *Rule B (Poverty exact idle preference)*: The rule states that an unlicensed user i selects exactly $\mathscr{PL}(i)$ available channels (i.e. the local poverty line). However, if the number of available channels is less than $\mathscr{PL}(i)$, this unlicensed user will obtain channel from richer users (i.e. unlicensed users with more available channels). If an unlicensed user has conflicts with a poorer user (i.e. unlicensed users with fewer numbers of available channels), this unlicensed user yields a channel and selects another channel. However, since this rule will instruct an unlicensed user to obtain the number of channels equal to the lower bound, the channel utilization could be low. This can be avoided by Rule D.
- *Rule C (Poverty guided idle preference)*: The number of channels that a poorer user can obtain from a richer user is defined as $\max(0, \min(A(i') - \mathscr{PL}(i), \mathscr{PL}(i) - A(i)))$ where $A(i)$ is the current number of channels allocated by user i. This rule prevents a poorer user from obtaining the channel from a richer user, if the number of channels allocated to the richer user will then become smaller than the poverty line.

Rules B and C require an unlicensed user to have information about the channel allocation of neighboring users. This information can be broadcast periodically by each user. However, to avoid information broadcasting, Rules D and E are introduced based on the multiple channel access protocol (e.g. CSMA). In this case, each unlicensed user performs contention detection, which allows multiple unlicensed users to access the same channel.

- *Rule D (Selfish spectrum contention)*: This rule states that each unlicensed user i can obtain a number of channels (i.e. based on the CSMA protocol) up to $\hat{N}$, where N is higher than $\mathscr{PL}(i)$. In this case, the number of neighboring users d_i can be estimated at the MAC layer.
- *Rule E (Poverty guided selfish spectrum contention)*: The rule states that the number of channels that each unlicensed user can use is limited by $\hat{N}_i = \max(\alpha \times \mathscr{PL}(i), 1)$, where $\alpha > 1$. In particular, an unlicensed user can switch to the channel with the

minimum contention (i.e. the channel which is accessed by the smallest number of unlicensed users).

It was proven that Rule A will result in conflict-free channel allocation. With Rules B and C, the channel allocation of all unlicensed users will reach an equilibrium point (i.e. the state with no channel switching) with the number of iterations smaller than or equal to $O(I^2)$. For the contention-based rule, it was proven that by using Rules D and E, the channel selection will reach an equilibrium point after at most $Z \times N$ iterations, where Z is bounded by $O(I^2)$. Also, with Rule D, the lower bound of the number of channels obtained by an unlicensed user can be obtained analytically according to the value of $\hat{N}$.

Several algorithms were proposed to implement the rules. The algorithm for Rules B and D assumes that the information of channel selection of neighboring users is available. Also, it assumes that neighboring users cannot perform channel allocation simultaneously. The algorithm is divided into two phases. In the first phase, the channel is classified as reserved, conflicting, or available. Then, in the second phase, channel allocation is performed based on Rules B and C. The algorithm to implement Rules D and E uses the busy time radio (BTR) [616] as the channel quality indicator. The details of both the algorithms can be found in [546].

Simulation results showed that the algorithm based on Rules B and C performs much better than that based on Rule A. In particular, as the user density increases, the performance of Rule A, which is based on the global poverty line, is limited by the unlicensed user with the lowest poverty line, while Rules B and C are not. Also, it was observed that Rule B performs better than Rule C, since some unlicensed users are allowed to obtain a higher number of channels than the poverty line. The algorithm corresponding to Rule E performs much better than that based on Rule D. Again, channel selection under Rule D is based on a global limit $\hat{N}$, while local limit $\hat{N}_i$ is applied in Rule E. In addition, a comparison with explicit coordination scheme in [545] and centralized assignment scheme in [374, 253] was performed. Although the performance of the proposed rule-based spectrum sharing is lower than the performances of explicit coordination and centralized assignment schemes, the rule-based spectrum sharing can converge to the equilibrium solution at a much faster rate and may not require global information of the network.

10.1.8 Game theory-based spectrum etiquette

The problem of channel selection in a cognitive radio network can be designed based on game theory. In [547], a learning game was developed by considering both cooperation and selfish behaviors of the unlicensed users. Also, a potential game was formulated for the channel selection problem. An underlay spectrum access model was considered in which multiple unlicensed users transmit data simultaneously on the selected channel. There are I unlicensed users and N available channels. The objective of the game formulation is to find the optimal channel for the transmission of each unlicensed user according to the utility functions. The **players** in the game are the unlicensed users. The **strategy** of each player is the selection of a channel. The **payoff** is given by a utility function. Two utility functions were considered for the payoff of an unlicensed user for selfish and cooperative behaviors. The utility function for a selfish user is defined as follows:

$$\mathscr{U}_i^{(s)}(n_i, \mathbf{n}_{-i}) = -\sum_{j \neq i, j=1}^{I} P_j h_{i,j} 1(n_j = n_i), \tag{10.50}$$

where n_i is the channel selected by an unlicensed user i, $\mathbf{n}_{-i}$ is a vector for the channels selected by all unlicensed users except user i, P_i is the transmit power, which is fixed for each unlicensed user, $h_{i,j}$ is the channel gain between unlicensed users i and j, and $1(\text{Cond})$ returns a value of 1 if the Cond is true and returns value of 0 otherwise. In particular, $1(n_j = n_i)$ indicates whether unlicensed users i and j select the same channel or not. The utility function defined in (10.50) determines the individual benefit of an unlicensed user i (i.e. the level of interference to an unlicensed user i). Therefore, this utility function is used for the case of selfish users. On the other hand, the utility function for a cooperative user is defined as follows:

$$\mathscr{U}_i^{(c)}(n_i, \mathbf{n}_{-i}) = -\sum_{j \neq i, j=1}^{I} P_j h_{i,j} 1(n_j = n_i) - \sum_{j \neq i, j=1}^{I} P_i h_{j,i} 1(n_i, n_j). \tag{10.51}$$

Note that the second term in (10.51) accounts for the interference which an unlicensed user i causes to the entire network. Therefore, with this utility function, an unlicensed user must choose a strategy so that not only the interference to itself is minimized but also the interference to other users is minimized.

To implement the game formulation, two approaches were considered, i.e. potential game (see Section 5.7.2) and Φ-no-regret learning (see Section 6.2.5). It was shown that with the cooperative utility function defined in (10.51), the channel selection game is a potential game for which the convergence property was well proven in the literature [261]. In a potential game, a potential function which determines any unilateral change in the payoff of any player is defined. This potential function can provide the information about the path to improve the payoff of the players in the game. Based on the cooperative utility function, the potential function of a channel selection game is defined as follows:

$$\mathscr{P}\mathscr{T}(n_i, {}_{-i}) = \sum_{i=1}^{I} \left(-\frac{1}{2} \sum_{j \neq i, j=1}^{I} P_j h_{i,j} 1(n_j = n_i) - \frac{1}{2} \sum_{j \neq i, j=1}^{I} P_i h_{j,i} 1(n_i = n_j) \right). \tag{10.52}$$

It was proven that (10.52) satisfies all requirements of the potential function in a potential game. An algorithm based on random access was proposed for this potential game. In particular, each unlicensed user randomly executes the channel selection algorithm with probability $\text{Pr}_a = 1/I$ in order to avoid multiple unlicensed users changing their strategies simultaneously. The algorithm requires all unlicensed users to have information about the payoff of other users.

A signaling protocol similar to that of RTS-CTS-ACK in IEEE 802.11 standard was proposed. This protocol is used to measure the interference component in each channel which can be used to compute the utility of other unlicensed users. Also, this protocol is used to broadcast the information about the current strategy to other unlicensed users. All the signaling messages are transmitted over a dedicated control channel. The protocol works as follows:

- START packet is sent by a transmitter of an unlicensed user. This packet includes an estimation for the interference to neighboring users on all channels.
- Upon reception of the START packet, a receiver computes interference and obtains utility for all channels. The channel with the highest utility is selected.
- The decision on selected channels is sent back to the transmitter by packet START_CH.
- The transmitter acknowledges by sending packet ACK_START_CH and starts transmitting on the selected channel.
- All other unlicensed users overhearing packets START_CH and ACK_START_CH update their information about strategy and interference accordingly.

Note that all packets of a signaling protocol are transmitted over a dedicated control channel.

Implementation of the game formulation is based on the no-external-regret learning algorithm with exponential update [617]. Note that this algorithm can converge to a set of equilibria which may not be the same as the Nash equilibrium [618]. Let $U_i^t(n_i)$ denote the cumulative utility obtained by unlicensed user i at time t by selecting channel n_i, which is obtained from

$$U_i^t(n_i) = \sum_{t'=1}^{t} \mathcal{U}_i(n_i, \mathbf{n}_{-i}^{t'}), \tag{10.53}$$

where $\mathcal{U}_i(\cdot)$ is normalized, i.e. $\mathcal{U}_i(\cdot) \in [0, 1]$. The probability of selecting channel n_i at time $t + 1$ can be obtained from

$$\rho_i^{t+1}(s_i) = \frac{(1+\beta)^{U_i^t(s_i)}}{\sum_{s_i'}(1+\beta)^{U_i^t(s_i')}}. \tag{10.54}$$

Note that this learning algorithm can be applied to both selfish and cooperative cases. Although there is no proof to guarantee the convergence of this learning algorithm, simulation studies (e.g. in [619]) showed that the algorithm can converge to the Nash equilibrium if it exists.

Simulation results revealed that with the cooperative utility function, both the potential game and the learning algorithm converge to the solution (i.e. within a few hundred iterations), and the final solutions provide similar performances. Since the learning algorithm does not use any information exchange as the potential game, it achieves slightly lower fairness than the potential game. Compared with the selfish utility function, the cooperative utility function achieves better performance. However, both the selfish and cooperative utility functions yield much better performance in terms of SIR than that of a random approach where a channel is selected by an unlicensed user randomly.

Note that most of the distributed dynamic spectrum access ignored the mobility of the users. However, this mobility information can be used to facilitate the data transmission in a mobile cognitive radio network.

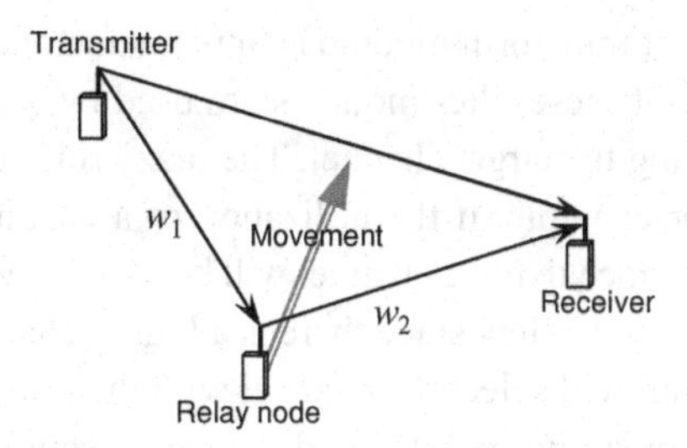

Figure 10.11 Mobility of relay node.

10.1.9 Opportunistic channel selection in cooperative mobility networks

In a cooperative network, the nodes help each other to transmit/relay data of other nodes, so that the performance (e.g. throughput or error) is improved [620, 621]. Specifically, in cooperative mobility, an autonomous node can change its location to provide traffic relay functions [622]. With location change, the transmission environment will change (e.g. shorter distance) and the communications performance can be enhanced. However, since the movement incurs a cost to a node, the mobility of a node will be constrained by the available resource. In [548], a channel selection algorithm was proposed for this cooperative mobility cognitive radio network. With this channel selection algorithm, a node can not only move to a better location, but also choose an available channel to improve the performance of traffic relaying.

With cooperative mobility, the movement planning of a node follows that in [622]. Let us consider a transmitter, a receiver, and a relay node as shown in Figure 10.11. The relay node will move towards the transmitter when the bit-error-rate (BER) in the link between the transmitter and the relay node is higher than or equal to that of the link between the relay node and the receiver. Otherwise, the relay node will move towards the receiver. In a more general scenario, weight w_l is assigned to link l (e.g. in Figure 10.11, w_1 and w_2 are the weights for the links between the transmitter and the relay node and between the relay node and the receiver, respectively). These weights can be chosen to be proportional to the BERs in the corresponding links. The direction of movement of the relay node will depend on the weights for the links corresponding to the transmitters and receivers that the relay node will cooperate with.

Each node has a cognitive radio capability. In particular, a node can choose a channel if the channel is not occupied by a licensed user. To make a proper decision, a node has to estimate the activity on the target channel. The parameters characterizing the activity of a channel are the channel utilization and correlation of channel access by a licensed user. The proposed channel selection algorithm uses these parameters for channel access by an unlicensed user/node. In this algorithm, first the node measures the activity on the channel. The outputs of this measurement are the durations that the channel is idle and occupied. The measurement data is processed by a flip-flop filter based on exponentially-weighted moving average (EWMA) [623], and a wavelet transform to obtain the utilization and the auto-correlation, respectively. The flip-flop filter is used to estimate the average and the variance of the durations that the channel remains idle and occupied. The wavelet transform module is implemented based on the Haar wavelet transform. The outputs of these filter modules are used by a fuzzy logic controller to obtain the cost of accessing the target channel. The fuzzy rules are defined based on the intuition of activity. For example, if the utilization of a channel is high, the cost is high since there is a small chance that the channel will be idle. However, if the degree of auto-correlation is high, the cost is low since there is a higher chance that the channel will remain idle. The algorithm will select the channel with the lowest cost.

There are two scenarios when the cooperative mobility and channel selection algorithm are used. In the first scenario, a node will minimize the number of channel switching, since channel switching incurs delay and energy overhead. In this case, a node will prefer to move to gain better channel quality, and the channel selection algorithm will be only executed by the nodes whose QoS performances are below the target level. Alternatively, in the second scenario, the nodes will minimize their mobility. In this case, all nodes execute the channel selection algorithm. Then, cooperative mobility is applied only to the node whose QoS performance is below the target level.

Simulations were performed assuming that routing in this cooperative network is performed using the standard Dijkstra's algorithm. The weight of a link is defined as $w_l = -\log(1 - BER(l))$, where $BER(l)$ is the BER for link l. It was shown that as the mobility resource increases (e.g. the distance of node movement is large), the performance in terms of BER improves, since a relay node can move towards the best point for traffic relaying. Also, as the number of relay nodes increases, the BER decreases. Due to the flip-flop and wavelet transform filters, the node can accurately estimate the state of the channel. The two different scenarios of combining cooperative mobility and channel selection algorithm achieve different objectives. In the scenario which minimizes the number of channel switching, a relay node consumes a large amount of resources to move to a good location for relaying traffic. However, in the scenario which minimizes mobility, a relay node switches among licensed bands frequently to search for an idle channel. These two scenarios can be applied in different situations depending on the application of this cognitive radio network.

10.2 Distributed signaling protocols for cognitive radio networks

10.2.1 Channel evacuation protocol

In a cognitive radio network, once an unlicensed user detects the presence of a licensed user, an unlicensed user has to stop channel access as soon as possible to avoid interference to a licensed use (Figure 10.12). This is referred to as the *evacuation process*. Also, an unlicensed user can inform other unlicensed users using *warning messages* to stop using the channel. This warning message will alleviate the interference problem for a licensed user since some of the unlicensed users may not be able to detect the presence of licensed users (e.g. due to fading). To support these two functionalities, the embedded spectrally agile radio protocol for evacuation (ESCAPE) was proposed

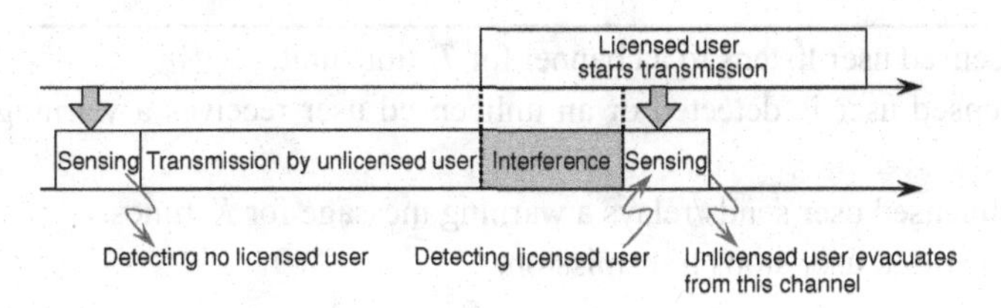

Figure 10.12 Channel evacuation.

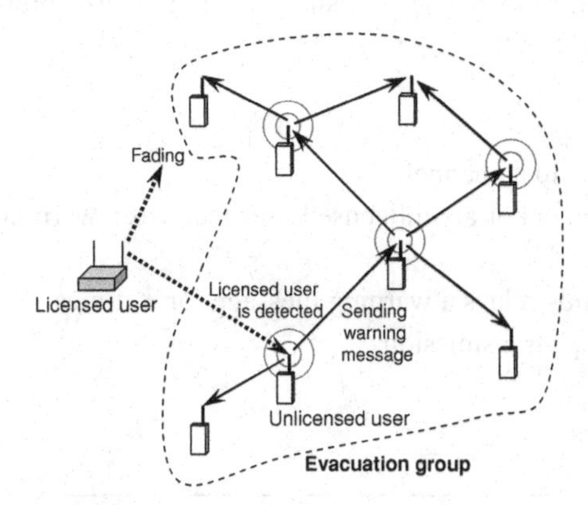

Figure 10.13 Flooding of warning messages.

in [549]. In this protocol, to minimize interference caused to a licensed user and maximize the probability of successful message reception, the warning message is transmitted using spread-spectrum modulation with a predefined spreading code. A simple flooding-based routing scheme is used. Therefore, unlicensed users do not require any coordination among each other and only a small amount of network information needs to be exchanged.

In the system model, there are licensed users and a group of unlicensed users sharing a channel using the spectrum overlay model (Figure 10.13). The unlicensed users communicate in an ad hoc mode, and there is no central controller to gather and report the presence of licensed users. An unlicensed user senses the channel periodically. If no licensed user is detected, the unlicensed user will access the channel. Otherwise, the unlicensed user stops transmission and broadcasts a warning message to other unlicensed users. An out-of-band control channel is used for propagating a warning message and exchanging other control information (e.g. traffic load and link condition). However, this ESCAPE protocol also supports in-band control channel if a dedicated channel for the unlicensed users is not available in the network.

During the initialization period of the ESCAPE protocol, the unlicensed users form an evacuation group which is a group of connectible unlicensed users. The unlicensed users in the same group will share the same warning message (e.g. stop transmission after warning message is received). After an initialization period, the ESCAPE protocol works as follows:

```
1: An unlicensed user listens to a channel for T_l time unit.
2: if a licensed user is detected or an unlicensed user receives a warning message
   then
3:     an unlicensed user sends/relays a warning message for K times.
4:     an unlicensed user stops transmission.
5: end if
6: if an unlicensed user has a packet to transmit then
7:     An unlicensed user transmits a packet using standard MAC protocol (e.g.
       CSMA/CA in IEEE 802.11).
8:     Go to step 1
9: else
10:    an unlicensed user listens to a channel
11:    if a licensed user is detected or an unlicensed user receives a warning message
       then
12:        an unlicensed user sends/relays a warning message for K times.
13:        an unlicensed user stops transmission.
14:    end if
15:    Go to step 1
16: end if
```

The warning messages are transmitted based on ALOHA protocol without backoff and retransmission. If there are overlapping transmissions (e.g. from different unlicensed users), the resolution will be handled by a spreading code.

Given the spreading code length, transmission power, the number of transmission repetitions of a warning message by an unlicensed user, the evacuation delay and failure probability of channel evacuation can be analyzed. Performance evaluation results with fixed and random packet lengths showed that a channel utilization by the licensed users is almost 50 percent while all unlicensed users can successfully evacuate the channel when a licensed user starts accessing the channel. In this case, the evacuation time was measured to be 1–2 ms (with a data rate of 11 Mbps and chip-length of 0.9 μs). Also, it was observed that as the number of transmissions of the warning message by an unlicensed user increases (i.e. K in lines 3 and 12 of the above algorithm), the evacuation delay increases while the evacuation failure probability decreases. Since there are more warning messages to transmit, the delay to finish the evacuation process increases. On the other hand, the chance of successful reception by an unlicensed receiver increases with the number of warning message transmissions.

10.2.2 Distributed channel management in uncoordinated wireless environment

In a commons-use spectrum access model, the spectrum is open for anyone to access. To avoid the *tragedy of commons*, a channel management mechanism can be used. This channel management mechanism can be efficiently designed if all nodes in a network can coordinate through a central controller [624, 625, 626]. However, in an uncoordinated

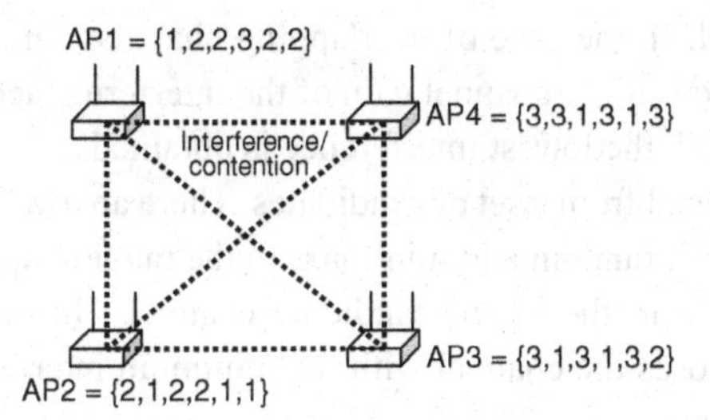

Figure 10.14 A sample network of APs with different channel hopping sequences.

environment which is common in a cognitive radio network, a distributed channel management mechanism would be required. In [550], a distributed channel management method, namely, MAXchop, was proposed by using channel hopping and partially overlapping channels [627]. For IEEE 802.11 networks operating in the unlicensed bands, the objective of this MAXchop channel management is to provide fair throughput for communication between access points and mobile stations.

The shortcoming of a static channel management/allocation method can be illustrated by the example network in Figure 10.14. There are four access points, and each of them is in the transmission range of each other. As a result, one channel cannot be used by two or more access points simultaneously without interference/contention. In the IEEE 802.11 standard, there are three non-overlapping channels. With static channel allocation, two channels can be allocated to the two access points, while the other channel has to be shared by two access points (e.g. using CSMA/CA MAC protocol). Consequently, two access points will gain higher throughput than that of the other two access points which share the same channel. To solve this fairness problem, the channel allocation can be performed periodically for which an access point hops to use different channels for different periods (i.e. channel hopping). In this channel hopping, the channel allocated to each access point is shown in Figure 10.14. In the first period, channels 1 and 2 are allocated to access points 1 and 2, respectively, while channel 3 is shared by access points 3 and 4. In the next period, channels 2 and 3 are allocated to access points 1 and 4, respectively, while channel 1 is shared by access points 2 and 3. In this way, there will be no access point which has to share the channel permanently, and the average number of allocated channels of each access point will be 0.75. Note that when an access point changes the channel, it will inform all associated stations so that in the next period, these stations will communicate with the same access point on the proper channel.

However, in an uncoordinated environment, each access point has to choose the channel hopping sequence to maximize the throughput independently. Equivalently, the objective of channel hopping in the proposed MAXchop algorithm is to minimize the number of channels to be shared (i.e. minimize interference/contention in the network). The proposed algorithm works as follows.

1. The channel hopping sequences of all access points are randomly initialized.
2. Each access point determines the interference for all channels. In the case of non-overlapping channels, this interference is determined from the number of access

points using the same channel. In the case of overlapping channels, interference is computed from the transmit power and channel gain of the interfering access point.
3. A set of candidate channels with the lowest interference is obtained.
4. An access point chooses a channel from a set of candidates. There are two approaches to choosing the channel, namely, random and min-max. In the random approach, the channel is randomly chosen from the set of candidate channels. In the min-max approach, an access point chooses the channel with the minimum interference from those with the highest interference.

The implementation issues for this MAXchop channel management algorithm are as follows:

- *Channel switching*: The stations associated with an access point have to be informed when the operating channel is changed due to the channel hopping. In this case, a periodic beacon message is broadcast to all stations. Similar beacon messages will be included in the IEEE 802.11k standard for radio resource management.
- *Determining interfering access points*: To obtain a hopping sequence, an access point must determine the interfering access points (i.e. Step 2 in the algorithm above). Two approaches can be used, namely, station driven and access point driven. An access point driven approach is easy to implement, since an access point can observe the channel and estimate the interference itself. On the other hand, even though a station driven approach requires a special signaling message to report interference, it can avoid the hidden access point problem. The reporting mechanism of interference from mobile stations to an access point will be included in the IEEE 802.11k standard.
- *Asynchronous hopping*: In the MAXchop algorithm, the length of a hopping sequence for different access points can be different. Therefore, all access points are not required to synchronize their channel hopping. Nonetheless, since the hopping is assumed to be performed infrequently, the impact of asynchronous channel hopping to the performance of an access point is minor.

To evaluate the performance of the MAXchop channel management algorithm, both simulations and experiments were performed. This algorithm was compared with the least congested channel search (LCCS) and centralized algorithm CFAssign-randomized compaction (RaC) [626] algorithms. With LCCS, the channel with the least congestion is chosen by an access point. The LCCS scheme is simple and is used in many commercial access points. In RaC, there is a central controller gathering and optimizing the channel assignment of all access points in a network. It was shown that channel hopping in the proposed MAXchop algorithm can achieve a much higher throughput than that of LCCS. Also, both the throughput and fairness of the MAXchop algorithm are close to those of RaC. However, the MAXchop algorithm does not require coordination and a central controller as in RaC. The performance of TCP with this MAXchop algorithm was also investigated. TCP suffers from frequent timeouts due to delays during channel switching. However, the performance gain in terms of throughput surpasses this overhead, and hence the overall performance of TCP with the MAXchop algorithm is better than that of a simple channel selection mechanism.

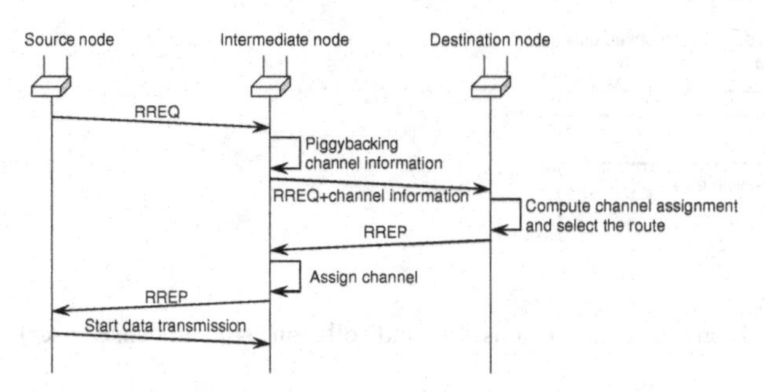

Figure 10.15 Spectrum-aware on-demand routing protocol.

10.2.3 Spectrum-aware on-demand routing protocol

In a multihop (e.g. ad hoc) cognitive radio network, the transmission of the unlicensed users/nodes is based on spectrum opportunity. Therefore, the routing protocol in such a network has to take into account the availability of the spectrum. In [551, 552], a spectrum-aware on-demand routing framework was proposed. The routing mechanism in this framework is based on the ad hoc on-demand distance vector (AODV) protocol [598]. The proposed protocol supports information exchange among unlicensed nodes. The channel switching and backoff delays are used as the metrics to obtain the best route. Here, if more than one node decides to use the same channel, they will experience backoff delays due to channel contention. However, if a node decides to change the channel, it will experience channel switching delay. To optimize the decision of the node, multi-flow multi-frequency scheduling was used. The scheduling mechanism will poll on the active channel (i.e. the channel that is used for data transmission of the unlicensed nodes) to minimize the switching delay.

The multi-flow multi-frequency scheduling first classifies the flows according to their channels. These flows are then allowed to transmit in a round-robin fashion. The delay of a node (DN) can be obtained from

$$\mathrm{DN} = D_{\mathrm{switch}} + D_{\mathrm{backoff}}, \tag{10.55}$$

where D_{switch} is the delay due to channel switching and D_{backoff} is the delay due to channel contention. While the channel switching delay can be obtained from the specifications of the transceiver interface (e.g. [628]), the backoff delay can be obtained from [629]

$$D_{\mathrm{backoff}}(I_n) = \frac{W_0}{(1 - P_{\mathrm{col}})(1 - (1 - P_{\mathrm{col}}))^{\frac{1}{(I_n - 1)}}}, \tag{10.56}$$

where I_n is the number of nodes using channel n, P_{col} is a collision probability, and W_0 is the minimum contention window size. Then, the node computes the delay along the route (i.e. the path delay) to the destination based on (10.55).

The routing protocol has two major steps, i.e. route discovery and route replay (Figure 10.15). In the route discovery step, a RREQ (route request) message is sent by a source node to acquire the possible route to the destination. Also, the channel information (i.e. spectrum opportunities) is piggybacked with this RREQ message. Once the

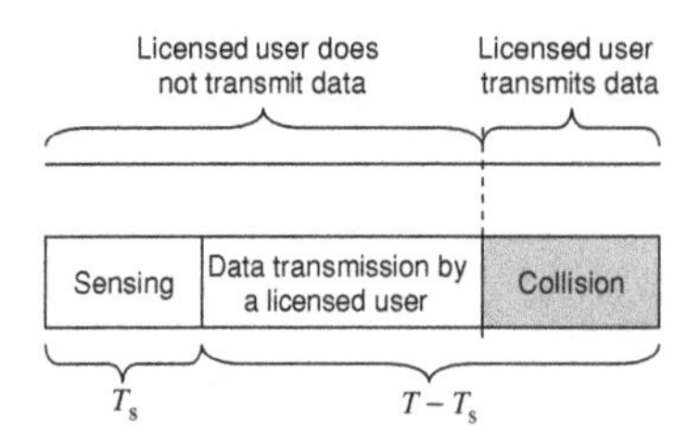

Figure 10.16 Channel sensing, data transmission, and collision (with a licensed user).

destination receives the RREQ message, it will have a full knowledge about the channel availability along the route from a source node. The destination then chooses the route with the lowest delay and also assigns a channel to each node along the route. Then, the destination node sends back a RREP (route reply) message to the source. This message also contains thc information on channel assignment so that the nodes along the route can adjust the channel allocation accordingly. Once this RREP is received by a source node, it starts data transmission.

Through simulations, this spectrum-aware on-demand routing protocol was compared with switch-aware [630, 436] and K-hop distinct [631] protocols. It was shown that the proposed protocol can achieve the lowest end-to-end delay since all switching and backoff delays are taken into account when selecting the route. Note that the switch-aware protocol considers only switching delays. The K-hop distinct protocol always switches to the best channel (e.g. with the lowest contention) without consideration of the switching delay.

10.2.4 Optimization of the transmission period

To access radio channels in a distributed cognitive radio network, the duration of transmission by an unlicensed user has to be optimized to minimize collision with the transmission of a licensed user. This optimization problem was considered in [632, 633, 634]. In [633], the transmission of an unlicensed user is frame-based with frame length T and the length of the sensing period is T_s (Figure 10.16). The average transmission period (i.e. the on state) and idle peiod (i.e. the off state) of a licensed user are denoted by $\overline{T}_{on}$ and $\overline{T}_{off}$, respectively. In this case, the transmission of a licensed user may collide with that of an unlicensed user. That is, a licensed user can start its transmission before an unlicensed user finishes its transmission (Figure 10.16). In this case, if the transmission period of an unlicensed user is long, the collision probability P_{col} increases. However, if this transmission period is short, the throughput of an unlicensed user is not maximized since an unlicensed user has to perform spectrum sensing frequently. To optimize the performance of an unlicensed user, its throughput (i.e. its rate) can be defined as a function of the frame length T and sensing period length T_s as follows:

$$\mathscr{R}(T, T_s) = \frac{T - T_s}{T}(1 - P_{col})r, \tag{10.57}$$

where r is the channel capacity. To maintain accuracy of channel sensing, the length of sensing period is kept fixed. Therefore, the normalized throughput of an unlicensed user can be expressed as $\mathscr{R}(T) = \frac{T-T_s}{T}(1 - P_{col})$.

When the channel is sensed to be unoccupied by a licensed user, an unlicensed user starts its transmission immediately. However, collision can occur during the transmission of an unlicensed user. Let t denote the time required for a licensed user to start transmission. It is assumed that the on and off periods of a licensed user are exponentially distributed. Therefore, the probability density function of t is given by

$$f(t) = \frac{1}{\overline{T}_{off}} \exp\left(-\frac{t}{\overline{T}_{off}}\right). \tag{10.58}$$

The time duration that the collision occurs can be modeled as a random variable $Y(t)$ which is defined as follows:

$$Y(t) = \begin{cases} T - T_s - t, & 0 \leq t \leq T - T_s \\ 0, & t > T - T_s. \end{cases} \tag{10.59}$$

Given the length of frame T, the average time that an unlicensed user experiences the collision can be computed as follows:

$$\begin{aligned} E(Y) &= \int_0^{T-T_s} (T - T_s - t) f(t) \mathrm{d}t \\ &= T - T_s - \overline{T}_{off}\left(1 - \exp\left(-\frac{T - T_s}{\overline{T}_{off}}\right)\right). \end{aligned} \tag{10.60}$$

The collision probability is obtained from

$$P_{col} = 1 - \frac{\overline{T}_{off}}{T - T_s}\left(1 - \exp\left(-\frac{T - T_s}{\overline{T}_{off}}\right)\right). \tag{10.61}$$

Then, the normalized throughput can be expressed as follows:

$$\mathscr{R}(t) = \frac{\overline{T}_{off}}{T}\left(1 - \exp\left(-\frac{T - T_s}{\overline{T}_{off}}\right)\right). \tag{10.62}$$

An optimization problem can be formulated to maximize this normalized throughput. In this case, the normalized throughput function is differentiated with respect to the frame length as follows:

$$\frac{d\mathscr{R}(T)}{dT} = \frac{1}{T^2}\left(-\overline{T}_{off} + (\overline{T}_{off} + T)\exp\left(-\frac{T - T_s}{\overline{T}_{off}}\right)\right) = 0. \tag{10.63}$$

The optimal frame length is

$$T^* = -\overline{T}_{off}\left(1 + \mathscr{W}\left(-\exp\left(-\frac{\overline{T}_{off} + T_s}{\overline{T}_{off}}\right)\right)\right), \tag{10.64}$$

where $\mathscr{W}(\cdot)$ is Lambert's W function [635]. By evaluating the second derivative of $\mathscr{R}(T)$, it can be verified that T^* yields the highest throughput (i.e. the maximum point).

In the case where $T_s = 1$ ms, the variations of collision probability and throughput are shown in Figures 10.17 and 10.18, respectively. As expected, when the frame length

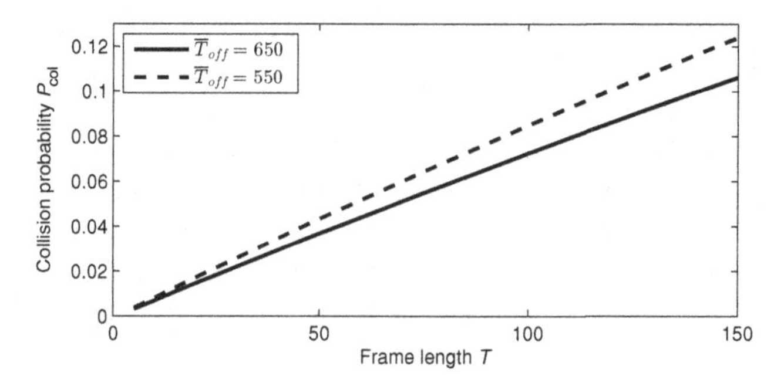

Figure 10.17 Collision probability under different frame lengths.

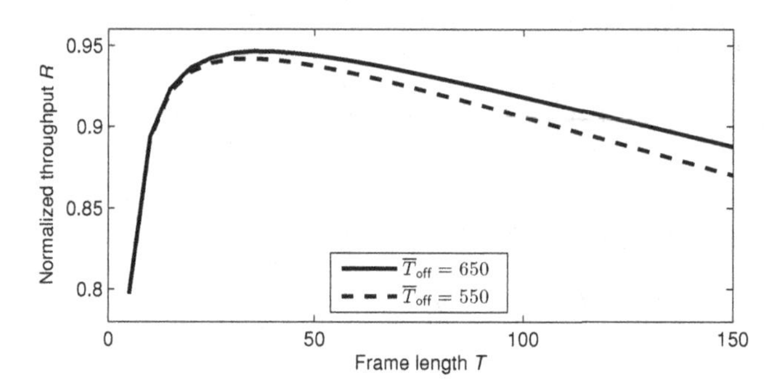

Figure 10.18 Throughput under different frame lengths.

increases, the collision probability increases, since there is a higher chance that a licensed user will start transmission before an unlicensed user finishes its transmission. Also, when a licensed user has a shorter off period, the collision probability increases, since a licensed user transmits data more often. Figure 10.18 shows the increase and decrease in throughput as the frame length increases. There is an optimal point for the frame length at which the highest throughput can be achieved by an unlicensed user. This optimal point can be obtained analytically from (10.64).

10.3 Summary

The unlicensed users in a cognitive radio network can adopt either collaborative or non-collaborative behaviors. If the unlicensed users are non-collaborative, a learning algorithm becomes crucial for an unlicensed user to estimate network information. In this chapter, distributed dynamic spectrum access schemes based on learning algorithms have been reviewed. The learning algorithms are used to build knowledge about both the ambient radio environment and the network based on the observations. This knowledge is then used by the cognitive radio users to make spectrum access decisions. In addition, to support distributed dynamic spectrum access, various signaling protocols are used for exchanging network information between the collaborative unlicensed users. Several of these protocols have been discussed.

11 Economics of dynamic spectrum access: spectrum trading

To design efficient and effective dynamic spectrum access techniques for a cognitive radio network, the related technical aspects (e.g. channel allocation, power control) as well as economic aspects (e.g. pricing, spectrum auction) need to be considered. The economic issues are crucial for cognitive radio networks operating under the exclusive-use spectrum access model, since they define the incentive for licensed users to yield the right of spectrum access to the unlicensed users. Economic issues are also important for dynamic spectrum access based on the shared-use and commons-use models, because they determine the competition and cooperation between the licensed and unlicensed users.

In this chapter, we describe the different economic aspects of dynamic spectrum access in cognitive radio networks. First, the concept of *spectrum trading* is presented, which involves spectrum selling by single or multiple licensed users and spectrum buying by unlicensed users. A taxonomy of the spectrum trading models is presented. The pricing issue for spectrum trading as well as authentication, authorization, and accounting (AAA) issues are discussed. Then, an overview of the economic theories for spectrum trading in a dynamic spectrum access environment is given. These include utility theory, the concept of market-equilibrium, competition in an oligopoly market, and auction theory. A survey on the spectrum trading models based on the above theories is then presented.

11.1 Introduction to spectrum trading

Generally, license for spectrum access is provided to a primary user or service provider through an auction process in a primary market (Figure 11.1). When the allocated spectrum is under-utilized, the licensed user can lease the spectrum in a secondary market to an unlicensed user which temporarily demands the spectrum for a particular service. In the primary market, the process of spectrum allocation is lengthy and inflexible due to the regulatory requirements. Therefore, spectrum leasing in the secondary market, which is not controlled by the government, appears to be an attractive tool to promote efficient use of the radio spectrum. Spectrum leasing (or trading) between the licensed and unlicensed users will be a significant issue in cognitive radio networks.

In economics, trading is defined as a process of exchanging commodity or service in a market. This process can be performed through direct exchange of commodity or

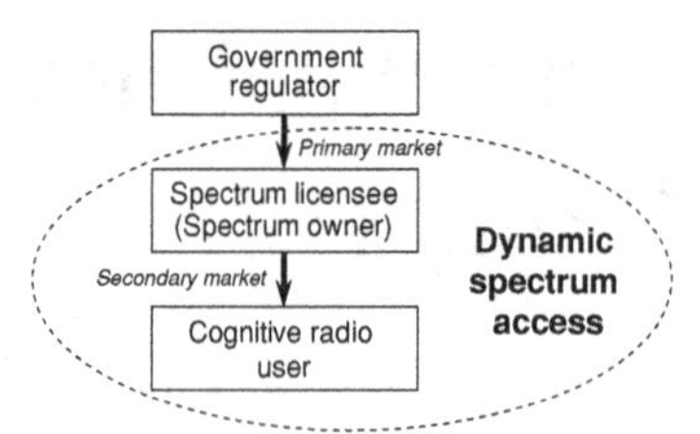

Figure 11.1 Government regulator, spectrum owner, and cognitive radio users.

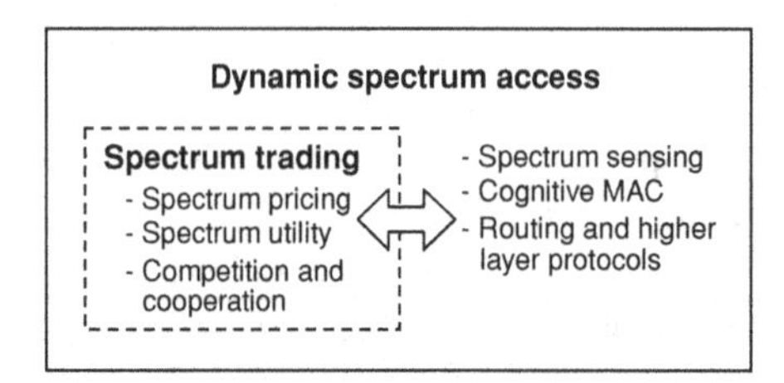

Figure 11.2 Dynamic spectrum access and spectrum trading.

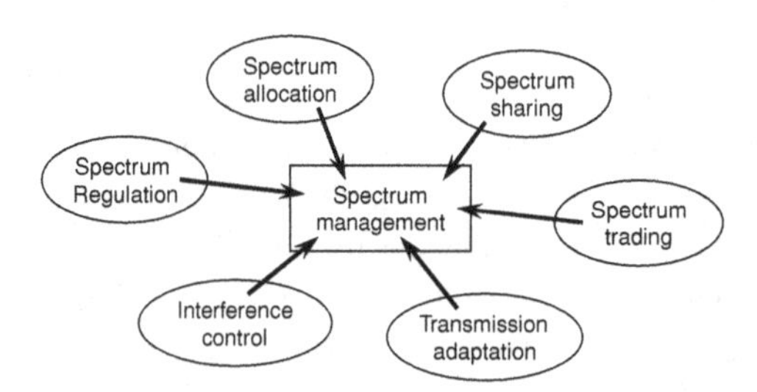

Figure 11.3 Spectrum management.

service (i.e. bartering) or through a medium of exchange which is generally the money. The concept of trading can be applied to spectrum leasing in the secondary market, which constitutes the concept of spectrum trading. Specifically, spectrum trading is defined as a process of selling and buying of spectrum resources in different dimensions (e.g. frequency band, time slot). Pricing is an important issue here for both the licensed user (or primary service provider) selling the spectrum and the unlicensed users (or secondary service provider or secondary user) buying the spectrum.

While dynamic spectrum access encompasses technical functionalities including spectrum sensing at the physical and MAC layers, channel access, routing, and higher layer protocols, spectrum trading can be regarded as its other component which deals with the economic aspects of dynamic spectrum access (Figure 11.2). Spectrum trading can be considered as a component of spectrum management [636] (Figure 11.3), and, therefore, it is required to be integrated with other components in a cognitive radio network. Recall that two major steps in spectrum sharing are spectrum exploration and spectrum exploitation (Figure 11.4). Spectrum trading is a process between spectrum exploration and exploitation. A spectrum seller has to perform spectrum exploration to identify spectrum opportunities. Then, these spectrum opportunities can be sold to the spectrum

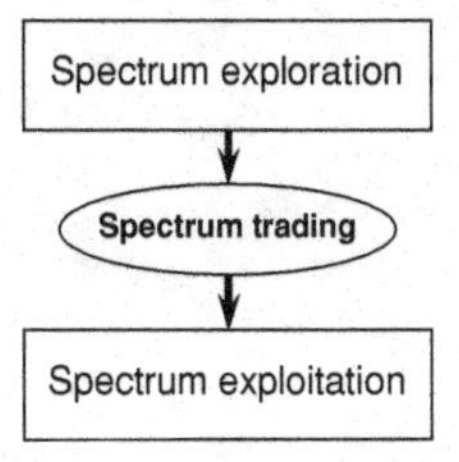

Figure 11.4 Spectrum exploration, spectrum trading, and spectrum exploitation.

buyer. After obtaining the right to access, the spectrum buyer performs the spectrum exploitation step to utilize the spectrum to achieve its objectives under the constraints defined by the spectrum seller.

Two major structures of spectrum trading are as follows:

- *Single seller – monopoly*: In this case, there is only a single seller in the system and this is referred to as a monopoly market. In this market, a seller can maximize its revenue given the demand from the buyers. In this case, the buyers can either compete or cooperate to buy spectrum opportunities from the seller. This behavior (i.e. competition, cooperation) will affect the revenue of a seller and depending on the behavior of the buyers, a seller may adapt the trading parameters accordingly.
- *Multiple sellers – oligopoly*: In this case, there are multiple sellers offering radio spectrum to the market and this is referred to as an oligopoly market. The buyer has a choice to choose the best offer to maximize its satisfaction in both performance and price. Since there is a competition among the sellers, the profit of a spectrum seller is always less than that in a single-seller case.

Alternatively, a spectrum trading market may not have a permanent seller (e.g. as in a commons-use spectrum access model). In this spectrum trading structure, all users have the right to access the spectrum. However, if a particular user (e.g. a renting user) requires to have a spectrum share more than another user (e.g. a leasing user), the other user will need to be compensated. For example, a renting user may provide an incentive to a leasing user and access the spectrum with special right for a certain period of time. When a renting user accesses the spectrum with special right, credit is given to the leasing user. Afterwards, when necessary, the leasing user may use this credit to access the spectrum as a renting user with special rights while another user becomes a leasing user.

In spectrum trading, the objective of a seller is to maximize the revenue/profit, while that of a buyer is to maximize the utility of spectrum usage. However, these objectives generally conflict with each other. This effect can be shown in Figure 11.5. As the seller increases the price to achieve higher revenue, the utility of a buyer decreases due to the higher cost. A similar effect on the QoS performance is observed when the amount of spectrum allocated to an unlicensed user is varied. In particular, as the spectrum size allocated to an unlicensed user increases, the utility of an unlicensed user increases, but the performance of a licensed user degrades. Therefore, an optimal and stable solution for spectrum trading in terms of price and allocated spectrum would be required so that

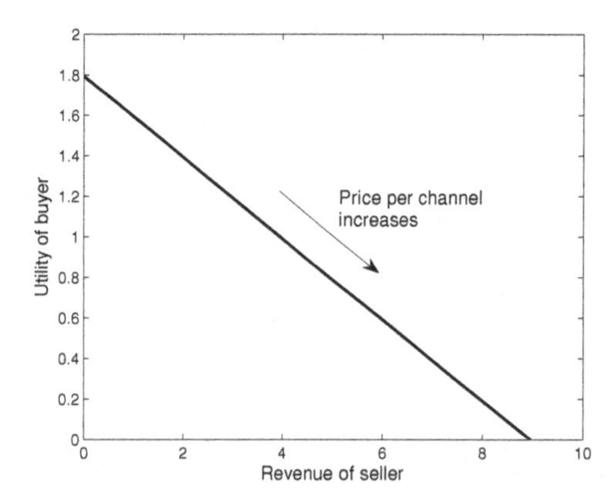

Figure 11.5 Revenue of a spectrum seller versus utility of a spectrum buyer.

the revenue and utility are maximized while both the seller and the buyer are satisfied and do not want to deviate from the solution.

Different techniques can be applied when designing a spectrum trading model to obtain an optimal and stable solution for the spectrum seller(s) and buyer(s) [637].

- *Microeconomic approach*: Microeconomic theory can be used to model spectrum trading in a cognitive radio network with two major entities, namely, the spectrum seller and the spectrum buyer. The solution of this approach is based on market-equilibrium, which gives the spectrum price and allocated spectrum size for which spectrum demand equals spectrum supply. At the market-equilibrium, the profit of the seller and the satisfaction of the buyer(s) are maximized.
- *Classical optimization approach*: Spectrum trading can be formulated as an optimization problem which consists of a single objective under a set of constraints. The objective can be, for example, to maximize profit of the spectrum owner or to maximize the utility of the cognitive radio user, under the constraint on limited performance degradation of primary users due to spectrum sharing or limited interference caused to the primary users.
- *Non-cooperative game approach*: A classical optimization formulation relies on a single objective function and the solution of such a formulation is "system-wise" optimal for the entire system. In a non-cooperative game model, several entities are involved and they have different (and possibly conflicting) interests. The solution of the game model (i.e. the equilibrium solutions) satisfy all of the entities involved.
- *Bargaining game approach*: A bargaining game formulation can be used in situations where cognitive radio entities can cooperate and each entity can influence the action of other entities during spectrum trading. In this bargaining game, the system entities can negotiate and bargain with each other so that a fair and efficient solution can be obtained.
- *Auction approach*: One efficient approach to spectrum trading is the auction approach. In an auction process, the buyers submit their spectrum bids and the profit of a spectrum seller is maximized by allocating spectrum to the buyer(s) submitting the highest bidding price.

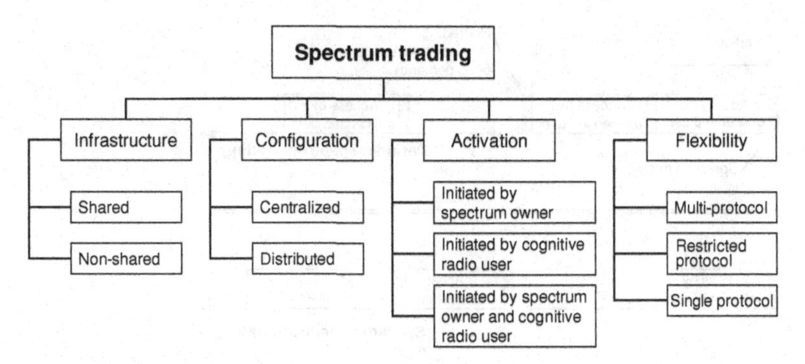

Figure 11.6 Classification of spectrum trading.

11.1.1 Classification and taxonomy of spectrum trading

When developing a spectrum trading model, the following three aspects need to be considered [638]:

- *Mode*: The mode refers to (1) the change of spectrum ownership and (2) the change of use (e.g. different wireless access technologies) due to spectrum trading. After spectrum trading is done, the spectrum ownership will be transferred from the spectrum seller to the appropriate buyer (i.e. change of spectrum ownership). Also, the type of wireless service operating on the traded spectrum can be changed (i.e. change of use).
- *Extent*: It defines the degree of a spectrum owner's rights and obligations which will be transferred to a buyer, i.e. (1) complete transfer and (2) shared. In a complete transfer case, all rights and obligations of spectrum access are completely transferred to the buyer. In contrast, in the shared case, both spectrum owner and buyer share the rights and obligations.
- *Duration*: The duration determines the length of time the traded spectrum can be accessed by a buyer. Different scales of duration can be defined as follows: (1) short-term lease, (2) long-term lease, (3) sale-and-buy-back, and (4) permanent. While the duration of short-term lease may be a few hours, that of long-term lease could be a few months. For the sale-and-buy-back case, the spectrum is sold and can be bought back by a spectrum owner if it is demanded. In the permanent case, the spectrum is permanently sold to a buyer. This buyer can access the spectrum until the license expires.

As shown in Figure 11.6, the spectrum trading models classified based on the different criteria are as follows [638]:

- *Infrastructure*: The spectrum obtained from trading can be used over a shared or a dedicated infrastructure. In the case of shared infrastructure, multiple unlicensed users share the same equipment, while in the case of dedicated infrastructure, each unlicensed user uses its own equipment to utilize the spectrum.
- *Configuration*: Spectrum trading can be centralized or distributed. In a centralized case, a spectrum broker is used to control spectrum trading (e.g. leasing duration

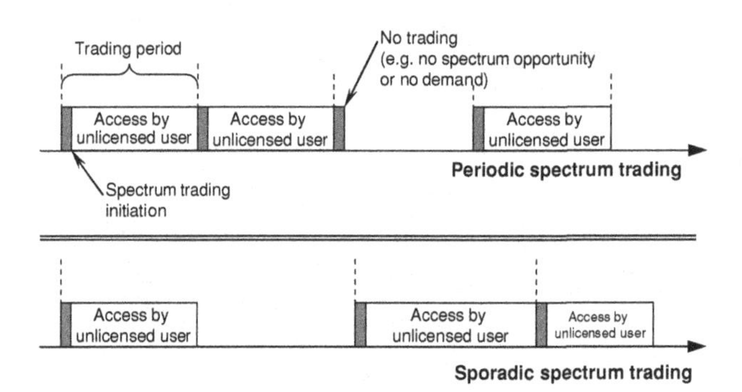

Figure 11.7 Periodic and sporadic spectrum trading.

and pricing) and the transmission parameters (e.g. transmit power). In a distributed case, each of the unlicensed users negotiates with a licensed user for spectrum trading independently. With this distributed model, the unlicensed users can either cooperate or compete to buy the spectrum from a licensed user.

- *Activation*: Spectrum trading can be initiated by a licensed user by observing its spectrum opportunities. In this case, a licensed user sends a request to a group of unlicensed users informing its intention for selling the spectrum. The unlicensed users with spectrum demand will participate in spectrum trading. In contrast, the spectrum trading can be initiated by an unlicensed user [639]. In this case, the request is sent to the licensed users, and the licensed users with spectrum opportunities will join the spectrum trading. Also, spectrum trading can be initiated by both sides. The activation of spectrum trading can be periodic or sporadic. In periodic spectrum trading, spectrum is traded for a fixed time interval. On the other hand, for the case of sporadic trading, spectrum trading can be initiated at any point in time (Figure 11.7).
- *Flexibility*: For spectrum trading, the type of wireless service and technology can be specified. A licensed user can determine a specific protocol or a set of protocols that an unlicensed user can use (i.e. single protocol and restricted protocol cases, respectively). On the other hand, in the multi-protocol case, there is no restriction on the protocol to be used by an unlicensed user.

Another criterion of classification is whether spectrum trading is performed in a non-real-time or real-time manner [640]. In the non-real-time case, spectrum trading is performed before the spectrum is needed or accessed by an unlicensed user. On the other hand, in the real-time case, spectrum trading can be performed in an on-demand basis. In general, the duration of non-real-time spectrum trading is several months long, while that of real-time trading is much shorter (e.g. a few days or hours).

Information management is an important issue in spectrum trading. Each of the sellers and buyers has to determine the information which needs to be exchanged among each other. The information flows required for spectrum trading include the following [638]:

- *Request and acknowledgement messages of spectrum trading from the licensed and unlicensed users*: These messages contain the details of spectrum demand (i.e. from a buyer's perspective) and spectrum supply (i.e. from a seller's perspective) for spectrum access as well as pricing information.

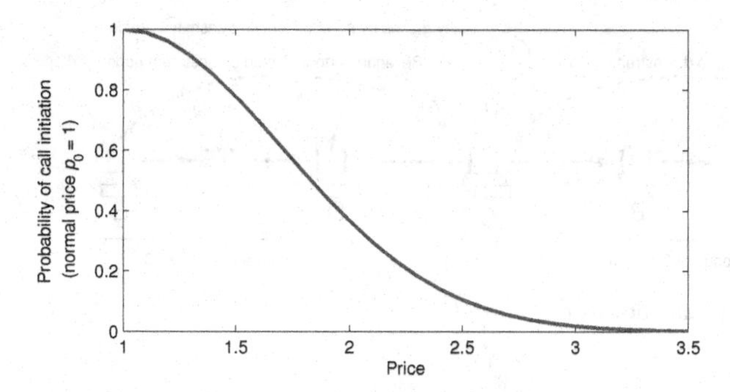

Figure 11.8 Example of demand function.

- *Spectrum access parameters*: This information is generally determined by a licensed user as public information. This set of parameters will be used by an unlicensed user to access the spectrum.
- *Spectrum occupancy information*: A licensed user uses this information to determine the spectrum supply which will be sold to an unlicensed user. This information can be also collected by an unlicensed user to identify the spectrum opportunities and initiate spectrum trading.
- *Report on spectrum access*: This information can be generated by either licensed or unlicensed users to be used by a licensed user. This may contain information such as spectrum utilization and interference level caused to the licensed users.

11.1.2 Radio resource pricing

The problem of radio resource pricing was considered in the literature. For example, pricing schemes were proposed for cellular networks [641, 642, 643, 644, 645] and for multihop relay networks [646, 647, 648, 649, 650, 651, 652, 653, 654]. These papers showed the significance of incorporating economic aspects into the radio resource management for wireless networks.

A pricing scheme can be integrated with the admission control mechanism, which is used to limit the number of users sharing a limited amount of radio resources. Admission control can be used to avoid network congestion which degrades the system performance. In a cellular mobile network, admission control can be used to prioritize handoff calls over new calls by limiting the number of new users. However, limiting the number of new users can lower the revenue of a service provider. Therefore, the admission control parameter(s) need to be optimized to achieve the desired system objective. The problem of pricing and admission control for a cellular network was considered jointly in [641]. In this work, pricing was used to control the user arrival rate. For example, if the price chosen by a service provider is higher than the normal price, a cellular user will be reluctant to initiate a call. As a result, the call initiation is delayed with an expectation from a user that the price in the future will be decreased. In particular, the probability of call initiation decreases as price increases (Figure 11.8). This integrated pricing and call admission control was formulated as a utility maximization problem. Here, the utility was defined as a decreasing function of blocking probability. Simulation results showed

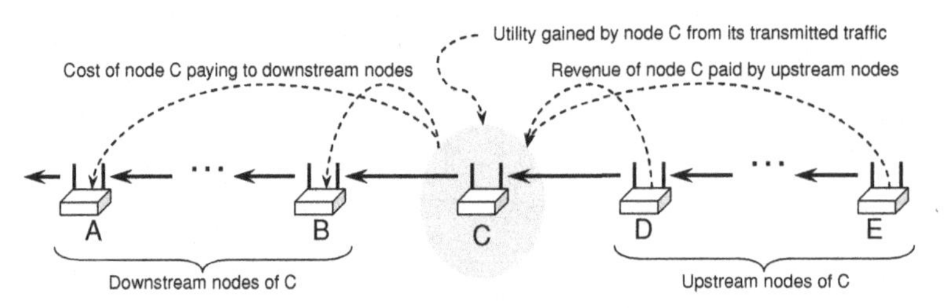

Figure 11.9 Upstream and downstream nodes.

that increasing the price can mitigate the problem of large call blocking rate. Also, when the traffic load (i.e. new call initiation) is low, the price can be decreased to attract more cellular users to access the system. With this adaptive pricing, the utility of a cellular service can be maximized over time-varying traffic load.

The problem of pricing for a wireless ad hoc network was considered in [646]. In particular, a market equilibrium in terms of price and bandwidth allocation of relaying traffic was studied. In the ad hoc network model, a node in an ad hoc network can transmit its own traffic and also relay the traffic from an upstream node. For example, as shown in Figure 11.9, nodes D–E are the upstream nodes of C. In this case, each node will charge its upstream node. Each node has to decide on an optimal set of parameters including the amount of bandwidth to be allocated to its own traffic and relayed traffic as well as the price to be charged to the relayed traffic (Figure 11.8). The payoff of each node is defined as a function of the utility obtained from transmitting its own traffic plus the revenue gained from relaying the traffic of other upstream nodes minus the price paid to downstream nodes along the route to the destination. The solution in terms of equilibrium was obtained which maximizes the total benefit of the node.

However, in a cognitive radio network, the problem of pricing is somewhat different from that in a traditional wireless network due to spectrum sharing and adaptability of the licensed and unlicensed users. A licensed user can charge a price to an unlicensed user for spectrum access. This price can be dynamically adjusted according to the availability of spectrum opportunity which is a function of traffic load in the licensed network and the demand from the unlicensed users. This demand depends on the current number of ongoing sessions and applications used by the unlicensed users. To support a secure pricing scheme, an authentication mechanism is required to verify the users to access the spectrum. An authorization mechanism is used to grant the access to the users. An accounting mechanism is used to record the usage statistics and calculate the price to be charged to the users.

11.1.3 Authentication, authorization, and accounting in spectrum trading

To support secure spectrum trading, an authentication, authorization, and accounting (AAA) framework is required. The AAA framework needs to support real-time information exchange and negotiation among the licensed and unlicensed users. Also, in a cognitive radio network, this framework has to be designed by considering the fact that

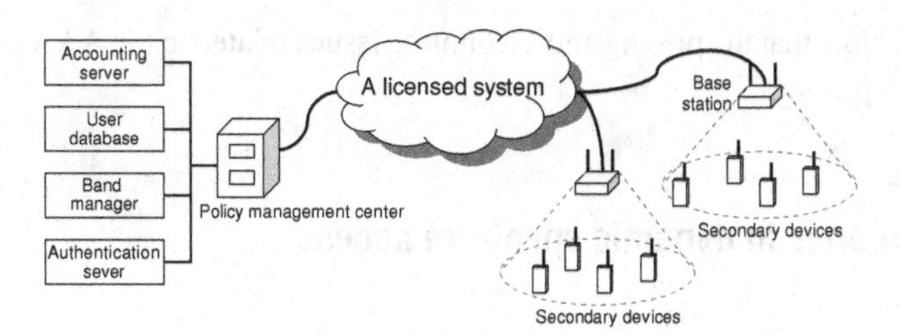

Figure 11.10 Authentication, authorization, and accounting architecture.

the communications among the unlicensed users cannot be fully controlled by a licensed user. Mechanisms are required to manage transmission from unlicensed users as well as to prevent unauthorized spectrum access.

In [655], an AAA framework was proposed for a cognitive radio system for which an unlicensed user buys radio spectrum from a licensed system. The architecture of this AAA framework is shown in Figure 11.10. In a licensed system, there is a policy management center (PMC) to manage spectrum access by unlicensed users. This policy management center is composed of an authentication server (AS), a user database, a band manager (BM), and an accounting server. The AAA framework works as follows. First, the authentication server verifies an unlicensed user. Then, the band manager allocates the available spectrum to a group of authorized unlicensed users. At each base station of a licensed system, an agent is installed to receive and forward control messages between the policy management center and an unlicensed user. Also, this agent monitors the spectrum usage, and reports to the band manager if necessary.

Multiple unlicensed users can be grouped together. A delegate device (DDev) is used as a representative of the unlicensed users in the same group. This delegate device negotiates with the authentication server through a secure control channel. After the authentication server has authorized the access, the unlicensed users are informed of this authorization through the control channel. While there are a few authentication methods (e.g. challenge/response interactive, authentication using synchronized data, one-way authentication using a password [656, 657]), the proposed AAA framework uses an authentication method based on an asymmetric crypto-system, in which a group of unlicensed users has a pair of public/private keys. When an unlicensed user wants to access the spectrum, the delegate device of that unlicensed user sends a request message which is encrypted by a public key to the authentication server. Then, an authentication algorithm decrypts the request message by using the private key and verifies information in the message. After authentication, the authentication server contacts the band manager to check the status of spectrum usage, and allocates free spectrum to the unlicensed users. A registration ticket is issued by a band manager and this ticket is broadcast through a control channel to the unlicensed users. To access the spectrum, an unlicensed user uses a received ticket to register to an agent at the base station. This registration is based on the authenticated Diffie–Hellman principle [658] to prevent an unauthorized user from accessing the spectrum. In addition, a service management mechanism for an unlicensed user was developed to support handoff, spectrum monitoring, and registration ticket renewal. Note that the pricing and accounting issues related to the AAA framework were studied in [659].

11.2 Economic theories in dynamic spectrum access

11.2.1 Utility theory

The concept of utility can generally be used to quantify the satisfaction of a cognitive radio entity. The general meaning of utility is "the usefulness of something, especially in a practical way" [660]. In economics, this term has a more specific definition, i.e. "utility is a consumer's perception of happiness or satisfaction [661]." In economics, the general term of usefulness is defined as the satisfaction of an individual. The concept of utility was first introduced by Jeremy Bentham in the early 1900s, in which pain and pleasure were used to explain the behavior of an individual. Also, D. Bernoulli provided the basis of economic behavior based on utility. In the 1870s, William S. Jevons and Leon Walras introduced the concept of measurable utility. While William S. Jevons used utility to represent the satisfaction from the exchange of goods and services, Leon Walras used the concept of utility for demand-curve analysis. In modern economics, utility is essentially used to express the personal feelings of consumers arising from the consumption of goods and services. This feeling includes pleasure, satisfaction, and lack of pain, which are all very subjective and difficult to measure quantitatively. Consumer demand for commodities and services is influenced by these feelings.

Utility and preference

The fundamental concept of utility is *preference*, which indicates a consumer's preference among different options. For example, a consumer may prefer cake to bread. In other words, this consumer has a higher level of satisfaction on cake than that on bread. When the utility is used to represent the level of satisfaction, cake yields higher utility than bread. It is possible that a consumer has the same preference for different choices. For example, a consumer may have the same level of satisfaction for cake and ice cream. In this case, cake and ice cream have the same utility. However, with the preference, we cannot quantify how much a consumer prefers one commodity to another. Namely, a consumer cannot say cake is preferred more than bread twice, and a consumer can only rank the satisfaction on the different choices. This type of utility is referred to as *ordinal utility*, and does not have any numerical meaning. For example, let the ordinal utility of cake be 2 while that of bread is 1. The numbers "2" and "1" do not imply that a consumer wants cake as twice as bread, but they indicate that cake is preferred to bread since 2 is bigger than 1. Clearly, the concept of ordinal utility is limited especially when a consumer has to purchase a commodity. For example, even though the utility of cake is higher than that of bread, a consumer may not always purchase cake if the price of bread is much cheaper. To analyze this situation, the concept of *cardinal utility* was introduced with a measurability property. For example, if the cardinal utilities of cake and bread are 2 and 1, then a consumer prefers cake twice as much as bread.

Based on cardinal utility, the concept of marginal utility was introduced. It is clear from the measurability of the cardinal utility that the more a commodity is consumed, the larger is the utility associated with it (e.g. the utility of two pieces of cake is higher than one piece). However, the rate of increase in utility can be different and this rate is quantified by marginal utility. For example, if one cake yields 5 units of utility, while two pieces yield 9 units of utility and three pieces yield 12 units, the marginal utilities of the second and third piece of cake are 4 and 3, respectively. This marginal utility is important for utility maximization.

Based on the measurable utility, a rational consumer will always try to maximize his utility by making the best decision. However, other factors such as the prices of the commodities and the level of consumer income also influence the decision which would maximize the utility. The income indicates the amount of money that a consumer can spend per unit of time. From a consumer's point of view, the price is defined as the amount of money that needs to be paid for a unit of the commodity. Clearly, given a particular level of income, fewer units of commodities will be bought if the price is higher. Specifically, both price and the level of income will constrain the consumption of a consumer.

However, the problem of utility maximization becomes difficult due to the limited available information. For example, a consumer may not be certain about the utility of each commodity. In this case, a consumer can learn and adapt the decision over time to reach the highest utility due to consumption. This heuristic learning and decision making process of a consumer is referred to as *bounded rationality*. Bounded rationality is also assumed for a consumer when the cost of gathering complete information is high for which a consumer may make decisions heuristically to maximize the utility.

Based on the concept of preference, a utility function can be used to quantify user satisfaction, and it can be defined as follows:

$$\mathscr{U} = f(\mathbf{x}), \quad \text{for} \quad f : \mathbb{R}_+^N \rightarrow \mathbb{R}, \tag{11.1}$$

where $\mathbf{x} = [x_1, \ldots, x_n, \ldots, x_N]$ denotes a vector containing different quantities of each of the commodities n (e.g. channel n). This $\mathbf{x}$ is also called the *consumption plan* of a consumer. One of the most important theorems for utility function relates to its existence. This theorem states that, given the proper preference relation, there is at least one continuous real-valued function to represent utility. A utility function has an additivity property if it can be expressed as $\mathscr{U}(\mathbf{x}) = \sum_{n=1}^{N} f_n(\mathbf{x}_n)$. In other words, each commodity affects the utility by the corresponding quantity. The utility function is generally differentiable, and its first derivative represents marginal utility.

When the utility function is defined over a consumption plan and represents the preferences of the consumer directly, it is called a *direct utility function*. In contrast, an *indirect utility function* is defined by the choice of a consumption plan that maximizes the utility based on the income of a consumer and the price. This indirect utility function can be obtained by solving a utility maximization problem defined as follows:

$$\mathscr{V}(\mathbf{p}, Inc) = \max_{\mathbf{x} \in \mathbb{R}_+^N} \mathscr{U}(\mathbf{x}), \quad \text{subject to} \quad \mathbf{px} \leq Inc, \tag{11.2}$$

where **p** and **x** are vectors of the price and consumption plan of the corresponding commodity, *Inc* is the level of income, and $\mathscr{V}(\cdot, \cdot)$ is an indirect utility function. This indirect utility function indicates the satisfaction of the consumer not only from the preference but also from the price of the commodity and the available budget. In general, an indirect utility function is non-increasing in price and non-decreasing in income. This indirect utility function plays an important role to obtain the commodity demand function (i.e. given the prices, how much of the commodities a consumer is willing to buy). That is, the demand function is the optimal consumption plan such that the utility is maximized. This is referred to as the *Marshallian or Walrasian demand function* $\mathbf{x}^m(\mathbf{p}, I)$.

An indirect utility function is defined with the objective of achieving the highest satisfaction. That is, given a level of income, a consumer searches for the consumption plan such that the utility is maximized. Alternatively, a consumer may want to obtain a consumption plan such that a certain level of the utility u is achieved while spending the smallest amount of money. Here, the objective is to minimize the monetary expenditure. This can be expressed mathematically as follows:

$$\mathscr{E}(\mathbf{p}, u) = \min_{\mathbf{x} \in \mathbb{R}_+^N} \mathbf{px}, \quad \text{subject to} \quad \mathscr{U}(\mathbf{x}) \geq u. \tag{11.3}$$

Here $\mathscr{E}(\cdot, \cdot)$ is called the *expenditure function*. Similarly, the optimal consumption plan for which the expenditure is minimized is referred to as the *Hicksian demand function* $\mathbf{x}^h(\mathbf{p}, u)$.

A consumer could have the same satisfaction with different consumption plans. This implies that one commodity can be substituted by other commodities. This substitutability impacts optimization of the utility, especially when the consumption plan is constrained by price and income. This substitutability can be indicated by the level of substitutability (i.e. the ratio between the amount of commodities which gives the same level of utility). The level of substitutability affects the consumer demand function significantly.

Application of utility in wireless networks

The concept of utility has been used widely in wireless networks to solve the radio resource management problem. Specifically, since a utility function can provide a layer of abstraction for the QoS performance (e.g. throughput, signal-to-interference ratio (SIR), or delay) in a wireless system, the concept of utility is used when formulating a radio resource allocation model (e.g. time slot in TDMA, subcarrier in OFDMA, or transmit power in a CDMA-based wireless system) to achieve the desired objectives. Based on utility functions, two different approaches, namely, classical optimization and game theory, have been used for radio resource allocation in wireless networks.

In a classical optimization approach, a single objective is defined under multiple constraints. This objective can be, for example, to maximize users' utility under resource constraints (e.g. limited ratio link capacity). Alternatively, the objective can be to maximize resource utilization while the utility of the users is maintained higher than the minimum level. Traditional optimization techniques (i.e. deterministic or stochastic, discrete or continuous, static or dynamic) can be used to obtain the optimal solution for resource allocation. Alternatively, game theory can be used for resource allocation in an environment consisting of multiple entities. These entities may have different interests which conflict with each other. Utility is used to describe the payoff (i.e. interest) of each entity in a game formulation. Different from a classical optimization approach with a single objective function, the aim of a game theory approach is to provide "stable" (i.e. equilibrium) solutions in presence of multiple objectives – one for each entity. The stable solution ensures that all the entities involved are satisfied with the solution, and none of the entities wants to deviate from the solution. Among different concepts of equilibrium, Nash equilibrium is the most popular one used as the solution of a game model for resource allocation in wireless networks.

Different utility functions are used to quantify the level of satisfaction on different QoS performances in wireless networks. Some of these utility functions are described below. Note that in this section we use capital letters to denote constants in the utility function and lower case letters to denote variables.

Transmission rate

Transmission rate or transmission bandwidth is a major QoS performance measure in wireless as well as in wired networks. In the seminal work by Kelly *et al.* [662], the concept of rate allocation to maximize utility was introduced. The network utility maximization problem can be formulated as follows [663]. A communication network has multiple links and each link has capacity C_l. There are multiple traffic sources and the transmission rate of each source is r. The utility corresponding to this transmission rate is $\mathcal{U}(r)$. Each link can be shared by multiple sources. The objective of this optimization is to maximize utility while the transmission rate allocated to each link is less than or equal to the capacity. In this case, the utility function is assumed to be concave. Various utility functions can be used to quantify the user satisfaction on the transmission rate in a wireless system [664]:

- The *exponential utility function* is given by

$$\mathcal{U}(r) = A + B(r + C)^D, \tag{11.4}$$

  where r is the transmission rate, $D < 1$, and B and A are constants of the utility function.

- The *logarithmic utility function* is given by

$$\mathcal{U}(r) = A + B\log(r + C). \tag{11.5}$$

- The *power utility function* is given by

$$\mathcal{U}(r) = A + B \times C^r, \tag{11.6}$$

  where $0 < C < 1$.

The above utility functions were considered for the transmission rate since the user's satisfaction, especially for elastic applications, becomes saturated when the transmission rate is large. It was observed that the empirical data obtained from the subjective surveys fit these utility functions quite well [664].

Transmission delay

The quality of real-time services (e.g. voice and video) is affected by transmission delay, which is generally composed of two components, namely, queueing delay and access delay. The queueing delay is the time period that a data packet waits in the transmission buffer until it is scheduled for transmission. Access delay is the time that a data packet waits since it has been scheduled for transmission until it is successfully transmitted. The delay performance depends on traffic scheduling and admission control methods as well as the wireless link quality.

A utility function can be used to quantify the quality of real-time services. In [665], a decreasing and concave utility function was applied to quantify the level of user satisfaction as a function of transmission delay d. This function is defined as follows:

$$\mathscr{U}(d) = -d^B, \tag{11.7}$$

where $B > 1$. A downlink packet scheduling method was proposed in [665] to maximize the utility.

Signal-to-interference ratio (SIR)

Since the transmit power of one user can significantly affect the quality of transmission for other users, power control is a critical issue in a CDMA system, The signal-to-interference ratio (SIR) at the receiver determines the bit-error-rate (BER) for transmission of a packet. A power control algorithm based on utility was proposed in [666]. The power control problem was modeled as a non-cooperative game in which each user determines the transmit power to maximize its individual utility. The utility is a function of the number of successfully transmitted bits per unit amount of energy consumption, and can be expressed as follows:

$$\mathscr{U}(r, \gamma, p) = r \frac{\mathscr{F}(\gamma)}{P} \frac{l}{L}, \tag{11.8}$$

where l and L denote the number of information bits and the total number of bits in a packet, r is the transmission rate, γ denotes the SIR, and P denotes the transmit power. Here $\mathscr{F}(\cdot)$ is an efficiency function indicating the probability of successful packet transmission. This function is assumed to be of sigmoid type (i.e. $\mathscr{F}(\gamma) = (1 - e^{-\gamma})^{\alpha}$) where α is the function parameter.

Connection blocking probability

Due to limited radio resources in a wireless access network, an admission control method is required to decide whether an incoming connection/session (i.e. new session or handoff session) can be accepted or not. One of the QoS performance metrics related to admission control is the new call blocking probability. The performance of an admission control method can be optimized based on the utility function indicating users' satisfaction. In [641], to quantify the level of satisfaction on call blocking probability b, the following utility function was defined:

$$\mathscr{U} = 1 - \exp(A(b - B)). \tag{11.9}$$

Note that this utility function is non-increasing with blocking probability.

Energy consumption

In a mobile wireless environment, the radio transmitter mostly operates on energy supplied from the battery. To reduce the energy consumption and hence to extend the lifetime of the wireless node, the transmit power needs to be controlled. Also, the radio transceiver can be turned off periodically (i.e. switched to sleep or standby mode). To design a resource allocation method in an energy-limited environment (e.g. ad hoc and sensor networks), the energy consumption performance needs to be considered. In [667], utility was used to quantify the level of user satisfaction on energy consumption and to optimize data access for a wireless node. The utility function was defined as follows:

$$\mathscr{U}(e) = E_t/e, \tag{11.10}$$

where E_t is the total amount of energy available to each node, and e is the energy required to complete the transmission.

Utility fairness

In an optimization-based resource allocation method, the objective is typically either to maximize the system capacity or to maintain fairness. These objectives could be in conflict with each other. For example, a base station may allocate a large portion of the radio resource to a user with the best channel quality. Even though this objective can maximize the capacity of the base station, this may not be fair to other users. In [668], utility fairness was used to study the rate performance in an energy-limited multihop wireless network. This utility function was defined as a function of transmission rate r as follows:

$$\mathscr{U}(r) = \begin{cases} (1-A)^{-1} r^{1-A}, & A \neq 1 \\ \log(r), & A = 1, \end{cases} \tag{11.11}$$

where A is an adjustable parameter to control fairness. Specifically, if $A = 0$, optimizing the utility will maximize the capacity of the system since the utility is simply the transmission rate r. If $A = 1$, it will compromise efficiency for the sake of fairness. This is referred to as *proportional fairness*. If $A \to \infty$, it will provide strict fairness. That is, extra resources will be allocated to other users only when the performance of the worst user cannot be improved by allocating more resources. This is referred to as the *max-min fairness*. With this utility function, therefore, the parameter A can be flexibly adjusted to achieve the desired fairness objective.

Application-related utility

In [669], a utility function was used to quantify users' satisfaction on the application performance (e.g. web browsing and file download). This utility was used as a payoff in a dynamic game model to determine optimal pricing for WiFi access points. For web browsing, the utility function was defined as follows:

$$\mathscr{U}(t, \tau) = A \min(t, \tau), \tag{11.12}$$

where t is the length of time of web browsing, A is a constant, and τ is the maximum session length. Note that the utility becomes saturated for the maximum session length

τ. For file transfer, the utility function was expressed as follows:

$$\mathscr{U}(t,\tau) = \begin{cases} 0, & t < \tau \\ A\tau, & t = \tau. \end{cases} \tag{11.13}$$

To complete file downloading, a user must be connected for the entire duration τ. Otherwise, the utility becomes zero.

In [670], a generic utility function was used to optimize video layering with advanced video encoding (e.g. MPEG-4) and bandwidth allocation so that the system utility is maximized. This generic utility was defined as a function of the receiver's capacity, received video bandwidth, and the total number of subscribed layers.

Utility and price

Price is one of the major factors which affects users' satisfaction on wireless access services. In particular, the concept of *net utility* can be used to indicate the satisfaction on both performance and price. This net utility [671] can be defined as follows:

$$\mathscr{N}(x,p) = \mathscr{U}(\mathscr{R}(x)) - p \times x, \tag{11.14}$$

where $\mathscr{U}(q)$ is the utility function for performance q, $\mathscr{R}(x)$ is the performance function for the given amount of allocated resource x, p is the price per unit of resource, and $p \times x$ is the total cost.

Alternatively, the utility function can be defined as follows [672]:

$$\mathscr{U}(q,p) = (A(q-Q) + (1-A)(p_{\max} - p))1(q - Q > 0)1(p_{\max} - p > 0), \tag{11.15}$$

where q and p denote, respectively, the quality and price of the wireless access service, Q and $p_{\max}$ denote the highest quality and the highest price, and 1(Cond) returns 1 if Cond is true and returns 0 otherwise. This utility function was used in [672] to express user satisfaction for dynamic spectrum access in a cognitive radio environment. Since competition exists among several service providers, a game formulation was used to determine users' strategies to access the radio spectrum based on the quality and the price.

Subjective survey of utility

A subjective survey was conducted to obtain the utility function corresponding to the performance of a wireless access service [664]. In particular, user experience was evaluated for data services in a cellular wireless network. The channel rates and the number of users sharing the system were varied to observe users' satisfaction. The respondents work with the network emulator and provide different scores for their subjective opinion. These scores are, for example, 5 for "excellent," 4 for "good," 3 for "moderate," 2 for "poor," and 1 for "terrible." This score information can be used as raw data to construct a utility function, for example, by using curve fitting techniques. That is, the surveyed data are fitted by a set of candidate utility functions. The best fitted function and the corresponding parameters are selected and then used for system optimization.

11.2.2 Market-equilibrium

Pricing is an important issue in economics. In a market environment, which includes a number of sellers and buyers, the concept of market-equilibrium price can be used so that the pricing satisfies all the sellers and buyers. On the seller side, the amount of commodity that the seller is willing to supply to the market is indicated by a *supply function*. The supplied quantity is a function of price p, i.e. $\mathscr{S}(p)$. On the buyer side, the amount of commodity that the buyer is willing to buy from the market is indicated by a *demand function*. The demand for a commodity is a function of price p, i.e. $\mathscr{D}(p)$. In general, the amount of supply from the seller is an increasing function of price. On the other hand, the demand for a commodity in the market is a decreasing function of price.

Given the demand and supply functions in a market, the market-equilibrium price p^* is given by the price for which the supply equals the demand, i.e.

$$\mathscr{S}(p^*) = \mathscr{D}(p^*). \tag{11.16}$$

This market-equilibrium price is the best possible strategy for both sellers and buyers. In particular, any other price, which is not a market-equilibrium price, is undesirable for both sellers and buyers. Therefore, there will be an incentive for the sellers or buyers to deviate from that price. For example, let us consider a price p such that $p < p^*$. In such a case, due to the low price, the demand will be larger than the supply. Consequently, some sellers will find that they could sell their commodity at a higher price due to the large demand from the buyers. As the price is increased by the seller, the market price will be driven until supply equals demand. On the other hand, for a price p', where $p' > p^*$, the demand will be less than the supply. As a result, some sellers will not be able to sell all of their commodities. Therefore, these sellers will reduce the price. The price will be decreased until all excess supply can be sold to the buyers, which is the location of the market-equilibrium price.

Let us consider the following linear demand and supply functions:

$$\mathscr{D}(p) = A - Bp, \quad \mathscr{S}(p) = E + Fp, \tag{11.17}$$

where A, B, E, and F are constants. The market-equilibrium price can be obtained by solving the following equation:

$$\mathscr{D}(p) = A - Bp = E + Fp = \mathscr{S}(p), \tag{11.18}$$

which gives

$$p^* = \frac{A - E}{F + B}. \tag{11.19}$$

The amount of supply (which is same as the demand) at the market-equilibrium price is given by

$$\begin{aligned} \mathscr{D}(p^*) &= \mathscr{S}(p^*) \\ &= A - Bp^* \\ &= \frac{AF + EB}{B + F}. \end{aligned} \tag{11.20}$$

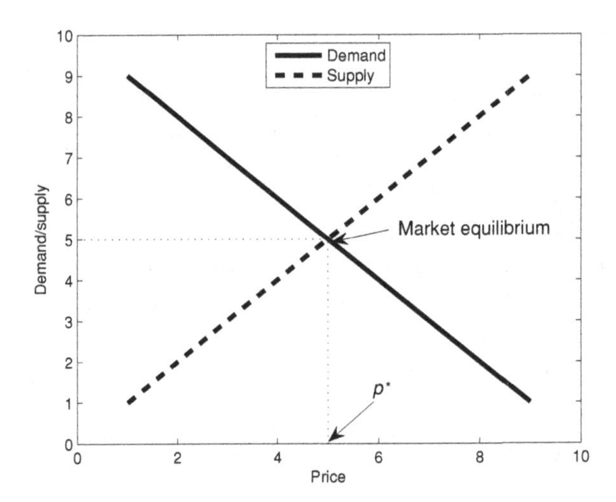

Figure 11.11 Linear demand function.

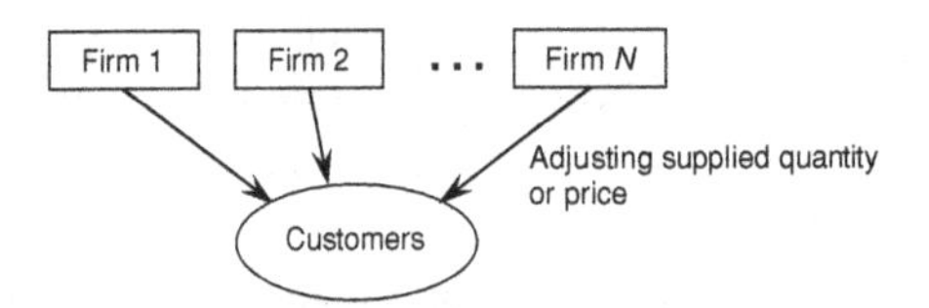

Figure 11.12 Oligopoly market model.

For the case of $A = 10$, $B = 1$, $E = 0$, and $F = 1$, the demand and supply functions as well as the market-equilibrium price are shown in Figure 11.11.

11.2.3 Oligopoly market

Spectrum trading in a cognitive radio network can be modeled as an oligopoly market where a small number of firms (i.e. oligopolists) dominate a particular market (Figure 11.12). In this market structure, the firms compete with each other independently to sell a commodity so that their profits are maximized. These firms control the amount of supply and/or the price. The amount of supply and/or the price offered by a firm will affect not only its own profit, but also the profits of other firms. For example, if one firm increases its supplied quantity to the market, the price for the entire market will decrease. As a result, profits of other firms tend to decrease. Note that in an oligopoly market, the commodity supplied by the firms can be either homogeneous or differentiated. In the former case, all firms produce the same commodity, and in the latter case, each firm produces different commodity for which each of them is substitutable. However, since each commodity may not be perfectly substitutable, the competition among the firms is affected by the level of substitutability.

In a cognitive radio network, a firm could be analogous to a primary user or a primary service provider. The primary service providers compete with each other to sell the spectrum opportunities to the secondary users or secondary service providers. The objective of a primary service provider is to maximize the profit of selling the spectrum opportunities (i.e. rationality behavior). The profit of a primary service provider depends not only on the cost of sharing the spectrum with the secondary service providers (e.g. due to performance degradation of primary users), but also on the strategy chosen by other primary service providers. In addition, the spectrum may not be perfectly substitutable due to the difference in quality of propagation and/or interference conditions. For example, a spectrum band at the lower frequency range could have better propagation characteristics than that in the higher frequency range.

The theory of non-cooperative games can be used to analyze and predict the behavior of the firms in an oligopoly market. A non-cooperative game model is composed of players, their strategies, and the payoffs of the players. In an oligopoly market, the players correspond to the firms. The strategy for each firm corresponds to the supplied quantity or offering price. The payoff for a firm is given by the profit. Three classical oligopoly models which we will analyze using game theory are the *Cournot*, *Bertrand*, and *Stackelberg* game models. These models have different market structures and the strategies used in these competitions are different. In particular, in a Cournot model, the firms compete in terms of the amount of supply to the market. All the firms make their decisions at the same time. On the other hand, in a Bertrand model, all firms make their decision simultaneously in terms of price. In a Stackelberg model, there is a firm (referred to as a leader) who is able to make decisions on the supplied quantity or price before other firms (i.e. followers). Then, these followers make their decisions by taking into account the decision of the leader. These different market structures result in different behaviors of the firms in achieving the best and stable decisions.

To illustrate the models for Cournot, Bertrand, and Stackelberg competitions, a market with two firms (i.e. a duopoly market) will be considered. A duopoly market is a special case of an oligopoly market, where the total number of players is two. The following linear demand function is assumed: $\mathscr{D}(p) = A - Bp$, and the inverse demand function is expressed as $p(d) = A' - B'd$, where $A' = A/B$ and $B' = 1/B$.

Cournot model

In Cournot competition, all firms make their decisions on the supplied quantity simultaneously. For this aggregated supplied quantity, the customers will react with the price that they are willing to pay for. Specifically, the price is determined from the inverse demand function as follows: $\mathscr{D}^{-1}(p) = p(d) = A' - B'd$. Here, the supplied quantity of one firm will affect the market price of the commodity (e.g. a larger quantity results in a lower price), and hence the profit of other firms. With the production cost of C per unit of commodity and fixed cost C_f, for the case of homogeneous commodity (i.e. all firms produce the same commodity), the profit of a firm i can be expressed as follows:

$$\begin{aligned}\mathscr{P}_i(s_i, s_j) &= p(s_i + s_j)s_i - Cs_i - C_f \\ &= (A' - B'(s_i + s_j))s_i - Cs_i - C_f,\end{aligned} \tag{11.21}$$

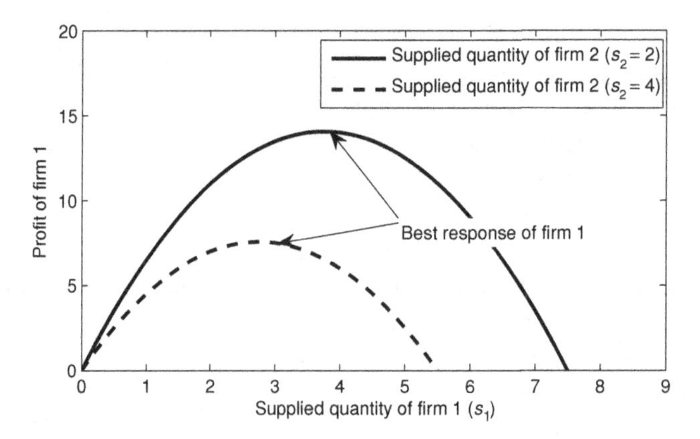

Figure 11.13 Profit of a firm in a Cournot duopoly market model.

where s_i and s_j are the supplied quantity to the duopoly market from firms i and j, respectively. Since the objective of each firm is to maximize its profit, the best response of each firm can be obtained by finding the optimal amount of supplied quantity which maximizes the profit given the amount of supply from other firms. For this, the profit function is differentiated with respect to the supplied quantity as follows:

$$\begin{aligned} \frac{\partial \mathscr{P}_i(s_i, s_j)}{\partial s_i} &= A' - B'(2s_i + s_j) - C \\ 0 &= A' - B'(2s_i + s_j) - C \\ s_i^* = \mathscr{B}_i(s_j) &= \frac{A' - B's_j - C}{2B'}. \end{aligned} \tag{11.22}$$

Similarly, the best response function of firm j is obtained as $s_j^* = \mathscr{B}_j(s_i) = (A' - B's_i - C)/2B'$. For example, with $A' = 10$, $B' = 1$, and $C = 0.5$, the profit of firm $i = 1$ is as shown in Figure 11.13 under different supplied quantity s_1 given the supplied quantity s_2 from firm 2. Here, as the supplied quantity s_i increases, since the firm i can sell more, the profit of firm i first increases. However, at a certain point, the profit decreases due to the decrease in price resulting from the abundance in supply of the commodity in the market. The optimal amount of supply s_i^*, which gives the highest profit to firm i, is referred to as the best response of firm i.

The solution of the Cournot game model, namely, the *Nash equilibrium*, gives the optimal amount of supply which maximizes the profits of the firms. The Nash equilibrium of a game is a set of strategies for all players with the property that no player can increase his payoff by choosing a different action, given the other players' actions. In the context of the Cournot model, this Nash equilibrium can be expressed as follows:

$$\mathscr{B}_i(s_j^*) = \mathscr{B}_j(s_i^*). \tag{11.23}$$

For the above example, the best responses of the two firms and the Nash equilibrium of the duopoly market are graphically shown in Figure 11.14. Here, the Nash equilibrium is the point where the best responses intersect with each other. The Nash equilibrium

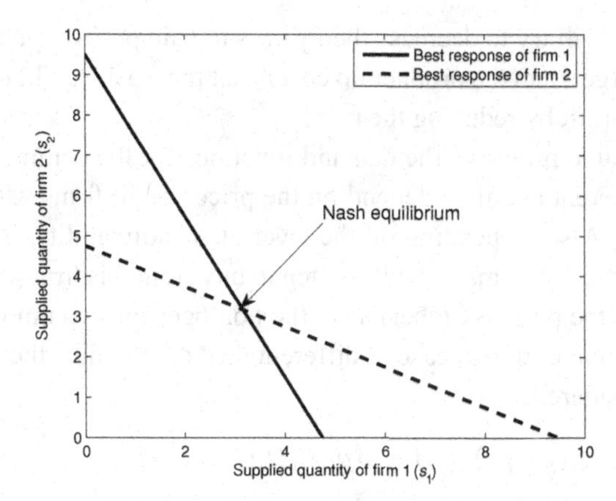

Figure 11.14 Best responses of the firms in a Cournot duopoly market model.

can be obtained from

$$(s_i^*, s_j^*) = \left(\frac{A' - C}{3B'}, \frac{A' - C}{3B'} \right). \tag{11.24}$$

At the Nash equilibrium, none of the firms can have better profit without adjustment in the supplied quantity from the other firms. For example, if firm i tries to increase its supply, the price will decrease. As a result, firm j must also increase its supply to gain higher profit. However, this will reduce the profit of both the firms due to the lower price. As a result, firm j is forced to reduce its supply. This process will repeat until the Nash equilibrium is reached.

Bertrand game

Instead of adjusting the amount of supply, a firm can change its offered price. In such a case, the firm first chooses the price of the commodity and announces it to the customer. The customer responds with its demand for the commodity. The firm can choose the price so that its profit is maximized given the price chosen by other firms. This competition through pricing is referred to as a Bertrand game. In a Bertrand game, the solution depends mainly on the substitutability of the commodity. If the commodities from the different firms are identical (i.e. the homogeneous case), then they are said to be fully substitutable. On the other hand, if the commodities are different, the products may be partly substitutable or may be completely unsubstitutable.

In the case of homogeneous commodity, the customer can choose to buy the commodity from any one of the firms. The consumer will always choose to buy from the firm which offers the lowest price. If there is no limit on the maximum supplied quantity, that firm will supply the commodity for the entire market. In this case, other firms offering higher prices will experience zero profit. It can be shown that in the case of homogeneous commodity, there is a unique Nash equilibrium in which the prices charged by all firms are identical. In particular, at the Nash equilibrium, the price is equal to the production cost. The interpretation of this solution is as follows. Given any initial price, if one firm decreases the price, the entire market will buy from that firm. Consequently, since other firms gain zero profit, they will try to decrease their prices to gain positive profit. In this case, any price which is larger than the production cost is not the Nash equilibrium since one firm can gain higher profit by reducing the price.

In the case of differentiated products, the demand functions for the commodity from the different firms are different and they depend on the prices of all firms whose commodities are substitutable. Also, depending on the level of substitutability, if the price from one firm is changed, the entire market will switch to buy from this firm (if the price is lower) or other firms (if the price is higher) as in the homogeneous commodity case. To describe the Bertrand competition in case of differentiated commodity, the following demand functions are considered:

$$s_i(p_i, p_j) = A - Bp_i + Dp_j, \tag{11.25}$$

$$s_j(p_i, p_j) = \tilde{A} - \tilde{B}p_j + \tilde{D}p_i, \tag{11.26}$$

where A, B, D, $\tilde{A}$, $\tilde{B}$, and $\tilde{D}$ are the constants of the demand functions. Constants D and $\tilde{D}$, in particular, indicate the level of substitutability. These demand functions indicate that if the price from firm i increases, the demand for the commodity from firm i will decrease. If the price from firm j decreases, the demand for the commodity from firm i will also decrease, since the commodity from firm i is substitutable by the commodity from firm j which has lower price. In this case, the profit of firm i can be expressed as follows:

$$\begin{aligned}\mathscr{P}_i(p_i, p_j) &= p_i s_i - Cs_i - C_f \\ &= (A - Bp_i + Dp_j)(p_i - C) - C_f.\end{aligned} \tag{11.27}$$

The profit of the firm j is similar to that of firm i. The best response of firm i (i.e. which maximizes its profit) can be derived by differentiating $\mathscr{P}_i(p_i, p_j)$ with respect to p_i as follows:

$$\begin{aligned}\frac{\partial \mathscr{P}_i(p_i, p_j)}{\partial p_i} &= B(C - p_i) + A - Bp_i + Dp_j = 0 \\ p_i^* &= \frac{BC + A + Dp_j}{2B}.\end{aligned} \tag{11.28}$$

Similarly, the best response function of firm j is

$$p_j^* = \frac{\tilde{B}C + \tilde{A} + \tilde{D}p_i}{2\tilde{B}}. \tag{11.29}$$

The Nash equilibrium price is found to be

$$(p_i^*, p_j^*) = \left(\frac{2B\tilde{B}C + 2A\tilde{B} + \tilde{B}CD + \tilde{A}D}{4B\tilde{B} - D\tilde{D}}, \frac{2B\tilde{B}C + 2\tilde{A}B + BC\tilde{D} + A\tilde{D}}{4B\tilde{B} - D\tilde{D}}\right). \tag{11.30}$$

For the Bertrand duopoly game, the best responses of firm $i = 1$ and $j = 2$ and the Nash equilibrium points are shown in Figure 11.15 for $A = \tilde{A} = 10$, $B = \tilde{B} = 1$, and $C = 0.5$. In this case, given the price of firm 2 (i.e. p_2), the best response for the price

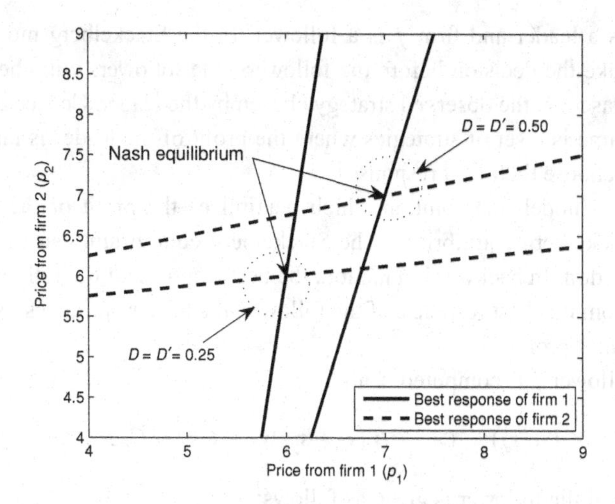

Figure 11.15 Best responses of the firms in a Bertrand duopoly game.

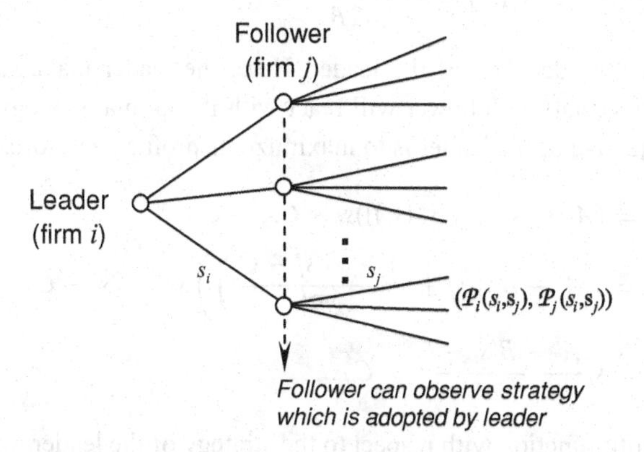

Figure 11.16 Best responses of the firms in a Bertrand duopoly game.

of firm 1 (i.e. p_1) can be obtained, and vice versa. Unlike the Cournot model, in the Bertrand game, if one firm increases the price, the other firms will increase the price as well to achieve the highest profit. The Nash equilibrium is located at the point where the best responses of both the firms intersect with each other. It is observed that the parameters D and $\tilde{D}$, which represent the substitutability of the commodity impacts the slope of the best response curves and hence the location of the Nash equilibrium.

Stackelberg model

In a Stackelberg model, the firms can compete with each other in terms of supplied quantity or price. In this competition, there is at least one firm (referred to as a leader) who can commit the chosen strategy (i.e. supplied quantity or price) before other firms (referred to as the followers). An extensive form [673] of this game is shown in Figure 11.16 for the case of competition in terms of supplied quantity similar to that in a Cournot model. This extensive form game shows the sequence of decision making in which firm i is a leader and firm j is a follower. In the Stackelberg model, since the leader will make the decision before the followers, the followers will choose their optimal strategy based on the observed strategy chosen by the leader. Consequently, the solution of this game is a set of strategies where the profit of the leader is maximized and the followers choose their best responses.

In a Stackelberg model, the solution which maximizes the profit of the leader is defined as the Stackelberg equilibrium. The Stackelberg equilibrium can be obtained by backward induction. In backward induction, the best response of the follower is first obtained. Then, from this best response of the follower, the leader optimizes its strategy to achieve the highest profit.

The profit of follower j is computed from

$$\mathscr{P}_j(s_i, s_j) = (A' - B'(s_i + s_j))s_j - Cs_j - C_f. \tag{11.31}$$

The best response of the follower is given as follows:

$$s_j^*(s_i) = \frac{A' - B's_i - C}{2B'}. \tag{11.32}$$

Then, we backtrack to the decision of the leader. Here, the leader makes a decision based on the assumption that the follower will react with its optimal strategy (i.e. best response), and the objective of the leader is to maximize its profit. Therefore, we have

$$\begin{aligned}\mathscr{P}_i(s_i, s_j) &= (A' - B'(s_i + s_j^*(s_i)))s_i - Cs_i - C_f. \\ &= \left(A' - B'\left(s_i + \frac{A' - B's_i - C}{2B'}\right)\right) s_i - Cs_i - C_f \\ &= s_i \frac{A' - B's_i - C}{2} - C_f.\end{aligned} \tag{11.33}$$

Differentiating this profit function with respect to the strategy of the leader, which is s_i, we obtain

$$\begin{aligned}\frac{\partial \mathscr{P}_i(s_i, s_j)}{\partial s_i} &= A' - \frac{B's_i}{2} - B'\left(s_i + \frac{A' - B's_i - C}{2B'}\right) - C \\ 0 &= \frac{A' - C}{2} - B's_i \\ s_i^* &= \frac{A - C}{2B'}.\end{aligned} \tag{11.34}$$

This is the subgame perfect Nash equilibrium or the optimal strategy for the leader if the leader can make a decision before the follower. Based on the optimal strategy of the leader, the optimal strategy for the follower is obtained as follows:

$$s_j^* = \frac{A' - C}{4B'}. \tag{11.35}$$

The Stackelberg equilibrium is shown graphically in Figure 11.17, where s_1 and s_2 are the supplied quantities of firms 1 and 2, respectively. The Stackelberg equilibrium can

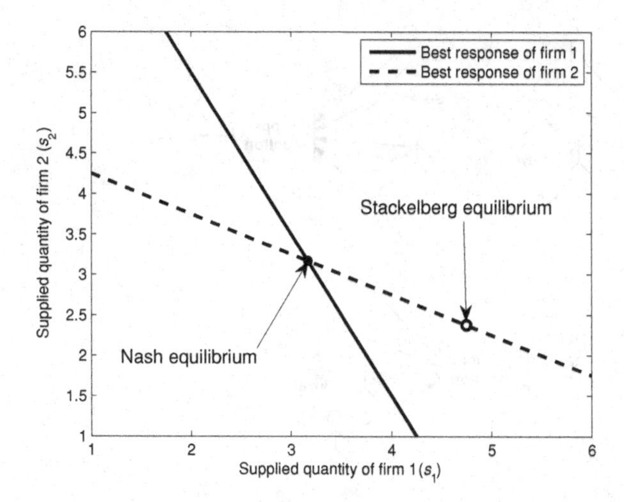

Figure 11.17 Best responses of the firms and Stackelberg equilibrium in a Stackelberg model.

be expressed mathematically as follows:

$$(s_1^*, s_2^*) = \left(\frac{A' - C}{2B'}, \frac{A' - C}{4B'}\right). \tag{11.36}$$

At the Stackelberg equilibrium, which is generally different from the Nash equilibrium, the leader will offer a larger supplied quantity than that of the follower. Consequently, the profit of the leader is higher. This higher profit of the leader in a Stackelberg model is also known as the *first-move advantage* in which the player of the game with the ability to make decisions before other players gains higher payoff.

The Stackelberg model was applied to model resource management problems in wireless networks [674, 675, 676, 677, 678, 679]. For example, the problem of pricing for bandwidth sharing between IEEE 802.16 WMAN and IEEE 802.11 WLAN was formulated as a Stackelberg game where IEEE 802.16 WMAN is the leader while IEEE 802.11 WLAN is a follower [674]. In [679], the problem of relay selection and power control in a cooperative communication environment was modeled as a Stackelberg game. In this case, the leader is a relay node, while the follower is a source node. The relay node maximizes its revenue by choosing the price, and the source node chooses the transmit power accordingly so that its utility is maximized.

11.2.4 Auction theory

An auction is a process used to obtain the price of a commodity with an undetermined value. There are three categories of auction, namely, supply auction, demand auction, and double auction (Figure 11.18). In a supply auction, multiple sellers offer their commodities to a buyer. In a demand auction, multiple buyers bid for a commodity being sold by a seller. In a double auction, multiple buyers bid to buy commodities from multiple sellers.

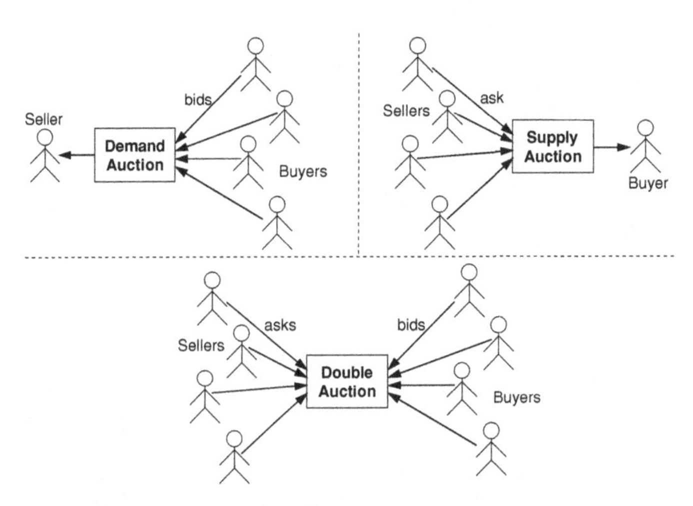

Figure 11.18 Three types of auction.

The components in an auction market are as follows:

- A *seller* is a market entity who wants to sell the commodity. A seller offers the price (i.e. the *asking* price) and the amount of commodity to be traded by auction.
- The *buyer* is an entity who wants to buy the commodity. A buyer submits a bid in terms of price and bidding quantity to buy through the auction.
- The *trading/clearing price* is the price of each commodity to be traded in an auction market. The trading price has to satisfy the asking price and the bidding price (e.g. it should be higher than or equal to the asking price but lower than or equal to the bidding price) from the seller and the buyer, respectively.

Single-side auction

A single-side auction is a situation where there is one auctioneer – which could be a seller or a buyer, in the case of a supply auction or a demand auction, respectively. In the single-side auction, the bidders submit their bids to the auctioneer. Then, the auctioneer decides to sell or buy from any bidder. The major types of single-side auction are the increasing-price auction (English auction), decreasing-price auction (Dutch auction), first-price sealed-bid auction, and second-price sealed-bid auction (Vickrey auction).

In an increasing-price or English auction, the minimum price is set. Then, a bidder submits a bid (i.e. a bidding price which is higher than the minimum price) to the auctioneer. Each bidder may observe the bids from other bidders and compete by increasing its bidding price. Thereafter, the bidding price is continuously increased until the auction is terminated (e.g. after a limited time duration or when all bidders stop submitting bids). The bidder with the highest bidding price wins the auction.

In a decreasing-price auction or Dutch auction, the maximum price is set, and a bidder submits a bid which is lower than the maximum price to the auctioneer. The first bidder to accept the price will win the auction. Different from an increasing-price auction, in a decreasing-price auction, the bidder information does not need to be revealed other than the winning bidder and its bidding price.

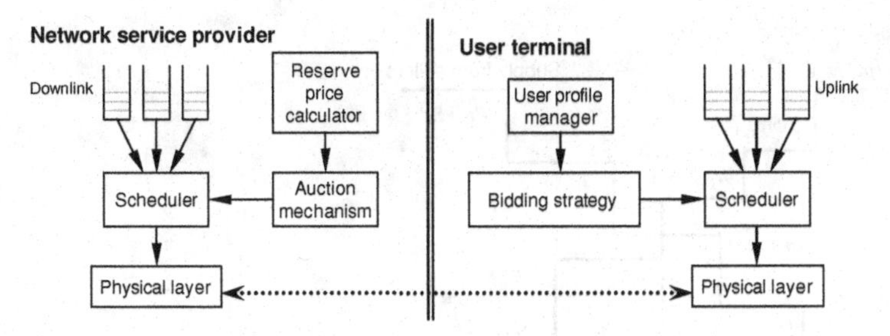

Figure 11.19 Joint scheduling and spectrum bidding architecture.

In a sealed-bid auction, all bidders independently submit sealed bids. The auctioneer opens the bids and determines the winning bidder whose bidding price is the highest. For the winning bidder, the price to pay the auctioneer could be its bidding price (i.e. first-price auction) or the second highest bidding price (i.e. second-price auction or Vickrey auction).

Note that spectrum auction may be jointly designed with a resource allocation framework (e.g. scheduling). An example of such a spectrum auction is shown in Figure 11.19 [680]. In particular, the downlink and uplink schedulers will use information from the auction mechanism from the network service provider and the user's bidding strategy. Therefore, the user can bid for the spectrum based on the QoS requirement, while the network service provider can charge a price according to the bids from all users.

Double auction

In a double auction [681], there are I buyers and N sellers. Each buyer i wants to purchase x_i items and each seller n wants to sell y_n items. The information about x_i and y_n are available publicly. In a double auction, a buyer i reports a price $p_i^{(b)}$ (i.e. the bidding price) per unit of the commodity, while a seller n reports a price $p_n^{(s)}$ (i.e. the asking price). Without loss of generality, we may assume $p_1^{(b)} > p_2^{(b)} > \cdots > p_I^{(b)}$ and $p_1^{(s)} < p_2^{(s)} < \cdots < p_N^{(s)}$. Note that if two prices are equal, their indexes are interchangeable. Also, each seller or each buyer can set different prices for different items, where the seller and buyer sells and buys each item separately.

To determine the trading price in double auction, the demanded quantities from all buyers are arranged according to the ascending order of price. Similarly, the supplied quantities from all sellers are arranged according to the descending order of price (Figure 11.20). At the trading point T^*, the aggregated demand and supply intersect, and hence n' sellers will sell T^* items to i' buyers. There are two cases used to determine the trading price and trading quantity:

- *Case 1*: The bidding and asking prices satisfy the condition $p_{i'}^{(b)} \geq p_{n'}^{(s)} \geq p_{i'+1}^{(b)}$ and aggregate demand and supply satisfy $\sum_{n=1}^{n'-1} y_n \leq \sum_{i=1}^{i'} x_i \leq \sum_{n=1}^{n'} y_n$. In this case, the sellers $n = \{1, \ldots, n'\}$ sell all their items y_n at price $p_{n'}^{(s)}$, and the buyers $i = \{1, \ldots, i'\}$ buy at price $p_{i'}^{(b)}$. The amount that each buyer buys is $\left\lfloor x_i - \frac{\sum_{j=1}^{i'-1} x_j - \sum_{j=1}^{n'-1} y_j}{i'-1} \right\rfloor$.

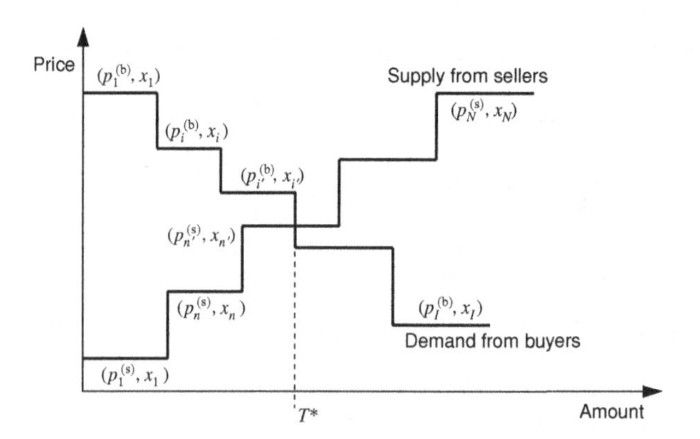

Figure 11.20 Example of ordered demand and supply in a double auction.

- *Case 2*: The bidding and asking prices satisfy the condition $p_{n'+1}^{(s)} \geq p_{i'}^{(b)} \geq p_{n'}^{(s)}$ and aggregate demand and supply satisfy $\sum_{i=1}^{i'-1} x_i \leq \sum_{n=1}^{n'} y_n \leq \sum_{i=1}^{i'} x_i$. In this case, the buyers $i = \{1, \ldots, i'\}$ buy at price $p_{i'}^{(b)}$, and the sellers $n = \{1, \ldots, n'\}$ sell at price $p_{n'}^{(s)}$. The amount that each seller sells is $\left\lfloor y_n - \frac{\sum_{j=1}^{n'-1} y_j - \sum_{j=1}^{i'-1} x_j}{n'-1} \right\rfloor$.

However, when a central controller is available in this double auction, an optimization problem can be formulated to obtain the quantity of items to be traded. Let the reserved price of each buyer and seller be fixed and respectively denoted by $\hat{p}_i^{(b)}$ and $\hat{p}_n^{(s)}$. Let $\hat{x}_{i,n}$ and $\hat{p}_{i,n}$ denote the solutions in terms of the quantity that buyer i buys from seller n and the trading price, respectively. The benefit of buyer i from double auction can be defined as follows:

$$B_i^{(b)} = \sum_{n=1}^{\hat{n}} (p_i^{(b)} - \hat{p}_{i,n})\hat{x}_{i,n}, \tag{11.37}$$

and that of seller n is defined as follows:

$$B_n^{(s)} = \sum_{i=1}^{\hat{i}} (\hat{p}_{i,n} - p_n^{(s)})\hat{x}_{i,n}. \tag{11.38}$$

To maximize the benefit of both the seller and the buyer, an optimization problem can be formulated as a linear programming problem as follows:

$$\text{maximize: } \sum_{i=1}^{\hat{i}} \sum_{n=1}^{\hat{n}} \hat{x}_{i,n} \left(p_i^{(b)} - p_n^{(s)} \right), \tag{11.39}$$

$$\text{subject to: } \sum_{i=1}^{\hat{i}} \hat{x}_{i,n} \leq y_n, \quad \forall n, \tag{11.40}$$

$$\sum_{n=1}^{\hat{n}} \hat{x}_{i,n} \leq x_i, \quad \forall i, \tag{11.41}$$

$$\hat{x}_{i,n} \geq 0, \quad \forall i, n. \tag{11.42}$$

The constraints limit the quantity of the item to be traded to the supplied and demanded quantity for the seller and the buyer, respectively.

11.3 Spectrum trading and dynamic spectrum access

In this section, different spectrum trading models are discussed, based on the microeconomic theories discussed earlier.

11.3.1 Double auction-based pricing for dynamic spectrum access

When multiple sellers (i.e. licensed users) and multiple buyers (i.e. unlicensed users) are involved in spectrum trading, a double auction can be applied to obtain the competitive equilibrium solution [526]. The model assumes that all sellers and buyers are rational to maximize their payoffs. However, sellers and buyers may not exchange information with each other. They can even lie to each other if this helps them to improve their payoffs. Therefore, in order to obtain an optimal solution which maximizes the payoff, each of the sellers and buyers has to develop its own belief on the information from others.

In [526], a static pricing game model was formulated. In this model, the payoff of licensed user n can be expressed as follows:

$$U_n = \sum_{k=1}^{K_n} \left(\phi_{a_n^k} - c_n^k\right) y_n^k, \tag{11.43}$$

where K_n is the total number of channels of licensed user n, $\phi_{a_n^k}$ is the payment received by licensed user n from selling channel a_n^k to an unlicensed user, c_n^k is the cost incurred to licensed user n due to selling channel k, and y_n^k is a binary variable indicating whether channel k is sold or not. Similarly, the payoff of unlicensed user i can be expressed as follows:

$$U_i = \sum_{k=1}^{K} \left(v_i^k - \phi_k\right) x_i^k, \tag{11.44}$$

where K is the total number of channels, v_i^k is the benefit gained from channel k by unlicensed user i, and x_i^k is a binary variable indicating whether channel k is obtained by unlicensed user i or not. Since both the licensed and the unlicensed users are rational, an optimization problem can be formulated to obtain the payments $\phi_{a_n^k}$ and ϕ_k such that the payoffs of the licensed and unlicensed users are maximized. To avoid the complexity due to the multi-objective nature of this optimization problem, competitive equilibrium from double auction theory can be used to obtain the solution. This competitive equilibrium denotes the price at which the number of channels that the unlicensed users want to buy is equal to the number of channels that the licensed users want to sell. In this case, the supply function defines the relationship between the number of channels to be sold and the cost of licensed users. The demand function defines the relationship between the number of channels to be obtained and the benefit of unlicensed users.

Since the payoff functions of licensed and unlicensed users depend on the decisions of both unlicensed and licensed users, these licensed and unlicensed users can cheat each other by posting incorrect information. In this case, the licensed and unlicensed users have to establish their own beliefs on the available information.

The static pricing game model was extended to a dynamic model for which there are multiple periods and in each period, the parameters in the payoff functions of both licensed and unlicensed users can be different. The objective function for both types of users now include the factor δ, which discounts the payoff in the future. These objective functions are in the form of Bellman's equation, which can be solved by using a standard dynamic programming method [592]. The budget of the unlicensed users to buy the channel can be integrated into the discounted objective function.

The performance of the proposed scheme was compared with the theoretical competitive equilibrium (i.e. when perfect knowledge is available for all users). When the number of unlicensed users increases, the total payoff increases. It was observed that the total payoff obtained from the proposed scheme is slightly smaller than that from the theoretic model. Since the belief update algorithm can reduce the amount of information exchange in a spectrum trading environment, the overhead of the proposed scheme was much smaller than that of the traditional continuous double auction.

While this immediately solves the auction problem when all the data from licensed and unlicensed users is available, in cognitive radio networks it is common for the data from licensed and unlicensed users to be not available simultaneously. Therefore, the auction problem may need to be solved sequentially according to the arriving request.

11.3.2 Sequential and concurrent auction for dynamic spectrum access

In [682], the problem of dynamic spectrum access was formulated as a multi-unit sealed-bid sequential and concurrent auction. In this case, the service providers bid for the spectrum from a spectrum broker. For a sequential auction, the channels are auctioned one by one, while for a concurrent auction, all channels are auctioned at the same time. In this spectrum auction environment, there are N channels and I service providers. Each of the service providers is rational to maximize its profit, which is calculated as the difference between the revenue gained from using the spectrum to serve the corresponding users and the price paid to the spectrum broker. In each round of the auction, service provider i submits bid b_i to the spectrum broker. Given the bids from all service providers, the spectrum broker determines the winning service provider (with the highest bid), and allocates a channel to that service provider. At the end of each round of the auction, the spectrum broker broadcasts the minimum bid during that round. In the next round of the auction, the winning service provider will decrease its bid to increase its profit. These steps are repeated until a steady state is reached.

The value of the channel for service provider i is denoted by v_i. In the case of a substitutable channel, this value is fixed for all channels. Then, the profit of service provider i is given by

$$\overline{\mathscr{P}}_i = \begin{cases} v_i - b_i, & \text{if service provider } i \text{ wins} \\ 0, & \text{if service provider } i \text{ loses.} \end{cases} \tag{11.45}$$

For the sequential auction, the probability density function of bid b can be expressed as

$$f(b) = \frac{1}{v_{\max} - b_{\min}} \tag{11.46}$$

where $v_{\max}$ is the maximum value of the channel, and $b_{\min}$ is the lowest among all submitted bids. If k channels are already auctioned, the probability that any bid b_j is less than b_i where $j \in (I-k-1)$, $j \neq i$, and $i \in (I-k)$, is $\int_{b_{\min}}^{b_i} f(b)\mathrm{d}b$, or

$$\Pr(b_j < b_i) = \frac{b_i - b_{\min}}{v_{\max} - b_{\min}}. \tag{11.47}$$

To win the auction, the condition $\forall b_j < b_i$ needs to be satisfied, and the probability of service provider i winning the sequential auction can be obtained from

$$P_{\text{win}}^{\text{seq}}(i) = \left(\frac{b_i - b_{\min}}{v_{\max} - b_{\min}}\right)^{(I-k-1)}. \tag{11.48}$$

The expected profit of service provider i is

$$\overline{\mathscr{P}}_i = (v_i - b_i)\left(\frac{b_i - b_{\min}}{v_{\max} - b_{\min}}\right)^{(I-k-1)}. \tag{11.49}$$

Then, the optimal bid to maximize this profit can be obtained by differentiating the expected profit $\overline{\mathscr{P}}_i$ with respect to bid b_i, which is

$$b_{i_{\text{seq}}}^{*} = \frac{(I-k-1)v_i + b_{\min}}{I-k}. \tag{11.50}$$

For the concurrent auction, the probability of winning the auction is the probability that $b_j < b_i$ for $j \neq i$, and $j \in (I-N)$. That is, bid b_i will win the auction if all other $I-N$ bids are smaller. The probability of winning the concurrent auction can be obtained from

$$P_{\text{win}}^{\text{con}}(i) = \left(\frac{b_i - b_{\min}}{v_{\max} - b_{\min}}\right)^{(I-N)}. \tag{11.51}$$

Similarly, the optimal bid for a service provider i in concurrent auction is

$$b_{i_{\text{con}}}^{*} = \frac{(I-N)v_i + b_{\min}}{I-N+1}. \tag{11.52}$$

A comparison between sequential and concurrent auctions was also presented in [682]. The comparison was divided into three cases. Two cases considered performance evaluations at the transient state, while the other considered performance evaluations at the steady state. Here, the steady state refers to the state where all the service providers reach fixed bids, and they cannot unilaterally change their bids to obtain higher profit. In contrast, the transient state refers to the state where the service providers adjust their bids before reaching the steady state. This transient state is divided into two cases, namely, when no channel is auctioned, and k channels are already auctioned. In all three cases of comparison, it was observed that the optimal bids obtained from a sequential auction are higher than those of a concurrent auction. Therefore, the spectrum broker will prefer sequential auction since it can achieve higher revenue from the service providers.

11.3.3 Dynamic spectrum allocation via spectrum server

Spectrum policy server (SPS) can be used as a broker to allocate spectrum to the service providers. Then, these service providers use the allocated spectrum to provide wireless access services to the cognitive radio users. A non-cooperative game was formulated to obtain the solution of spectrum allocation in this spectrum policy server [671]. In this system, a user can accept service from the service provider based on the price and its satisfaction with the performance. In particular, the service acceptance probability of users for service from provider i is quantified by

$$\mathscr{A}(\mathscr{U}(r), p) = 1 - \mathrm{e}^{-C(\mathscr{U}(r))^{\mu} p^{-\epsilon}}, \quad \text{where} \quad \mathscr{U}(r) = \frac{(r/k)^{\beta}}{1+(r/k)^{\beta}}, \tag{11.53}$$

where $\mathscr{U}(r)$ is the utility defined as a function of transmission rate r, p is the price of the service, C, μ, ϵ, k, and β are constants. As the transmission rate increases, the utility of a user increases. As a result, the user will accept the service with higher probability. However, when the price is high, this acceptance probability becomes low due to the larger cost for the user. If there are more than one service providers, the user will choose the service provider with the higher acceptance probability.

For service provider i, the profit $\mathscr{P}_i$ is defined as

$$\mathscr{P}_i(r_i, p_i) = p_i - F_i - V_i b_i, \tag{11.54}$$

where F_i is the fixed cost of service provider and V_i is the price per unit of bandwidth b_i charged by the spectrum server. This bandwidth is a function of transmission rate r_i and spectral efficiency c_i, i.e. $b_i = r_i/c_i$. The acceptance probability of the service provider is $\overline{\mathscr{P}}_i(r_i, p_i) = \mathscr{A}(\mathscr{U}(r_i), p_i) \times \mathscr{P}_i(r_i, p_i)$.

Each service provider adjusts the service price and the transmission rate to achieve the highest profit. In this case, a non-cooperative game can be formulated as follows. The players of this game are the service providers. The strategy of each player is the price and the offered transmission rate, and the payoff of each player is the resulting expected profit defined as follows:

$$\phi_i((r_i, p_i), (r_j, p_j)) = \begin{cases} 0, & \mathscr{A}(\mathscr{U}(r_i), p_i) < \mathscr{A}(\mathscr{U}(r_j), p_j) \\ 1/2\overline{\mathscr{P}}_i(r_i, p_i), & \mathscr{A}(\mathscr{U}(r_i), p_i) = \mathscr{A}(\mathscr{U}(r_j), p_j) \\ \overline{\mathscr{P}}_i(r_i, p_i), & \mathscr{A}(\mathscr{U}(r_i), p_i) > \mathscr{A}(\mathscr{U}(r_j), p_j). \end{cases} \tag{11.55}$$

The Nash equilibrium of this non-cooperative game is defined as follows:

$$(r_i^*, p_i^*) = \arg\max_{r_i, p_i} \phi_i((r_i, p_i), (r_j, p_j)). \tag{11.56}$$

Since there could be multiple service providers in the system, the service selection of the user can be performed by the spectrum server to reduce complexity and overhead of

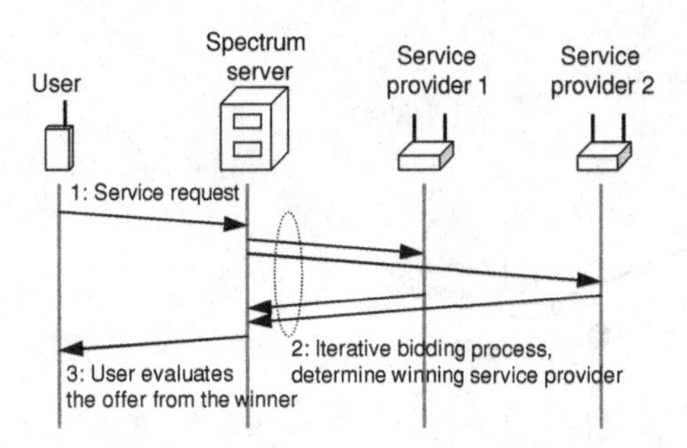

Figure 11.21 Iterative bidding through spectrum server.

the system. This service selection is referred to as the bidding process and it works as follows:

1. A new user connects to the spectrum server and also submits the acceptance probability function.
2. Service providers compete with each other to provide service to the users. This competition is iterated until the Nash equilibrium is achieved for the service price and transmission rate.
3. The winning service provider reports the winning bid in terms of service price and transmission rate to the user. Then, the user makes a decision on whether to accept the service or not based on the acceptance probability function.

These steps are shown in Figure 11.21.

The multiple-user case was also considered. In particular, service providers compete with each user individually using the same steps for iterative bidding in Figure 11.21. The spectrum server can maximize its revenue by adjusting the bandwidth allocation for each user, and an optimization problem can be formulated as follows:

$$\text{maximize: } R_{\text{server}}(\mathbf{b}) = \sum_{m=1}^{M} V_m A_m(\cdot,\cdot) B_m(\mathbf{b}), \tag{11.57}$$

$$\text{subject to: } \sum_{m=1}^{M} b_m \geq b_{\max}, \tag{11.58}$$

where $\mathbf{b}$ is a vector of bandwidth allocated to each user (i.e. $\mathbf{b} = [\cdots b_m \cdots]$), M is the total number of users, $b_{\max}$ is the maximum size of the spectrum that can be allocated to all service providers, and B_m is the transmission rate offered by the winning service provider.

With two service providers, the acceptance probabilities of both providers in case of competition are higher than those without competition. The competition benefits the users since the price becomes lower. However, the expected profits of the service providers are lower when there is a competition. In the case of multiple users, the proposed scheme was compared with the equal bandwidth partition scheme (i.e. same amount of bandwidth is allocated to all users). When the number of users is varied,

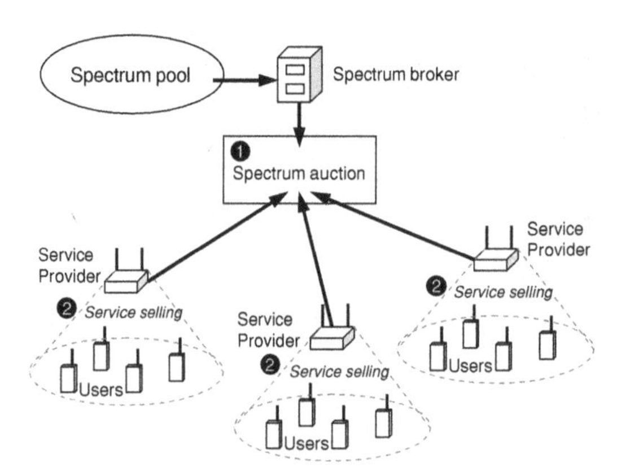

Figure 11.22 Spectrum pool and secondary market model.

the proposed spectrum server-based scheme was observed to achieve a higher expected bandwidth utilization and a higher average acceptance probability.

11.3.4 Techno-economic model for dynamic spectrum access

In [683], a dynamic spectrum access framework was proposed as a part of the *end-to-end reconfigurability* project [684] considering both economic and technical issues in cognitive radio networks. The spectrum trading model is divided into two levels. In the first level, the spectrum is auctioned among the service providers [685] (Figure 11.22). This spectrum auction is controlled by a spectrum broker. Then, in the second level, the spectrum is sold to the end-users in the MAC layer.

Given the bid b_i from service provider i, the size of the spectrum share is determined from $\Delta_b = b_i / \sum_{i=1}^{I} b_i$ which is based on the concept of weighted proportional fairness [468]. The bid submitted by each service provider is obtained by estimating expected load (i.e. service demand) from end-users. From this expected load, a service provider calculates the size of the required spectrum which gives the optimal bid in spectrum auction. This optimal bid depends not only on the QoS requirement of the end-users but also on the transmission and physical layer parameters (e.g. data rate, modulation and coding mode, and channel gain). In addition, since the expected load will depend on the service price, a service provider must choose the bid and service price simultaneously to maximize its profit. On the end-user side, negotiation between the user terminal and the service provider is used to adapt the transmission parameters and the service price accordingly. This negotiation is performed through a MAC protocol between the economic manager (EM) at the base station of a service provider and the radio resource management auction agent (RAA) at the end-user's terminal. For an end-user, the utility is used to quantify the satisfaction on the received service under a budget constraint. If the QoS requirement of both ongoing and incoming users cannot be achieved, an incoming user will be blocked.

The performance of the proposed framework was analyzed and compared with that of traditional static channel allocation. It was observed that the proposed framework can

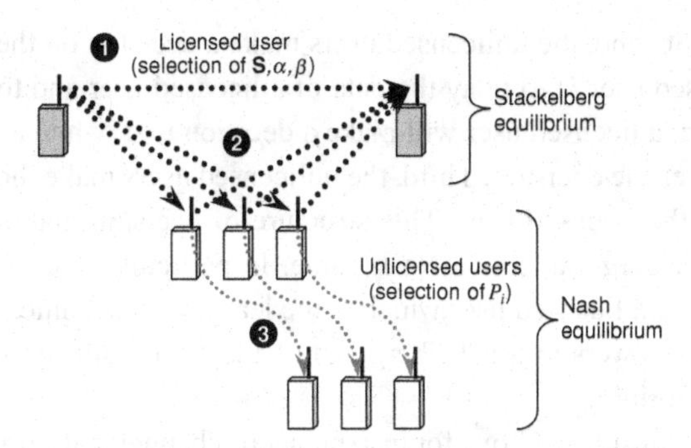

Figure 11.23 Spectrum leasing through cooperation.

accommodate a higher number of users under the same blocking probability constraint. Also, comparison of the proposed scheme with a non-shared spectrum allocation scheme revealed that the proposed framework can achieve lower blocking probability when the traffic load is small. However, since the proposed framework has to maintain the QoS guarantee to the ongoing users, the blocking probability becomes higher than that of the non-shared spectrum allocation scheme under normal and heavy load cases.

In the above, we have only considered the problem of spectrum trading through pricing. However, there are other forms of incentive for the licensed user to allow unlicensed users to access the spectrum (e.g. cooperation to relay the data). This situation was considered in [686].

11.3.5 Spectrum leasing through cooperation

Spectrum leasing from a licensed user can be implemented without charging any price to the unlicensed users. However, in this case, the unlicensed users can offer cooperation (i.e. cooperation through relaying) to licensed users. A spectrum leasing framework of this type was presented in [686]. In this system, a licensed user will lease the spectrum access to a group of unlicensed users as long as this group of unlicensed users agrees to cooperate with the licensed user to relay its transmission. With help from the unlicensed users, the transmission performance of a licensed user can be improved. The technique used for relay transmission in this work is based on distributed space-time coding [687] (i.e. decode-and-forward). While a licensed user gains benefit from the data relayed by the unlicensed users, the unlicensed users gain benefit from their own transmissions allowed by the licensed users. Three major steps in this spectrum leasing are shown in Figure 11.23. In the first step, the transmitter of a licensed user transmits data to a licensed receiver. At the same time, a group of unlicensed users can receive this data. In the second step, a selected subgroup of unlicensed users (i.e. denoted by set $\mathbb{S}$) relays this received data to the licensed receiver. In the third step, the licensed user allows a subgroup of unlicensed users to communicate. In a time slot, the duration of step one, two, and three are denoted by $(1-\alpha)$, $\alpha\beta$, and $\alpha(1-\beta)$, respectively.

There are two decisions involved in this spectrum leasing. First, a licensed user makes a decision on the duration of transmission and the subgroup of unlicensed users to relay its data (i.e. α, β, and $\mathbb{S}$). Second, the unlicensed users make a decision on the transmit power (i.e. P_i for unlicensed user i) to relay the data of a licensed user and to transmit their own data. In this case, a licensed user will make a decision first, while a subgroup of unlicensed users observes the decision. Third, the unlicensed users make the decision based on the decision of the licensed user. This structure of decision making can be modeled either as a Stackelberg game or a Nash game. In particular, the Stackelberg game models the decision of a licensed user which is the leader of this game. Then, the unlicensed users (i.e. the followers in Stackelberg game) adapt the transmission power obtained as the Nash equilibrium.

In [686], two cases were considered – one for instantaneous channel-state-information and the other for long-term channel-state-information (CSI). In the instantaneous-CSI case, the transmitter is aware of all fading power gains, namely, $h^{(l)}$ between the transmitter and the receiver of a licensed user, $h_i^{(lu)}$ between the transmitter of a licensed user and the receiver of the unlicensed user i, $h_i^{(ul)}$ between the transmitter of an unlicensed user i and the receiver of a licensed user, and $h_{i,j}$ between the transmitter of an unlicensed user i and the receiver of an unlicensed user j. In the long-term-CSI case, a licensed user knows the fading channel statistics for the entire system, while an unlicensed user knows the channel fading statistics only between its transmitter and receiver.

The transmission rate between a transmitter and receiver of a licensed user without any cooperation from an unlicensed user is given by

$$\mathscr{R}_{\text{dir}} = \log_2\left(1 + \frac{|h^{(l)}|^2 P_{\text{L}}}{\sigma^2}\right), \tag{11.59}$$

where P_{L} denotes the fixed amount of transmit power of a licensed user, and σ^2 is the noise power. The transmission rate of a licensed user with cooperation from the unlicensed users is denoted by

$$\mathscr{R}_{\text{coop}}(\alpha, \beta, \mathbb{S}) = \min((1-\alpha)\mathscr{R}_{\text{LU}}(\mathbb{S}), \alpha\beta\mathscr{R}_{\text{UL}}(\alpha, \beta, \mathbb{S})), \tag{11.60}$$

where

$$\mathscr{R}_{\text{LU}}(\mathbb{S}) = \log_2\left(1 + \frac{\min_{i\in\mathbb{S}} |h_i^{(\text{lu})}|^2 P_{\text{L}}}{\sigma^2}\right), \tag{11.61}$$

and

$$\mathscr{R}_{\text{UL}}(\alpha, \beta, \mathbb{S}) = \log_2\left(1 + \sum_{i\in\mathbb{S}} \frac{|h_i^{(\text{ul})}|^2 P_i^*(\alpha, \beta, \mathbb{S})}{\sigma^2}\right). \tag{11.62}$$

In this case, $P_i^*(\alpha, \beta, \mathbb{S})$ denotes the Nash equilibrium of the transmit power of the unlicensed users. For the Stackelberg game, an optimization problem can be formulated as follows:

$$\text{maximize:} \quad \mathscr{R}_{\text{P}}(\alpha, \beta, \mathbb{S}), \tag{11.63}$$

$$\text{subject to:} \quad 0 \leq \alpha, \beta \leq 1, \quad \mathbb{S} \subset \mathbb{S}_{\text{tot}}, \tag{11.64}$$

where $\mathbb{S}_{\text{tot}}$ is the set of all unlicensed users. To obtain the transmit power of the unlicensed users, a non-cooperative game was formulated where the utility of an unlicensed user is defined as follows [688]:

$$\mathcal{U}_i(P_i, \mathbf{P}_{-i}) = (1-\beta)\mathcal{R}_i(P_i, \mathbf{P}_{-i}) - cP_i, \tag{11.65}$$

where $\mathbf{P}_{-i}$ is a vector of transmit power of all unlicensed users except user i, and c is a cost constant. The transmission rate of an unlicensed user i can be obtained from

$$\mathcal{R}_i(P_i, \mathbf{P}_{-i}) = \log_2\left(1 + \frac{|h_{i,i}|^2 P_i}{\sigma^2 + \sum_{j \neq i} |h_{i,j}|^2 P_j}\right). \tag{11.66}$$

The Nash equilibrium of the transmit power game of the unlicensed users was observed to be unique. Also, the condition for a licensed user to allow the unlicensed users to cooperate and access the spectrum was established. An algorithm was proposed to select the unlicensed users for relaying and transmitting the data. This algorithm has linear time complexity with respect to the total number of unlicensed users. The solutions of α and β can be obtained from

$$\beta^* = \arg\max_{\beta} \beta\mathcal{R}_{\mathrm{UL}}(\beta, \mathbb{S}), \quad \alpha^* = \frac{1}{1 + \frac{\beta\mathcal{R}_{\mathrm{UL}}(\beta,\mathbb{S})}{\mathcal{R}_{\mathrm{LU}}(\mathbb{S})}}. \tag{11.67}$$

Simulation results showed that as the distance between the transmitter and the receiver increases, a licensed user reduces the sharing time (i.e. smaller α) to the unlicensed users since the licensed user has to consider its utility by maximizing its rate due to the direct transmission. However, in this case, the unlicensed users attempt to contribute more towards cooperation (i.e. larger β). A larger value of β will persuade a licensed user not to reduce the value of α too much. In other words, the unlicensed users try to contribute more towards cooperation so that a licensed user allows them to transmit over a longer time duration.

11.3.6 Improving the efficiency of spectrum access by using pricing

In an underlay spectrum access scenario, the power control problem for the unlicensed users with a constraint on the power mask to limit the interference caused to the licensed users can be modeled by using non-cooperative games. However, it was shown that the solution of such a non-cooperative game (i.e. Nash equilibrium) is not efficient [673, 689]. In the context of a cognitive radio network this implies that the payoffs of the unlicensed users may not be maximized. Also, the Nash equilibrium may not be Pareto optimal, implying that at least two players can improve their payoffs by changing their strategies. Figure 11.24 illustrates this phenomenon in which there is a region such that both players 1 and 2 can improve their payoffs. The inefficiency of Nash equilibrium arises due to the fact that each player in a non-cooperative game is rational, and hence the player will choose the best strategy which maximizes only its own payoff given the strategies of other players.

To solve this inefficiency problem of Nash equilibrium, pricing was used to provide incentives to the players [238, 690]. In [511], pricing was used to solve the problem of power control and spectrum access in a cognitive radio network. In particular, when the price charged to the transmitter is determined based on the amount of transmit power,

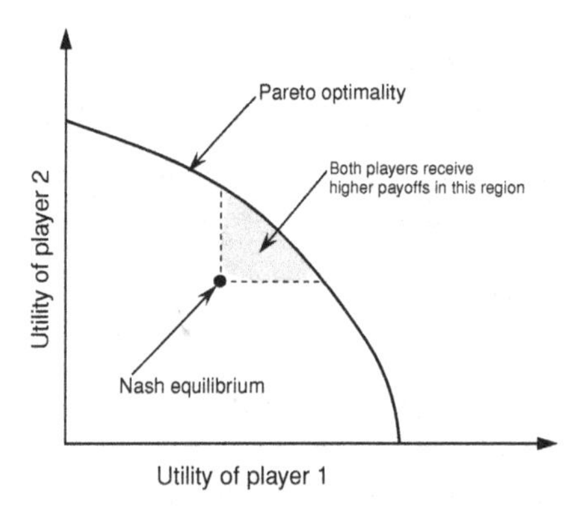

Figure 11.24 Inefficiency of Nash equilibrium.

all of the players will have incentive to transmit at low power. Consequently, the total interference in a network decreases and a higher transmission rate can be achieved for the unlicensed users. In particular, the sum of transmission rate would be the highest when the optimal price is used.

For the non-cooperative game model for a spectrum underlay scenario considered in [511], the payoff of unlicensed user i for a set of radio spectrum $\mathbb{N}$ is obtained as

$$\mathscr{U}_i(n) = \sum_{n \in \mathbb{N}} \left(\log_2 \left(1 + \frac{h_{i,i,n} P_i(n)}{\sum_{j \neq i} h_{i,j,n} P_j(n) + h_{n,i}^{(\mathrm{l})} P_n^{(\mathrm{l})} + \sigma^2} \right) \right) - p_i(n) P_i(n), \tag{11.68}$$

where $h_{i,j,n}$ is the channel gain between the transmitter of the unlicensed user i and receiver of the unlicensed user j on channel n, $P_i(n)$ is the transmit power, $h_{n,i}^{(\mathrm{l})} P_n^{(\mathrm{l})}$ denotes the interference from licensed user n (i.e. $h_{n,i}^{(\mathrm{l})}$ and $P_n^{(\mathrm{l})}$ are the channel gain and the transmit power of the licensed user, respectively), and σ^2 is the noise power. Note that, in this payoff function, a linear pricing function is applied in which $p_i(n)$ is the price per unit transmit power over channel n charged to an unlicensed use i. The Nash equilibrium can be obtained from the best response by establishing the Lagrangian function and the Karush–Kuhn–Tucker (KKT) conditions [220].

In order to improve the efficiency of the Nash equilibrium, an optimization problem was formulated to maximize the sum of transmission rates. The objective function for this optimization formulation is as follows:

$$\max_{P_1(1),\ldots,P_I(N)} \sum_{i=1}^{I} w_i \sum_{n \in \mathbb{N}} \mathscr{U}_i(n), \tag{11.69}$$

where I is the total number of unlicensed users, and w_i is the weight of the payoff from the unlicensed user i. Again, to obtain the optimal solution in terms of transmit power, the Lagrangian function and the KKT conditions were established for this optimization problem. It was shown that the price $p_i(n)$ can be chosen so that the Nash equilibrium of the transmit power game is identical to the optimal solution of (11.69).

To achieve the improved Nash equilibrium under pricing, three iterative distributed algorithms were proposed, i.e. sequential and parallel water-filling algorithms, and the

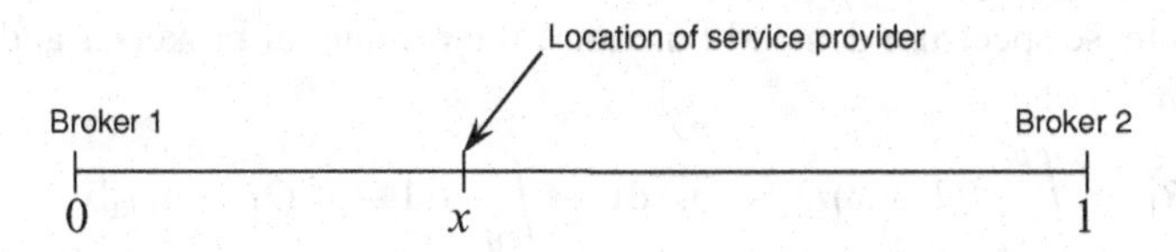

Figure 11.25 Location of service provider and the spectrum brokers.

relaxation algorithm. In the sequential algorithm, which is based on the *Gauss–Seidel* procedure, one player changes its strategy according to the best response, while the other players observe. In the parallel algorithm, which is based on the *Jacobi* procedure, all unlicensed users can simultaneously change their strategies based on the previous observation. However, these sequential and parallel algorithms require correct information of the system, and their performances are susceptible to the error. Therefore, a relaxation algorithm was proposed [691], which is robust to the estimation error and channel fluctuation in a wireless communications environment. A similar approach was considered in [692, 89] where non-linear and exponential pricing functions were applied for power control and pricing, respectively.

11.3.7 Spectrum pricing competition based on Hotelling's model

While the spectrum brokers can sell the spectrum to the service providers, the spectrum demand of the service providers depends on the price charged by the brokers. To quantify this demand, different functions can be used, and one of the most popular demand functions used in economics is Hotelling's model [693]. This model is based on the location of the buyer, and, therefore, it is used widely for spatial competition. Hotelling's model was used for pricing competition between spectrum brokers in [694].

In the system model considered in [694], there are two brokers – broker 1 and broker 2 – and two service providers, A and B. Brokers 1 and 2 are located at points 0 and 1, respectively, on the line of unit length (Figure 11.25). The service providers are uniformly distributed along this unit interval, and the locations of these service providers determine the value (e.g. utility) from buying spectrum from the broker. Let the location of service provider A be at point x (Figure 11.25). Then, for service provider A, the utility due to buying spectrum from broker 1 can be expressed as follows:

$$V_{1,\mathrm{A}} = (1 - x)Q_1 - p_1 \tag{11.70}$$

where Q_1 is the quality and p_1 is the price of spectrum from broker 1. Similarly, the utility of service provider A due to buying spectrum from broker 2 is obtained from $V_{2,\mathrm{A}} = xQ_2 - p_2$. The spectrum demand of the service providers A and B from brokers 1 and 2 can then be expressed as follows:

$$D_{1,\mathrm{A}} = \frac{Q_1 - p_1 + p_2}{Q_1 + Q_2}, \quad D_{2,\mathrm{A}} = \frac{Q_2 + p_1 - p_2}{Q_1 + Q_2}$$
$$D_{2,\mathrm{B}} = \frac{Q_2 + p_1 - p_2}{Q_1 + Q_2}, \quad D_{1,\mathrm{B}} = \frac{Q_1 - p_1 + p_2}{Q_1 + Q_2}. \tag{11.71}$$

According to these spectrum demand functions, the revenue of brokers 1 and 2 can be obtained from

$$R_1 = \int_{D_{1,A}} ((1-x)Q_1 - p_1)\,dx + \int_{D_{1,B}} ((1-x)Q_1 - p_1)\,dx, \tag{11.72}$$

$$R_2 = \int_{D_{2,A}} (xQ_2 - p_2)\,dx + \int_{D_{2,B}} (xQ_2 - p_2)\,dx. \tag{11.73}$$

Simulation results showed that, as the spectrum quality becomes better, the spectrum price of the corresponding broker increases. Also, it is observed that there are multiple Nash equilibria for the price of both brokers.

11.3.8 Pricing and admission control

The problem of admission control and pricing in cognitive radio networks was considered in [695]. In the system model, a licensed system (e.g. a cellular system) has N channels for serving the licensed users. However, if these N channels are not all occupied, some of them can be sold and allocated to unlicensed users. In this case, when a new licensed user initiates a connection, a licensed service allocates a channel to that user. The channel will be freed when the connection is terminated. Similarly, when an unlicensed user initiates the connection, a licensed service checks whether the connection can be accepted or not. This decision depends on the admission control policy employed by the licensed system. If a licensed system decides to accept an unlicensed user, that unlicensed user is charged with price p. The unlicensed user is sensitive to price p. For pricing, the following decreasing demand function was assumed [696]:

$$\lambda_u(p) = Ae^{-Bp^2}, \tag{11.74}$$

where A and B are constants, and λ_u is the connection arrival rate for the unlicensed users. Since the channel can be occupied by an unlicensed user, and the connection is non-interruptable, connections from licensed users will be blocked if all channels are occupied. This blocking effect is considered to be a cost to the licensed system. The revenue of the licensed service can be obtained from

$$R = \sum_{n=0}^{N-1} \pi_n \lambda_u(p_n) p_n - (\pi_N - P_{bl}(\lambda_l, N))\lambda_l C_{bl}, \tag{11.75}$$

where λ_l and λ_u are the connection arrival rates for licensed and unlicensed users, respectively, and π_n is the steady state probability that n channels are occupied. $P_{bl}(\lambda_l, N)$ is the blocking probability of a licensed user in the absence of unlicensed user arrivals, and C_{bl} is the cost of blocking connections from a licensed user.

The licensed system seeks an optimal policy for price p_n to maximize its revenue. Stochastic dynamic programming [697] was applied to obtain the solution. In this case, the price is discretized so that the action space is finite, and the optimal pricing policy can be obtained by the policy iteration algorithm. However, this optimal price depends on the number of occupied channels n. As a result, from an unlicensed user's viewpoint, the price is dynamic and difficult to predict. Therefore, single-price policies were also proposed in which the price charged to the unlicensed users is independent of the system state (i.e. the number of occupied channels). Two single-price policies, namely, static and threshold pricing, were considered. In both the pricing policies, connections from the unlicensed users are charged at the same price. However, for the static scheme, a connection is accepted unless all channels are occupied. In the threshold scheme, a connection is accepted only if the number of occupied channels is less than the given threshold T. It was shown that the threshold pricing policy can achieve revenue close to that of optimal pricing policy. Also, these pricing policies provide much higher revenue compared to a static pricing policy.

11.4 Summary

We have considered the economic aspects of dynamic spectrum access in a cognitive radio network. Since the radio spectrum is a very valuable resource and it requires efficient allocation to satisfy both licensed and unlicensed users, the economic issues involved in dynamic spectrum sharing are important. In this chapter, the concept of spectrum trading between primary and secondary users has been introduced. Spectrum trading is a part of the spectrum management framework which could be integrated with other components (e.g. power and admission control). One of the most important issues in spectrum trading, namely, spectrum pricing, has been discussed. To provide a background and basic understanding of spectrum trading, some basic economics theories (i.e. utility, market-equilibrium, oligopoly market, and auction) have been briefly described, and work on spectrum trading based on these theories has been reviewed.

12 Economics of dynamic spectrum access: applications of spectrum trading models

In this chapter, a number of different spectrum trading models based on economic theory are presented. In the first model, dynamic competitive spectrum sharing is modeled as a Cournot competition, which is formulated as static and dynamic non-cooperative games. From this competition, given the pricing function adopted by the primary user, the optimal amount of spectrum for secondary users needs to be determined so that the utility of each of the secondary users is maximized. In the second model, competitive spectrum pricing among primary users (or service providers) is modeled as a Bertrand competition where multiple primary service providers sell the available spectrum opportunities to a secondary service provider. The third model is a cooperative pricing model for which spectrum pricing can be obtained as the solution of an optimization model solved by a central controller. Another model is the market-equilibrium pricing model in which there is neither competition nor cooperation among primary service providers. A comparison between market-equilibrium, competitive, and cooperative spectrum pricing is presented. The characteristics of these pricing schemes are qualitatively and quantitatively compared. Also, competitive spectrum pricing in the Bertrand model is formulated as a repeated game to investigate the long-term behavior of the primary service providers. In this case, if a punishment mechanism is used and the primary service providers properly weigh their profits in the future, a collusion can be maintained to achieve the highest profit for all primary service providers. To this end, a hierarchical framework for spectrum trading in IEEE 802.22 WRANs is presented. This framework consists of a double auction model, a non-cooperative game model, and an evolutionary game model. The double auction is used for spectrum trading among multiple spectrum sellers (i.e. TV broadcasters) and multiple spectrum buyers (i.e. WRAN service providers). The evolutionary game is used to model the selection of service providers by the WRAN users. The non-cooperative game is used to obtain the solution in terms of the number of bidding TV bands and the price charged to WRAN users. A number of open research issues on spectrum trading in a dynamic spectrum access environment are then outlined.

12.1 Dynamic competitive spectrum sharing: Cournot model

In this section, the problem of spectrum sharing among a primary user and multiple secondary users is considered [698]. This problem can be formulated as an oligopoly

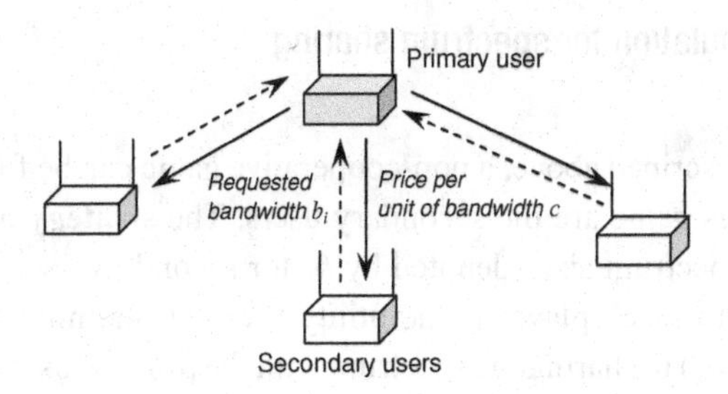

Figure 12.1 System model for spectrum sharing.

market competition and a non-cooperative game is used to obtain the spectrum allocation for secondary users. First, the formulation of a static game for the case where all secondary users have current information of the adopted strategies and the payoff of each other is presented. However, this assumption may not be realistic in some cognitive radio systems. The secondary users may gradually and iteratively adjust their strategies based on the observations on their previous strategies. The stability condition of the dynamic behavior for this spectrum sharing scheme is analytically investigated.

12.1.1 System model

We consider a spectrum overlay-based cognitive radio wireless system with one primary user and I secondary users (Figure 12.1). The primary user is willing to share a portion of the spectrum (b_i) with secondary user i. The primary user charges a secondary user for the spectrum at a rate of c per unit bandwidth, where c is a function of the total size of spectrum available for sharing by the secondary users. To enhance the transmission performance, adaptive modulation and coding (AMC) is used so that the transmission rate can be dynamically adjusted based on channel quality. The revenue of secondary user i is denoted by r_i per unit of achievable transmission rate. With AMC, the signal-to-noise ratio (SNR) γ at the receiver is partitioned into $S+1$ non-overlapping intervals with threshold denoted by γ_s, where S denotes the number of transmission modes [10]. Based on the model in [10], the probability of using transmission mode s, i.e. $\Pr(s)$ and average packet error rate $\overline{\text{PER}}_s$ for a given average SNR $\overline{\gamma}$ can be obtained. Then, the average transmission rate (i.e. spectral efficiency) k_i can be obtained for user i as follows: $k_i = \sum_{s=1}^{S} e_s \Pr(s)(1 - \overline{\text{PER}}_s)$, where e_s is the spectral efficiency of transmission mode s.

Both centralized and distributed decision making of the secondary users are considered. In the former case, each secondary user is assumed to be able to observe the strategies adopted by other users. In the latter case, the adaptation for spectrum sharing is performed in a distributed fashion based on communication between each of the secondary users and the primary user only (i.e. the secondary users are unable to observe the strategies and payoffs of each other).

12.1.2 Non-cooperative game formulation for spectrum sharing

Static game

Based on the system model described above, a non-cooperative game can be formulated as follows. The **players** in this game are the secondary users. The **strategy** of each of the players is the requested spectrum size (denoted by b_i for secondary user i), which is non-negative. The **payoff** for each player is the utility (i.e. revenue minus cost) of secondary user i (denoted by $\mathcal{U}_i$) in sharing the spectrum with the primary user and other secondary users. The **commodity** of this oligopoly market is the frequency spectrum.

The pricing function, which is used by the primary user to charge the secondary users, is given by

$$c(\mathbf{b}) = w_1 + w_2 \left(\sum_{b_j \in \mathbf{b}} b_j \right)^{\kappa}, \tag{12.1}$$

where w_1, w_2, and κ are non-negative constants, $\kappa \geq 1$ (so that this pricing function is convex), and $\mathbf{b}$ denotes a vector of strategies of all secondary users (i.e. $\mathbf{b} = [b_1 \ \ldots \ b_I]$). Let ω denote the worth of the spectrum for the primary user. Then, the condition $c(\mathbf{b}) > \omega \times \sum_{b_j \in \mathbf{b}} b_j$ is necessary to ensure that the primary user is willing to share a spectrum of size $b = \sum_{b_j \in \mathbf{b}} b_j$ with the secondary users. Note that the primary user charges all of the secondary users the same price.

The revenue of secondary user i can be obtained from $r_i k_i b_i$, while the cost of spectrum sharing is $b_i c(\mathbf{b})$. Therefore, the utility of secondary user i can be obtained as follows: $\mathcal{U}_i(\mathbf{b}) = r_i k_i b_i - b_i c(\mathbf{b})$.

Assume that the guard band used to separate the spectrum allocated to different users is fixed and small. Then, the utility can be rewritten as follows:

$$\mathcal{U}_i(\mathbf{b}) = r_i k_i b_i - b_i \left[w_1 + w_2 \left(\sum_{b_j \in \mathbf{b}} b_j \right)^{\kappa} \right]. \tag{12.2}$$

The marginal utility function for secondary user i can be obtained from

$$\frac{\partial \mathcal{U}_i(\mathbf{b})}{\partial b_i} = r_i k_i - w_1 - w_2 \left(\sum_{b_j \in \mathbf{b}} b_j \right)^{\kappa} - w_2 b_i \kappa \left(\sum_{b_j \in \mathbf{b}} b_j \right)^{\kappa - 1}. \tag{12.3}$$

Let $\mathbf{b}_{-i}$ denote a vector of strategies adopted by all except secondary user i (i.e. $\mathbf{b}_{-i} = \{b_j | j = 1, \ldots, I; j \neq i\}$ and $\mathbf{b} = \mathbf{b}_{-i} \cup \{b_i\}$). In this case, the optimal size of allocated spectrum to one secondary user depends on the strategies of other secondary users. Nash equilibrium is considered as the solution of the game to ensure that all secondary users are satisfied with the solution.

The *Nash equilibrium* of a game is a strategy profile (list of strategies, one for each player) with the property that no player can increase his payoff by choosing a different strategy, given the other players' strategies. In this case, the Nash equilibrium is obtained by using the best response function which is the best strategy of one player given others' strategies. The best response function of secondary user i given the size of the shared spectrum by other secondary users b_j, where $j \neq i$, is defined as follows:

$$\mathscr{B}_i\left(\mathbf{b}_{-i}\right) = \arg\max_{b_i} \mathscr{U}_i(\mathbf{b}_{-i} \cup \{b_i\}). \tag{12.4}$$

A vector $\mathbf{b}^* = \{b_1^*, \ldots b_I^*\}$ denotes the Nash equilibrium of this game if the following condition is satified, i.e.

$$b_i^* = \mathscr{B}_i(\mathbf{b}_{-i}^*), \quad \forall i, \tag{12.5}$$

where $\mathbf{b}_{-i}^*$ denotes a vector of best responses for secondary users j for $j \neq i$. Mathematically, to obtain the Nash equilibrium, the following set of equations is solved:

$$\begin{aligned}\frac{\partial \mathscr{U}_i(\mathbf{b})}{\partial b_i} &= r_i k_i - w_1 - w_2 \left(\sum_{b_j \in \mathbf{b}} b_j\right)^{\kappa} - w_2 b_i \kappa \left(\sum_{b_j \in \mathbf{b}} b_j\right)^{\kappa-1} \\ &= 0.\end{aligned} \tag{12.6}$$

A numerical method can be used to obtain the Nash equilibrium (i.e. allocated spectrum size b_i^*) by solving (12.6). For this, an optimization problem is formulated, with the objective defined as follows:

$$\text{minimize} \sum_{i=1}^{I} \left|b_i - \mathscr{B}_i\left(\mathbf{b}_{-i}\right)\right|. \tag{12.7}$$

This objective is to minimize the sum of the differences between decision variables b_i and the corresponding best response functions. Note that the minimum value of the objective function in (12.7) is zero if the algorithm reaches the Nash equilibrium.

12.1.3 Dynamic game model

In a distributed cognitive radio environment, secondary users may only be able to observe the pricing information from the primary user but not the strategies (i.e. the requested spectrum size) and utility of other secondary users. Therefore, the Nash equilibrium for each secondary user has to be obtained based on the observation of the price offered by the primary user only. Since all secondary users are rational, to maximize their utility they can adjust the size of the requested spectrum b_i based on the marginal utility function. In this case, each secondary user can communicate with the primary user to obtain the differentiated pricing function for different strategies. The adjustment of the requested/allocated spectrum size can be modeled as a dynamic game as follows:

$$b_i[t+1] = \mathscr{Q}(b_i[t]) = b_i[t] + \alpha_i b_i[t] \frac{\partial \mathscr{U}_i(\mathbf{b})}{\partial b_i[t]}, \tag{12.8}$$

where $b_i[t]$ is the allocated spectrum size at time t and α_i is the adjustment speed parameter of secondary user i. Note that $\mathscr{Q}(\cdot)$ here denotes the self-mapping function.

This dynamic game can be expressed as follows:

$$b_i[t+1] = b_i[t] + \alpha_i b_i[t] \left(r_i k_i - w_1 - y \left(\sum_{b_j \in \mathbf{b}} b_j[t] \right)^{\kappa} - b_i[t] y \kappa \left(\sum_{b_j \in \mathbf{b}} b_j[t] \right)^{\kappa - 1} \right). \tag{12.9}$$

Since the strategies of the players are adjusted iteratively in a dynamic game, ensuring the stability/convergence of the algorithm is crucial. The stability analysis can be performed for the case of $\kappa = 1$. This case corresponds to a linear pricing function adopted by the primary user. With this linear pricing function, the primary user will charge a higher price if the demand for the spectrum is higher. At the Nash equilibrium, the following condition holds, i.e. $\mathbf{b}[t+1] = \mathbf{b}[t] = \mathbf{b}$, namely, $\mathbf{b} = \mathscr{Q}(\mathbf{b})$, where $\mathscr{Q}(\mathbf{b})$ is the self-mapping function of the fixed point $\mathbf{b}$. With the linear pricing function, the fixed point can be obtained by solving the following set of equations

$$\alpha_i b_i \left(r_i k_i - w_1 - 2 b_i w_2 - w_2 \sum_{j \neq i} b_j \right) = 0, \quad i = \{1, \ldots, I\}. \tag{12.10}$$

The Jacobian matrix of the fixed point of the Nash equilibrium can be expressed as follows:

$$\mathbf{J} = \begin{bmatrix} \frac{\partial b_1[t+1]}{\partial b_1} & \cdots & \frac{\partial b_1[t+1]}{\partial b_I} \\ \vdots & & \vdots \\ \frac{\partial b_I[t+1]}{\partial b_1} & \cdots & \frac{\partial b_I[t+1]}{\partial b_I} \end{bmatrix}. \tag{12.11}$$

The local stability of the dynamic game spectrum sharing can be analyzed based on localization by considering the eigenvalues of the Jacobian matrix of the mapping. The fixed point is stable if and only if the eigenvalues λ_i are all inside the unit circle of the complex plane (i.e. $|\lambda_i| < 1$) [699].

With two secondary users in the cognitive radio environment, four fixed points are defined as $\mathbf{b}_0$, $\mathbf{b}_1$, $\mathbf{b}_2$, and $\mathbf{b}_3$, which can be expressed as follows:

$$\mathbf{b}_1 = \left(\frac{r_1 k_1 - w_1}{2 w_2}, 0 \right), \quad \mathbf{b}_2 = \left(0, \frac{r_2 k_2 - w_1}{2 w_2} \right)$$

$$\mathbf{b}_3 = \left(\frac{r_2 k_2 - 2(r_1 k_1) + w_1}{-3 w_2}, \frac{r_1 k_1 - 2(r_2 k_2) + w_1}{-3 w_2} \right) \tag{12.12}$$

and $\mathbf{b}_0 = (0, 0)$, where $\mathbf{b}_3$ is the Nash equilibrium.

With two secondary users, there are two eigenvalues, and the Jacobian matrix can be expressed as in (12.13).

$$\mathbf{J}(b_1, b_2) = \begin{bmatrix} 1 + \alpha_1 (r_1 k_1 - w_1 - 4 w_2 b_1 - w_2 b_2) & -w_2 \alpha_1 b_1 \\ -w_2 \alpha_2 b_2 & 1 + \alpha_2 (r_2 k_2 - w_1 - 4 w_2 b_2 - w_2 b_1) \end{bmatrix}. \tag{12.13}$$

Then, the stability condition at each fixed point is investigated. For $\mathbf{b}_0$, the Jacobian matrix becomes

$$\mathbf{J}(0,0)=\begin{bmatrix}1+\alpha_1(r_1k_1-w_1) & 0\\ 0 & 1+\alpha_2(r_2k_2-w_1)\end{bmatrix}. \tag{12.14}$$

The eigenvalues are given by the diagonal elements of $\mathbf{J}(\cdot)$ if matrix $\mathbf{J}(\cdot)$ is diagonal or triangular. The coordinate (0, 0) will be stable if $|1+\alpha_1(r_1k_1-w_1)|<1$ and $|1+\alpha_2(r_2k_2-w_1)|<1$. First, the following cases are considered

$$1+\alpha_1(r_1k_1-w_1)>-1,\ 1+\alpha_2(r_2k_2-w_1)>-1, \tag{12.15}$$

or, equivalently,

$$\alpha_1(r_1k_1-w_1)>-2,\ \alpha_2(r_2k_2-w_1)>-2, \tag{12.16}$$

which are possible. Now, the following cases are considered: $1+\alpha_1(r_1k_1-w_1)<1$ and $1+\alpha_2(r_2k_2-w_1)<1$, from which the stability condition for $\mathbf{b}_0$ can be expressed as follows:

$$r_1k_1<w_1 \quad \text{and} \quad r_2k_2<w_1. \tag{12.17}$$

These conditions imply that when none of the secondary users is willing to share the spectrum with the primary user (which corresponds to fixed point (0, 0)) the system is stable. That is, when the cost of the spectrum (as offered by the primary user) is higher than the revenue gained from allocated spectrum, a secondary user will not be willing to share the spectrum.

For fixed point $\mathbf{b}_1$, the Jacobian matrix can be expressed as follows:

$$\mathbf{J}\left(\frac{r_1k_1-w_1}{2w_2},0\right)=\begin{bmatrix}1+\alpha_1(-r_1k_1+w_1) & \alpha_1\frac{r_1k_1-w_1}{2}\\ 0 & 1+\alpha_2\frac{1}{2}(2r_2k_2-r_1k_1-w_1)\end{bmatrix}. \tag{12.18}$$

For the first eigenvalue, it can be shown in a way similar to that of $\mathbf{b}_0$ that the condition $|1+\alpha_1(-r_1k_1+w_1)|<1$ or $r_1k_1>w_1$ is satisfied. However, for the second eigenvalue, the following condition is evaluated

$$\left|1+\alpha_2\frac{1}{2}(2r_2k_2-r_1k_1-w_1)\right|<1. \tag{12.19}$$

Only the case where $1+\alpha_2\frac{1}{2}(2r_2k_2-r_1k_1-x)<1$ or $2r_2k_2-r_1k_1-x<0$ is considered.

The possible conditions are:

- $2r_2k_2<r_1k_1+w_1$: In this case, the eigenvalue can be smaller than one. As a result, fixed point $\mathbf{b}_1$ is stable. In other words, fixed point $\mathbf{b}_1$ can be obtained by *map* $\mathcal{Q}(\mathbf{b}(t))$ if the value of r_1k_1 in the revenue function for the first secondary user is much larger than that for the other secondary user (i.e. r_2k_2).
- $2r_2k_2\geq r_1k_1+w_1$: In this case, fixed point $\left(\frac{r_1k_1-w_1}{2w_2},0\right)$ is never stable, since the Nash equilibrium obtained from (12.12) gives a solution which is non-negative.

For fixed point $\mathbf{b}_2$, the Jacobian matrix can be expressed as follows:

$$\mathbf{J}\left(\frac{r_1k_1 - w_1}{2w_2}, 0\right) = \begin{bmatrix} 1 + \alpha_1(-r_1k_1 + w_1) & 0 \\ \alpha_2 \frac{r_2k_2 - w_1}{2} & 1 + \alpha_2 \frac{1}{2}(2r_2k_2 - r_1k_1 - w_1) \end{bmatrix}. \tag{12.20}$$

Similar to the case for fixed point $\mathbf{b}_1$, the possible conditions are

- $2r_1k_1 < r_2k_2 + w_1$: In this case, the eigenvalue can be smaller than one. As a result, fixed point $\mathbf{b}_2$ is stable. In other words, fixed point $\mathbf{b}_2$ can be obtained by *map* $\mathcal{Q}(\mathbf{b}(t))$ if the value of r_2k_2 in the revenue function of the second user is much larger than that of the first secondary user (i.e. r_1k_1).
- $2r_1k_1 \geq r_2k_2 + w_1$: In this case, fixed point $\left(\frac{r_2k_2 - w_1}{2w_2}, 0\right)$ is never stable, since the Nash equilibrium obtained from (12.12) gives the solution which is non-negative.

For fixed point $\mathbf{b}_3$, which is the Nash equilibrium, the Jacobian matrix can be expressed as follows:

$$\mathbf{J}\left(\frac{r_2k_2 - 2(r_1k_1) + w_1}{-3w_2}, \frac{r_1k_1 - 2(r_2k_2) + w_1}{-3w_2}\right) = \begin{bmatrix} j_{1,1} & j_{1,2} \\ j_{2,1} & j_{2,2} \end{bmatrix}, \tag{12.21}$$

where

$$j_{1,1} = 1 + \alpha_1\left(\frac{r_1k_1 - 2w_2r_2k_2 + 3w_2r_1k_1 - w_2w_1 - w_1}{-3w_2}\right),$$

$$j_{1,2} = -w_2\left(\frac{r_2k_2 - 2r_1k_1 + w_1}{-3w_2}\right),$$

$$j_{2,1} = -w_2\left(\frac{r_1k_1 - 2r_2k_2 + w_1}{-3w_2}\right),$$

$$j_{2,2} = 1 + \alpha_2\left(\frac{r_2k_2 - 2w_2r_1k_1 + 3w_2r_2k_2 - w_2w_1 - w_1}{-3w_2}\right).$$

Since this Jacobian matrix is neither diagonal nor triangular, the characteristic equation to obtain the eigenvalues is given as follows: $\lambda^2 - \lambda(j_{1,1} + j_{2,2}) + (j_{1,1}j_{2,2} - j_{1,2}j_{2,1}) = 0$, and hence

$$(\lambda_1, \lambda_2) = \frac{(j_{1,1} + j_{2,2}) \pm \sqrt{4j_{1,2}j_{2,1} + (j_{1,1} - j_{2,2})^2}}{2}. \tag{12.22}$$

Basically, given r_1, r_2, k_1, k_2, w_1, and w_2, the relationship between α_1 and α_2 for which the fixed point of Nash equilibrium is stable can be obtained. When the Nash equilibrium is stable, the profit of the secondary users cannot be increased by altering the allocated spectrum size (i.e. the marginal profit is zero).

12.1.4 Performance evaluation results

The numerical results obtained for the above static and dynamic game formulations of spectrum sharing between two secondary users can be summarized as follows [698]:

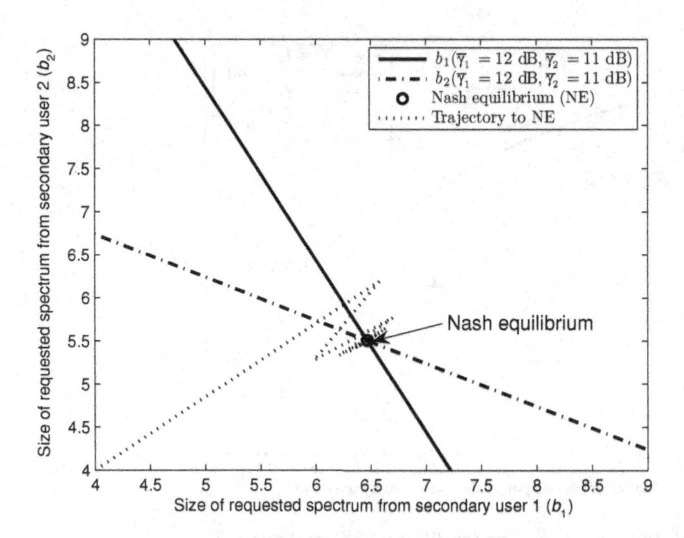

Figure 12.2 Best response, Nash equilibrium, and convergence of dynamic game.

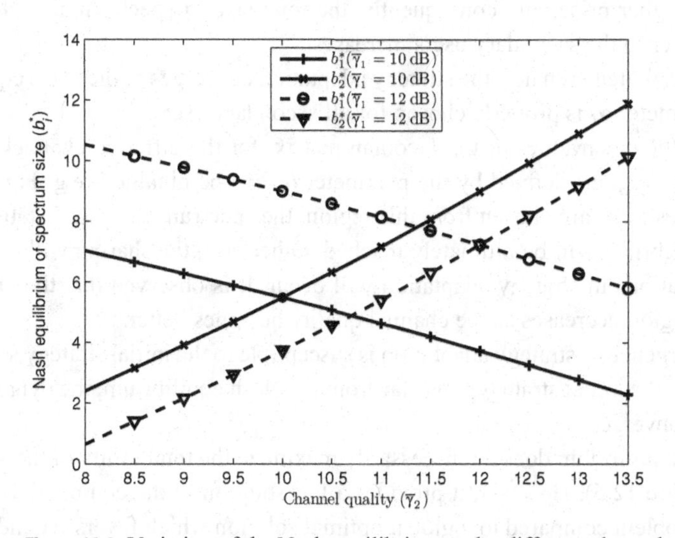

Figure 12.3 Variation of the Nash equilibrium under different channel quality.

- The Nash equilibrium is located at the point where all best responses of the secondary users intersect (Figure 12.2). This Nash equilibrium depends on the channel quality. As the channel quality improves, a secondary user prefers to share a larger spectrum size with the primary user (Figure 12.3). Also, the channel quality of one secondary user impacts the allocated spectrum size for the other secondary user. Similar results are expected for a larger number of secondary users.
- Due to the fixed spectrum size of the primary user, the spectrum share for each of the secondary users decreases as the number of users increases. However, since there is more competition (i.e. with a higher number of secondary users), the primary user can

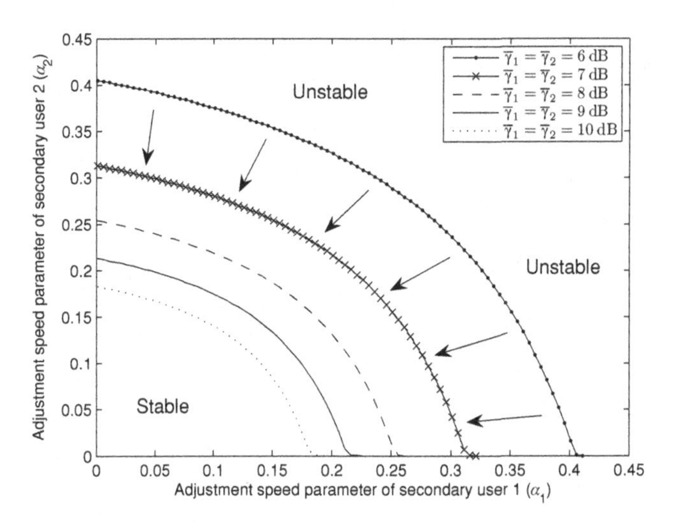

Figure 12.4 Stability region of α_i under different channel quality.

charge a higher price, and consequently, the total size of spectrum allocated by the primary user to the secondary users increases.

- For the distributed scenario, the strategy adaptation converges to the Nash equilibrium if the parameter α_i is properly chosen for all secondary users.
- Based on the eigenvalues of the Jacobian matrix, for the different channel qualities, the stability regions defined by the parameter α_i can be obtained (e.g. Figure 12.4). If the values of α_i are chosen from this region, the spectrum sharing is stable and the Nash equilibrium will be ultimately reached. Otherwise, the sharing will be unstable and fluctuations in strategy adaptation will occur. It is observed that the area of the stability region decreases as the channel quality becomes better.
- The convergence of strategy adaptation is susceptible to the initial strategy setting. For example, if the initial strategy is too far from the Nash equilibrium, the dynamic game may not converge.
- The Nash equilibrium does not necessarily maximize the total utility of the secondary users (Figure 12.5). However, it provides a fair solution to the competitive spectrum sharing problem compared to a global optimal solution which favors secondary users with larger transmission rates.

12.2 Bertrand competition, market-equilibrium pricing, and cooperative spectrum pricing

In this section, a spectrum trading formulation is presented for multiple licensed service providers to sell spectrum opportunities to a group of secondary users (i.e. a secondary service provider) [3]. Since spectrum pricing depends not only on the demand from the secondary service providers (or secondary users), but also on the behavior of the licensed service providers, the following different cases for the behaviors of licensed

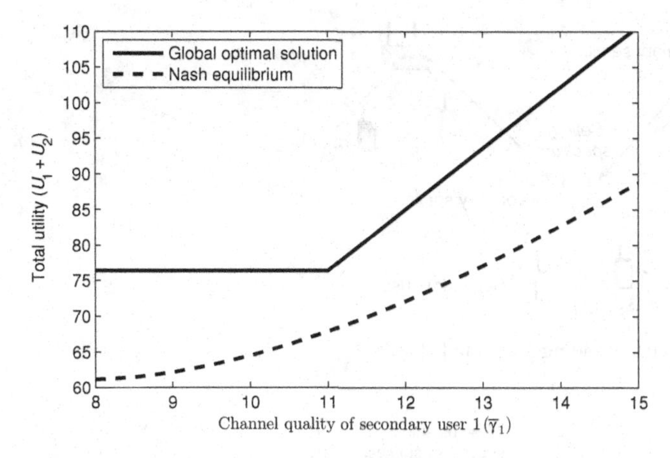

Figure 12.5 Total utility obtained from global optimal solution compared with that from Nash equilibrium.

service providers are considered: (1) neither competition nor cooperation among service providers, (2) competition among service providers, and (3) cooperation among service providers. The pricing schemes corresponding to these behaviors are referred to as market-equilibrium pricing, competitive pricing, and cooperative pricing, respectively. In a market-equilibrium pricing scheme, the objective of spectrum trading is only to satisfy spectrum demand from the secondary service providers (or secondary users). The licensed service provider in this case is not aware of the existence of other licensed service providers. In a competitive pricing scheme, since the licensed service providers are aware of the existence of other licensed service providers, the objective of each licensed service provider is to maximize its own profit. In a cooperative pricing scheme, a licensed service provider is not only aware of the existence of other licensed services, but also cooperates with other service providers. In this case, the objective of spectrum trading is to maximize the total profit. To achieve the solutions of these pricing schemes, distributed algorithms are also presented. Performance evaluation results show that while the cooperative pricing can achieve the highest profit for the licensed service providers, in terms of convergence, the market-equilibrium pricing is the most stable.

12.2.1 System model and pricing schemes

In the system model considered in [3], there are N licensed service providers where service provider n serves M_n local licensed users (Figure 12.6). A licensed service provider may sell portions of its spectrum to a secondary service provider (or secondary users). The spectrum sharing between licensed and secondary users can be performed based on a time-division multiple access (TDMA)-based wireless access scheme as shown in Figure 12.7. In such a spectrum trading market, the seller and the buyer correspond to the licensed and secondary service providers, respectively. The price (per unit spectrum) is denoted by p_n. Both licensed users and secondary users exploit adaptive modulation for wireless transmission. The spectrum demand of the secondary

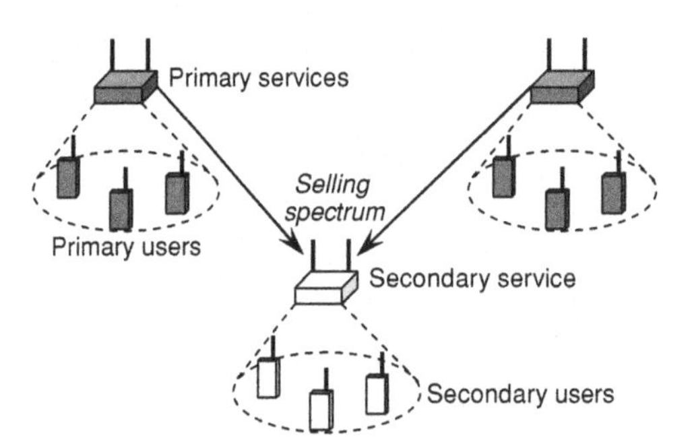

Figure 12.6 Spectrum trading model in [3].

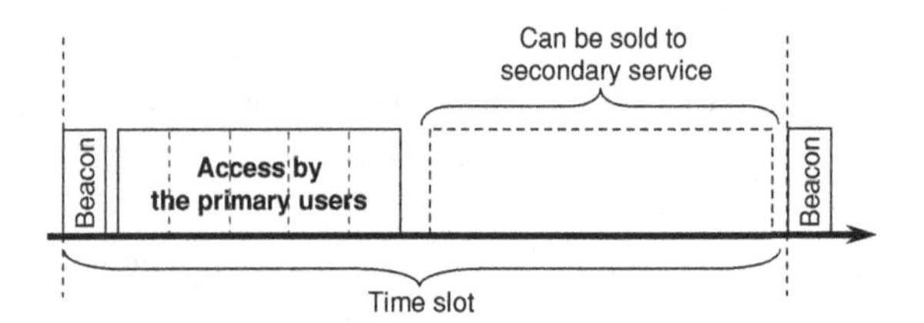

Figure 12.7 Time-slot based spectrum sharing.

users depends on the transmission rate achieved through adaptive modulation on the allocated frequency spectrum and the price charged by the licensed service providers. The price p_n is set based on three different pricing schemes, namely, market-equilibrium, competitive, and cooperative pricing schemes:

- *Market-equilibrium pricing scheme:* In this pricing scheme, a primary service provider is not aware of other primary service providers. This phenomenon could be due to the lack of any central controller or information exchange among primary service providers. As a result, at the seller side, a primary service provider naively sets the price according to the spectrum demand from the secondary service providers (or secondary users). This price setting is based on the willingness of the primary service provider to sell spectrum which is generally determined by the *supply function*. For a given price, the supply function indicates the size of radio spectrum to be shared by a primary user with secondary users. At the buyer side, the willingness of a secondary service provider to buy spectrum is determined by the *demand function*. Again, for a given price, the demand function determines the size of radio spectrum required by secondary users. The market-equilibrium price denotes the price for which spectrum supply from the primary service provider is equal to the spectrum demand from the secondary users. This market-equilibrium price ensures that there is no excess supply in the market and all spectrum demand is satisfied.
- *Competitive pricing scheme based on the Bertrand model:* In this pricing scheme, a primary service provider is aware of the existence of other primary service providers. These primary service providers compete with each other to achieve the highest individual profit. The primary service providers compete through price adjustment.

That is, given the spectrum prices offered by other primary service providers, one primary service provider chooses the price for its own spectrum so that its individual profit is maximized. This competitive spectrum pricing is similar to a price war in an oligopoly market where a smaller number of sellers dominate a particular market.

- *Cooperative pricing scheme:* In this pricing scheme, all of the primary service providers know each other and they fully cooperate to obtain the highest total profit by selling spectrum to the secondary users. In an actual environment, to achieve this full cooperation, all the primary service providers would be required to communicate with each other.

12.2.2 Utility and spectrum demand of secondary users

If the spectrum available to a secondary user creates high utility (e.g. due to larger size and/or higher quality), the spectrum demand by a secondary service provider becomes high. To quantify the satisfaction of spectrum access by secondary users, a utility function is used. The utility function used in [3] is a quadratic utility function [700] which can be expressed as follows:

$$\mathscr{U}(\mathbf{b}) = \sum_{n=1}^{N} b_n k_n^{(s)} - \frac{1}{2}\left(\sum_{n=1}^{N} b_n^2 + 2\nu \sum_{n \neq n'} b_n b_{n'}\right) - \sum_{n=1}^{N} p_n b_n \tag{12.23}$$

where $\mathbf{b}$ is a vector of shared spectrum sizes from all the primary services, i.e. $\mathbf{b} = [b_1 \ \ldots \ b_n \ \ldots \ b_N]$, p_n is the price offered by primary service provider n, and N is the total number of primary service providers. This utility function takes the quality of spectrum into account through parameter $k_n^{(s)}$, which denotes the spectral efficiency of wireless transmission by the secondary users using frequency spectrum from primary service provider n, and ν denotes the spectrum substitutability factor where $(0 \leq \nu \leq 1)$. In this instance, if a secondary user uses a multi-interface network adaptor, it is able to switch among the frequency spectra freely depending on the price offered by primary service providers. This spectrum substitutability factor ν is defined as follows: when $\nu = 0.0$, a secondary user cannot switch among the frequency spectra, while for $\nu = 1.0$ a secondary user can switch among the operating frequency spectra freely.

The demand function for spectrum from primary service provider n can be obtained from $\partial\mathscr{U}(\mathbf{b})/\partial b_n = 0$ as follows:

$$\mathscr{D}_n(\mathbf{p}) = \frac{(k_n^{(s)} - p_n)(\nu(N-2)+1) - \nu \sum_{n' \neq n}(k_{n'}^{(s)} - p_{n'})}{(1-\nu)(\nu(N-1)+1)} \tag{12.24}$$

where $\mathbf{p}$ denotes a vector of prices offered by all primary service providers in the market (i.e. $\mathbf{p} = \begin{bmatrix} p_1 & \ldots & p_n & \ldots & p_N \end{bmatrix}^{\mathrm{T}}$). To simplify the expression, the demand function in (12.24) can be rewritten as follows: $\mathscr{D}_n(\mathbf{p}) = D_1(\mathbf{p}_{-n}) - D_2 p_n$, where $\mathbf{p}_{-n}$ denotes the vector of prices of all primary service providers except service provider n, $D_1(\mathbf{p}_{-n})$

and D_2 are constants for given $p_{n'}$ for $n \neq n'$, which are given as follows:

$$D_1(\mathbf{p}_{-n}) = \frac{k_n^{(s)}(\nu(N-2)+1) - \nu \sum_{n \neq n'} (k_{n'}^{(s)} - p_{n'})}{(1-\nu)(\nu(N-1)+1)},$$
$$D_2 = \frac{(\nu(N-2)+1)}{(1-\nu)(\nu(N-1)+1)}. \tag{12.25}$$

12.2.3 Revenue and cost functions for a primary service provider

A primary service provider can earn revenue from two sources, i.e. from local primary users and secondary users. While the spectrum can be sold to the secondary users to receive higher revenue to the primary service provider, the performance of primary users can be degraded. In this case, the performance degradation is considered as a cost for a primary service provider. In particular, the local primary users are charged by a primary service provider at a flat rate for a guaranteed amount of bandwidth. However, if the required bandwidth cannot be provided, the primary service provider offers a "discount" to the users, and this is considered as the cost of sharing spectrum with the secondary service provider.

Let $\mathscr{R}_n^{\mathrm{l}}$ denote the revenue gained from primary users served by primary service provider n, $\mathscr{R}_n^{\mathrm{s}}$ denote the revenue gained from sharing spectrum with secondary users, and $\mathscr{C}_n$ denote the cost due to QoS performance degradation of all primary users. The revenue and cost functions can be defined as follows:

$$\mathscr{R}_n^{\mathrm{s}} = p_n b_n, \quad \mathscr{R}_n^{\mathrm{l}} = c_1 M_n,$$
$$\mathscr{C}_n(b_n) = c_2 M_n \left(B_n^{\mathrm{req}} - k_n^{(p)} \frac{W_n - b_n}{M_n} \right)^2, \tag{12.26}$$

where b_n and p_n denote the spectrum size shared with the secondary users and the corresponding price, respectively, c_1 and c_2 are constants for the revenue and cost functions at the primary service, respectively, B_n^{req} is the bandwidth requirement per user, W_n is the spectrum size, M_n is the number of ongoing primary users, and $k_n^{(\mathrm{p})}$ is the spectral efficiency of the wireless transmission for primary service provider n. In this case, given the spectrum price, the revenue from secondary users is a linear function of the shared spectrum size. The cost is proportional to the square of the difference between bandwidth requirement and allocated bandwidth to a primary user.

12.2.4 Spectrum supply, spectrum demand, and market-equilibrium

The spectrum supply indicates the size of the spectrum to be sold by a primary service provider given price p_n. This spectrum supply function can be derived based on a profit maximization problem. The solution of this optimization formulation is the optimal spectrum size b_n to be shared with the secondary users. Based on revenue (i.e. $\mathscr{R}_n^{\mathrm{l}}$ and $\mathscr{R}_n^{\mathrm{s}}$) and cost (i.e. $\mathscr{C}_n$), the profit $\mathscr{P}_n$ of a particular primary service provider n can be expressed as follows:

$$\mathscr{P}_n = p_n b_n + c_1 M_n - c_2 M_n \left(B_n^{\text{req}} - k_n^{(\text{p})} \frac{W_n - b_n}{M_n} \right)^2. \qquad (12.27)$$

The optimal spectrum size to be sold can be obtained by differentiating the profit function with respect to b_n (when p_n is given) as follows:

$$\frac{\partial \mathscr{P}_n}{\partial b_n} = 0 = -p_n + 2c_2 M_n \left(B_n^{\text{req}} - k_n^{(\text{p})} \frac{W_n - b_n}{M_n} \right) \frac{k_n^{(\text{p})}}{M_n}. \qquad (12.28)$$

The optimal value of b_n^* gives the supply function as follows:

$$b_n^* = \mathscr{S}_n(p_n) = W_n - \frac{M_n}{k_n^{(\text{p})}} \left(B_n^{\text{req}} - \frac{p_n}{2c_2 k_n^{(\text{p})}} \right). \qquad (12.29)$$

The market-equilibrium (i.e. solution) is defined as the price p_n^* at which spectrum supply equals spectrum demand, i.e.

$$\mathscr{S}_n(p_n^*) = \mathscr{D}_n(\mathbf{p}^*), \quad \forall n, \qquad (12.30)$$

where the vector $\mathbf{p}^* = [\cdots \quad p_n^* \quad \cdots]^{\text{T}}$ consists of the market-equilibrium prices for all primary service providers.

It can be easily observed that the supply function does not depend on the prices offered by other primary service providers, since a primary service provider is not aware of the existence of other primary service providers.

12.2.5 Non-cooperative game formulation for Bertrand competition

To model the price competition among primary service providers, a non-cooperative game is applied. The players in this game are the primary service providers. The strategy of each of the players is the price offered per unit of spectrum. The payoff for primary service provider n (denoted by $\mathscr{P}_n$) is the individual profit due to selling spectrum under competition to the secondary users.

Based on the demand, revenue, and cost functions, the individual profit of each primary service provider can be expressed as follows:

$$\mathscr{P}_n(\mathbf{p}) = \mathscr{R}_n^{\text{s}} + \mathscr{R}_n^{\text{l}} - \mathscr{C}_n, \qquad (12.31)$$

where $\mathbf{p}$ denotes a vector of prices offered by all of the players (i.e. the primary service providers) in the game.

The solution of this non-cooperative game for price competition is the Nash equilibrium. In this case, the Nash equilibrium is obtained by using the best response function of the players. The best response of a player is the best strategy of this player given other players' strategies. The best response function of primary service provider n, given the vector of prices offered by other primary service providers $\mathbf{p}_{-n}$, is defined as follows:

$$\mathscr{B}_n(\mathbf{p}_{-n}) = \arg\max_{p_n} \mathscr{P}_n(p_n, \mathbf{p}_{-n}). \qquad (12.32)$$

The vector $\mathbf{p}^\star = \begin{bmatrix} \cdots & p_n^\star & \cdots \end{bmatrix}^{\mathrm{T}}$ denotes the Nash equilibrium of this game if

$$p_n^\star = \mathscr{B}_n(\mathbf{p}_{-n}^\star), \quad \forall n, \tag{12.33}$$

where $\mathbf{p}_{-n}^\star$ denotes the vector of best responses for player n' for $n' \neq n$. Mathematically, to obtain the Nash equilibrium, the following set of equations is solved: $\partial \mathscr{P}_n(\mathbf{p})/\partial p_n = 0$ for all n. In this case, the size of the shared bandwidth b_n in the individual profit function is replaced with the spectrum demand $\mathscr{D}_n(\mathbf{p})$, so the profit function can be expressed as follows:

$$\mathscr{P}_n(\mathbf{p}) = p_n \mathscr{D}_n(\mathbf{p}) + c_1 M_n - c_2 M_n \left(B_n^{\text{req}} - k_n^{(p)} \frac{W_n - \mathscr{D}_n(\mathbf{p})}{M_n} \right)^2. \tag{12.34}$$

Then, using $\frac{\partial \mathscr{P}_n(\mathbf{p})}{\partial p_n} = 0$, we obtain

$$\begin{aligned} 0 = {} & 2c_2 k_n^{(\mathrm{p})} D_2 \left(B_n^{\text{req}} - k_n^{(p)} \frac{W_n - (D_1(\mathbf{p}_{-n}) - D_2 p_n)}{M_n} \right) \\ & + D_1(\mathbf{p}_{-n}) - 2D_2 p_n. \end{aligned} \tag{12.35}$$

Recall that the demand function can be expressed as $\mathscr{D}_n(\mathbf{p}) = D_1(\mathbf{p}_{-n}) - D_2 p_n$.

The solution $p_n^\star$, which is the Nash equilibrium, can be obtained by solving the above set of linear equations. Then, given the vector of prices $\mathbf{p}^\star$, the size of the shared spectrum can be obtained from the spectrum demand function $\mathscr{D}_n(\mathbf{p}^\star)$.

12.2.6 Cooperative spectrum pricing based on optimization

When all primary service providers cooperate, the highest profit can be achieved through an optimal price which is obtained from an optimization problem formulation. This optimization problem can be expressed as follows:

$$\text{maximize:} \quad \sum_{n=1}^{N} \mathscr{P}_n(\mathbf{p}), \tag{12.36}$$

$$\text{subject to:} \quad W_n \geq b_n \geq 0, \tag{12.37}$$

$$p_n \geq 0, \tag{12.38}$$

where the total profit for all the primary service providers is given by $\sum_{n=1}^{N} \mathscr{P}_n(\mathbf{p})$. Since the constraint in (12.37) can be written as $W_n \geq \mathscr{D}(\mathbf{p}) \geq 0$, the Lagrangian can be expressed as follows:

$$\begin{aligned} \mathscr{L}(\mathbf{p}) = {} & \sum_{n=1}^{N} \mathscr{P}_n(\mathbf{p}) - \sum_{j=1}^{N} \lambda_j(-p_j) - \sum_{k=1}^{N} \mu_k(\mathscr{D}_k(\mathbf{p}) - W_k) \\ & - \sum_{l=1}^{N} \sigma_l(-\mathscr{D}_l(\mathbf{p})), \end{aligned} \tag{12.39}$$

where λ_j, μ_k, and σ_l are Lagrange multipliers for the constraints in (12.37), and (12.38), respectively. Using Karush–Kuhn–Tucker (KKT) conditions, the vector of optimal prices $\mathbf{p}^\bullet$ can be obtained such that the total profit of all of the primary services is maximized.

12.2.7 Distributed algorithms for the pricing schemes

In a cognitive radio environment, the primary service providers (or primary users) may not have the complete network information. For example, a primary service may not know the form of the demand function and the channel quality for the secondary users. In case of competitive and cooperative pricing schemes, the prices and profits of other primary service providers may not be observable. From a practical perspective, distributed algorithms would be required to reach the solutions of the above pricing schemes. A primary service provider may observe and learn the behavior of secondary users and other primary service providers from the historical observations.

Distributed algorithm for market-equilibrium pricing

Since the market-equilibrium gives a price for which spectrum supply from a primary service provider is equal to spectrum demand from the secondary users, the price offered by each primary service provider is gradually adjusted in a direction that minimizes the difference between spectrum demand and spectrum supply. This distributed market-equilibrium pricing algorithm works as follows:

1. The spectrum price $p_n[0]$ is initialized and this price is sent to the secondary service provider (or secondary service controller). Note that this initial price should be chosen to be neither too small nor too large so that the algorithm converges to the final solution quickly.
2. The secondary service provider replies with the size of spectrum demand which is computed from the demand function $\mathscr{D}_n(\mathbf{p}[t])$ for spectrum from primary service provider n.
3. The primary service provider computes the size of the supplied spectrum $\mathscr{S}_n(p_n[t])$.
4. To obtain the price in iteration $t+1$, the difference between spectrum demand and supply at iteration t is computed, weighted by the learning rate α_n, and added to the price in the current iteration.
5. The above steps are repeated until the termination criterion is satisfied (i.e. the difference of prices in current iteration t and next iteration $t+1$ becomes less than the threshold ϵ, e.g. $\epsilon = 10^{-5}$).

This price adjustment in each iteration (i.e. Step 4 of the algorithm above) can be expressed as follows:

$$p_n[t+1] = p_n[t] + \alpha_n \left(\mathscr{D}_n(\mathbf{p}[t]) - \mathscr{S}_n(p_n[t])\right), \tag{12.40}$$

where $\mathbf{p}[t]$ is the vector of prices at iteration t, i.e.

$$\mathbf{p}[t] = \begin{bmatrix} p_1[t] & \cdots & p_n[t] & \cdots & p_N[t] \end{bmatrix}^{\mathrm{T}}. \tag{12.41}$$

Distributed algorithm for Bertrand pricing competition

In a competitive pricing scheme, the Nash equilibrium is a solution (i.e. set of prices) that satisfies all of the primary service providers. If a primary service provider cannot observe the prices of other primary service providers, each primary service provider can use only its price information and spectrum demand information from the secondary service provider to adjust its strategy. This distributed competitive pricing algorithm works as follows:

1. The spectrum price is initialized to $p_n[0]$ and this price is sent to the secondary service provider (or controller).
2. The secondary service provider replies with the size of spectrum demand.
3. The primary service provider estimates marginal individual profit $\frac{\partial \mathscr{P}_n(\mathbf{p}[t])}{\partial p_n[t]}$.
4. The marginal individual profit is used together with spectrum demand information to compute the spectrum price in the next iteration.
5. The above steps are repeated until a termination criterion is satisfied.

The relationship between the prices in the current and the next iterations (i.e. Step 4 of the algorithm above) is as follows:

$$p_n[t+1] = p_n[t] + \alpha_n \left(\frac{\partial \mathscr{P}_n(\mathbf{p}[t])}{\partial p_n[t]} \right), \tag{12.42}$$

where α_n is the learning rate.

Let $\mathbf{p}_{-n}[t]$ denote the vector of prices of all primary services except service n at iteration t. To estimate the marginal individual profit (i.e. Step 3 of the algorithm above), a primary service can observe the marginal spectrum demand for a small variation in price ξ (e.g. $\xi = 10^{-4}$). That is,

$$\frac{\partial \mathscr{P}_n(\mathbf{p}[t])}{\partial p_n[t]} \approx \frac{\mathscr{P}_n([\cdots p_n[t] + \xi \cdots]) - \mathscr{P}_n([\cdots p_n[t] - \xi \cdots])}{2\xi}. \tag{12.43}$$

Distributed algorithm for cooperative pricing

The optimal price obtained from the cooperative pricing scheme achieves the highest total profit of all primary service providers. In this case, a primary service provider can observe the spectrum demand from the secondary service provider and also can exchange profit information with all primary service providers. This distributed cooperative pricing algorithm works as follows:

1. The spectrum price is initialized to $p_n[0]$ and then it is sent to the secondary service provider.
2. The secondary service provider replies with the size of spectrum demand.
3. A primary service provider estimates marginal total profit $\frac{\partial \sum_{n'=1}^{N} \mathscr{P}_{n'}(\mathbf{p}[t])}{\partial p_n[t]}$ by exchanging information with the rest of the primary service providers.
4. The marginal total profit is used together with the spectrum demand from a secondary service provider to compute the spectrum price in the next iteration.
5. The above steps are repeated until a termination criterion is satisfied.

The relationship between the prices in the current and next iterations (i.e. Step 4 in the algorithm above) can be expressed as follows:

$$p_n[t+1] = p_n[t] + \alpha_n \left(\frac{\partial \sum_{n'=1}^{N} \mathscr{P}_{n'}(\mathbf{p}[t])}{\partial p_n[t]} \right). \tag{12.44}$$

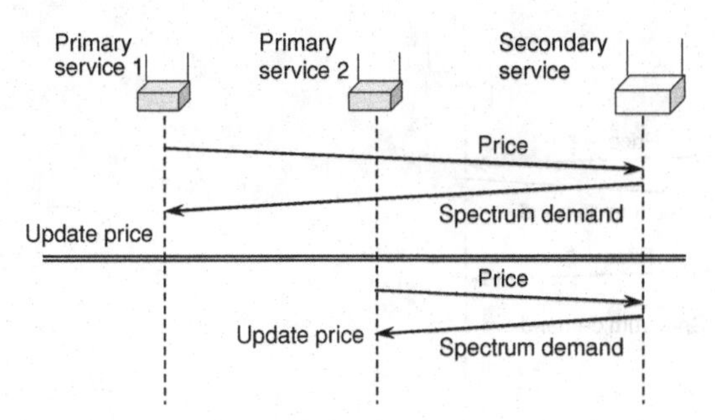

Figure 12.8 Message exchange in market-equilibrium pricing scheme.

Again, similar to that in (12.43), to estimate the marginal total profit (i.e. Step 3 in the algorithm above), a primary service provider can observe the marginal total profit (e.g. through information exchange among all primary service providers) for a small variation in price ξ as in (12.45):

$$\frac{\partial \sum_{n'=1}^{N} \mathscr{P}_{n'}(\mathbf{p}[t])}{\partial p_n[t]} \approx \frac{1}{2\xi} \left(\sum_{n'=1}^{N} \mathscr{P}_{n'}([\quad \cdots \quad p_n[t] + \xi \quad \cdots \quad]) - \sum_{n'=1}^{N} \mathscr{P}_{n'}([\quad \cdots \quad p_n[t] - \xi \quad \cdots \quad]) \right). \tag{12.45}$$

12.2.8 Information exchange protocol

For distributing pricing competition, the signaling protocols to exchange information among primary service providers and secondary service providers (or secondary users) can be described as follows:

- *Market-equilibrium pricing:* In one iteration, each primary service provider chooses the price and sends it to the secondary service provider. The secondary service provider responds with spectrum demand to each primary service provider (Figure 12.8). Therefore, in one iteration, two messages are exchanged between a primary service provider and the secondary service provider.
- *Competitive pricing:* In one iteration, each primary service provider estimates marginal profit by using two messages. One message is for $p_n[t] - \xi$ and another is for $p_n[t] + \xi$. The secondary service provider also responds with two messages (Figure 12.9). After the marginal profit is obtained, each primary service provider adjusts its price and continues to the next iteration. Therefore, in one iteration, four messages are exchanged between a primary service provider and the secondary service provider.
- *Cooperative pricing:* In one iteration, each primary service provider estimates marginal total profit by using two messages (Figure 12.10). Similar to that of competitive pricing, the first message is for $p_n[t] - \xi$. Then, the secondary service provider responds with spectrum demand to all primary service providers. Each primary service provider computes its own profit and exchanges this information with others. The second

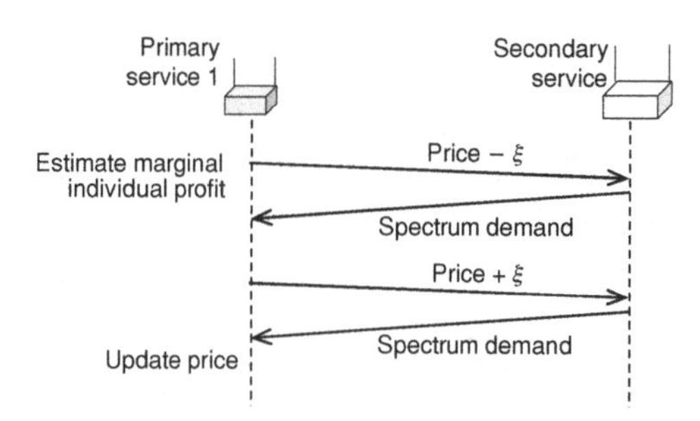

Figure 12.9 Message exchange in competitive pricing scheme.

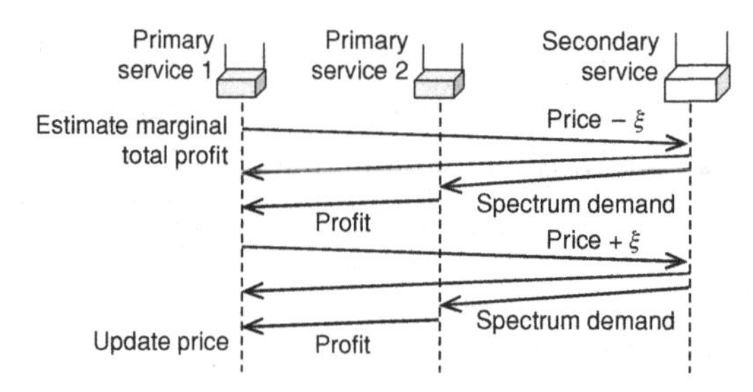

Figure 12.10 Message exchange in cooperative pricing scheme.

message is for $p_n[t] + \xi$. The same process is repeated. Therefore, in one iteration, $4N$ messages are exchanged between primary service providers and the secondary service provider, where N is the total number of primary service providers.

12.2.9 Performance evaluation results

The performance evaluation results for the three different pricing schemes with two primary service providers can be summarized as follows [3]:

- A primary service provider sharing spectrum with secondary service providers (or secondary users) can gain higher profit than a service provider which does not share.
- In general, cooperative pricing can maximize the total profit of all primary service providers, while the market-equilibrium pricing achieves the lowest total profit (e.g. in Figure 12.11).
- When the channel quality of the spectrum owned by one primary service provider becomes better (i.e. higher SNR), the price charged to the corresponding secondary users increases while the price offered by the other primary service provider decreases (Figure 12.12). Since the secondary users achieve higher spectral efficiency, the secondary users can gain higher transmission rates from the same amount of spectrum, which results in higher utility, and, therefore, the primary service provider can increase the spectrum price to earn more revenue.
- As the value of the spectrum substitutability factor increases (i.e. the secondary users can switch among the spectra from primary service providers freely),

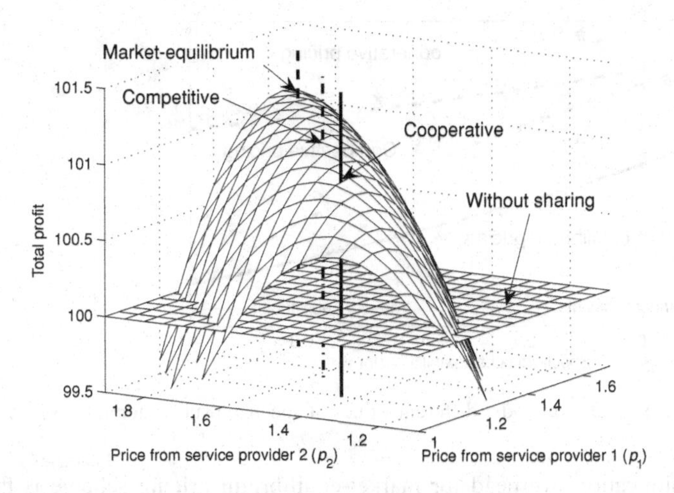

Figure 12.11 Total profit of the service providers.

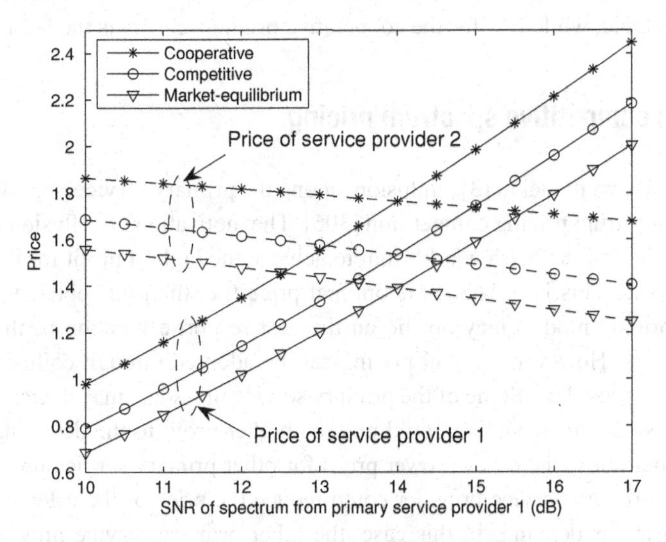

Figure 12.12 Adaptation of price under different SNR.

spectrum price decreases for all the pricing schemes. In addition, when the spectrum is not substitutable, the competitive price becomes identical to the cooperative price. When $\nu = 0$, there is no competition since none of the secondary users can change its operating spectrum and each of the primary service providers can set the spectrum price to gain the highest profit. However, when the spectrum is perfectly substitutable or $\nu = 1$, the degree of competition becomes high since a secondary user can switch to the primary service provider offering the lowest price. As a result, the primary service providers decrease the spectrum price to attract more demand, and this decrease in price converges to the market-equilibrium price. These effects are shown in Figure 12.13.

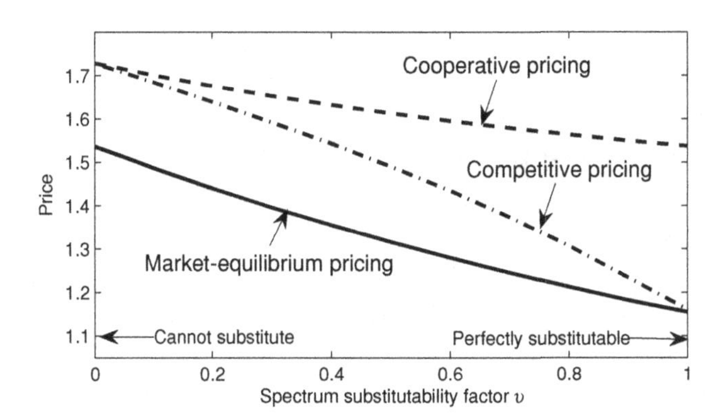

Figure 12.13 Price adaptation under different values of spectrum substitutability factor.

- The communication overhead for market-equilibrium pricing scheme is the lowest while that for the cooperative pricing scheme is the largest.
- It is observed that the distributed algorithm for the market-equilibrium pricing scheme is the most stable, while that for the cooperative pricing scheme is the least stable.

12.3 Collusion in competitive spectrum pricing

Based on the system model in [3], collusion among the primary service providers can be analyzed for spectrum pricing competition [306]. The motivation of collusion arises due to the inefficiency of the Nash equilibrium to achieve the highest profit for the primary service providers. This is because, the optimal price (i.e. the price obtained from the cooperative pricing model) may not lie on the best response function of the primary service providers. However, optimal pricing can be adopted through collusion among primary service providers. Some of the primary service providers may deviate from this optimal price since this deviation could provide higher profit to the deviating primary service provider but at the cost of lower profit for other primary service providers. For example, one primary service provider could reduce the price of the offered spectrum to generate a larger demand. In this case, the other primary service providers could punish the deviating primary service (e.g. by reducing the spectrum price to generate higher demand) to decrease the profit of the deviating primary service provider. A repeated game is formulated to analyze this situation. It can be shown that if the primary service providers are aware of the punishment due to deviation, and if the profit in the future is properly weighted, a collusion among the primary service providers can be maintained to choose the optimal price so that the highest profit can be achieved in the long run.

12.3.1 Formulation of a repeated game

Since the optimal prices (i.e. obtained from the cooperative pricing model) to maximize the total profit of the primary service providers are different from those at the Nash equilibrium, in a non-cooperative (or competitive) environment, some primary service

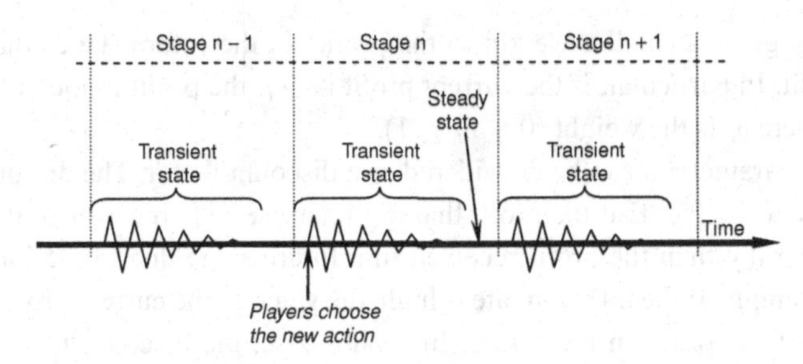

Figure 12.14 Repeated game.

providers could unilaterally deviate from optimal price, especially when the game is played only once (i.e. a one-shot game). In other words, the optimal pricing does not give a stable equilibrium. However, if the game is repeated, it is desirable for the players in the game to achieve an efficient solution to earn a high long-term profit. This can be achieved if all of the primary service providers establish a collusion. Since the game is repeated, a punishment mechanism can be implemented to deter any primary service provider from deviating from the collusion. To model this, a repeated game is formulated which captures the behavior of the primary service providers when the pricing game is infinitely repeated.

In a repeated game, the players play the game multiple times, and the outcome of the previous play can be observed. As a result, the players can learn and coordinate their actions so that the desired result is achieved. If the game is repeated, each step in playing the game can be defined as a stage. In general, one stage lasts for the period starting at the time when the players choose their actions and ending at the time when the outcome is observed. In a distributed implementation of competitive pricing, since the price adaptation can be in a transient state before it reaches the steady state, the outcome of the game cannot be instantaneously observed. In this case, the transient state exists for a time period during which the difference between prices in two consecutive iterations is larger than the threshold ϵ (otherwise, the price adaptation is in steady state). This dynamic Bertrand game can be defined as a repeated game for which each stage is defined from the time that the players change the parameters of price adaptation to the time that the steady state is reached. The stages of a repeated dynamic game can be shown as in Figure 12.14.

In this repeated dynamic game of spectrum pricing competition, the set of actions consists of *maintaining collusion*, *deviating from collusion*, and *punishment actions*. When the punishment action is used, all of the primary service providers choose the Nash equilibrium strategy from which none of the primary service providers wants to deviate. Also, the total profit at the Nash equilibrium is less than or equal to that due to the optimal pricing.

A *trigger strategy* is used in this repeated game. With this *trigger strategy*, any primary service provider will maintain collusion as long as other service providers agree to do so. However, if a primary service provider deviates from the collusion (i.e. triggers), then all of the other primary service providers will use the punishment action permanently. In this case, the primary service providers will consider the long-term profit. However, since part of the long-term profit will be gained in future stages, a primary service provider usually gives a smaller weight to the profits in the future stages than that to the current profit. In particular, if the current profit is $\mathscr{P}_i$, the profit in the next stage is worth $\delta_i \mathscr{P}_i$, where δ_i is the weight ($0 \leq \delta_i \leq 1$).

Note that the parameter δ can be considered as a discount factor. The discount factor is able to consider the fact that the profit that a primary service receives in the current stage is more worthy than the profit received in a future stage due to the interest and inflation. For example, if the inflation rate is high, the value of the current profit is higher than the value of the profit in the future. In such a case, the discount factor is small since the primary service will be interested in the current profit more than the future profit. Similarly, if the interest rate is high, the current profit can be invested to earn more benefit. Therefore, the current value of the profit is higher, and again the discount factor is small.

For primary service provider i, let $\mathscr{P}_i^{\mathrm{o}}$, $\mathscr{P}_i^{\mathrm{n}}$, and $\mathscr{P}_i^{\mathrm{d}}$ denote the profits due to the optimal price (i.e. cooperative pricing), the price at the Nash equilibrium, and the price due to deviation, respectively. Then, for the case that the collusion is maintained forever, the long-term profit of primary service provider i can be expressed as follows:

$$\mathscr{P}_i^{\mathrm{o}} + \delta_i \mathscr{P}_i^{\mathrm{o}} + \delta_i^2 \mathscr{P}_i^{\mathrm{o}} + \delta_i^3 \mathscr{P}_i^{\mathrm{o}} + \cdots = \frac{1}{1-\delta_i} \mathscr{P}_i^{\mathrm{o}}. \tag{12.46}$$

In other words, the long-term profit is the weighted-sum of the current profit and the profit in all future stages. On the other hand, if one primary service provider deviates from the optimal price, that service provider will gain deviated profit in the first stage, while during the rest of the stages, this primary service provider will gain the profit corresponding to the Nash equilibrium. Therefore, the long-term profit can be expressed as follows:

$$\mathscr{P}_i^{\mathrm{d}} + \delta_i \mathscr{P}_i^{\mathrm{n}} + \delta_i^2 \mathscr{P}_i^{\mathrm{n}} + \delta_i^3 \mathscr{P}_i^{\mathrm{n}} + \cdots = \mathscr{P}_i^{\mathrm{d}} + \frac{\delta_i}{1-\delta_i} \mathscr{P}_i^{\mathrm{n}}. \tag{12.47}$$

The collusion will be maintained if the long-term profit due to adopting collusion is higher than that due to deviation, i.e.

$$\frac{1}{1-\delta_i} \mathscr{P}_i^{\mathrm{o}} \geq \mathscr{P}_i^{\mathrm{d}} + \frac{\delta_i}{1-\delta_i} \mathscr{P}_i^{\mathrm{n}}. \tag{12.48}$$

Therefore, the value of δ_i needs to be chosen such that the following condition is satisfied:

$$\delta_i \geq \frac{\mathscr{P}_i^{\mathrm{d}} - \mathscr{P}_i^{\mathrm{o}}}{\mathscr{P}_i^{\mathrm{d}} - \mathscr{P}_i^{\mathrm{n}}}. \tag{12.49}$$

For the values of δ_i obtained as above, a collusion will be maintained in which the optimal price will be used to gain the highest profit in each state, and consequently, the highest long-term profit.

12.3.2 Performance evaluation results

The numerical results obtained for the repeated game formulation with two primary service providers can be summarized as follows [306]:

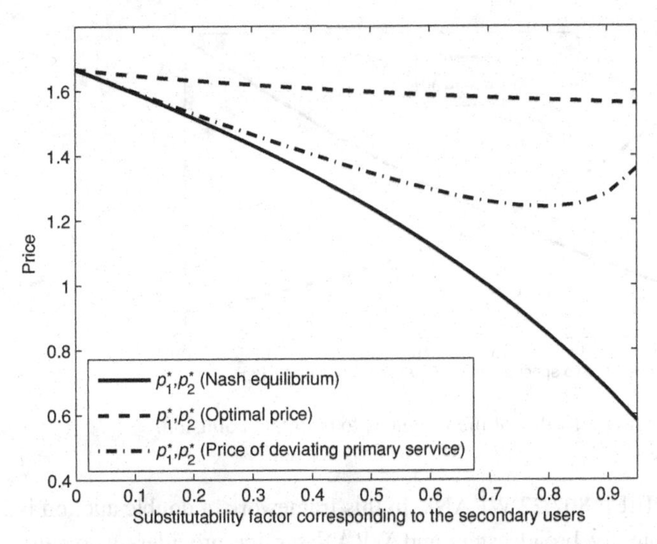

Figure 12.15 Price at the Nash equilibrium, the optimal price, and the price of deviating primary service provider.

- If all primary service providers make an agreement to establish a collusion, the optimal price will be offered to the secondary users. However, one primary service provider can unilaterally deviate to achieve a higher profit than that it can gain at the optimal price. This price due to deviation is observed to be always lower than the optimal price and higher than that at the Nash equilibrium (Figure 12.15).
- In order to maintain the collusion, both of the primary service providers must consider the outcome (i.e. profit) in the future. If one of the primary service providers deviates from the optimal price, the future outcome is affected by the punishment mechanism. To maintain the collusion, a proper value of weight δ_i needs to be used in the punishment mechanism.
- When the channel quality corresponding to the spectrum offered by primary service provider 2 becomes higher, the corresponding weight δ_i becomes smaller while that of primary service provider 1 becomes larger. Since a better channel quality may result in higher benefit (i.e. profit) due to deviation from the optimal price, primary service provider 2 has more motivation to deviate. As a result, the value of δ_i can be reduced to maintain the collusion (Figure 12.16).

12.4 Economic model for spectrum sharing in IEEE 802.22 WRANs

Spectrum trading among TV broadcasters, WRAN service providers, and WRAN users will be a significant issue in an IEEE 802.22-based WRAN. An IEEE 802.22-based WRAN will operate on TV bands which are licensed to the TV broadcasters. Therefore, radio resource allocation and pricing need to be considered between the service providers and WRAN users, as well as between the spectrum owner (i.e. TV broadcasters) and WRAN service providers. In this section, a hierarchical spectrum trading framework is

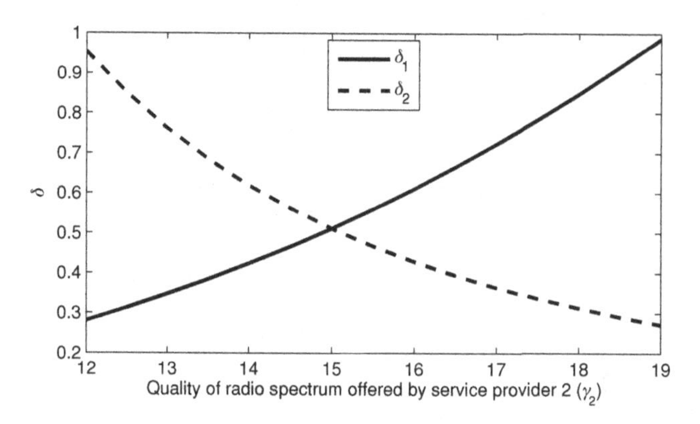

Figure 12.16 The highest value of the weight δ_n to maintain collusion.

presented for IEEE 802.22 WRANs. In this framework, a double auction is established among multiple TV broadcasters and WRAN service providers who sell and buy the TV bands, respectively. Again, multiple WRAN service providers compete with each other by adjusting the price charged to the WRAN users. This framework considers the problem of spectrum bidding and pricing jointly for the WRAN service providers to maximize their profits.

12.4.1 System model for spectrum trading in IEEE 802.22 WRANs

A WRAN system with N TV broadcasters, each of which owns one TV band (i.e. a total of N bands) is considered. There are I WRAN BSs each of which is operated by a different competitive service provider. Without loss of generality, it is assumed that there are I service providers in total. The radio spectrum used by these BSs is procured from the TV broadcasters. All of these TV bands are assumed to be non-overlapping to avoid co-channel interference among cognitive radio transmissions controlled by different BSs in a service region. The service region under consideration is composed of multiple service areas, and each service area contains at least one WRAN BS. The number of WRAN users in area a is denoted by U_a. It is possible that one area can have multiple services from multiple WRAN BSs (e.g. area 3 in Figure 12.17). In each service area, a WRAN user can choose the service provider which provides the best payoff in terms of allocated bandwidth and service price. An example of a service region with three service areas is shown in Figure 12.17.

A double auction [681] is used to model this multiple-seller and multiple-buyer spectrum trading (i.e. multiple TV broadcasters owning multiple TV bands and multiple WRAN service providers, respectively). After a WRAN service provider obtains the TV band through a double auction, the price charged to the WRAN users is determined. In a service area where multiple WRAN service providers are available, WRAN users can churn among the service providers when the performances and/or prices offered by the WRANs change. Therefore, the WRAN service providers must competitively determine their spectrum bidding for a double auction and price for WRAN users optimally so

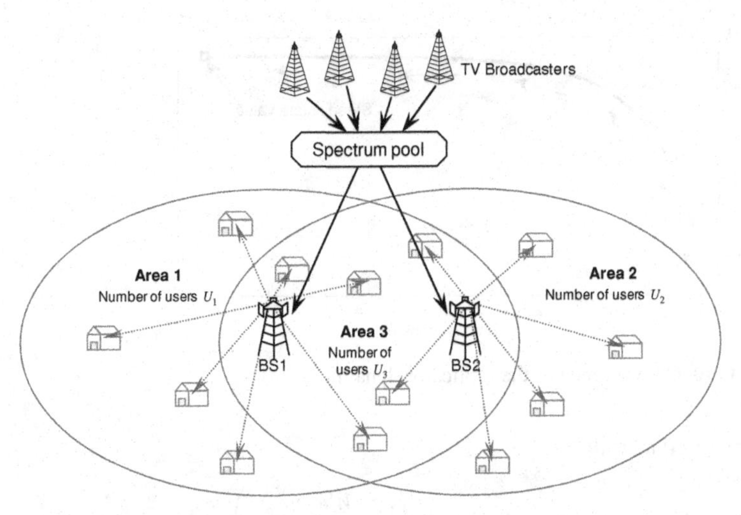

Figure 12.17 A service region covered by the IEEE 802.22 WRAN BSs.

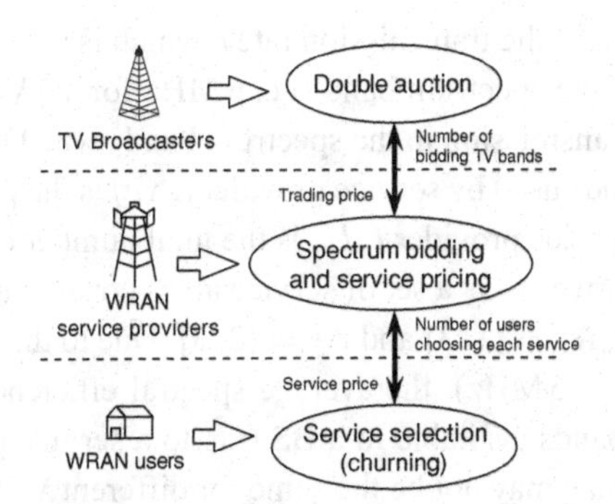

Figure 12.18 Joint spectrum bidding and service pricing for spectrum trading in IEEE 802.22 WRAN.

that their profits are maximized. The structure of this joint spectrum bidding and pricing scheme is shown in Figure 12.18.

12.4.2 Selection of service providers

Among multiple available WRAN service providers, a WRAN user will choose the service provider which returns the highest benefit. In particular, if one service provider offers higher transmission rate and/or lower service price, a user will churn to that service provider. To select and change the service provider, a user does not have global information of all service providers. Therefore, a user will observe the performances and prices offered by other wireless service providers, and the decision of a user to choose a particular service provider will be made locally and independently without considering any effect to other users. The benefit of a user is the net utility, which can be defined as the difference between the utility and the price charged by the WRAN service providers. That is, the benefit of a user in service area a selecting service provider i can

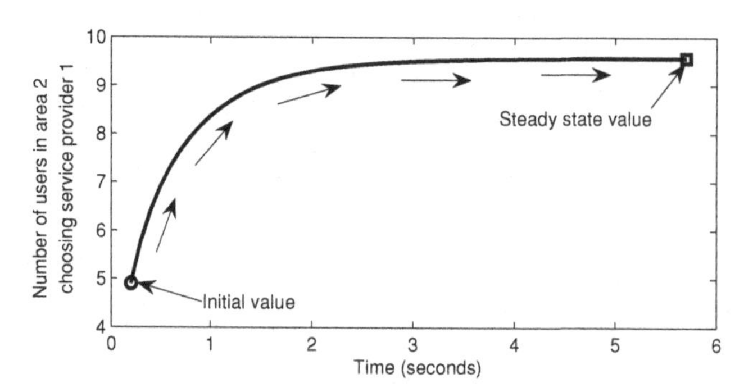

Figure 12.19 Convergence of the replicator dynamics.

be expressed as follows:

$$\mathcal{N}_{i,a} = \mathcal{U}\left(\frac{Wk_i c_i}{\sum_{a \in \mathbb{A}_i} x_{i,a} U_a}\right) - p_i, \tag{12.50}$$

where $\mathcal{U}(r)$ is the utility as a function of the transmission rate r which is assumed to be $\mathcal{U}(r) = \log(1 + r)$, W is the size of the spectrum band (i.e. 6 MHz for a TV band), k_i is the average spectral efficiency of transmission in the spectrum band owned by service provider i, c_i is the number of TV bands used by service provider i, $x_{i,a}$ is the proportion of users in service area a choosing service provider i, U_a is the total number of users in service area a, and p_i is the service price. $\mathbb{A}_i$ is a set of service areas covered by service provider i. As shown in Figure 12.17, $\mathbb{A}_1 = \{1, 3\}$ and $\mathbb{A}_2 = \{2, 3\}$. Due to the relatively large bandwidth of a TV channel (i.e. 6 MHz), the average spectral efficiency can be assumed to be the same for all TV bands available to a BS (i.e. to a service provider). However, the average spectral efficiency may not be the same for different WRAN BSs, for instance, due to different locations of the BSs.

Based on the benefit of a user defined in (12.50), the change of proportion of users in area a choosing service provider i can be expressed as follows:

$$\frac{\mathrm{d}x_{i,a}}{\mathrm{d}t} = \dot{x}_{i,a} = x_{i,a}(\mathcal{N}_{i,a} - \overline{\mathcal{N}}_a), \tag{12.51}$$

where $\overline{\mathcal{N}}_a$ is the average benefit of users in service area a, obtained from

$$\overline{\mathcal{N}}_a = \sum_{i=1}^{I} x_{i,a} \mathcal{N}_{i,a}. \tag{12.52}$$

The convergence of $x_{i,a}$ defined in (12.51) is shown in Figure 12.19. Given an initial value, the user will switch to the service provider which offers better performance and/or lower price. At the steady state, the number of users choosing each service provider remains constant.

Note that (12.51) above is referred to as the replicator dynamics [701] in the evolutionary game theory. The replicator dynamics are obtained based on the fact that a population of users will tend to evolve to choose a strategy (i.e. service provider) which provides a higher than the average benefit. From the evolution of the users in (12.51),

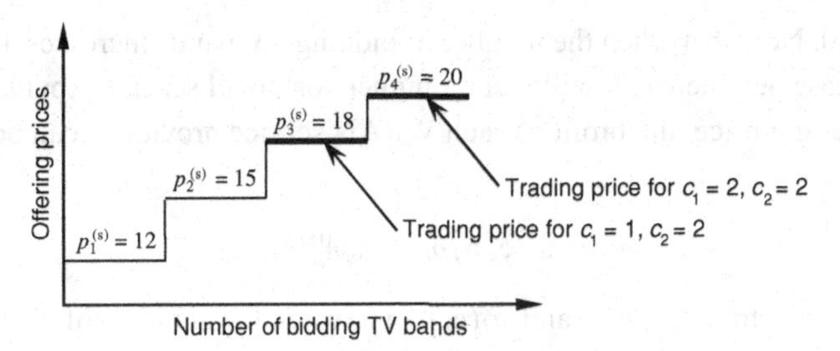

Figure 12.20 Example of trading price in double auction for spectrum trading in IEEE 802.22 WRAN.

the solution of the replicator dynamics which gives the total number of users choosing service provider i can be obtained from

$$u_i = \sum_{a \in \mathbf{A}_i} x_{i,a} U_a. \tag{12.53}$$

12.4.3 Non-cooperative game formulation of joint spectrum bidding and service pricing

According to the hierarchical spectrum trading shown in Figure 12.18, a WRAN service provider has to determine (1) the number of TV bands to bid for (referred to as bidding bands here) given the offered price from the TV broadcasters, and (2) the service price to a WRAN user. For spectrum trading among multiple TV broadcasters and multiple WRAN service providers, the double auction model [681] (see Section 11.2.4) from microeconomics can be naturally applied.

A sealed-bid double auction model is assumed for trading TV bands. In this sealed-bid double auction, all sellers (i.e. TV broadcasters) and buyers (i.e. WRAN service providers) submit offering and bidding prices without knowing similar information of others. Then, given all offering and bidding prices, the trading price is determined. Also, the assignment of a TV band from a seller to a buyer is obtained. In this sealed-bid double auction, there are N sellers and I buyers. A seller sells TV band n by offering price $p_n^{(s)}$ (i.e. the offering price). Without the loss of generality, it is assumed that $p_1^{(s)} < \cdots < p_n^{(s)} < \cdots < p_N^{(s)}$. Service provider i bids for c_i TV bands, and the total number of bidding TV bands is denoted by $C = \sum_{i=1}^{I} c_i$. It is assumed that $C \leq N$. From this offering price and the number of bidding TV bands, the trading price, which is defined as the price of TV band to be sold by each TV broadcaster, can be obtained from $p^{(t)} = p_C^{(s)}$. In particular, this trading price is the highest offering price at which C TV bands are sold to the WRAN service providers.

For example (Figure 12.20), assume that there are four TV broadcasters selling four TV bands and their offering prices are $p_n^{(s)} = \{12, 15, 18, 20\}$ for $n = 1, \ldots, 4$. If there are two WRAN service providers and the number of bidding bands for them are $c_1 = 1$ and $c_2 = 2$ (i.e. the first and second service providers require one and two TV bands, respectively), the trading price is $p^{(t)} = p_3^{(s)} = 18$. However, if the number of bidding TV bands from the service providers is $c_1 = 2$ and $c_2 = 2$, the trading price becomes

$p^{(t)} = p_4^{(s)} = 20$. Note that when the number of bidding TV bands increases, the trading price will increase, and hence, it will incur a higher cost to all service providers.

Given the trading price, the profit of each WRAN service provider i can be obtained as follows:

$$\mathscr{P}_i = u_i(\mathbf{c}, \mathbf{p})p_i - c_i p^{(t)}(\mathbf{c}), \tag{12.54}$$

where $p^{(t)}(\mathbf{c})$ is the trading price and $u_i(\mathbf{c}, \mathbf{p})$ indicates the number of WRAN users choosing service provider i. Note that $u_i(\mathbf{c}, \mathbf{p})$ can be expressed as a function of the set of number of TV bands the different service providers bid for, i.e. $\mathbf{c} = [c_1, \ldots, c_i, \ldots, c_I]$ and the set of prices offered by the service providers to the WRAN users, i.e. $\mathbf{p} = [p_1, \ldots, p_i, \ldots, p_I]$. It is assumed that the TV broadcasters fix their offering prices. Therefore, the trading price can be expressed as a function of the set of TV bands that all of the WRAN service providers bid for (i.e. $p^{(t)}(\mathbf{c})$).

Since the WRAN service providers are rational to maximize their own profits, a non-cooperative game can be formulated to obtain the number of TV bands to bid for and the service price for each of the WRAN service providers. The players of this game are the WRAN service providers. The strategy of service provider i is the number of TV bands c_i that it bids for and service price p_i that it charges to the WRAN users. The payoff of service provider i is given by its profit $\mathscr{P}_i$.

A *solution* of this game is given by the Nash equilibrium and the best responses of the players are used to obtain the Nash equilibrium. This best response of a player is defined as its best strategy given the strategies of other players, i.e.

$$\mathscr{B}_i(\mathbf{c}_{-i}, \mathbf{p}_{-i}) = \arg\max_{c_i, p_i} u_i(\mathbf{c}, \mathbf{p})p_i - c_i p^{(t)}, \tag{12.55}$$

in which $\mathbf{c}_{-i} = [\ldots, c_j, \ldots]$, $j \neq i$ and $\mathbf{p}_{-i} = [\ldots, p_j, \ldots]$, $j \neq i$ denote, respectively, the sets corresponding to the number of bidding TV bands that the service providers bid for, and the prices offered by all service providers except service provider i. The Nash equilibrium can be obtained from the sets $\mathbf{c}^*$ and $\mathbf{p}^*$ such that

$$(c_i^*, p_i^*) = \mathscr{B}_i(\mathbf{c}_{-i}^*, \mathbf{p}_{-i}^*). \tag{12.56}$$

As an example, consider the case where two service providers exist. To obtain the Nash equilibrium, all possible strategy pairs corresponding to the number of bidding TV bands (e.g. (c_1, c_2)) have to be determined for these service providers. For each of these strategy pairs, the optimal price which maximizes the profit of each service provider, is then obtained (Figure 12.21). This optimal price can be obtained based on the service selection behavior of the WRAN users. Then, based on the profit calculated for each strategy pair, the Nash equilibrium for the spectrum bidding strategy is determined. The exchange of information involved in this competitive spectrum bidding and service pricing scheme is shown in Figure 12.20.

12.4.4 Performance evaluation results

Figure 12.22 shows the best responses of two service providers in terms of prices. In this case, when one service provider increases the price, the other service provider can

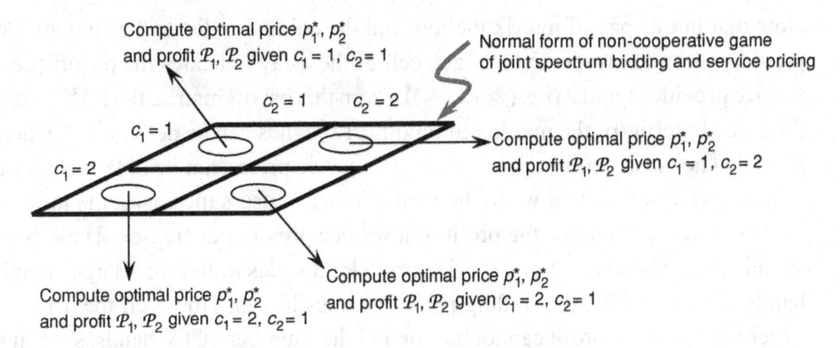

Figure 12.21 Procedure to obtain Nash equilibrium for the non-cooperative game of joint spectrum bidding and service pricing for IEEE 802.22 WRAN.

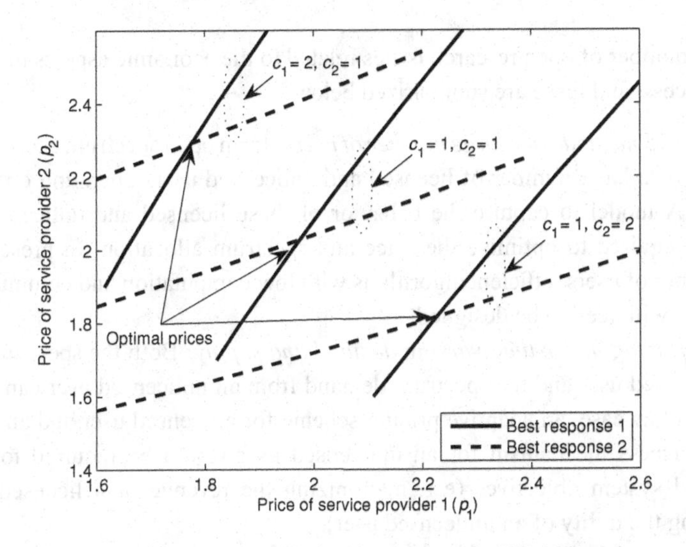

Figure 12.22 Best responses and optimal prices for WRAN service providers.

improve its profit by also increasing the price. The location where the best responses of prices intersect gives the optimal service prices for the service providers. These optimal prices depend on the number of bidding TV bands. In particular, when one service provider obtains a larger number of TV bands, the transmission rate available to the users increases. Consequently, that service provider can increase the price to achieve a higher profit.

Given the optimal service prices, the Nash equilibrium in terms of the number of bidding TV bands can be obtained. For example, if the total number of users is $U = 21$, the payoff matrix of the non-cooperative game can be expressed as follows:

	$c_2 = 1$	$c_2 = 2$	$c_2 = 3$
$c_1 = 1$	(28.82, 31.22)	**(23.81, 33.75)***	(16.69, 19.48)
$c_1 = 2$	(26.51, 23.61)	(18.42, 21.62)	–
$c_1 = 3$	(11.32, 16.47)	–	–

(12.57)

Note that in this payoff matrix the row and the column indicate the strategies of service providers 1 and 2, respectively. Each cell of the array indicates the payoff (i.e. profit) of service provider 1 and 2 (i.e. $(\mathscr{P}_1, \mathscr{P}_2)$). From this payoff matrix, the Nash equilibrium of the game in terms of the number of bidding TV bands can be determined, where $c_1^* = 1$, $c_2^* = 2$. The Nash equilibrium in terms of service prices can be obtained accordingly.

It is also observed that when the total number of users increases, the profit increases. However, at some points, the profit of a service provider decreases. These points correspond to the cases where one service provider decides to bid for a larger number of TV bands. Consequently, the trading price becomes higher. However, the service provider experiencing lower profit cannot bid for a higher number of TV bands, since it will incur a high cost. As a result, the profit will decrease.

12.5 Open research issues

There are a number of open research issues related to the economic aspects of dynamic spectrum access, and these are summarized below:

- *Spectrum trading model for large number of users:* In an open spectrum access market, there will be a large number of licensed and unlicensed users operating on multiple channels. A model to capture the behavior of these licensed and unlicensed users would be required to optimize the price and spectrum allocation. In presence of a large number of users, efficient algorithms with low computation and communication complexity will need to be designed.
- *Spectrum pricing under time-varying demand and supply:* Both the spectrum supply from a licensed user and the spectrum demand from an unlicensed user can be time-varying. In this case, an adaptive pricing scheme for a licensed user and an adaptive spectrum access mechanism for an unlicensed user would be required to achieve the desired system objectives (e.g. maximizing the revenue of a licensed user or maximizing the utility of an unlicensed user).
- *Risk–return model of dynamic spectrum access:* The risk–return model [702, 703, 704, 705] from microeconomics can be applied to dynamic spectrum access. In particular, by accessing radio spectrum an unlicensed user experiences an uncertain risk of interfering with the transmissions of licensed users due to imperfect spectrum sensing. In this case, an unlicensed user can choose the level of risk and return by adjusting the sensing period. If the sensing period is small, the transmission period will be larger, but there is a high risk for an unlicensed user to have incorrect sensing results. An optimal decision based on the risk–return model for the spectrum access will be required to optimize the performance of unlicensed users under the constraints set by the licensed users.

12.6 Summary

Four spectrum trading models have been presented. In the first model, the dynamic competitive spectrum sharing among secondary users is formulated as a non-cooperative game. The secondary users choose the requested spectrum sizes (i.e. competition by adjusting quantity) to achieve the highest utility. In the second model, three spectrum pricing schemes are considered and compared, i.e. market-equilibrium, competitive pricing based on Bertrand competition, and cooperative pricing based on profit optimization. The differences in the various aspects of these pricing schemes (e.g. the amount of profit and complexity of implementation) are stated. In the third model, based on competitive spectrum pricing, the long-term behavior of the primary service providers is investigated. A repeated game is formulated for Bertrand pricing competition in which the primary service providers can collude with each other to set the price. A punishment strategy is used to ensure that if the future profit of primary service providers is properly weighted, the collusion can be maintained. In the fourth model, the spectrum trading of the IEEE 802.22 WRAN is presented. This hierarchical spectrum trading model is formulated as a non-cooperative game for the competition among WRAN service providers and as an evolutionary game for the service selection of WRAN users.

References

[1] R. V. Prasad, P. Pawlczak, J. A. Hoffmeyer, and H. S. Berger, "Cognitive functionality in next generation wireless networks: standardization efforts," *IEEE Communications Magazine*, vol. 46, no. 4, pp. 72–78, 2008.

[2] Z. Zhang and X. Xie, "Intelligent cognitive radio: research on learning and evaluation of CR based on neural network," in *Proceedings of International Conference on Information and Communications Technology (ICICT)*, December 2007, pp. 33–37.

[3] D. Niyato and E. Hossain, "Market-equilibrium, competitive, and cooperative pricing for spectrum sharing in cognitive radio networks: analysis and comparison," *IEEE Transactions on Wireless Communications*, vol. 7, no. 11 (Part 1), pp. 4273–4283, 2008.

[4] "Federal communications commission: spectrum policy task force report," *Federal Communications Commission ET Docket 02-135*, November 2002.

[5] M. M. Buddhikot, "Understanding dynamic spectrum access: models, taxonomy and challenges," in *Proceedings of IEEE International Symposium on New Frontiers in Dynamic Spectrum Access Networks (DySPAN)*, April 2007, pp. 649–663.

[6] T. S. Rappaport, *Wireless Communications*. Prentice Hall, 1996.

[7] M. Schwartz, *Mobile Wireless Communications*. Cambridge University Press, 2005.

[8] A. Chockalingam and M. Zorzi, "Adaptive ARQ with energy efficient backoff on Markov fading links," *IEEE Transactions on Wireless Communications*, vol. 7, no. 5, pp. 1445–1449, 2008.

[9] M. Zorzi, A. Chockalingam, and R. R. Rao, "Throughput analysis of TCP on channels with memory," *IEEE Journal on Selected Areas in Communications*, vol. 18, no. 7, pp. 1289–1300, 2000.

[10] Q. Liu, S. Zhou, and G. B. Giannakis, "Queuing with adaptive modulation and coding over wireless links: cross-layer analysis and design," *IEEE Transactions on Wireless Communications*, vol. 4, no. 3, pp. 1142–1153, 2005.

[11] J. Misic and V. B. Misic, *Performance Modeling and Analysis of Bluetooth Networks: Polling, Scheduling, and Traffic Control*. Auerbach, 2005.

[12] S. Farahani, *ZigBee Wireless Networks and Transceivers*. Newnes, 2008.

[13] Y. Xiao, "IEEE 802.11n: enhancements for higher throughput in wireless lans," *IEEE Wireless Communications*, vol. 12, no. 6, pp. 82–91, 2005.

[14] J. Proakis, *Digital Communications*. McGraw-Hill, 2000.

[15] K. Etemad, *CDMA2000 Evolution: System Concepts and Design Principles*. Wiley, 2004.

[16] H. Holma and A. Toskala, *WCDMA for UMTS: HSPA Evolution and LTE*. Wiley, 2007.

[17] M. S. Kuran and T. Tugcu, "A survey on emerging broadband wireless access technologies," *Computer Networks*, vol. 51, no. 11, pp. 3013–3046, 2007.

[18] A. Ghosh, D. R. Wolter, J. G. Andrews, and R. Chen, "Broadband wireless access with wimax/802.16: current performance benchmarks and future potential," *IEEE Communications Magazine*, vol. 43, no. 2, pp. 129–136, 2005.

[19] W. Bolton, Y. Xiao, and M. Guizani, "IEEE 802.20: mobile broadband wireless access," *IEEE Wireless Communications*, vol. 14, no. 1, pp. 84–95, 2007.

[20] C. Cordeiro, K. Challapali, D. Birru, and S. Shankar, "IEEE 802.22: the first worldwide wireless standard based on cognitive radios," in *Proceedings of IEEE International Symposium on New Frontiers in Dynamic Spectrum Access Networks (DySPAN)*, November 2005, pp. 328–337.

[21] Y. G. Li and G. L. Stuber, *Orthogonal Frequency Division Multiplexing for Wireless Communications*. Springer, 2006.

[22] V. Kuhn, *Wireless Communications over MIMO Channels: Applications to CDMA and Multiple Antenna Systems*. Wiley, 2006.

[23] S. Wood and R. Aiello, *Essentials of UWB*. Cambridge University Press, 2008.

[24] A. Sendonaris, E. Erkip, and B. Aazhang, "User cooperation diversity – Part I. System description," *IEEE Transactions on Communications*, vol. 51, no. 11, pp. 1927–1938, 2003.

[25] ——, "User cooperation diversity – Part II. Implementation aspects and performance analysis," *IEEE Transactions on Communications*, vol. 51, no. 11, pp. 1939–1948, 2003.

[26] J. Zheng, M. J. Lee, and M. Anshel, "Toward secure low rate wireless personal area networks," *IEEE Transactions on Mobile Computing*, vol. 5, no. 10, pp. 1361–1373, 2006.

[27] W.-Y. Chung, V. K. Singh, R. Myllyla, and H. Lim, "Security enhanced indoor location tracking system for ubiquitous home healthcare," in *Proceedings of IEEE Conference on Sensors*, October 2006, pp. 522–525.

[28] R.-G. Lee, C.-C. Lai, S.-S. Chiang, H.-S. Liu, C.-C. Chen, and G.-Y. Hsieh, "Design and implementation of a mobile-care system over wireless sensor network for home healthcare applications," in *Proceedings of International Conference of the IEEE Engineering in Medicine and Biology Society (EMBS)*, August–September 2006, pp. 6004–6007.

[29] D. Trossen and D. Pavel, "Sensor networks, wearable computing, and healthcare applications," *IEEE Pervasive Computing*, vol. 6, no. 2, pp. 58–61, 2007.

[30] I. F. Akyildiz, W. Su, Y. Sankarasubramaniam, and E. Cayirci, "Wireless sensor networks: a survey," *Computer Networks*, vol. 38, no. 4, pp. 393–422, 2002.

[31] Y. Wang, G. Attebury, and B. Ramamurthy, "A survey of security issues in wireless sensor networks," *IEEE Communications Surveys and Tutorials*, vol. 8, no. 2, pp. 2–23, 2006.

[32] R. Rajagopalan and P. K. Varshney, "Data-aggregation techniques in sensor networks: a survey," *IEEE Communications Surveys and Tutorials*, vol. 8, no. 4, pp. 48–63, 2006.

[33] A. A. Abbasi and M. Younis, "A survey on clustering algorithms for wireless sensor networks," *Computer Communications*, vol. 30, no. 14–15, pp. 2826–2841, 2007.

[34] K. Akkaya and M. Younis, "A survey on routing protocols for wireless sensor networks," *Ad Hoc Networks*, vol. 3, no. 3, pp. 325–349, 2005.

[35] M. Younis and K. Akkaya, "Strategies and techniques for node placement in wireless sensor networks: a survey," *Ad Hoc Networks*, vol. 6, no. 4, pp. 621–655, 2008.

[36] B. Sundararaman, U. Buy, and A. D. Kshemkalyani, "Clock synchronization for wireless sensor networks: a survey," *Ad Hoc Networks*, vol. 3, no. 3, pp. 281–323, 2005.

[37] Y. Xiao, V. K. Rayi, B. Sun, X. Du, F. Hu, and M. Galloway, "A survey of key management schemes in wireless sensor networks," *Computer Communications*, vol. 30, no. 11–12, pp. 2314–2341, 2007.

[38] I. F. Akyildiz, T. Melodia, and K. R. Chowdhury, "A survey on wireless multimedia sensor networks," *Computer Networks*, vol. 51, no. 4, pp. 921–960, 2007.

[39] P. Baronti, P. Pillai, V. W. Chook, S. Chessa, A. Gotta, and Y. F. Hu, "Wireless sensor networks: a survey on the state of the art and the 802.15.4 and ZigBee standards," *Computer Communications*, vol. 30, no. 7, pp. 1655–1695, 2007.

[40] J. Jubin and J. D. Tornow, "The DARPA packet radio network protocols," in *Proceedings of the IEEE*, vol. 75, no. 1, January 1987, pp. 21–32.

[41] G. S. Lauer, *Routing in Communications Networks*. Prentice-Hall, 1995, ch. "Packet-radio routing."

[42] X. Zhang and H.-H. Chen, "Cluster-based multi-channel communications protocols in vehicle ad hoc networks," *IEEE Wireless Communications*, vol. 13, no. 5, pp. 44–51, 2006.

[43] T. Taleb, E. Sakhaee, A. Jamalipour, K. Hashimoto, N. Kato, and Y. Nemoto, "A stable routing protocol to support its services in VANET networks," *IEEE Transactions on Vehicular Technology*, vol. 56, no. 6, pp. 3337–3347, 2007.

[44] B. Ducourthial, Y. Khaled, and M. Shawky, "Conditional transmissions: performance study of a new communication strategy in VANET," *IEEE Transactions on Vehicular Technology*, vol. 56, no. 6, pp. 3348–3357, 2007.

[45] N. Wisitpongphan, F. Bai, P. Mudalige, V. Sadekar, and O. Tonguz, "Routing in sparse vehicular ad hoc wireless networks," *IEEE Journal on Selected Areas in Communications*, vol. 25, no. 8, pp. 1538–1556, 2007.

[46] D. Henkel, C. Dixon, J. Elston, and T. X. Brown, "A reliable sensor data collection network using unmanned aircraft," in *Proceedings of 2nd International Workshop on Multi-Hop Ad Hoc Networks: From Theory to Reality (REALMAN)*. ACM, 2006, pp. 125–127.

[47] S. Kumar, V. S. Raghavan, and J. Deng, "Medium access control protocols for ad hoc wireless networks: a survey," *Ad Hoc Networks*, vol. 4, no. 3, pp. 326–358, 2006.

[48] T. B. Reddy, I. Karthigeyan, B. Manoj, and C. S. R. Murthy, "Quality of service provisioning in ad hoc wireless networks: a survey of issues and solutions," *Ad Hoc Networks*, vol. 4, no. 1, pp. 83–124, 2006.

[49] J. Y. Yu and P. H. J. Chong, "A survey of clustering schemes for mobile ad hoc networks," *IEEE Communications Surveys and Tutorials*, vol. 7, no. 1, pp. 32–48, 2005.

[50] L. Hanzo-II and R. Tafazolli, "A survey of QoS routing solutions for ad hoc networks," *IEEE Communications Surveys and Tutorials*, vol. 9, no. 2, pp. 50–70, 2005.

[51] T.-Y. Wu, C.-Y. Huang, and H.-C. Chao, "A survey of mobile IP in cellular and mobile ad-hoc network environments," *Ad Hoc Networks*, vol. 3, no. 3, pp. 351–370, 2005.

[52] A. A. Hanbali, E. Altman, and P. Nain, "A survey of security issues in mobile ad hoc and sensor networks," *IEEE Communications Surveys and Tutorials*, vol. 7, no. 3, pp. 22–36, 2005.

[53] S. Guo and O. W. W. Yang, "Energy-aware multicasting in wireless ad hoc networks: a survey and discussion," *Computer Communications*, vol. 30, no. 9, pp. 2129–2148, 2007.

[54] D. Djenouri, L. Khelladi, and A. N. Badache, "A survey of security issues in mobile ad hoc and sensor networks," *IEEE Communications Surveys and Tutorials*, vol. 7, no. 4, pp. 2–28, 2005.

[55] V. Srivastava, J. Neel, A. B. Mackenzie *et al.*, "Using game theory to analyze wireless ad hoc networks," *IEEE Communications Surveys and Tutorials*, vol. 7, no. 4, pp. 46–56, 2005.

[56] J. He, K. Yang, K. Guild, and H.-H. Chen, "Application of IEEE 802.16 mesh networks as the backhaul of multihop cellular networks," *IEEE Communications Magazine*, vol. 45, no. 9, pp. 82–90, 2007.

[57] H. Wu, F. Yang, K. Tan, J. Chen, Q. Zhang, and Z. Zhang, "Distributed channel assignment and routing in multiradio multichannel multihop wireless networks," *IEEE Journal on Selected Areas in Communications*, vol. 24, no. 11, pp. 1972–1983, 2006.

[58] A. H. M. Rad and V. W. S. Wong, "Joint channel allocation, interface assignment and MAC design for multi-channel wireless mesh networks," in *Proceedings of IEEE International Conference on Computer Communications (INFOCOM)*, May 2007, pp. 1469–1477.

[59] I. F. Akyildiz, X. Wang, and W. Wang, "Wireless mesh networks: a survey," *Computer Networks*, vol. 47, no. 4, pp. 445–487, 2005.

[60] D. T. Chen, "On the analysis of using 802.16e WiMAX for point-to-point wireless backhaul," in *Proceedings of IEEE Radio and Wireless Symposium 2007*, January 2007, pp. 507–510.

[61] M. H. Ahmed, "Call admission control in wireless networks: a comprehensive survey," *IEEE Communications Surveys and Tutorials*, vol. 7, no. 1, pp. 49–68, 2005.

[62] M. Ghaderi and R. Boutaba, "Call admission control in mobile cellular networks: a comprehensive survey," *Wireless Communications and Mobile Computing*, vol. 6, no. 1, pp. 69–93, 2006.

[63] D. Hong and S. S. Rappaport, "Traffic model and performance analysis for cellular mobile radio telephone systems with prioritized and nonprioritized handoff procedures," *IEEE Transactions on Vehicular Technology*, vol. 35, no. 3, pp. 77–92, 1986.

[64] A. H. M. Rad and V. W. S. Wong, "On optimal call admission control in cellular networks," in *Proceedings of IEEE International Conference on Computer Communications (INFOCOM)*, vol. 1, March 1996, pp. 43–50.

[65] Y. Fang and Y. Zhang, "Call admission control schemes and performance analysis in wireless mobile networks," *IEEE Transactions on Vehicular Technology*, vol. 51, no. 2, pp. 371–382, 2002.

[66] S. Chatziperis, P. Koutsakis, and M. Paterakis, "A new call admission control mechanism for multimedia traffic over next-generation wireless cellular networks," *IEEE Transactions on Mobile Computing*, vol. 7, no. 1, pp. 95–112, 2008.

[67] S. Jiang and X. Ling, "A CAC considering both intracell and intercell handoffs for measurement-based DCA," *IEEE Transactions on Vehicular Technology*, vol. 56, no. 2, pp. 789–800, 2007.

[68] T.-C. Chau, K. Y. M. Wong, and B. Li, "Optimal call admission control with QoS guarantee in a voice/data integrated cellular network," *IEEE Transactions on Wireless Communications*, vol. 5, no. 5, pp. 1133–1141, 2006.

[69] D.-S. Lee and Y.-H. Hsueh, "Bandwidth-reservation scheme based on road information for next-generation cellular networks," *IEEE Transactions on Vehicular Technology*, vol. 53, no. 1, pp. 243–252, 2004.

[70] L. L. H. Andrew, S. V. Hanly, and R. G. Mukhtar, "Active queue management for fair resource allocation in wireless networks," *IEEE Transactions on Mobile Computing*, vol. 7, no. 2, pp. 231–246, 2008.

[71] W.-C. Feng, K. G. Shin, D. D. Kandlur, and D. Saha, "The blue active queue management algorithms," *IEEE/ACM Transactions on Networking*, vol. 10, no. 4, pp. 513–528, 2002.

[72] L. Le, J. Aikat, K. Jeffay, and F. D. Smith, "The effects of active queue management and explicit congestion notification on web performance," *IEEE/ACM Transactions on Networking*, vol. 15, no. 6, pp. 1217–1230, 2007.

[73] D. Niyato and E. Hossain, "Queue-aware uplink bandwidth allocation and rate control for polling service in IEEE 802.16 broadband wireless networks," *IEEE Transactions on Mobile Computing*, vol. 5, no. 6, pp. 668–679, 2006.

[74] T. Issariyakul and E. Hossain, "ORCA-MRT: an optimization-based approach for fair scheduling in multirate TDMA wireless networks," *IEEE Transactions on Wireless Communications*, vol. 4, no. 6, pp. 2823–2835, 2005.

[75] M. Andrews, "A survey of scheduling theory in wireless data networks," in *Proceedings of IMA Summer Workshop on Wireless Communications*, 2005.

[76] T. Ng, I. Stoica, and H. Zhang, "Packet fair queueing algorithms for wireless networks with location-dependent errors," in *Proceedings of IEEE International Conference on Computer Communications (INFOCOM)*, vol. 3, March–April 1998, pp. 1103–1111.

[77] S. Lu, V. Bharghavan, and R. Srikant, "Fair scheduling in wireless packet networks," *IEEE/ACM Transactions on Networking*, vol. 7, no. 4, pp. 473–537, 1999.

[78] T. Issariyakul, E. Hossain, and D. I. Kim, "Medium access control protocols for wireless mobile ad hoc networks: Issues and approaches," *Wireless Communications and Mobile Computing*, vol. 3, no. 8, pp. 935–958, 2003.

[79] P. Roshan and J. Leary, *802.11 Wireless LAN Fundamentals*. Cisco Press, 2004.

[80] A. C. V. Gummalla and J. O. Limb, "Wireless medium access control protocols," *IEEE Communications Surveys and Tutorials*, vol. 3, no. 2, 2000.

[81] W. Ye and J. Heidemann, "Medium access control in wireless sensor networks," USC/Information Sciences Institute, Tech. Rep. ISI-TR-580, October 2003, to appear as a chapter in *Wireless Sensor Networks*, T. Znati, K. M. Sivalingam and C. Raghavendra (eds.), Kluwer Academic Publishers. [Online]. Available: http://www.isi.edu/johnh/PAPERS/Ye03c.html

[82] A. Gupta and P. Mohapatra, "A survey on ultra wide band medium access control schemes," *Computer Networks*, vol. 51, no. 11, pp. 2976–2993, 2007.

[83] I. F. Akyildiz, J. McNair, L. Carrasco, and R. Puigjaner, "Medium access control protocols for multimedia traffic in wireless networks," *IEEE Network*, vol. 13, no. 4, pp. 39–47, 1999.

[84] H. W. Kuhn, "The Hungarian method for the assignment problem," *Naval Research Logistic Quarterly*, vol. 2, pp. 83–97, 1955.

[85] J. Cai, X. Shen, and J. Mark, "Downlink resource management for packet transmission in OFDM wireless communication systems," *IEEE Transactions on Wireless Communications*, vol. 4, no. 4, pp. 1688–1703, 2005.

[86] D. Niyato, E. Hossain, and V. Bhargava, "Scheduling and admission control in power-constrained OFDM wireless mesh routers: analysis and optimization," *IEEE Transactions on Wireless Communications*, vol. 6, no. 10, pp. 3738–3748, 2007.

[87] D. Niyato and E. Hossain, "Adaptive fair subcarrier/rate allocation in multirate OFDMA networks: radio link level queuing performance analysis," *IEEE Transactions on Vehicular Technology*, vol. 55, no. 6, pp. 1897–1907, 2006.

[88] S.-J. Oh, D. Zhang, and K. Wasserman, "Optimal resource allocation in multiservice CDMA networks," *IEEE Transactions on Wireless Communications*, vol. 2, no. 4, pp. 811–821, 2003.

[89] C. W. Sung and W. S. Wong, "A noncooperative power control game for multirate CDMA data networks," *IEEE Transactions on Wireless Communications*, vol. 2, no. 1, pp. 186–194, 2003.

[90] S. Koskie and Z. Gajic, "A Nash game algorithm for SIR-based power control in 3G wireless CDMA networks," *IEEE/ACM Transactions on Networking*, vol. 13, no. 5, pp. 1017–1026, 2005.

[91] S. Shakkottai, T. S. Rappaport, and P. C. Karlsson, "Cross-layer design for wireless networks," *IEEE Communications Magazine*, vol. 41, no. 10, pp. 74–80, 2003.

[92] G. Carneiro, J. Ruela, and M. Ricardo, "Cross-layer design in 4G wireless terminals," *IEEE Wireless Communications*, vol. 11, no. 2, pp. 7–13, 2004.

[93] F. Bari and V. C. M. Leung, "Automated network selection in a heterogeneous wireless network environment," *IEEE Network*, vol. 21, no. 1, pp. 34–40, 2007.

[94] D. Niyato and E. Hossain, "A noncooperative game-theoretic framework for radio resource management in 4G heterogeneous wireless access networks," *IEEE Transactions on Mobile Computing*, vol. 7, no. 3, pp. 332–345, 2008.

[95] R. Skehill, M. Barry, W. Kent, M. Ocallaghan, N. Gawley, and S. Mcgrath, "The common RRM approach to admission control for converged heterogeneous wireless networks," *IEEE Wireless Communications*, vol. 14, no. 2, pp. 48–56, 2007.

[96] Z. Chen, "A customizable QoS strategy for convergent heterogeneous wireless communications," *IEEE Wireless Communications*, vol. 14, no. 2, pp. 20–27, 2007.

[97] K. Yang, J. Zhang, and H.-H. Chen, "A flexible QoS-aware service gateway for heterogeneous wireless networks," *IEEE Network*, vol. 21, no. 2, pp. 6–12, 2007.

[98] L. Min, L. Zhongcheng, G. Xiaobing, and D. Eryk, "Performance analysis and optimization of handoff algorithms in heterogeneous wireless networks," *IEEE Transactions on Mobile Computing*, vol. 7, no. 7, pp. 846–857, 2008.

[99] E. Stevens-Navarro, Y. Lin, and V.-W.-S. Wong, "An MDP-based vertical handoff decision algorithm for heterogeneous wireless networks," *IEEE Transactions on Vehicular Technology*, vol. 57, no. 2, pp. 1243–1254, 2008.

[100] A. Assouma, R. Beaubrun, and S. Pierre, "Mobility management in heterogeneous wireless networks," *IEEE Journal on Selected Areas in Communications*, vol. 24, no. 3, pp. 638–648, 2006.

[101] L. Badia, M. Miozzo, M. Rossi, and M. Zorzi, "Routing schemes in heterogeneous wireless networks based on access advertisement and backward utilities for QoS support," *IEEE Communications Magazine*, vol. 45, no. 2, pp. 67–73, 2007.

[102] S. Haykin, "Cognitive radio: brain-empowered wireless communications," *IEEE Journal on Selected Areas in Communications*, vol. 23, no. 2, pp. 201–220, 2005.

[103] M. Heddebaut, V. Deniau, and K. Adouane, "In-vehicle WLAN radio-frequency communication characterization," *IEEE Transactions on Intelligent Transportation Systems*, vol. 5, no. 2, pp. 114–121, 2004.

[104] K. Yang, S. Ou, H.-H. Chen, and J. He, "A multihop peer communication protocol with fairness guarantee for IEEE 802.16 based vehicular networks," *IEEE Transactions on Vehicular Technology*, vol. 56, no. 6, pp. 3358–3370, 2007.

[105] A. Soomro and D. Cavalcanti, "Opportunities and challenges in using WPAN and WLAN technologies in medical environments," *IEEE Communications Magazine*, vol. 45, no. 2, pp. 114–122, 2007.

[106] D. Niyato, E. Hossain, and J. Diamond, "IEEE 802.16/wimax-based broadband wireless access and its application for telemedicine/e-health services," *IEEE Wireless Communications*, vol. 14, no. 1, pp. 72–83, 2007.

[107] C. Salvador, M. P. Carrasco, M. A. G. de Mingo *et al.*, "Airmed-cardio: A GSM and Internet services-based system for out-of-hospital follow-up of cardiac patients," *IEEE*

Transactions on Information Technology in Biomedicine, vol. 9, no. 1, pp. 73–85, 2005.

[108] Y. Chu and A. Ganz, "A mobile teletrauma system using 3G networks," *IEEE Transactions on Information Technology in Biomedicine*, vol. 8, no. 4, pp. 456–462, 2004.

[109] S. Pavlopoulos, E. Kyriacou, A. Berler, S. Dembeyiotis, and D. Koutsouris, "A novel emergency telemedicine system based on wireless communication technology-ambulance," *IEEE Transactions on Information Technology in Biomedicine*, vol. 2, no. 4, pp. 261–267, 1998.

[110] "Report of the spectrum efficiency group," FCC Spectrum Policy Task Force, Report, November 2002.

[111] M. A. McHenry, P. A. Tenhula, D. McCloskey, D. A. Roberson, and C. S. Hood, "Chicago spectrum occupancy measurements & analysis and a long-term studies proposal," in *Proceedings of First International Workshop on Technology and Policy for Accessing Spectrum (TAPAS)*. ACM, 2006, p. 1.

[112] F. K. Jondral, "Software-defined radio: basics and evolution to cognitive radio," *EURASIP Journal on Wireless Communications and Networking*, vol. 5, no. 3, pp. 275–283, 2005.

[113] U. Ramacher, "Software-defined radio prospects for multistandard mobile phones," *IEEE Computer*, vol. 40, no. 10, pp. 62–69, 2007.

[114] G. J. Minden, J. B. Evans, L. S. Searl *et al.*, "An agile radio for wireless innovation," *IEEE Communications Magazine*, vol. 45, no. 5, pp. 113–121, 2007.

[115] R. Bagheri, A. Mirzaei, M. E. Heidari, S. Chehrazi, M. Lee, M. Mikhemar, W. K. Tang, and A. A. Abidi, "Software-defined radio receiver: dream to reality," *IEEE Communications Magazine*, vol. 44, no. 8, pp. 111–118, 2006.

[116] K. G. Shin, H. Kim, C. Cordeiro, and K. Challapali, "An experimental approach to spectrum sensing in cognitive radio networks with off-the-shelf IEEE 802.11 devices," in *Proceedings of IEEE Consumer Communications and Networking Conference (CCNC)*, January 2007, pp. 1154–1158.

[117] H. Harada, "Software defined radio prototype toward cognitive radio communication systems," in *Proceedings of IEEE International Symposium on New Frontiers in Dynamic Spectrum Access Networks (DySPAN)*, November 2005, pp. 539–547.

[118] R. J. DeGroot, D. P. Gurney, K. Hutchinson *et al.* "A cognitive-enabled experimental system," in *Proceedings of IEEE International Symposium on New Frontiers in Dynamic Spectrum Access Networks (DySPAN)*, November 2005, pp. 556–561.

[119] Y. Tachwali and H. Refai, "Implementation of a BPSK transceiver on hybrid software defined radio platforms," in *Proceedings of IEEE International Conference on Information and Communication Technologies: From Theory to Applications (ICTTA)*, April 2008.

[120] A. Tribble, "The software defined radio: fact and fiction," in *Proceedings of IEEE Radio and Wireless Symposium*, January 2008, pp. 5–8.

[121] J. Glossner, D. Iancu, M. Moudgill *et al.*, "Implementing communications systems on an SDR SOC," in *Proceedings of IEEE International Conference on Acoustics, Speech and Signal Processing (ICASSP)*, March–April 2008, pp. 5380–5383.

[122] A. Saha and A. Sinha, "Radio processor – a new reconfigurable architecture for software defined radio," in *Proceedings of International Conference on Computer Science and Information Technology (ICCSIT)*, August–September 2008, pp. 709–713.

[123] Q. Zhao and B. M. Sadler, "A survey of dynamic spectrum access," *IEEE Signal Processing Magazine*, vol. 24, no. 3, pp. 79–89, 2007.

[124] W. Lehr and J. Crowcroft, "Managing shared access to a spectrum commons," in *Proceedings of IEEE International Symposium on New Frontiers in Dynamic Spectrum Access Networks (DySPAN)*, November 2005, pp. 420–444.

[125] D. Hatfield and P. Weiser, "Property rights in spectrum: taking the next step," in *Proceedings of IEEE International Symposium on New Frontiers in Dynamic Spectrum Access Networks (DySPAN)*, November 2005, pp. 43–55.

[126] Q. Pang and V. C. M. Leung, "Channel clustering and probabilistic channel visiting techniques for WLAN interference mitigation in Bluetooth devices," *IEEE Transactions on Electromagnetic Compatibility*, vol. 49, no. 4, pp. 914–923, 2007.

[127] S. Jeng, C. Tsung, and F. Chang, "WLAN smart antenna with Bluetooth interference reduction," *IET Communications*, vol. 2, no. 8, pp. 1098–1107, 2008.

[128] L. Angrisani, M. Bertocco, D. Fortin, and A. Sona, "Experimental study of coexistence issues between IEEE 802.11b and IEEE 802.15.4 wireless networks," *IEEE Transactions on Instrumentation and Measurement*, vol. 57, no. 8, pp. 1514–1523, 2008.

[129] S. Pollin, I. Tan, B. Hodge, C. Chun, and A. Bahai, "Harmful coexistence between 802.15.4 and 802.11: a measurement-based study," in *Proceedings of International Conference on Cognitive Radio Oriented Wireless Networks and Communications (CrownCom)*, May 2008.

[130] W. Yuan, X. Wang, and J.-P. Linnartz, "A coexistence model of IEEE 802.15.4 and IEEE 802.11b/g," in *Proceedings of IEEE Symposium on Communications and Vehicular Technology in the Benelux*, November 2007.

[131] B. A. Fette and B. Fette, *Cognitive Radio Technology*. Newnes, 2006.

[132] T. Clancy and W. Arbaugh, "Measuring interference temperature," in *Proceedings of Virginia Tech Wireless Personal Communications Symposium*, June 2006.

[133] T. C. Clancy, "Achievable capacity under the interference temperature model," in *Proceedings of IEEE International Conference on Computer Communications (INFOCOM)*, May 2007, pp. 794–802.

[134] B. Wild and K. Ramchandran, "Detecting primary receivers for cognitive radio applications," in *Proceedings of IEEE International Symposium on New Frontiers in Dynamic Spectrum Access Networks (DySPAN)*, November 2005, pp. 124–130.

[135] A. Ghasemi and E. S. Sousa, "Collaborative spectrum sensing for opportunistic access in fading environment," in *Proceedings of IEEE International Symposium on New Frontiers in Dynamic Spectrum Access Networks (DySPAN)*, November 2005, pp. 131–136.

[136] A. Sahai, N. Hoven, and R. Tandra, "Some fundamental limits in cognitive radio," in *Proceedings of Allerton Conference on Communications, Control and Computing*, October 2004.

[137] J. Vartiainen, H. Sarvanko, J. Lehtomaki, M. Juntti, and M. Latva-aho, "Spectrum sensing with LAD-based methods," in *Proceedings of IEEE International Symposium on Personal, Indoor and Mobile Radio Communications (PIMRC)*, September 2007.

[138] F. F. Digham, M.-S. Alouini, and M. K. Simon, "On the energy detection of unknown signals over fading channels," in *Proceedings of International Conference on Communications (ICC)*, vol. 5, May 2003, pp. 3575–3579.

[139] D. Cabric, S. M. Mishra, and R. W. Brodersen, "Implementation issues in spectrum sensing for cognitive radios," in *Proceedings of the Thirty-Eighth Asilomar Conference on Signals, Systems and Computers*, vol. 1, November 2004, pp. 772–776.

[140] A. Fehske, J. Gaeddert, and J. H. Reed, "A new approach to signal classification using spectral correlation and neural networks," in *Proceedings of IEEE International Symposium*

on New Frontiers in Dynamic Spectrum Access Networks (DySPAN), November 2005, pp. 144–150.

[141] Y. Hur, J. Park, W. Woo *et al.*, "A wideband analog multi-resolution spectrum sensing (MRSS) technique for cognitive radio (CR) systems," in *Proceedings of IEEE International Symposium on Circuits and Systems (ISCAS)*, May 2006, pp. 1912–1916.

[142] S. Y. Shin, I. Ramachandran, S. Roy, and W. H. Kwon, "Cascaded clear channel assessment: enhanced carrier sensing for cognitive radios," in *Proceedings of IEEE International Conference on Communications (ICC)*, June 2007, pp. 6532–6537.

[143] G. Ganesan and Y. G. Li, "Cooperative spectrum sensing in cognitive radio networks," in *Proceedings of IEEE International Symposium on New Frontiers in Dynamic Spectrum Access Networks (DySPAN)*, November 2005, pp. 137–143.

[144] J. Zhao, H. Zheng, and G.-H. Yang, "Distributed coordination in dynamic spectrum allocation networks," in *Proceedings of IEEE International Symposium on New Frontiers in Dynamic Spectrum Access Networks (DySPAN)*, November 2005, pp. 259–268.

[145] S. Shankar, "Spectrum agile radios: utilization and sensing architecture," in *Proceedings of IEEE International Symposium on New Frontiers in Dynamic Spectrum Access Networks (DySPAN)*, November 2005, pp. 160–169.

[146] FCC, "Et docket no 03-237 notice of inquiry and notice of proposed rulemaking," FCC, Tech. Rep. 03-237, November 2003.

[147] E. Altman, T. Boulogne, R. El-Azouzi, T. Jiménez, and L. Wynter, "A survey on networking games in telecommunications," *Computers and Operations Research*, vol. 33, no. 2, pp. 286–311, 2006.

[148] K. Challapali, C. Cordeiro, and D. Birru, "Evolution of spectrum-agile cognitive radios: first wireless Internet standard and beyond," in *Proceedings of 2nd Annual International Workshop on Wireless Internet (WICON)*. ACM, 2006, p. 27.

[149] V. R. Petty, R. Rajbanshi, D. Datla *et al.*, "Feasibility of dynamic spectrum access in underutilized television bands," in *Proceedings of IEEE International Symposium on New Frontiers in Dynamic Spectrum Access Networks (DySPAN)*, April 2007, pp. 331–339.

[150] J. F. Hauris, "Genetic algorithm optimization in a cognitive radio for autonomous vehicle communications," in *Proceedings of IEEE International Symposium on Computational Intelligence in Robotics and Automation (CIRA)*, June 2007, pp. 427–431.

[151] R. D. Hinman, "Application of cognitive radio technology to legacy military waveforms in a JTRS (joint tactical radio system) radio," in *Proceedings of IEEE Military Communications Conference (MILCOM)*, October 2006.

[152] H. Kim and K. G. Shin, "Efficient discovery of spectrum opportunities with MAC-layer sensing in cognitive radio networks," *IEEE Transactions on Mobile Computing*, vol. 7, no. 5, pp. 533–545, 2008.

[153] Q. Wang and H. Zheng, "Route and spectrum selection in dynamic spectrum networks," in *Proceedings of 3rd IEEE Consumer Communications and Networking Conference (CCNC)*, vol. 1, January 2006, pp. 625–629.

[154] C. Xin, "A novel layered graph model for topology formation and routing in dynamic spectrum access networks," in *Proceedings of IEEE International Symposium on New Frontiers in Dynamic Spectrum Access Networks (DySPAN)*, November 2005, pp. 308–317.

[155] A. M. R. Slingerland, P. Pawelczak, R. V. Prasad, A. Lo, and R. Hekmat, "Performance of transport control protocol over dynamic spectrum access links," in *Proceedings of*

IEEE International Symposium on New Frontiers in Dynamic Spectrum Access Networks (DySPAN), April 2007, pp. 486–495.

[156] X. Chen, H. Zhai, J. Wang, and Y. Fang, *Resource Management in Wireless Networking*. Springer, 2005, ch. "A survey on improving TCP performance over wireless networks," pp. 657–695.

[157] A. Maharshi, L. Tong, and A. Swami, "Cross-layer designs of multichannel reservation MAC under Rayleigh fading," *IEEE Transactions on Signal Processing*, vol. 51, no. 8, pp. 2054–2067, 2003.

[158] J. Wang, L. Li, S. H. Low, and J. C. Doyle, "Cross-layer optimization in TCP/IP networks," *IEEE/ACM Transactions on Networking*, vol. 13, no. 3, pp. 582–595, 2005.

[159] V. T. Raisinghani and S. Iyer, "Cross-layer design optimizations in wireless protocol stacks," *Computer Communications*, vol. 27, no. 8, pp. 720–724, 2004.

[160] L. Galluccio, F. Licandro, G. Morabito, and G. Schembra, "An analytical framework for the design of intelligent algorithms for adaptive-rate MPEG video encoding in next-generation time-varying wireless networks," *IEEE Journal on Selected Areas in Communications*, vol. 23, no. 2, pp. 369–384, 2005.

[161] H. Arslan and S. Yarkan, *Cognitive Radio, Software Defined Radio, and Adaptive Wireless Systems*. Springer Netherlands, 2007, ch. "Cross-layer adaptation and optimization for cognitive radio."

[162] L. Osadciw and K. Veeramachaneni, "Sensor network management through fitness function design in multi-objective optimization," in *Proceedings of Asilomar Conference on Signals, Systems and Computers (ACSSC)*, November 2007, pp. 1648–1651.

[163] N. Baldo and M. Zorzi, "Fuzzy logic for cross-layer optimization in cognitive radio networks," in *Proceedings of Consumer Communications and Networking Conference (CCNC)*, January 2007, pp. 1128–1133.

[164] C. P. Fu and S. C. Liew, "TCP Veno: TCP enhancement for transmission over wireless access networks," *IEEE Journal on Selected Areas in Communications*, vol. 21, no. 2, pp. 216–228, 2003.

[165] C. Clancy, J. Hecker, E. Stuntebeck, and T. O'Shea, "Applications of machine learning to cognitive radio networks," *IEEE Wireless Communications*, vol. 14, no. 4, pp. 47–52, 2007.

[166] G. Xu and Y. Lu, "Channel and modulation selection based on support vector machines for cognitive radio," in *Proceedings of IEEE International Conference on Wireless Communications, Networking and Mobile Computing (WiCOM)*, September 2006.

[167] A. Ginsberg, J. D. Poston, and W. D. Horne, "Toward a cognitive radio architecture: integrating knowledge representation with software defined radio technologies," in *Proceedings of IEEE Military Communications Conference (MILCOM)*, October 2006.

[168] P. Jackson, *Introduction to Expert Systems*. Addison Wesley, 1998.

[169] L. R. Rabiner, "A tutorial on hidden Markov models and selected applications in speech recognition," in *Proceedings of the IEEE*, vol. 77, no. 2, February 1989, pp. 257–286.

[170] S. Haykin, *Neural Networks: A Comprehensive Foundation*. Prentice Hall, 1998.

[171] R. S. Sutton and A. G. Barto, *Reinforcement Learning: An Introduction*. MIT Press, 1998.

[172] D. E. Goldberg, *Genetic Algorithms in Search, Optimization, and Machine Learning*. Addison-Wesley, 1989.

[173] D. A. Levine, I. F. Akyildiz, and M. Naghshineh, "A resource estimation and call admission algorithm for wireless multimedia networks using the shadow cluster concept," *IEEE/ACM Transactions on Networking*, vol. 5, no. 1, pp. 1–12, 1997.

[174] A. Aljadhai and T. F. Znati, "Predictive mobility support for QoS provisioning in mobile wireless environments," *IEEE Journal on Selected Areas in Communications*, vol. 19, no. 10, pp. 1915–1930, 2001.

[175] H. Celebi and H. Arslan, "Adaptive positioning systems for cognitive radios," in *Proceedings of IEEE International Symposium on New Frontiers in Dynamic Spectrum Access Networks (DySPAN)*, April 2007, pp. 78–84.

[176] ——, "Cognitive positioning systems," *IEEE Transactions on Wireless Communications*, vol. 6, no. 12, pp. 4475–4483, 2007.

[177] ——, "Utilization of location information in cognitive wireless networks," *IEEE Wireless Communications*, vol. 14, no. 4, pp. 6–13, 2007.

[178] S. Yarkan and H. Arslan, "Exploiting location awareness toward improved wireless system design in cognitive radio," *IEEE Communications Magazine*, vol. 46, no. 1, pp. 128–136, 2008.

[179] M. J. Marcus, P. Kolodzy, and A. Lippman, "Reclaiming the vast wasteland: why unlicensed use of the white space in the TV bands will not cause interference to DTV viewers," New America Foundation: Wireless Future Program, Report, 2005.

[180] D. X. WG, "The XG architectural framework v1.0," Tech. Rep.

[181] I. F. Akyildiz, W.-Y. Lee, M. C. Vuran, and S. Mohanty, "Next generation/dynamic spectrum access/cognitive radio wireless networks: a survey," *Computer Networks*, vol. 50, no. 13, pp. 2127–2159, 2006.

[182] D. Cabric, S. M. Mishra, D. Willkomm, R. Brodersen, and A. Wolisz, "A cognitive radio approach for usage of virtual unlicensed spectrum," in *Proceedings of 14th IST Mobile and Wireless Communications Summit*, June 2005.

[183] M. M. Buddhikot, P. Kolody, S. Miller, K. Ryan, and J. Evans, "DIMSUMnet: new directions in wireless networking using coordinated dynamic spectrum," in *Proceedings of IEEE International Symposium on a World of Wireless, Mobile, and Multimedia Networks (WoWMoM)*, June 2005, pp. 78–85.

[184] L. Xu, R. Tonjes, T. Paila, W. Hansmann, M. Frank, and M. Albrecht, "DRiVE-ing to the Internet: dynamic radio for IP services in vehicular environments," in *Proceedings of 25th Annual IEEE Conference on Local Computer Networks (LCN)*, November 2000, pp. 281–289.

[185] D. Grandblaise, D. Bourse, K. Moessner, and P. Leaves, "Dynamic spectrum allocation (DSA) and reconfigurability," in *Software-Defined Radio (SDR) Forum*, November 2002.

[186] I. F. Akyildiz and Y. Li, "OCRA: OFDM-based cognitive radio networks," Broadband and Wireless Networking Laboratory, Tech. Rep., 2006.

[187] "IEEE 1900 standards committee, IEEE SCC 41." [Online]. Available: http://www.scc41.org

[188] C. Cordeiro, K. Challapali, and M. Ghosh, "Cognitive PHY and MAC layers for dynamic spectrum access and sharing of TV bands," in *Proceedings of International Workshop on Technology and Policy for Accessing Spectrum (TAPAS)*. ACM, 2006, p. 3.

[189] M. H. Hayes, *Statistical Digital Signal Processing and Modeling*. Wiley, 1996.

[190] H. Cramer, *Mathematical Methods of Statistics*. Princeton University Press, 1946.

[191] M. J. Schervish, *Theory of Statistics*. Springer, 1996.

[192] Z. Tian and G. B. Giannakis, "A wavelet approach to wideband spectrum sensing for cognitive radios," in *Proceedings of 1st International Conference on Cognitive Radio Oriented Wireless Networks and Communications (CROWNCOM)*, June 2006.

[193] K. B. Letaief and W. Zhang, *Cognitive Wireless Communication Networks*. Springer, 2007, ch. "Cooperative spectrum sensing," pp. 115–138.

[194] C. Sun, W. Zhang, and K. B. Letaief, "Cluster-based cooperative spectrum sensing in cognitive radio systems," in *Proceedings of IEEE International Conference on Communications (ICC)*, June 2007, pp. 2511–2515.

[195] ——, "Cooperative spectrum sensing for cognitive radios under bandwidth constraints," in *Proceedings of IEEE Wireless Communications and Networking Conference (WCNC)*, March 2007, pp. 1–5.

[196] C.-H. Lee and W. Wolf, "Energy efficient techniques for cooperative spectrum sensing in cognitive radios," in *Proceedings of IEEE Consumer Communications and Networking Conference (CCNC)*, January 2008, pp. 968–972.

[197] G. Ganesan and Y. Li, "Cooperative spectrum sensing in cognitive radio, part I: two user networks," *IEEE Transactions on Wireless Communications*, vol. 6, no. 6, pp. 2204–2213, 2007.

[198] ——, "Cooperative spectrum sensing in cognitive radio, part II: multiuser networks," *IEEE Transactions on Wireless Communications*, vol. 6, no. 6, pp. 2214–2222, 2007.

[199] Z. Han and H. Jiang, "Replacement of spectrum sensing and avoidance of hidden terminal for cognitive radio," in *Proceedings of IEEE Wireless Communications and Networking Conference (WCNC)*, March–April 2008, pp. 1448–1452.

[200] W. Z. H. Jiang, P. Wang and X. Shen, "An interference aware distributed resource management scheme for CDMA-based wireless mesh backbone," *IEEE Transactions on Wireless Communications*, vol. 6, no. 12, pp. 4558–4567, December 2007.

[201] T. W. Parks and C. S. Burrus, *Digital Filter Design*. John Wiley and Sons Canada, 1987.

[202] Wiener, N., *Extrapolation, Interpolation, and Smoothing of Stationary Time Series*. Wiley, 1949.

[203] D. Donoho, "Compressed sensing," *IEEE Transactions on Information Theory*, vol. 52, no. 4, pp. 1289–1306, 2006.

[204] E. J. Candès, J. Romberg, and T. Tao, "Robust uncertainty principles: exact signal reconstruction from highly incomplete frequency information," *IEEE Transactions on Information Theory*, vol. 52, no. 2, pp. 489–509, 2006.

[205] E. Candès and T. Tao, "Near optimal signal recovery from random projections: universal encoding strategies," *IEEE Transactions on Information Theory*, vol. 52, no. 12, pp. 5406–5425, 2006.

[206] E. Candès and J. Romberg, "Encoding the l_p ball from limited measurements," in *Proceedings of IEEE Data Compression Conference (DCC)*, March 2006, pp. 33–42.

[207] S. Sarvotham, D. Baron, and R. Baraniuk, "Measurements vs. bits: compressed sensing meets information theory," in *Proceedings of Allerton Conference on Communications, Control and Computing*, September 2006.

[208] P. Boufounos and R. Baraniuk, "Quantization of sparse representations," in *Proceedings of IEEE Data Compression Conference (DCC)*, March 2007.

[209] W. Bajwa, J. Haupt, A. Sayeed, and R. Nowak, "Compressive wireless sensing," in *Proceedings of International Conference on Information Processing in Sensor Networks (IPSN)*, April 2006, pp. 134–142.

[210] W. Wang, M. Garofalakis, and K. Ramchandran, "Distributed sparse random projections for refinable approximation," in *Proceedings of International Conference on Information Processing in Sensor Networks (IPSN)*, April 2006, pp. 331–339.

[211] D. Waagen, N. Shah, M. Ordaz, and M. Cassabaum, "Random subspaces and SAR classification efficacy," in *Proceedings of SPIE Algorithms for Synthetic Aperture Radar Imagery XII*, May 2005.

[212] M. Duarte, M. Davenport, M. Wakin, and R. Baraniuk, "Sparse signal detection from incoherent projections," in *Proceedings of IEEE International Conference on Acoustics, Speech and Signal Processing (ICASSP)*, vol. 3, May 2006.

[213] M. Davenport, M. Wakin, and R. Baraniuk, "Detection and estimation with compressive measurements," Rice ECE Department Tech. Rep. TREE 0610, November 2006.

[214] J. Haupt, R. Castro, R. Nowak, G. Fudge, and A. Yeh, "Compressive sampling for signal classification," in *Proceedings of Asilomar Conference on Signals, Systems and Computers (ACSSC)*, October–November 2006, pp. 1430–1434.

[215] M. Davenport, M. Duarte, M. Wakin *et al.*, "The smashed filter for compressive classification and target recognition," in *Proceedings of Computational Imaging V at SPIE Electronic Imaging*, January 2007.

[216] N. Alon, P. B. Gibbons, Y. Matias, and M. Szegedy, "Tracking join and self-join sizes in limited storage," in *Proceedings of Eighteenth ACM SIGMOD–SIGACT–SIGART Symposium on Principles of Database Systems (PODS)*. ACM, 1999, pp. 10–20.

[217] N. Thaper, S. Guha, P. Indyk, and N. Koudas, "Dynamic multidimensional histograms," in *Proceedings of 2002 ACM SIGMOD International Conference on Management of Data (SIGMOD)*. ACM, June 2002, pp. 428–439.

[218] A. C. Gilbert, S. Guha, P. Indyk, Y. Kotidis, S. Muthukrishnan, and M. J. Strauss, "Fast, small-space algorithms for approximate histogram maintenance," in *Proceedings of Annual ACM Symposium on Theory of Computing (STOC)*. ACM, May 2002, pp. 389–398.

[219] Z. Tian and G. Giannakis, "Compressed sensing for wideband cognitive radios," in *Proceedings of IEEE International Conference on Acoustics, Speech and Signal Processing (ICASSP)*, vol. 4, April 2007, pp. 1357–1360.

[220] S. Boyd and L. Vandenberghe, *Convex Optimization*. Cambridge University Press, 2004.

[221] M. K. Wood and G. B. Dantzig, "Programming of interdependent activities: I, general discussion," *Econometrica*, vol. 17, no. 3–4, pp. 193–199, 1949.

[222] G. B. Dantzig, "Programming of interdependent activities: II, mathematical model," *Econometrica*, vol. 17, no. 3–4, pp. 200–211, 1949.

[223] L. G. Khachian, "A polynomial algorithm in linear programming," *Doklady Akademii Nauk SSSR*, vol. 244, no. 5, pp. 1093–1096, 1979.

[224] N. Karmarkar, "A new polynomial-time algorithm for linear programming," *Combinatorica*, vol. 4, no. 4, pp. 373–395, 1984.

[225] M. W. Cooper and K. Farhangian, "Multi-criteria optimization for nonlinear integer-variable problems," *Large Scale Systems*, vol. 9, pp. 73–78, 1985.

[226] S. Martello and P. Toth, *Knapsack Problems*. John Wiley, 1990.

[227] M. L. Fisher, "The Lagrangian relaxation method for solving integer programming problems," *Management Science*, vol. 27, no. 1, pp. 1–18, 1981.

[228] M. Guignard and S. Kim, "Lagrangian decomposition: a model yielding stronger Lagrangian bounds," *Mathematical Programming*, vol. 39, pp. 215–228, 1987.

[229] J. F. Benders, "Partitioning procedures for solving mixed-variables programming problems," *Numerische Mathematik*, vol. 4, no. 1, pp. 238–252, December 1962.

[230] H. Weyl, "Elementare theorie der konvexen polyheder," *Commentarii Mathematici Helvetici*, vol. 7, pp. 290–306, 1935.

[231] R. E. Gomory, "Outline of an algorithm for integer solution to linear programs," *Bulletin of the American Mathematical Society*, vol. 64, no. 5, pp. 275–278, 1958.

[232] D. P. Bertsekas, *Dynamic Programming and Optimal Control*. Athena Scientific, 2001.

[233] L. S. Shapley, "Stochastic games," *Proceedings of the National Academy of Sciences of the United States of America*, vol. 39, no. 10, pp. 1095–1100, 1953.

[234] J. F. Mertens and A. Neyman, "Stochastic games," *International Journal of Game Theory*, vol. 10, no. 2, pp. 53–66, 1981.

[235] D. Fudenberg and J. Tirole, *Game Theory*. MIT Press, 1991.

[236] G. Owen, *Game Theory*. Academic Press, 2001.

[237] V. Krishna, *Auction Theory*. Academic Press, 2002.

[238] C. U. Saraydar, N. B. Mandayam, and D. J. Goodman, "Efficient power control via pricing in wireless data networks," *IEEE Transactions on Communications*, vol. 50, no. 2, pp. 291–303, 2002.

[239] H. Yaïche, R. R. Mazumdar, and C. Rosenberg, "A game theoretic framework for bandwidth allocation and pricing in broadband networks," *IEEE/ACM Transactions on Networking*, vol. 8, no. 5, pp. 667–678, 2000.

[240] Z. Han, Z. Ji, and K. J. R. Liu, "Power minimization for multi-cell OFDM networks using distributed non-cooperative game approach," in *Proceedings of IEEE Global Telecommunications Conference (GLOBECOM)*, vol. 6, November–December 2004, pp. 3742–3747.

[241] ——, "Non-cooperative resource competition game by virtual referee in multi-cell OFDMA networks," *IEEE Journal on Selected Areas in Communications*, vol. 25, no. 6, pp. 1079–1090, 2007.

[242] V. Srinivasan, P. Nuggehalli, C. F. Chiasserini, and R. R. Rao, "Cooperation in wireless ad hoc networks," in *Proceedings of IEEE International Conference on Computer Communications (INFOCOM)*, vol. 2, March–April 2003, pp. 808–817.

[243] E. Altman, A. A. Kherani, P. Michiardi, and R. Molva, "Non-cooperative forwarding in ad-hoc networks," in *Proceedings of IEEE International IFIP-TC6 Networking Conference (NETWORK)*, May 2005, pp. 486–498.

[244] R. H. Porter, "Optimal cartel trigger price strategies," *Journal of Economic Theory*, vol. 29, no. 2, pp. 313–318, 1983.

[245] N. Vieille, "Stochastic games: recent results," *Handbook of Game Theory with Economic Applications*, vol. 3, pp. 1833–1850, 2002.

[246] A. Neyman and S. Sorin, *Stochastic Games and Applications*. Springer, 2003.

[247] J. Filar and K. Vrieze, *Competitive Markov Decision Processes*. Springer, 1996.

[248] E. Altman, "Applications of dynamic games in queues," *Advances in Dynamic Games*, vol. 7, pp. 309–342, 2005.

[249] E. Altman, T. Jimenez, R. N. Queija, and U. Yechiali, "Optimal routing among */m/1 queues with partial information," *Stochastic Models*, vol. 20, no. 2, pp. 149–172, 2004.

[250] D. Grosu, A. T. Chronopoulos, and M.-Y. Leung, "Load balancing in distributed systems: an approach using cooperative games," in *Proceedings of IEEE International Parallel and Distributed Processing Symposium (IPDPS)*, 2002, pp. 52–61.

[251] W. Rhee and J. M. Cioffi, "Increase in capacity of multiuser OFDM system using dynamic subchannel allocation," in *Proceedings of IEEE Vehicular Technology Conference (VTC) Spring*, vol. 2, 2000, pp. 1085–1089.

[252] Z. Han, Z. Ji, and K. J. R. Liu, "Fair multiuser channel allocation for OFDMA networks using Nash bargaining solutions and coalitions," *IEEE Transactions on Communications*, vol. 53, no. 8, pp. 1366–1376, August 2005.

[253] C. Peng, H. Zheng, and B. Y. Zhao, "Utilization and fairness in spectrum assignment for opportunistic spectrum access," *Mobile Networks and Applications*, vol. 11, no. 4, pp. 555–576, 2006.

[254] J. E. Suris, L. A. DaSilva, Z. Han, and A. B. MacKenzie, "Cooperative game theory for distributed spectrum sharing," in *Proceedings of IEEE International Conference on Communications (ICC)*, June 2007, pp. 5282–5287.

[255] K.-D. Lee and V. C. M. Leung, "Fair allocation of subcarrier and power in an OFDMA wireless mesh network," *IEEE Journal on Selected Areas in Communications*, vol. 24, no. 11, pp. 2051–2060, 2006.

[256] H. Park and M. van der Schaar, "Bargaining strategies for networked multimedia resource management," *IEEE Transactions on Signal Processing*, vol. 55, no. 7, pp. 3496–3511, 2007.

[257] K. Apt and A. Witzel, "A generic approach to coalition formation," in *arXiv:0709.0435v1 [cs.GT]*, September 2007.

[258] K. Apt and T. Radzik, "Stable partitions in coalitional games," in *arXiv:cs/0605132v1 [cs.GT]*, May 2006.

[259] K. Apt and A. Witzel, "A generic approach to coalition formation," in *Proceedings of First International Workshop on Computational Social Choice (COMSOC)*, December 2006, pp. 21–34.

[260] D. M. Kreps and R. Wilson, "Sequential equilibria," *Econometrica*, vol. 50, no. 4, pp. 863–894, 1982.

[261] D. Monderer and L. S. Shapley, "Potential game," *Games and Economic Behavior*, vol. 14, pp. 124–143, 1996.

[262] A. MacKenzie, *Game Theory for Wireless Engineers*. Morgan & Claypool, 2006.

[263] G. Scutari, S. Barbarossa, and D. P. Palomar, "Potential games: a framework for vector power control problems with coupled constraints," in *Proceedings of IEEE International Conference on Acoustics, Speech and Signal Processing (ICASSP)*, vol. 4, May 2006.

[264] J. Neel, J. Reed, and R. Gilles, "Game models for cognitive radio analysis," in *Proceedings of SDR Forum Technical Conference*, November 2004.

[265] J. O. Neel, J. H. Reed, and R. P. Gilles, "Convergence of cognitive radio networks," in *Proceedings of IEEE Wireless Communications and Networking Conference (WCNC)*, vol. 4, March 2004, pp. 2250–2255.

[266] J. Neel, J. Reed, and R. Gilles, "The role of game theory in the analysis of software radio networks," in *Proceedings of SDR Forum Technical Conference*, November 2002.

[267] A. R. Fattahi and F. Paganini, "New economic perspectives for resource allocation in wireless networks," in *Proceedings of American Control Conference*, vol. 6, June 2005, pp. 3960–3965.

[268] E. Altman and Z. Altman, "S-modular games and power control in wireless networks," *IEEE Transactions on Automatic Control*, vol. 48, no. 5, pp. 839–842, 2003.

[269] R. Menon, A. MacKenzie, R. Buehrer, and J. Reed, "Game theory and interference avoidance in decentralized networks," in *Proceedings of SDR Forum Technical Conference*, November 2004.

[270] J. E. Hicks, A. B. MacKenzie, J. A. Neel, and J. H. Reed, "A game theory perspective on interference avoidance," in *Proceedings of IEEE Global Telecommunications Conference (GLOBECOM)*, vol. 1, November–December 2004, pp. 257–261.

[271] J. Hicks and A. B. MacKenzie, "A convergence result for potential games," in *Proceedings of Symposium on Dynamic Games and Applications*, December 2004.

[272] R. J. Aumann, "Subjectivity and correlation in randomized strategies," *Journal of Mathematical Economics*, vol. 1, no. 1, pp. 67–96, 1974.

[273] ——, "Correlated equilibrium as an expression of Bayesian rationality," *Econometrica*, vol. 55, no. 1, pp. 1–18, 1987.

[274] S. Hart and A. Mas-Colell, "A simple adaptive procedure leading to correlated equilibrium," *Econometrica*, vol. 68, no. 5, pp. 1127–1150, 2000. [Online]. Available: http://ideas.repec.org/a/ecm/emetrp/v68y2000i5p1127-1150.html

[275] C. J. C. H. Watkins, "Learning from delayed rewards," Ph.D. dissertation, Cambridge University, 1989.

[276] J. Hu and M. P. Wellman, "Nash Q-learning for general-sum stochastic games," *Journal of Machine Learning Research*, vol. 4, no. 6, pp. 1039–1070, 2004.

[277] A. M. Fink, "Equilibrium in a stochastic *n*-person game," *Journal of Science in Hiroshima University Series A–I*, vol. 28, no. 1, pp. 89–93, 1964.

[278] N. Baldo and M. Zorzi, "Learning and adaptation in cognitive radios using neural networks," in *Proceedings of IEEE Consumer Communications and Networking Conference (CCNC)*, January 2008, pp. 998–1003.

[279] A. F. Cattoni, M. Ottonello, M. Raffetto, and C. S. Regazzoni, "Neural networks mode classification based on frequency distribution features," in *Proceedings of International Conference on Cognitive Radio Oriented Wireless Networks and Communications (CrownCom)*, August 2007, pp. 251–257.

[280] A. Fehske, J. Gaeddert, and J. Reed, "A new approach to signal classification using spectral correlation and neural networks," in *Proceedings of IEEE International Symposium on New Frontiers in Dynamic Spectrum Access Networks (DySPAN)*, November 2005, pp. 144–150.

[281] D. J. C. MacKay, *Information Theory, Inference and Learning Algorithms*. Cambridge University Press, June 2002.

[282] M. M. Gupta, L. Jin, and N. Homma, *Static and Dynamic Neural Networks: From Fundamentals to Advanced Theory*. Wiley-IEEE Press, 2003.

[283] G. Bianchi, "Performance analysis of the IEEE 802.11 distributed coordination function," *IEEE Journal on Selected Areas in Communications*, vol. 18, no. 3, pp. 535–547, 2000.

[284] A. Kamerman and L. Monteban, "WaveLAN II: A high-performance wireless LAN for the unlicensed band," *Bell Labs Technical Journal*, vol. 2, no. 3, pp. 278–292, 1997.

[285] D. Qiao, S. Choi, and K. G. Shin, "Goodput analysis and link adaptation for IEEE 802.11a wireless LANS," *IEEE Transactions on Mobile Computing*, vol. 1, no. 4, pp. 278–292, 2002.

[286] J. Nie and S. Haykin, "A Q-learning-based dynamic channel assignment technique for mobile communication systems," *IEEE Transactions on Vehicular Technology*, vol. 48, no. 5, pp. 1676–1687, 1999.

[287] M. Zhang and T. S. Yum, "Comparisons of channel assignment strategies in cellular mobile telephone systems," in *Proceedings of IEEE International Conference on Communications (ICC)*, June 1989, pp. 467–471.

[288] K. N. Sivarajan, R. J. McEliece, and J. W. Ketchum, "Dynamic channel assignment in cellular radio," in *Proceedings of IEEE Vehicular Technology Conference (VTC)*, May 1990, pp. 631–637.

[289] R. Farha, N. Abji, O. Sheikh, and A. Leon-Garcia, "Market-based resource management for cognitive radios using machine learning," in *Proceedings of IEEE Global Telecommunications Conference (GLOBECOM)*, November 2007, pp. 4630–4635.

[290] Y. Zhao, J. Gaeddert, L. M. abd Kyung Bae, J.-S. Um, and J. H. Reed, "Development of radio environment map enabled case- and knowledge-based learning algorithms for IEEE 802.22 WRAN cognitive engines," in *Proceedings of International Conference on Cognitive Radio Oriented Wireless Networks and Communications (CrownCom)*, August 2007, pp. 44–49.

[291] Y. Zhao, B. Le, and J. H. Reed, *Cognitive Radio Technology*. Elsevier/Newnes, 2006, ch. "Network support: the radio environment map."

[292] Y. Zhao, L. Morales, J. Gaeddert, K. K. Bae, J.-S. Um, and J. H. Reed, "Applying radio environment maps to cognitive wireless regional area networks," in *Proceedings of IEEE International Symposium on New Frontiers in Dynamic Spectrum Access Networks (DySPAN)*, April 2007, pp. 115–118.

[293] A. Aamodt, "The case-based reasoning paradigm: perspective, technology and impacts (abstract)," in *Proceedings of 13th Brazilian Symposium on Artificial Intelligence (SBIA)*. Springer-Verlag, 1996, p. 236.

[294] H.-S. T. Le and Q. Liang, "An efficient power control scheme for cognitive radios," in *Proceedings of IEEE Wireless Communications and Networking Conference (WCNC)*, March 2007, pp. 2559–2563.

[295] N. Baldo and M. Zorzi, "Cognitive network access using fuzzy decision making," in *Proceedings of IEEE International Conference on Communications (ICC)*, June 2007, pp. 6504–6510.

[296] R. J. Berger, "Open spectrum: a path to ubiquitous connectivity," *Queue*, vol. 1, no. 3, pp. 60–68, 2003.

[297] P. Kolodzy, "Spectrum policy task force report," in *FCC*, December 2002.

[298] J. M. Peha and S. Panichpapiboon, "Real-time secondary markets for spectrum," *Telecommunications Policy*, vol. 28, no. 7–8, pp. 603–618, 2004.

[299] G. Hardin, "The tragedy of the commons," *Science*, vol. 162, pp. 1243–1248, 1968.

[300] G. Li, S. Srikanteswara, and C. Maciocco, "Interference mitigation for WLAN devices using spectrum sensing," in *Proceedings of IEEE Consumer Communications and Networking Conference (CCNC)*, January 2008, pp. 958–962.

[301] S. Srikanteswara, G. Li, and C. Maciocco, "Cross layer interference mitigation using spectrum sensing," in *Proceedings of IEEE Global Telecommunications Conference (GLOBECOM)*, November 2007, pp. 3553–3557.

[302] J. Brito, "The spectrum commons in theory and practice," *Stanford Technology Law Review*, 2007.

[303] P. Kyasanur and N. H. Vaidya, "Selfish MAC layer misbehavior in wireless networks," *IEEE Transactions on Mobile Computing*, vol. 4, no. 5, pp. 502–516, 2005.

[304] A. L. Toledo and X. Wang, "Robust detection of selfish misbehavior in wireless networks," *IEEE Journal on Selected Areas in Communications*, vol. 25, no. 6, pp. 1124–1134, 2007.

[305] J. Konorski, "A game-theoretic study of CSMA/CA under a backoff attack," *IEEE/ACM Transactions on Networking*, vol. 14, no. 6, pp. 1167–1178, 2006.

[306] D. Niyato and E. Hossain, "Competitive pricing for spectrum sharing in cognitive radio networks: dynamic game, inefficiency of Nash equilibrium, and collusion," *IEEE Journal on Selected Areas in Communications*, vol. 26, no. 1, pp. 192–202, 2008.

[307] S. Almeida, J. Queijo, and L. Correia, "Spatial and temporal traffic distribution models for GSM," in *Proceedings of IEEE Vehicular Technology Conference (VTC) Fall*, vol. 1, September 1999, pp. 131–135.

[308] B. Kiefl, "What will we watch? A forecast of TV viewing habits in 10 years," *The Advertising Research Foundation*, 1998.

[309] P. Leaves, S. Ghaheri-Niri, R. Tafazolli *et al.*, "Dynamic spectrum allocation in a multi-radio environment: concept and algorithm," in *Proceedings of International Conference on 3G Mobile Communication Technologies*, March 2001, pp. 53–57.

[310] S. Anand, A. Sridharan, and K. N. Sivarajan, "Performance analysis of channelized cellular systems with dynamic channel allocation," *IEEE Transactions on Vehicular Technology*, vol. 52, no. 4, pp. 847–859, 2003.

[311] Y. Argyropoulos, S. Jordan, and S. P. R. Kumar, "Dynamic channel allocation in interference-limited cellular systems with uneven traffic distribution," *IEEE Transactions on Vehicular Technology*, vol. 48, no. 1, pp. 224–232, 1999.

[312] G. Boggia and P. Camarda, "Modeling dynamic channel allocation in multicellular communication networks," *IEEE Journal on Selected Areas in Communications*, vol. 19, no. 11, pp. 2233–2242, 2001.

[313] J. Jiang, T.-H. Lai, and N. Soundarajan, "On distributed dynamic channel allocation in mobile cellular networks," *IEEE Transactions on Parallel and Distributed Systems*, vol. 13, no. 10, pp. 1024–1037, 2002.

[314] A. Baiocchi, F. D. Priscoli, F. Grilli, and F. Sestini, "The geometric dynamic channel allocation as a practical strategy in mobile networks with bursty user mobility," *IEEE Transactions on Vehicular Technology*, vol. 44, no. 1, pp. 14–23, 1995.

[315] M. Bublin, M. Konegger, and P. Slanina, "A cost-function-based dynamic channel allocation and its limits," *IEEE Transactions on Vehicular Technology*, vol. 55, no. 4, pp. 2286–2295, 2007.

[316] O. Lazaro and D. Girma, "A hopfield neural-network-based dynamic channel allocation with handoff channel reservation control," *IEEE Transactions on Vehicular Technology*, vol. 49, no. 5, pp. 1578–1587, 2000.

[317] X. Tian and C. Ji, "Bounding the performance of dynamic channel allocation with QoS provisioning for distributed admission control in wireless networks," *IEEE Transactions on Vehicular Technology*, vol. 50, no. 2, pp. 388–397, 2001.

[318] A. Ghasemi and E. S. Sousa, "Asymptotic performance of collaborative spectrum sensing under correlated log-normal shadowing," *IEEE Communications Letters*, vol. 11, no. 1, pp. 34–36, 2007.

[319] ——, "Interference aggregation in spectrum-sensing cognitive wireless networks," *IEEE Journal of Selected Topics in Signal Processing*, vol. 2, no. 1, pp. 41–56, 2008.

[320] Y. Zhang, W.-D. Yang, and Y.-M. Cai, "Cooperative spectrum sensing technique," in *Proceedings of IEEE International Conference on Wireless Communications, Networking and Mobile Computing (WiCom)*, September 2007, pp. 1167–1170.

[321] G. Ganesan, Y. Li, B. Bing, and S. Li, "Spatiotemporal sensing in cognitive radio networks," *IEEE Journal on Selected Areas in Communications*, vol. 26, no. 1, pp. 5–12, 2008.

[322] P. Flajolet and G. N. Martin, "Probabilistic counting algorithms for data base applications," *Journal of Computer and System Sciences*, vol. 31, no. 2, pp. 182–209, 1985.

[323] J. Ma and Y. G. Li, "Soft combination and detection for cooperative spectrum sensing in cognitive radio networks," in *Proceedings of IEEE Global Telecommunications Conference (GLOBECOM)*, November 2007, pp. 3139–3143.

[324] Z. Quan, S. Cui, and A. H. Sayed, "Optimal linear cooperation for spectrum sensing in cognitive radio networks," *IEEE Journal of Selected Topics in Signal Processing*, vol. 2, no. 1, pp. 28–40, 2008.

[325] G. Atia, E. Ermis, and V. Saligrama, "Robust energy efficient cooperative spectrum sensing in cognitive radios," in *Proceedings of IEEE Workshop on Statistical Signal Processing*, August 2007, pp. 502–506.

[326] E. Visotsky, S. Kuffner, and R. Peterson, "On collaborative detection of TV transmissions in support of dynamic spectrum sharing," in *Proceedings of IEEE International Symposium on New Frontiers in Dynamic Spectrum Access Networks (DySPAN)*, November 2005, pp. 338–345.

[327] X. Liu and S. Shankar, "Sensing-based opportunistic channel access," *Mobile Networks and Applications*, vol. 11, no. 4, pp. 577–591, 2006.

[328] P. Qihang, Z. Kun, W. Jun, and L. Shaoqian, "A distributed spectrum sensing scheme based on credibility and evidence theory in cognitive radio context," in *Proceedings of IEEE International Symposium on Personal, Indoor and Mobile Radio Communications (PIMRC)*, September 2006, pp. 1–5.

[329] J. Unnikrishnan and V. V. Veeravalli, "Cooperative spectrum sensing and detection for cognitive radio," in *Proceedings of IEEE Global Telecommunications Conference (GLOBECOM)*, November 2007, pp. 2972–2976.

[330] Z. Quan, S. Cui, and A. H. Sayed, "An optimal strategy for cooperative spectrum sensing in cognitive radio networks," in *Proceedings of IEEE Global Telecommunications Conference (GLOBECOM)*, November 2007, pp. 2947–2951.

[331] A. Taherpour, M. Nasiri-Kenari, and A. Jamshidi, "Efficient cooperative spectrum sensing in cognitive radio networks," in *Proceedings of IEEE International Symposium on Personal, Indoor and Mobile Radio Communications (PIMRC)*, September 2007.

[332] J. N. Laneman, D. N. C. Tse, and G. W. Wornell, "Cooperative diversity in wireless networks: efficient protocols and outage behavior," *IEEE Transactions on Information Theory*, vol. 50, no. 12, pp. 3062–3080, 2004.

[333] H. V. Poor, *An Introduction to Signal Detection and Estimation*. Springer-Verlag, 1994.

[334] N. Ahmed, D. Hadaller, and S. Keshav, "GUESS: gossiping updates for efficient spectrum sensing," in *Proceedings of International Workshop on Decentralized Resource Sharing in Mobile Computing and Networking (MobiShare)*, 2006, pp. 12–17.

[335] H. Tang, "Some physical layer issues of wide-band cognitive radio system," in *Proceedings of IEEE International Symposium on New Frontiers in Dynamic Spectrum Access Networks (DySPAN)*, November 2005, pp. 151–159.

[336] M. Gandetto and C. Regazzoni, "Spectrum sensing: a distributed approach for cognitive terminals," *IEEE Journal on Selected Areas in Communications*, vol. 25, no. 3, pp. 546–557, 2007.

[337] W. Wang, L. Zhang, W. Zou, and Z. Zhou, "On the distributed cooperative spectrum sensing for cognitive radio," in *Proceedings of IEEE International Symposium on Communications and Information Technologies (ISCIT)*, October 2007, pp. 1496–1501.

[338] D. Kempe, A. Dobra, and J. Gehrke, "Gossip-based computation of aggregate information," in *Proceedings of IEEE Symposium on Foundations of Computer Science (FOC)*, October 2003, pp. 482–491.

[339] A. D. Spaulding and G. H. Hagn, "On the definition and estimation of spectrum occupancy," *IEEE Transactions on Electromagnetic Compatibility*, vol. 19, no. 3, pp. 269–280, 1977.

[340] P. J. Laycock, M. Morrell, G. F. Gott, and A. R. Ray, "A model for HF spectral occupancy," in *Proceedings of IEE International Conference on HF Radio Systems and Techniques*, April 1988, pp. 165–171.

[341] D. J. Percival, M. Kraetzl, and M. S. Britton, "A Markov model for HF spectral occupancy in central Australia," in *Proceedings of IEE International Conference on HF Radio Systems and Techniques*, July 1997, pp. 14–18.

[342] S. Yarkan and H. Arslan, "Binary time series approach to spectrum prediction for cognitive radio," in *Proceedings of IEEE Vehicular Technology Conference (VTC) Fall*, September–October 2007, pp. 1563–1567.

[343] S. Kaneko, S. Nomoto, T. Ueda, S. Nomura, K. Takeuchi, and K. Sugiyama, "Experimental verification on the prediction of the trend in radio resource availability in cognitive radio," in *Proceedings of IEEE Vehicular Technology Conference (VTC) Fall*, September–October 2007, pp. 1568–1572.

[344] H. Lutkepohl, *Introduction to Multiple Time Series Analysis*. Springer-Verlag, 1993.

[345] T. Clancy and B. Walker, "Predictive dynamic spectrum access," in *Proceedings of SDR Forum Technical Conference*, November 2006.

[346] S. Mangold and L. Berlemann, "IEEE 802.11k: improving confidence in radio resource measurements," in *Proceedings of IEEE International Symposium on Personal, Indoor and Mobile Radio Communications (PIMRC)*, vol. 2, September 2005, pp. 1009–1013.

[347] E. G. Villegas, R. V. Ferre, and J. P. Aspas, "Load balancing in WLANS through IEEE 802.11k mechanisms," in *Proceedings of IEEE Symposium on Computers and Communications (ISCC)*, June 2006, pp. 844–850.

[348] S. Hermann, M. Emmelmann, O. Belaifa, and A. Wolisz, "Investigation of IEEE 802.11k-based access point coverage area and neighbor discovery," in *Proceedings of IEEE Conference on Local Computer Networks (LCN)*, October 2007, pp. 949–954.

[349] N. Golmie, O. Rebala, and N. Chevrollier, "Bluetooth adaptive frequency hopping and scheduling," in *Proceedings of IEEE Military Communications Conference (MILCOM)*, vol. 2, October 2003, pp. 1138–1142.

[350] O. A. Bamahdi and S. A. Zummo, "An adaptive frequency hopping technique with application to Bluetooth–WLAN coexistence," in *Proceedings of International Conference on Networking (ICN), International Conference on Systems (ICONS) and International Conference on Mobile Communications and Learning Technologies (MCL)*, vol. 2, April 2006.

[351] N. Han, S. Shon, J. H. Chung, and J. M. Kim, "Spectral correlation based signal detection method for spectrum sensing in IEEE 802.22 WRAN systems," in *Proceedings of IEEE International Conference Advanced Communication Technology (ICACT)*, vol. 3, February 2006, pp. 1765–1770.

[352] H.-S. Chen, W. Gao, and D. G. Daut, "Spectrum sensing using cyclostationary properties and application to IEEE 802.22 WRAN," in *Proceedings of IEEE Global Telecommunications Conference (GLOBECOM)*, November 2007, pp. 3133–3138.

[353] J.-K. Lee, J.-H. Yoon, and J.-U. Kim, "A new spectral correlation approach to spectrum sensing for 802.22 WRAN system," in *Proceedings of IEEE International Conference on Intelligent Pervasive Computing (IPC)*, October 2007, pp. 101–104.

[354] Y. Youn, H. Jeon, J. H. Choi, and H. Lee, "Fast spectrum sensing algorithm for 802.22 WRAN systems," in *Proceedings of IEEE International Symposium on Communications and Information Technologies (ISCIT)*, October–September 2006, pp. 960–964.

[355] G. Zheng, N. Han, X. Huang, S. H. Sohn, and J. M. Kim, "Enhanced energy detector for IEEE 802.22 WRAN systems using maximal-to-mean power ratio," in *Proceedings of IEEE International Symposium on Wireless Communication Systems (ISWCS)*, October 2007, pp. 370–374.

[356] H.-S. Chen, W. Gao, and D. G. Daut, "Signature based spectrum sensing algorithms for IEEE 802.22 WRAN," in *Proceedings of IEEE International Conference on Communications (ICC)*, June 2007, pp. 6487–6492.

[357] F. Sheikh and B. Bing, "Cognitive spectrum sensing and detection using polyphase DFT filter banks," in *Proceedings of IEEE Consumer Communications and Networking Conference (CCNC)*, January 2008, pp. 973–977.

[358] Q. Zhao, L. Tong, A. Swami, and Y. Chen, "Decentralized cognitive MAC for opportunistic spectrum access in ad hoc networks: a POMDP framework," *IEEE Journal on Selected Areas in Communications*, vol. 25, no. 3, pp. 589–600, April 2007.

[359] J. Jia, Q. Zhang, and X. Shen, "HC-MAC: a hardware-constrained cognitive MAC for efficient spectrum management," *IEEE Journal on Selected Areas in Communications*, vol. 26, no. 1, pp. 106–117, 2008.

[360] A. Motamedi and A. Bahai, "MAC protocol design for spectrum-agile wireless networks: stochastic control approach," in *Proceedings of IEEE International Symposium on New Frontiers in Dynamic Spectrum Access Networks (DySPAN)*, April 2007, pp. 448–451.

[361] P. Papadimitratos, S. Sankaranarayanan, and A. Mishra, "A bandwidth sharing approach to improve licensed spectrum utilization," *IEEE Communications Magazine*, vol. 43, no. 12, pp. S10–S14, 2005.

[362] C. Cordeiro and K. Challapali, "C-MAC: a cognitive MAC protocol for multi-channel wireless networks," in *Proceedings of IEEE International Symposium on New Frontiers in Dynamic Spectrum Access Networks (DySPAN)*, April 2007, pp. 147–157.

[363] L. Ma, X. Han, and C.-C. Shen, "Dynamic open spectrum sharing MAC protocol for wireless ad hoc networks," in *Proceedings of IEEE International Symposium on New Frontiers in Dynamic Spectrum Access Networks (DySPAN)*, November 2005, pp. 203–213.

[364] L. B. Le and E. Hossain, "OSA-MAC: a multi-channel MAC protocol for opportunistic spectrum access in cognitive radio networks," in *Proceedings of IEEE Wireless Communications and Networking Conference (WCNC)*, March–April 2008.

[365] A. Sabharwal, A. Khoshnevis, and E. Knightly, "Opportunistic spectral usage: bounds and a multi-band CSMA/CA protocol," *IEEE/ACM Transactions on Networking*, vol. 15, no. 3, pp. 533–545, 2007.

[366] T. Chen, H. Zhang, G. M. Maggio, and I. Chlamtac, "Topology management in CogMesh: a cluster-based cognitive radio mesh network," in *Proceedings of IEEE International Conference on Communications (ICC)*, June 2007, pp. 6516–6521.

[367] L.-C. Wang, C.-W. Wang, Y.-C. Lu, and C.-M. Liu, "A concurrent transmission MAC protocol for enhancing throughput and avoiding spectrum sensing in cognitive radio," in *Proceedings of IEEE Wireless Communications and Networking Conference (WCNC)*, March 2007, pp. 105–110.

[368] G. Auer, H. Haas, and P. Omiyi, "Interference aware medium access for dynamic spectrum sharing," in *Proceedings of IEEE International Symposium on New Frontiers in Dynamic Spectrum Access Networks (DySPAN)*, April 2007, pp. 399–402.

[369] H. Nan, T.-I. Hyon, and S.-J. Yoo, "Distributed coordinated spectrum sharing MAC protocol for cognitive radio," in *Proceedings of IEEE International Symposium on New Frontiers in Dynamic Spectrum Access Networks (DySPAN)*, April 2007, pp. 240–249.

[370] T. Shu, S. Cui, and M. Krunz, "Medium access control for multi-channel parallel transmission in cognitive radio networks," in *Proceedings of IEEE Global Telecommunications Conference (GLOBECOM)*, November 2006.

[371] A. C.-C. Hsu, D. S. L. Weit, and C.-C. J. Kuo, "A cognitive MAC protocol using statistical channel allocation for wireless ad-hoc networks," in *Proceedings of IEEE Wireless Communications and Networking Conference (WCNC)*, March 2007, pp. 105–110.

[372] S. Sengupta, S. Brahma, M. Chatterjee, and S. Shankar, "Enhancements to cognitive radio based IEEE 802.22 air-interface," in *Proceedings of IEEE International Conference on Communications (ICC)*, June 2007, pp. 5155–5160.

[373] M. Thoppian, S. Venkatesan, R. Prakash, and R. Chandrasekaran, "MAC-layer scheduling in cognitive radio based multi-hop wireless networks," in *Proceedings of IEEE International Symposium on a World of Wireless, Mobile, and Multimedia Networks (WoWMoM)*, June 2006.

[374] H. Zheng and C. Peng, "Collaboration and fairness in opportunistic spectrum access," in *Proceedings of IEEE International Conference on Communications (ICC)*, vol. 5, May 2005, pp. 3132–3136.

[375] W. Wang and X. Liu, "List-coloring based channel allocation for open-spectrum wireless networks," in *Proceedings of IEEE Vehicular Technology Conference (VTC) Fall*, September 2005, pp. 690–694.

[376] C. Doerr, M. Neufeld, J. Fifield, T. Weingart, D. Sicker, and D. Grunwald, "MultiMAC – an adaptive MAC framework for dynamic radio networking," in *Proceedings of IEEE International Symposium on New Frontiers in Dynamic Spectrum Access Networks (DySPAN)*, November 2005, pp. 548–555.

[377] D. Grandblaise, K. Moessner, G. Vivier, and R. Tafazolli, "Credit token based rental protocol for dynamic channel allocation," in *Proceedings of International Conference on Cognitive Radio Oriented Wireless Networks and Communications (CrownCom)*, June 2006, pp. 1–5.

[378] V. Brik, E. Rozner, S. Banerjee, and P. Bahl, "DSAP: a protocol for coordinated spectrum access," in *Proceedings of IEEE International Symposium on New Frontiers in Dynamic Spectrum Access Networks (DySPAN)*, November 2005, pp. 611–614.

[379] D. Raychaudhuri and X. Jing, "A spectrum etiquette protocol for efficient coordination of radio devices in unlicensed bands," in *Proceedings of IEEE International Symposium on Personal, Indoor and Mobile Radio Communications (PIMRC)*, vol. 1, September 2003, pp. 172–176.

[380] M. M. Rashid, J. Hossain, E. Hossain, and V. K. Bhargava, "Opportunistic spectrum access in cognitive radio networks: a queueing analytic model and admission controller design," in *Proceedings of IEEE Global Telecommunications Conference (GLOBECOM)*, November 2007, pp. 4647–4652.

[381] S. N. Shankar, "Squeezing the most out of cognitive radio: a joint MAC/PHY perspective," in *Proceedings of IEEE International Conference on Acoustics, Speech and Signal Processing (ICASSP)*, vol. 4, April 2007, pp. 1361–1364.

[382] L.-C. Wang, Y.-C. Lu, C.-W. Wang, and D. S. L. Wei, "Latency analysis for dynamic spectrum access in cognitive radio: dedicated or embedded control channel?," in *Proceedings of IEEE International Symposium on Personal, Indoor and Mobile Radio Communications (PIMRC)*, September 2007.

[383] P. Pawelczak, G. J. Janssen, and R. Prasad, "Performance measures of dynamic spectrum access networks," in *Proceedings of IEEE Global Telecommunications Conference (GLOBECOM)*, November 2006.

[384] S. Keshavamurthy and K. Chandra, "Multiplexing analysis for dynamic spectrum access," in *Proceedings of IEEE Military Communications Conference (MILCOM)*, October 2006.

[385] M. Raspopovic, C. Thompson, and K. Chandra, "Performance models for wireless spectrum shared by wideband and narrowband sources," in *Proceedings of IEEE Military Communications Conference (MILCOM)*, October 2005, pp. 1642–1647.

[386] S. Tang and B. L. Mark, "Performance analysis of a wireless network with opportunistic spectrum sharing," in *Proceedings of IEEE Global Telecommunications Conference (GLOBECOM)*, November 2007, pp. 4636–4640.

[387] M. Raspopovic and C. Thompson, "Finite population model for performance evaluation between narrowband and wideband users in the shared radio spectrum," in *Proceedings of IEEE International Symposium on New Frontiers in Dynamic Spectrum Access Networks (DySPAN)*, April 2007, pp. 340–346.

[388] D. V. Djonin and V. Krishnamurthy, "MIMO transmission control in fading channels – a constrained Markov decision process formulation with monotone randomized policies," *IEEE Transactions on Signal Processing*, vol. 55, no. 10, pp. 5069–5083, 2007.

[389] D. V. Djonin, A. K. Karmokar, and V. K. Bhargava, "Joint rate and power adaptation for type-I hybrid ARQ systems over correlated fading channels under different buffer-cost constraints," *IEEE Transactions on Vehicular Technology*, vol. 57, no. 1, pp. 421–435, 2008.

[390] L. A. Johnston and V. Krishnamurthy, "Opportunistic file transfer over a fading channel: a POMDP search theory formulation with optimal threshold policies," *IEEE Transactions on Wireless Communications*, vol. 5, no. 2, pp. 394–405, 2006.

[391] K. Yang and X. Wang, "Battery-aware adaptive modulation with QoS constraints," *IEEE Transactions on Communications*, vol. 54, no. 10, pp. 1797–1805, 2006.

[392] Y. S. Chow, H. Robbins, and D. Siegmund, *Great Expectations: The Theory of Optimal Stopping*. Houghton Mifflin Company, 1971.

[393] J. C. Gittins, *Multi-Armed Bandit Allocation Indices*. Wiley, 1989.

[394] J. Mo, H.-S. W. So, and J. Walrand, "Comparison of multichannel MAC protocols," *IEEE Transactions on Mobile Computing*, vol. 7, no. 1, pp. 50–65, 2008.

[395] J. Tang, G. Xue, and W. Zhang, "Cross-layer design for end-to-end throughput and fairness enhancement in multi-channel wireless mesh networks," *IEEE Transactions on Wireless Communications*, vol. 6, no. 10, pp. 3482–3486, 2007.

[396] H. Su and X. Zhang, "Clustering-based multichannel MAC protocols for QoS provisionings over vehicular ad hoc networks," *IEEE Transactions on Vehicular Technology*, vol. 56, no. 6, pp. 3309–3323, 2007.

[397] L. Zhang, B.-H. Soong, and W. Xiao, "Location-aware two-phase coding multi-channel MAC protocol (LA-TPCMMP) for MANETs," *IEEE Transactions on Wireless Communications*, vol. 6, no. 5, pp. 1656–1669, 2007.

[398] F. Capar, I. Martoyo, T. Weiss, and F. Jondral, "Comparison of bandwidth utilization for controlled and uncontrolled channel assignment in a spectrum pooling system," in *Proceedings of IEEE Vehicular Technology Conference (VTC) Spring*, May 2002, pp. 1069–1073.

[399] L.-C. Wang and K.-J. Shieh, "Spectrum sharing for frequency hopped CDMA systems with overlaying cellular structures," in *Proceedings of IEEE Vehicular Technology Conference (VTC) Spring*, vol. 3, April 2003, pp. 1945–1949.

[400] B. Aazhang, J. Lilleberg, and G. Middleton, "Spectrum sharing in a cellular system," in *Proceedings of IEEE International Symposium on Spread Spectrum Techniques and Applications*, August–September 2004, pp. 355–359.

[401] M. K. Pereirasamy, J. Luo, M. Dillinger, and C. Hartmann, "Dynamic inter-operator spectrum sharing for UMTS FDD with displaced cellular networks," in *Proceedings of IEEE Wireless Communications and Networking Conference (WCNC)*, vol. 3, March 2005, pp. 1720–1725.

[402] G. Middleton, K. Hooli, A. Tolli, and J. Lilleberg, "Inter-operator spectrum sharing in a broadband cellular network," in *Proceedings of IEEE International Symposium on Spread Spectrum Techniques and Applications*, August 2006, pp. 376–380.

[403] D. Poppen, "Spectrum sharing between a fixed-service microwave system and a cellular CDMA mobile radio system," in *Proceedings of IEEE Vehicular Technology Conference (VTC)*, May 1993, pp. 564–567.

[404] R. S. Sahota and P. A. Whiting, "On the feasibility of spectrum sharing between GSM and IS-95," in *Proceedings of IEEE International Conference on Personal Wireless Communications*, December 1997, pp. 439–443.

[405] H. H. Xia, F. Rico, and A. Herrera, "Spectrum sharing feasibility for CDMA PCS," in *Proceedings of International Conference on Universal Personal Communications*, September–October 1994, pp. 267–271.

[406] T. Kamakaris, M. M. Buddhikot, and R. Iyer, "A case for coordinated dynamic spectrum access in cellular networks," in *Proceedings of IEEE International Symposium on New Frontiers in Dynamic Spectrum Access Networks (DySPAN)*, November 2005, pp. 289–298.

[407] H.-J. Ju and I. Rubin, "Mobile backbone synthesis for ad hoc wireless networks," *IEEE Transactions on Wireless Communications*, vol. 6, no. 12, pp. 4285–4298, 2007.

[408] S. Leng, L. Zhang, H. Fu, and J. Yang, "A novel location-service protocol based on *k*-hop clustering for mobile ad hoc networks," *IEEE Transactions on Vehicular Technology*, vol. 56, no. 2, pp. 810–817, 2007.

[409] L. Zhang, B.-H. Soong, and W. Xiao, "An integrated cluster-based multi-channel MAC protocol for mobile ad hoc networks," *IEEE Transactions on Wireless Communications*, vol. 6, no. 11, pp. 3964–3974, 2007.

[410] G. Venkataraman, S. Emmanuel, and S. Thambipillai, "Size-restricted cluster formation and cluster maintenance technique for mobile ad hoc networks," *International Journal of Network Management*, vol. 17, no. 2, pp. 171–194, 2007.

[411] G.-J. Yu and C.-Y. Chang, "An efficient cluster-based multi-channel management protocol for wireless ad hoc networks," *Computer Communications*, vol. 30, no. 8, pp. 1742–1753, 2007.

[412] A. D. Amis and R. Prakash, "Load-balancing clusters in wireless ad hoc networks," in *Proceedings of 3rd IEEE Symposium on Application-Specific Systems and Software Engineering Technology (ASSET)*. IEEE Computer Society, 2000, p. 25.

[413] H. Artail, H. Safa, H. Hamze, and K. Mershad, "A cluster based service discovery model for mobile ad hoc networks," in *Proceedings of Third IEEE International Conference on Wireless and Mobile Computing, Networking and Communications (WIMOB)*. IEEE Computer Society, 2007, p. 57.

[414] T. Chen, H. Zhang, G. M. Maggio, and I. Chlamtac, "CogMesh: a cluster-based cognitive radio network," in *Proceedings of IEEE International Symposium on New Frontiers in Dynamic Spectrum Access Networks (DySPAN)*, April 2007, pp. 168–178.

[415] Z. J. Haas and J. Deng, "Dual busy tone multiple access (DBTMA) – a multiple access control scheme for ad hoc networks," *IEEE Transactions on Communications*, vol. 50, no. 6, pp. 975–985, 2002.

[416] H. Su and X. Zhang, "Opportunistic MAC protocols for cognitive radio based wireless networks," in *Proceedings of Annual Conference on Information Sciences and Systems (CISS)*, March 2007, pp. 363–368.

[417] ——, "Cognitive radio based multi-channel MAC protocols for wireless ad hoc networks," in *Proceedings of IEEE Global Telecommunications Conference (GLOBECOM)*, November 2007, pp. 4857–4861.

[418] ——, "Cross-layer based opportunistic MAC protocols for QoS provisionings over cognitive radio wireless networks," *IEEE Journal on Selected Areas in Communications*, vol. 26, no. 1, pp. 118–129, 2008.

[419] W. Hu, D. Willkomm, M. Abusubaih *et al.*, "Dynamic frequency hopping communities for efficient IEEE 802.22 operation," *IEEE Communications Magazine*, vol. 45, no. 5, pp. 80–87, 2007.

[420] M. Bublin, M. Konegger, and P. Slanina, "A cost-function-based dynamic channel allocation and its limits," *IEEE Transactions on Vehicular Technology*, vol. 56, no. 4, pp. 2286–2295, 2007.

[421] D. Niyato and E. Hossain, "Service differentiation in broadband wireless access networks with scheduling and connection admission control: a unified analysis," *IEEE Transactions on Wireless Communications*, vol. 6, no. 1, pp. 293–301, 2007.

[422] J. Yang, Q. Jiang, and D. Manivannan, "A fault-tolerant channel-allocation algorithm for cellular networks with mobile base stations," *IEEE Transactions on Vehicular Technology*, vol. 56, no. 1, pp. 349–361, 2007.

[423] W. P. Siriwongpairat, Z. Han, and K. J. R. Liu, "Power controlled channel allocation for multiuser multiband UWB systems," *IEEE Transactions on Wireless Communications*, vol. 6, no. 2, pp. 583–592, 2007.

[424] P. Chaporkar, K. Kar, X. Luo, and S. Sarkar, "Throughput and fairness guarantees through maximal scheduling in wireless networks," *IEEE Transactions on Information Theory*, vol. 54, no. 2, pp. 572–594, 2008.

[425] K. W. Choi, D. G. Jeong, and W. S. Jeon, "Packet scheduler for mobile communications systems with time-varying capacity region," *IEEE Transactions on Wireless Communications*, vol. 6, no. 3, pp. 1034–1045, 2007.

[426] X. Wu, R. Srikant, and J. R. Perkins, "Scheduling efficiency of distributed greedy scheduling algorithms in wireless networks," *IEEE Transactions on Mobile Computing*, vol. 6, no. 6, pp. 595–605, 2007.

[427] C. Cicconetti, A. Erta, L. Lenzini, and E. Mingozzi, "Performance evaluation of the IEEE 802.16 MAC for QoS support," *IEEE Transactions on Mobile Computing*, vol. 6, no. 1, pp. 26–38, 2007.

[428] A. T. Hoang and Y.-C. Liang, "Maximizing spectrum utilization of cognitive radio networks using channel allocation and power control," in *Proceedings of IEEE Vehicular Technology Conference (VTC) Fall*, September 2006.

[429] H. Islam, Y.-C. Liang, and A. T. Hoang, "Joint beamforming and power control in the downlink of cognitive radio networks," in *Proceedings of IEEE Wireless Communications and Networking Conference (WCNC)*, March 2007, pp. 21–26.

[430] W. Wang, T. Peng, and W. Wang, "Optimal power control under interference temperature constraints in cognitive radio network," in *Proceedings of IEEE Wireless Communications and Networking Conference (WCNC)*, March 2007, pp. 116–120.

[431] H.-S. T. Le and Q. Liang, "An efficient power control scheme for cognitive radios," in *Proceedings of IEEE Wireless Communications and Networking Conference (WCNC)*, March 2007, pp. 2559–2563.

[432] K. Hamdi, W. Zhang, and K. B. Letaief, "Power control in cognitive radio systems based on spectrum sensing side information," in *Proceedings of IEEE International Conference on Communications (ICC)*, June 2007, pp. 5161–5165.

[433] D. Zhang and Z. Tian, "Adaptive games for agile spectrum access based on extended Kalman filtering," *IEEE Journal of Selected Topics in Signal Processing*, vol. 1, no. 1, pp. 79–90, 2007.

[434] N. Jain, S. R. Das, and A. Nasipuri, "A multichannel CSMA MAC protocol with receiver-based channel selection for multihop wireless networks," in *Proceedings of International Conference on Computer Communications and Networks*, October 2001, pp. 432–439.

[435] P. Kyasanur, J. Padhye, and P. Bahl, "On the efficacy of separating control and data into different frequency bands," in *Proceedings of International Conference on Broadband Networks*, vol. 1, October 2005, pp. 602–611.

[436] S. Krishnamurthy, M. Thoppian, S. Venkatesan, and R. Prakash, "Control channel based MAC-layer configuration, routing and situation awareness for cognitive radio networks," in *Proceedings of IEEE Military Communications Conference (MILCOM)*, vol. 1, October 2005, pp. 455–460.

[437] T. H. Cormen, C. E. Leiserson, R. L. Rivest, and C. Stein, *Introduction to Algorithms*. MIT Press and McGraw-Hill, 2001.

[438] D. Grandblaise, "Proposal for credit tokens based co-existence resolution and negotiation protocol," in *Contribution IEEE C802.16h-05/020rl*, July 2005.

[439] D. Niyato and E. Hossain, "A novel analytical framework for integrated cross-layer study of call-level and packet-level QoS in wireless mobile multimedia networks," *IEEE Transactions on Mobile Computing*, vol. 6, no. 3, pp. 322–335, 2007.

[440] ——, "Sleep and wakeup strategies in solar-powered wireless sensor/mesh networks: performance analysis and optimization," *IEEE Transactions on Mobile Computing*, vol. 6, no. 2, pp. 221–236, 2007.

[441] ——, "Call-level and packet-level quality of service and user utility in rate-adaptive cellular CDMA networks: a queuing analysis," *IEEE Transactions on Mobile Computing*, vol. 5, no. 12, pp. 1749–1763, 2006.

[442] G. Latouche and V. Ramaswami, *Introduction to Matrix Analytic Methods in Stochastic Modeling*. Society for Industrial Mathematics, 1987.

[443] P. Chatzimisios, A. C. Boucouvalas, and V. Vitsas, "IEEE 802.11 packet delay – a finite retry limit analysis," in *Proceedings of IEEE Global Telecommunications Conference (GLOBECOM)*, vol. 2, December 2003, pp. 950–954.

[444] E. A. Yavuz and V. C. M. Leung, "Computationally efficient method to evaluate the performance of guard-channel-based call admission control in cellular networks," *IEEE Transactions on Vehicular Technology*, vol. 55, no. 4, pp. 1412–1424, 2006.

[445] S. K. Das, S. K. Sen, K. Basu, and H. Lin, "A framework for bandwidth degradation and call admission control schemes for multiclass traffic in next-generation wireless networks," *IEEE Journal on Selected Areas in Communications*, vol. 21, no. 10, pp. 1790–1802, 2003.

[446] W. S. Jeon and D. G. Jeong, "Call admission control for mobile multimedia communications with traffic asymmetry between uplink and downlink," *IEEE Transactions on Vehicular Technology*, vol. 50, no. 1, pp. 59–66, 2001.

[447] C. W. Leong, W. Zhuang, Y. Cheng, and L. Wang, "Call admission control for integrated on/off voice and best-effort data services in mobile cellular communications," *IEEE Transactions on Communications*, vol. 52, no. 5, pp. 778–790, 2004.

[448] L. Kleinrock, *Queueing System, Volume 1: Theory*. John Wiley and Sons, 1975.

[449] S. Srinivasa and S. A. Jafar, "The throughput potential of cognitive radio: a theoretical perspective," in *Proceedings of Asilomar Conference on Signals, Systems and Computers (ACSSC)*, October–November 2006, pp. 221–225.

[450] ——, "The throughput potential of cognitive radio: a theoretical perspective," *IEEE Communications Magazine*, vol. 45, no. 5, pp. 73–79, 2007.

[451] S. A. Jafar and S. Srinivasa, "Capacity limits of cognitive radio with distributed and dynamic spectral activity," *IEEE Journal on Selected Areas in Communications*, vol. 25, no. 3, pp. 529–537, 2007.

[452] Y. Xing, C. N. Mathur, M. A. Haleem, R. Chandramouli, and K. P. Subbalakshmi, "Dynamic spectrum access with QoS and interference temperature constraints," *IEEE Transactions on Mobile Computing*, vol. 6, no. 4, pp. 423–433, 2007.

[453] B. Wang, Z. Ji, and K. J. R. Liu, "Primary-prioritized Markov approach for dynamic spectrum access," in *Proceedings of IEEE International Symposium on New Frontiers in Dynamic Spectrum Access Networks (DySPAN)*, April 2007, pp. 507–515.

[454] O. Yu, E. Saric, and A. Li, "Dynamic control of open spectrum management," in *Proceedings of IEEE Wireless Communications and Networking Conference (WCNC)*, March 2007, pp. 127–132.

[455] L. Zhang, Y.-C. Liang, and Y. Xin, "Joint admission control and power allocation for cognitive radio networks," in *Proceedings of IEEE International Conference on Acoustics, Speech and Signal Processing (ICASSP)*, vol. 3, April 2007, pp. 673–676.

[456] L. Gao, P. Wu, and S. Cui, "Power and rate control with dynamic programming for cognitive radios," in *Proceedings of IEEE Global Telecommunications Conference (GLOBECOM)*, November 2007, pp. 1699–1703.

[457] L. Kovács and A. Vidács, "Interference-tolerant spatio-temporal dynamic spectrum allocation," in *Proceedings of IEEE International Symposium on New Frontiers in Dynamic Spectrum Access Networks (DySPAN)*, April 2007, pp. 403–411.

[458] H. Kim, Y. Lee, and S. Yun, "A dynamic spectrum allocation between network operators with priority-based sharing and negotiation," in *Proceedings of IEEE International Symposium on Personal, Indoor and Mobile Radio Communications (PIMRC)*, vol. 2, September 2005, pp. 1004–1008.

[459] A. P. Subramanian, H. Gupta, S. R. Das, and B. M. Milind, "Fast spectrum allocation in coordinated dynamic spectrum access based cellular networks," in *Proceedings of IEEE International Symposium on New Frontiers in Dynamic Spectrum Access Networks (DySPAN)*, April 2007, pp. 320–330.

[460] M. Pan, J. Chen, R. Liu, Z. Feng, Y. Wang, and P. Zhang, "Dynamic spectrum access and joint radio resource management combining for resource allocation in cooperative

networks," in *Proceedings of IEEE Wireless Communications and Networking Conference (WCNC)*, March 2007, pp. 2746–2751.

[461] C. Raman, R. D. Yates, and N. B. Mandayam, "Scheduling variable rate links via a spectrum server," in *Proceedings of IEEE International Symposium on New Frontiers in Dynamic Spectrum Access Networks (DySPAN)*, November 2005, pp. 110–118.

[462] A. T. Hoang and Y.-C. Liang, "Adaptive scheduling of spectrum sensing periods in cognitive radio networks," in *Proceedings of IEEE Global Telecommunications Conference (GLOBECOM)*, November 2007, pp. 3128–3132.

[463] S. Gandhi, C. Buragohain, L. Cao, H. Zheng, and S. Suri, "A general framework for wireless spectrum auctions," in *Proceedings of IEEE International Symposium on New Frontiers in Dynamic Spectrum Access Networks (DySPAN)*, April 2007, pp. 22–33.

[464] L. Kovács and A. Vidács, "One-shot multi-bid auction method in dynamic spectrum allocation networks," in *Proceedings of IST Mobile and Wireless Communications Summit*, July 2007, pp. 1–5.

[465] S. Sengupta and M. Chatterjee, *Distributed Computing and Networking*. Springer, 2006, ch. "Synchronous and asynchronous auction models for dynamic spectrum access."

[466] T. Feng and Y. Zhen, "A new algorithm for weighted proportional fairness based spectrum allocation of cognitive radios," in *Proceedings of Eighth ACIS International Conference on Software Engineering, Artificial Intelligence, Networking, and Parallel/Distributed Computing (SNPD)*, vol. 1, July 2007, pp. 531–536.

[467] D. Grandblaise, C. Kloeck, T. Renk *et al.*, "Microeconomics inspired mechanisms to manage dynamic spectrum allocation," in *Proceedings of IEEE International Symposium on New Frontiers in Dynamic Spectrum Access Networks (DySPAN)*, April 2007, pp. 452–461.

[468] F. Kelly, "Charging and rate control for elastic traffic," *European Transactions on Telecommunications*, vol. 8, pp. 33–37, 1997.

[469] O. T. W. Yu and V. C. M. Leung, "Adaptive resource allocation for prioritized call admission over an ATM-based wireless PCN," *IEEE Journal on Selected Areas in Communications*, vol. 15, no. 7, pp. 1208–1225, September 1997.

[470] H. Chen, S. Kumar, and C.-C. J. Kuo, "Dynamic call admission control and resource reservation with interference guard margin (IGM) for CDMA systems," in *Proceedings of IEEE Wireless Communications and Networking Conference (WCNC)*, vol. 3, March 2003, pp. 1568–1572.

[471] O. Yu, E. Saric, and A. Li, "Fairly adjusted multimode dynamic guard bandwidth admission control over CDMA systems," *IEEE Journal on Selected Areas in Communications*, vol. 24, no. 3, pp. 579–592, 2006.

[472] T. Shu and Z. Niu, "Uplink capacity optimization by power allocation for multimedia CDMA networks with imperfect power control," *IEEE Journal on Selected Areas in Communications*, vol. 21, no. 10, pp. 1585–1594, 2003.

[473] M. Elmusrati, R. Jantti, and H. N. Koivo, "Multiobjective distributed power control algorithm for CDMA wireless communication systems," *IEEE Transactions on Vehicular Technology*, vol. 56, no. 2, pp. 779–788, 2007.

[474] N. Benvenuto, G. Carnevale, and S. Tomasin, "Joint power control and receiver optimization of CDMA transceivers using successive interference cancellation," *IEEE Transactions on Communications*, vol. 55, no. 3, pp. 563–573, 2007.

[475] C. Zhou, M. L. Honig, and S. Jordan, "Utility-based power control for a two-cell CDMA data network," *IEEE Transactions on Wireless Communications*, vol. 4, no. 6, pp. 2764–2776, 2005.

[476] C. W. Sung and W. S. Wong, "Power control and rate management for wireless multimedia CDMA systems," *IEEE Transactions on Communications*, vol. 49, no. 7, pp. 1215–1226, 2001.

[477] Y. Lu and R. W. Brodersen, "Integrating power control, error correction coding, and scheduling for a CDMA downlink system," *IEEE Journal on Selected Areas in Communications*, vol. 17, no. 5, pp. 978–989, 1999.

[478] A. Abrardo, G. Giambene, and D. Sennati, "Optimization of power control parameters for DS-CDMA cellular systems," *IEEE Transactions on Communications*, vol. 49, no. 8, pp. 1415–1424, 2001.

[479] G. Caire, R. R. Muller, and T. Tanaka, "Iterative multiuser joint decoding: optimal power allocation and low-complexity implementation," *IEEE Transactions on Information Theory*, vol. 50, no. 9, pp. 1950–1973, 2004.

[480] C. Comaniciu and H. V. Poor, "Jointly optimal power and admission control for delay sensitive traffic in CDMA networks with LMMSE receivers," *IEEE Transactions on Signal Processing*, vol. 51, no. 8, pp. 2031–2042, 2003.

[481] J. Zou and V. K. Bhargava, "Design issues in a CDMA cellular system with heterogeneous traffic types," *IEEE Transactions on Vehicular Technology*, vol. 47, no. 3, pp. 871–884, 1998.

[482] S. Zhao, Z. Xiong, and X. Wang, "Joint error control and power allocation for video transmission over CDMA networks with multiuser detection," *IEEE Transactions on Circuits and Systems for Video Technology*, vol. 12, no. 6, pp. 425–437, 2002.

[483] J.-F. Chamberland and V. V. Veeravalli, "Decentralized dynamic power control for cellular CDMA systems," *IEEE Transactions on Wireless Communications*, vol. 2, no. 3, pp. 549–559, 2003.

[484] M. Andersin, Z. Rosberg, and J. Zander, "Gradual removals in cellular PCS with constrained power control and noise," vol. 1, pp. 56–60, 1995.

[485] J. Zander and S. L. Kim, *Radio Resource Management for Wireless Networks*. Artech House, 2001.

[486] Y. Xing, C. N. Mathur, M. A. Haleem, R. Chandramouli, and K. P. Subbalakshmi, "Priority based dynamic spectrum access with QoS and interference temperature constraints," in *Proceedings of IEEE International Conference on Communications (ICC)*, June 2006, pp. 4420–4425.

[487] M. E. Steenstrup, "Opportunistic use of radio-frequency spectrum: a network perspective," in *Proceedings of IEEE International Symposium on New Frontiers in Dynamic Spectrum Access Networks (DySPAN)*, November 2005, pp. 638–641.

[488] S. Bahramian and B. H. Khalaj, "A novel low-complexity dynamic frequency selection algorithm for cognitive radios," in *Proceedings of IEEE International Symposium on Wireless Communication Systems (ISWCS)*, October 2007, pp. 558–562.

[489] S. Ramanathan, "A unified framework and algorithm for channel assignment in wireless networks," *Wireless Networks*, vol. 5, no. 2, pp. 81–94, 1999.

[490] L. Kovács and A. Vidács, "Spatio-temporal spectrum management model for dynamic spectrum access networks," in *Proceedings of First International Workshop on Technology and Policy for Accessing Spectrum (TAPAS)*. ACM, 2006, p. 10.

[491] M. Pan, S. Liang, H. Xiong, J. Chen, and G. Li, "A novel bargaining based dynamic spectrum management scheme in reconfigurable systems," in *Proceedings of International Conference on Systems and Networks Communication (ICSNC)*, October 2006, p. 54.

[492] Y. Zhang, K. Zhang, C. Chi, Y. Ji, Z. Feng, and P. Zhang, "An adaptive threshold load balancing scheme for the end-to-end reconfigurable system," *Springer Wireless Personal Communications*, vol. 46, no. 1, 2008.

[493] D. Bertsekas and R. Gallager, *Data Networks*. Prentice-Hall, 1992.

[494] J. A. V. Mieghem, "Due-date scheduling: asymptotic optimality of generalized longest queue and generalized largest delay rules," *Operations Research*, vol. 51, no. 1, pp. 113–122, 2003.

[495] M. Andrews, K. Kumaran, K. Ramanan, A. Stolyar, R. Vijayakumar, and P. Whiting, "Scheduling in a queueing system with asynchronous varying service rates," *Probability in the Engineering and Informational Sciences*, vol. 18, no. 2, pp. 191–217, 2004.

[496] M. Ma and D. H. K. Tsang, "Joint spectrum sharing and fair routing in cognitive radio networks," in *Proceedings of IEEE Consumer Communications and Networking Conference (CCNC)*, January 2008, pp. 978–982.

[497] Y. T. Hou, Y. Shi, and H. D. Sherali, "Optimal spectrum sharing for multi-hop software defined radio networks," in *Proceedings of IEEE International Conference on Computer Communications (INFOCOM)*, vol. 2, May 2007, pp. 1–9.

[498] ——, "Spectrum sharing for multi-hop networking with cognitive radios," *IEEE Journal on Selected Areas in Communications*, vol. 26, no. 1, pp. 146–155, 2008.

[499] C. Zeng, L. M. C. Hoo, and J. M. Cioff, "Efficient water-filling algorithms for a Gaussian multiaccess channel with ISI," in *Proceedings of IEEE Vehicular Technology Conference (VTC) Fall*, vol. 3, 2000, pp. 1072–1077.

[500] L. Lai and H. E. Gamal, "The water-filling game in fading multiple-access channels," *IEEE Transactions on Information Theory*, vol. 54, no. 5, pp. 2415–2425, 2008.

[501] X. Wang and G. B. Giannakis, "Power-efficient resource allocation for time-division multiple access over fading channels," *IEEE Transactions on Information Theory*, vol. 54, no. 3, pp. 1225–1240, 2008.

[502] J. Tang and X. Zhang, "Quality-of-service driven power and rate adaptation for multichannel communications over wireless links," *IEEE Transactions on Wireless Communications*, vol. 6, no. 12, pp. 4349–4360, 2007.

[503] G. Wunder and T. Michel, "Optimal resource allocation for parallel Gaussian broadcast channels: minimum rate constraints and sum power minimization," *IEEE Transactions on Information Theory*, vol. 53, no. 12, pp. 4817–4822, 2007.

[504] C. Liang and K. R. Dandekar, "Power management in MIMO ad hoc networks: a game-theoretic approach," *IEEE Transactions on Wireless Communications*, vol. 6, no. 4, pp. 1164–1170, 2007.

[505] S. Sanayei and A. Nosratinia, "Capacity of MIMO channels with antenna selection," *IEEE Transactions on Information Theory*, vol. 53, no. 11, pp. 4356–4362, November 2007.

[506] D. Rajan, "Towards universal power efficient scheduling in Gaussian channels," *IEEE Journal on Selected Areas in Communications*, vol. 25, no. 4, pp. 808–818, 2007.

[507] G. Arslan, M. F. Demirkol, and Y. Song, "Equilibrium efficiency improvement in MIMO interference systems: a decentralized stream control approach," *IEEE Transactions on Wireless Communications*, vol. 6, no. 8, pp. 2984–2993, 2007.

[508] D. S. W. Hui, V. K. N. Lau, and W. H. Lam, "Cross-layer design for OFDMA wireless systems with heterogeneous delay requirements," *IEEE Transactions on Wireless Communications*, vol. 6, no. 8, pp. 2872–2880, 2007.

[509] Z. Han and K. J. R. Liu, "Joint link quality and power management over wireless networks with fairness constraint and space-time diversity," *IEEE Transactions on Vehicular Technology*, vol. 53, no. 4, pp. 1138–1148, 2004.

[510] L. Zhang, Y.-C. Liang, and Y. Xin, "Joint beamforming and power allocation for multiple access channels in cognitive radio networks," *IEEE Journal on Selected Areas in Communications*, vol. 26, no. 1, pp. 38–51, 2008.

[511] F. Wang, M. Krunz, and S. Cui, "Price-based spectrum management in cognitive radio networks," *IEEE Journal on Selected Areas in Communications*, vol. 2, no. 1, pp. 74–87, 2008.

[512] G. Bansal, M. J. Hossain, and V. K. Bhargava, "Adaptive power loading for OFDM-based cognitive radio systems," in *Proceedings of IEEE International Conference on Communications (ICC)*, June 2007, pp. 5137–5142.

[513] M. Haddad, A. M. Hayar, and M. Debbah, "Spectral efficiency of cognitive radio systems," in *Proceedings of IEEE Global Telecommunications Conference (GLOBECOM)*, November 2007, pp. 4165–4169.

[514] T. Peng, W. Wang, Q. Lu, and W. Wang, "Subcarrier allocation based on water-filling level in OFDMA-based cognitive radio networks," in *Proceedings of International Conference on Wireless Communications, Networking and Mobile Computing (WiCom)*, September 2007, pp. 196–199.

[515] Q. Lu, W. Wang, T. Peng, and W. Wang, "Efficient multiuser water-filling algorithm under interference temperature constraints in OFDMA-based cognitive radio networks," in *Proceedings of International Symposium on Microwave, Antenna, Propagation and EMC Technologies for Wireless Communications*, August 2007, pp. 174–177.

[516] P. Wang, M. Zhao, L. Xiao, S. Zhou, and J. Wang, "Power allocation in OFDM-based cognitive radio systems," in *Proceedings of IEEE Global Telecommunications Conference (GLOBECOM)*, November 2007, pp. 4061–4065.

[517] X. Zhou, H. Zhang, and I. Chlamtac, "Transmit power allocation among PSWF-based pulse wavelets in cognitive UWB radio," in *Proceedings of International Conference on Cognitive Radio Oriented Wireless Networks and Communications*, June 2006.

[518] K. Jain, J. Padhye, V. N. Padmanabhan, and L. Qiu, "Impact of interference on multi-hop wireless network performance," in *Proceedings of 9th Annual International Conference on Mobile Computing and Networking (MobiCom)*, September 2003, pp. 66–80.

[519] F. S. Hillier and G. J. Lieberman, *Introduction to Operations Research*. McGraw-Hill, 2002.

[520] S. Gandhi, C. Buragohain, L. Cao, H. Zheng, and S. Suri, "A general framework for wireless spectrum auctions," UCSB, Tech. Rep., 2007.

[521] C.-Y. Wang, K.-T. Hong, and H.-Y. Wei, "Nash bargaining solution for cooperative shared-spectrum WLAN networks," in *Proceedings of IEEE International Symposium on Personal, Indoor and Mobile Radio Communications (PIMRC)*, September 2007.

[522] M. Nokleby, A. L. Swindlehurst, Y. Rong, and Y. Hua, "Cooperative power scheduling for wireless MIMO networks," in *Proceedings of IEEE Global Telecommunications Conference (GLOBECOM)*, November 2007, pp. 2982–2986.

[523] Q. Duan, L. Wang, C. D. Knutson, and M. A. Goodrich, "Axiomatic multi-transport bargaining: a quantitative method for dynamic transport selection in heterogeneous multi-transport

wireless environments," in *Proceedings of IEEE Wireless Communications and Networking Conference (WCNC)*, vol. 1, April 2006, pp. 98–105.

[524] L. Shang, R. P. Dick, and N. K. Jha, "DESP: a distributed economics-based subcontracting protocol for computation distribution in power-aware mobile ad hoc networks," *IEEE Transactions on Mobile Computing*, vol. 3, no. 1, pp. 33–45, 2004.

[525] T. K. Forde and L. E. Doyle, "Exclusivity, externalities & easements: dynamic spectrum access and Coasean bargaining," in *Proceedings of IEEE International Symposium on New Frontiers in Dynamic Spectrum Access Networks (DySPAN)*, April 2007, pp. 303–315.

[526] Z. Ji and K. J. R. Liu, "Belief-assisted pricing for dynamic spectrum allocation in wireless networks with selfish users," in *Proceedings of IEEE Communications Society on Sensor and Ad Hoc Communications and Networks (SECON)*, vol. 1, September 2006, pp. 119–127.

[527] H. Islam, Y.-C. Liang, and A. T. Hoang, "Distributed power and admission control for cognitive radio networks using antenna arrays," in *Proceedings of IEEE International Symposium on New Frontiers in Dynamic Spectrum Access Networks (DySPAN)*, April 2007, pp. 250–253.

[528] D. H. Friend, M. Y. ElNainay, Y. Shi, and A. B. MacKenzie, "Architecture and performance of an island genetic algorithm-based cognitive network," in *Proceedings of IEEE Consumer Communications and Networking Conference (CCNC)*, January 2008, pp. 993–997.

[529] K. R. Chowdhury and I. F. Akyildiz, "Cognitive wireless mesh networks with dynamic spectrum access," *IEEE Journal on Selected Areas in Communications*, vol. 26, no. 1, pp. 168–181, 2008.

[530] R. Etkin, A. Parekh, and D. Tse, "Spectrum sharing for unlicensed bands," *IEEE Journal on Selected Areas in Communications*, vol. 25, no. 3, pp. 517–528, 2007.

[531] L. Berlemann, G. R. Hiertz, B. Walke, and S. Mangold, "Strategies for distributed QoS support in radio spectrum sharing," in *Proceedings of IEEE International Conference on Communications (ICC)*, vol. 5, May 2005, pp. 3271–3277.

[532] A. Larcher, H. Sun, M. van der Schaar, and Z. Ding, "Decentralized transmission strategy for delay-sensitive applications over spectrum agile network," in *Proceedings of International Packet Video Workshop*, December 2004.

[533] G. Alyfantis, G. Marias, S. Hadjiefthymiades, and L. Merakos, "Non-cooperative dynamic spectrum access for CDMA networks," in *Proceedings of IEEE Global Telecommunications Conference (GLOBECOM)*, November 2007, pp. 3574–3578.

[534] O. Simeone and Y. Bar-Ness, "A game-theoretic view on the interference channel with random access," in *Proceedings of IEEE International Symposium on New Frontiers in Dynamic Spectrum Access Networks (DySPAN)*, April 2007, pp. 13–21.

[535] M. Sharma, A. Sahoo, and K. D. Nayak, "Channel selection under interference temperature model in multi-hop cognitive mesh networks," in *Proceedings of IEEE International Symposium on New Frontiers in Dynamic Spectrum Access Networks (DySPAN)*, April 2007, pp. 133–136.

[536] N. Clemens and C. Rose, "Intelligent power allocation strategies in an unlicensed spectrum," in *Proceedings of IEEE International Symposium on New Frontiers in Dynamic Spectrum Access Networks (DySPAN)*, November 2005, pp. 37–42.

[537] S. Geirhofer, L. Tong, and B. M. Sadler, "A measurement-based model for dynamic spectrum access in WLAN channels," in *Proceedings of IEEE Military Communications Conference (MILCOM)*, October 2006, pp. 1–7.

[538] ——, "Cognitive medium access: constraining interference based on experimental models," *IEEE Journal on Selected Areas in Communications*, vol. 26, no. 1, pp. 95–105, 2008.

[539] M. Musku and P. Cotae, "Cognitive radio: time domain spectrum allocation using game theory," in *Proceedings of IEEE International Conference on System of Systems Engineering (SoSE)*, April 2007, pp. 1–6.

[540] H.-P. Shiang and M. van der Schaar, "Distributed resource management in multihop cognitive radio networks for delay sensitive transmission," *IEEE Transactions on Vehicular Technology*, to appear.

[541] D. Niyato and E. Hossain, "Cognitive radio for next generation wireless networks: an approach to opportunistic channel selection in IEEE 802.11-based wireless mesh," *IEEE Wireless Communications*, to appear.

[542] Z. Han, C. Pandana, and K. J. R. Liu, "Distributive opportunistic spectrum access for cognitive radio using correlated equilibrium and no-regret learning," in *Proceedings of IEEE Wireless Communications and Networking Conference (WCNC)*, March 2007, pp. 11–15.

[543] A. Tonmukayakul and M. B. H. Weiss, "An agent-based model for secondary use of radio spectrum," in *Proceedings of IEEE International Symposium on New Frontiers in Dynamic Spectrum Access Networks (DySPAN)*, November 2005, pp. 467–475.

[544] B. Atakan and O. Akan, "Biologically-inspired spectrum sharing in cognitive radio networks," in *Proceedings of IEEE Wireless Communications and Networking Conference (WCNC)*, March 2007, pp. 43–48.

[545] H. Zheng and L. Cao, "Device-centric spectrum management," in *Proceedings of IEEE International Symposium on New Frontiers in Dynamic Spectrum Access Networks (DySPAN)*, November 2005, pp. 56–65.

[546] L. Cao and H. Zheng, "Distributed rule-regulated spectrum sharing," *IEEE Journal on Selected Areas in Communications*, vol. 26, no. 1, pp. 130–145, 2008.

[547] N. Nie and C. Comaniciu, "Adaptive channel allocation spectrum etiquette for cognitive radio networks," in *Proceedings of IEEE International Symposium on New Frontiers in Dynamic Spectrum Access Networks (DySPAN)*, November 2005, pp. 269–278.

[548] A. Al-Fuqaha, B. Khan, A. Rayes, M. Guizani, O. Awwad, and G. B. Brahim, "Opportunistic channel selection strategy for better QoS in cooperative networks with cognitive radio capabilities," *IEEE Journal on Selected Areas in Communications*, vol. 26, no. 1, pp. 156–167, 2008.

[549] X. Liu and Z. Ding, "ESCAPE: a channel evacuation protocol for spectrum-agile networks," in *Proceedings of IEEE International Symposium on New Frontiers in Dynamic Spectrum Access Networks (DySPAN)*, April 2007, pp. 292–302.

[550] A. Mishra, V. Shrivastava, D. Agrawal, S. Banerjee, and S. Ganguly, "Distributed channel management in uncoordinated wireless environments," in *Proceedings of 12th Annual International Conference on Mobile Computing and Networking (MobiCom)*. ACM, 2006, pp. 170–181.

[551] G. Cheng, W. Liu, Y. Li, and W. Cheng, "Spectrum aware on-demand routing in cognitive radio networks," in *Proceedings of IEEE International Symposium on New Frontiers in Dynamic Spectrum Access Networks (DySPAN)*, April 2007, pp. 571–574.

[552] ——, "Joint on-demand routing and spectrum assignment in cognitive radio networks," in *Proceedings of IEEE International Conference on Communications (ICC)*, June 2007, pp. 6499–6503.

[553] F. Rashid-Farrokhi, L. Tassiulas, and K. J. R. Liu, "Joint optimal power control and beamforming in wireless networks using antenna arrays," *IEEE Transactions on Communications*, vol. 46, no. 10, pp. 1313–1324, 1998.

[554] A. Yener, R. D. Yates, and S. Ulukus, "Joint power control, multiuser detection and beamforming for CDMA systems," in *Proceedings of IEEE Vehicular Technology Conference (VTC)*, vol. 2, July 1999, pp. 1032–1036.

[555] N. D. Bambos, S. C. Chen, and G. J. Pottie, "Radio link admission algorithms for wireless networks with power control and active link quality protection," in *Proceedings of IEEE International Conference on Computer Communications (INFOCOM)*, vol. 1, April 1995, pp. 97–104.

[556] G. Chakraborty and B. Chakraborty, "A genetic algorithm approach to solve channel assignment problem in cellular radio networks," in *Proceedings of IEEE Midnight-Sun Workshop on Soft Computing Methods in Industrial Applications*, June 1999, pp. 34–39.

[557] S. Matsui, I. Watanabe, and K.-I. Tokoro, "Application of the parameter-free genetic algorithm to the fixed channel assignment problem," *Systems and Computers in Japan*, vol. 36, no. 4, pp. 71–81, 2005.

[558] X. Fu, A. G. Bourgeois, P. Fan, and Y. Pan, "Using a genetic algorithm approach to solve the dynamic channel-assignment problem," *International Journal of Mobile Communications*, vol. 4, no. 3, pp. 333–353, 2006.

[559] Q. Zhang, G. J. M. Smit, L. T. Smit, A. Kokkeler, F. W. Hoeksema, and M. Heskamp, "A reconfigurable platform for cognitive radio," in *Proceedings of International Conference on Mobile Technology, Applications and Systems*, November 2005.

[560] W. H. Press, S. A. Teukolsky, W. T. Vetterling, and B. P. Flannery, *Numerical Recipes in C++: The Art of Scientific Computing*. Cambridge University Press, 2002.

[561] W. Yiu, G. Ginis, and J. Cioffi, "Distributed multiuser power control for digital subscriber lines," *IEEE Journal on Selected Areas in Communications*, vol. 20, no. 5, pp. 1105–1115, 2002.

[562] J. Friedman, "A non-cooperative equilibrium for supergames," *Review of Economic Studies*, vol. 38, no. 113, pp. 1–12, 1971.

[563] ——, *Oligopoly and the Theory of Games*. North-Holland, 1977.

[564] M. Cagalj, S. Ganeriwal, I. Aad, and J.-P. Hubaux, "On selfish behavior in CSMA/CA networks," in *Proceedings of IEEE International Conference on Computer Communications (INFOCOM)*, vol. 4, March 2005, pp. 2513–2524.

[565] L. Berlemann, G. Hiertz, B. Walke, and S. Mangold, "Radio resource sharing games: enabling QoS support in unlicensed bands," *IEEE Network*, vol. 19, no. 4, pp. 59–65, 2005.

[566] J. Konorski, "A game-theoretic study of CSMA/CA under a backoff attack," *IEEE/ACM Transactions on Networking*, vol. 14, no. 6, pp. 1167–1178, 2006.

[567] Y. Jin and G. Kesidis, "Distributed contention window control for selfish users in IEEE 802.11 wireless LANs," *IEEE Journal on Selected Areas in Communications*, vol. 25, no. 6, pp. 1113–1123, 2007.

[568] S. Mangold, L. Berlemann, and B. Walke, "Equilibrium analysis of coexisting IEEE 802.11e wireless LANs," in *Proceedings of IEEE International Symposium on Personal, Indoor and Mobile Radio Communications (PIMRC)*, September 2003.

[569] L. Berlemann, G. R. Hiertz, B. Walke, and S. Mangold, "Cooperation in radio resource sharing games of adaptive strategies," in *Proceedings of IEEE Vehicular Technology Conference (VTC) Fall*, September 2004.

[570] M. J. Osborne, *A Course in Game Theory*. MIT Press, 1994.

[571] Q. Qu, L. Milstein, and D. Vaman, "Cognitive radio based multi-user resource allocation in mobile ad hoc networks using multi-carrier CDMA modulation," *IEEE Journal on Selected Areas in Communications*, vol. 26, no. 1, pp. 70–82, 2008.

[572] G. Alyfantis, S. Hadjiefthymiades, and L. Merakos, "A cooperative uplink power control scheme for elastic data services in wireless CDMA systems," *ACM SIGCOMM Computer Communication Review*, vol. 36, no. 3, pp. 5–14, 2006.

[573] T. Grenager, R. Powers, and Y. Shoham, "Dispersion games: general definitions and some specific learning results," in *Eighteenth National Conference on Artificial Intelligence*. American Association for Artificial Intelligence, 2002, pp. 398–403.

[574] E. Even-Dar and Y. Mansour, "Fast convergence of selfish rerouting," in *Proceedings of Sixteenth annual ACM-SIAM Symposium on Discrete Algorithms (SODA)*. Society for Industrial and Applied Mathematics, 2005, pp. 772–781.

[575] E. Altman, R. E. Azouzi, and T. Jiménez, "Slotted ALOHA as a game with partial information," *Computer Networks*, vol. 45, no. 6, pp. 701–713, 2004.

[576] Y. E. Sagduyu and A. Ephremides, "A game-theoretic look at simple relay channel," *Wireless Networks*, vol. 12, no. 5, pp. 545–560, 2006.

[577] A. B. MacKenzie and S. B. Wicker, "Selfish users in Aloha: a game-theoretic approach," in *Proceedings of IEEE Vehicular Technology Conference (VTC) Fall*, vol. 3, October 2001, pp. 1354–1357.

[578] ——, "Stability of multipacket slotted Aloha with selfish users and perfect information," in *Proceedings of IEEE International Conference on Computer Communications (INFOCOM)*, vol. 3, March–April 2003, pp. 1583–1590.

[579] A. Federgruen, "On n-person stochastic games with denumerable state space," *Advances in Applied Probability*, vol. 10, no. 2, pp. 452–471, 1978.

[580] P. C. Fishburn, *Utility Theory for Decision Making*. Wiley, 1970.

[581] D. P. Bertsekas, *Parallel and Distributed Computation: Numerical Methods*. Athena Scientific, 1997.

[582] T. C. Clancy, "Formalizing the interference temperature model," *Wireless Communications and Mobile Computing*, vol. 7, no. 9, pp. 1077–1086, 2007.

[583] P. Kyasanur and N. H. Vaidya, "Routing and link-layer protocols for multi-channel multi-interface ad hoc wireless networks," *ACM SIGMOBILE Mobile Computing and Communications Review*, vol. 10, no. 1, pp. 31–43.

[584] J. H. Holland, *Adaptation in Natural and Artificial Systems: An Introductory Analysis with Applications to Biology, Control, and Artificial Intelligence*. MIT Press, 1992.

[585] R. Axelrod, *The Evolution Of Cooperation*. Basic Books, 1985.

[586] S. Geirhofer, L. Tong, and B. M. Sadler, "Dynamic spectrum access in WLAN channels: empirical model and its stochastic analysis," in *Proceedings of First International Workshop on Technology and Policy for Accessing spectrum (TAPAS)*. ACM, 2006, p. 14.

[587] B. S. I. Group, "Specification of the Bluetooth system," November 2004.

[588] N. Golmie, R. E. V. Dyck, A. Soltanian, A. Tonnerre, and O. Rébala, "Interference evaluation of Bluetooth and IEEE 802.11b systems," *Wireless Networks*, vol. 9, no. 3, pp. 201–211, 2003.

[589] S. Geirhofer, L. Tong, and B. M. Sadler, "Dynamic spectrum access in the time domain: modeling and exploiting white space," *IEEE Communications Magazine*, vol. 45, no. 5, pp. 66–72, 2007.

[590] Q. Zhao, S. Geirhofer, L. Tong, and B. M. Sadler, "Optimal dynamic spectrum access via periodic channel sensing," in *Proceedings of IEEE Wireless Communications and Networking Conference (WCNC)*, March 2005, pp. 33–37.

[591] E. Altman, *Constrained Markov Decision Processes*. Chapman and Hall/CRC Press, 1999.

[592] M. L. Puterman, *Markov Decision Processes: Discrete Stochastic Dynamic Programming*. Wiley, 2005.

[593] D. Grosu and A. T. Chronopoulos, "Noncooperative load balancing in distributed systems," *Journal of Parallel and Distributed Computing*, vol. 65, no. 9, pp. 1022–1034, 2005.

[594] R. Draves, J. Padhye, and B. Zill, "Routing in multi-radio, multi-hop wireless mesh networks," in *Proceedings of 10th Annual International Conference on Mobile Computing and Networking (MobiCom)*. ACM, 2004, pp. 114–128.

[595] D. Krishnaswamy, "Network-assisted link adaptation with power control and channel reassignment in wireless networks," in *Proceedings of 3G Wireless Conference*, May 2002, pp. 165–170.

[596] D. Fudenberg and D. K. Levine, *The Theory of Learning in Games*. MIT Press, 1998.

[597] H. P. Young, *Strategic Learning and its Limits*. Oxford University Press, 2005.

[598] C. E. Perkins and E. M. Royer, "Ad-hoc on-demand distance vector routing," in *Proceedings of IEEE Workshop on Mobile Computing Systems and Applications (WMCSA)*, February 1999, pp. 90–100.

[599] G. D. Kondylis and G. J. Pottie, "Dynamic channel allocation strategies for wireless packet access," in *Proceedings of IEEE Vehicular Technology Conference (VTC) Fall*, vol. 5, 1999, pp. 2819–2824.

[600] A. Nasipuri and S. R. Das, "Multichannel CSMA with signal power-based channel selection for multihop wireless networks," in *Proceedings of IEEE Vehicular Technology Conference (VTC) Fall*, vol. 1, September 2000, pp. 211–218.

[601] P. Bahl, R. Chandra, and J. Dunagan, "SSCH: slotted seeded channel hopping for capacity improvement in IEEE 802.11 ad-hoc wireless networks," in *Proceedings of 10th Annual International Conference on Mobile Computing and Networking (MobiCom)*. ACM, 2004, pp. 216–230.

[602] R. Chandra and P. Bahl, "MultiNet: connecting to multiple IEEE 802.11 networks using a single wireless card," in *Proceedings of IEEE International Conference on Computer Communications (INFOCOM)*, vol. 2, March 2004, pp. 882–893.

[603] D. Malone, P. Clifford, D. Reid, and D. J. Leith, "Experimental implementation of optimal WLAN channel selection without communication," in *Proceedings of IEEE International Symposium on New Frontiers in Dynamic Spectrum Access Networks (DySPAN)*, April 2007, pp. 316–319.

[604] H. Ishibuchi, R. Sakamoto, and T. Nakashima, "Learning fuzzy rules from iterative execution of games," *Fuzzy Sets Systems*, vol. 135, no. 2, pp. 213–240, 2003.

[605] M. Maskery, V. Krishnamurthy, and Q. Zhao, *Cognitive Wireless Communication Networks*. Springer, 2007, ch. "Game theoretic learning and pricing for dynamic spectrum access in cognitive radio," pp. 303–325.

[606] L. Busoniu, R. Babuska, and B. D. Schutter, "Multi-agent reinforcement learning: a survey," in *Proceedings of IEEE International Conference on Control, Automation, Robotics and Vision (ICARCV)*, December 2006, pp. 1–6.

[607] D. Mladenic, "Text-learning and related intelligent agents: a survey," *IEEE Intelligent Systems*, vol. 14, no. 4, pp. 44–54, 1999.

[608] R. Kowalczyk, M. Ulieru, and R. Unland, *Agent Technologies, Infrastructures, Tools, and Applications for E-Services*. Springer, 2003, ch. "Integrating mobile and intelligent agents in advanced e-commerce: a survey," pp. 295–313.

[609] J. P. Muller, "Architectures and applications of intelligent agents: a survey," *The Knowledge Engineering Review*, vol. 13, no. 4, pp. 353–380, 1999.

[610] I. E. Alvin and E. Roth, "Learning in extensive-form games: experimental data and simple dynamic models in the intermediate term," *Games and Economic Behavior*, vol. 8, pp. 164–212, 1995.

[611] T. B. Klos and B. Nooteboom, "Agent-based computational transaction cost economics," *Journal of Economic Dynamics and Control*, vol. 25, no. 3–4, pp. 503–526, 2001.

[612] E. Bonabeau, M. Dorigo, and G. Theraulaz, *Swarm Intelligence: From Natural to Artificial Systems*. Oxford University Press, 1999.

[613] A. A. Daoud, M. Alanyali, and D. Starobinski, "Secondary pricing of spectrum in cellular CDMA networks," in *Proceedings of IEEE International Symposium on New Frontiers in Dynamic Spectrum Access Networks (DySPAN)*, April 2007, pp. 535–542.

[614] F. P. Kelly, "Loss networks," *The Annals of Applied Probability*, vol. 1, no. 3, pp. 319–378, 1991.

[615] C. Reis, R. Mahajan, M. Rodrig, D. Wetherall, and J. Zahorjan, "Measurement-based models of delivery and interference in static wireless networks," *ACM SIGCOMM Computer Communication Review*, vol. 36, no. 4, pp. 51–62, 2006.

[616] G.-H. Yang, H. Zheng, J. Zhao, and V. Li, "Adaptive channel selection through collaborative sensing," in *Proceedings of IEEE International Conference on Communications (ICC)*, vol. 8, June 2006, pp. 3753–3758.

[617] Y. Freund and R. E. Schapire, "A decision-theoretic generalization of on-line learning and an application to boosting," in *Proceedings of the Second European Conference on Computational Learning Theory (EuroCOLT)*. Springer-Verlag, 1995, pp. 23–37.

[618] A. Greenwald and A. Jafari, "A general class of no-regret learning algorithms and game-theoretic equilibria," in *Proceedings of Computational Learning Theory Conference*, August 2003, pp. 1–11.

[619] A. Jafari, A. R. Greenwald, D. Gondek, and G. Ercal, "On no-regret learning, fictitious play, and Nash equilibrium," in *Proceedings of Eighteenth International Conference on Machine Learning (ICML)*. Morgan Kaufmann, 2001, pp. 226–233.

[620] A. Nosratinia, T. E. Hunter, and A. Hedayat, "Cooperative communication in wireless networks," *IEEE Communications Magazine*, vol. 42, no. 10, pp. 74–80, 2004.

[621] C. Politis, T. Oda, S. Dixit, *et al.*, "Cooperative networks for the future wireless world," *IEEE Communications Magazine*, vol. 42, no. 9, pp. 70–79, 2004.

[622] P. Basu and J. Redi, "Movement control algorithms for realization of fault-tolerant ad hoc robot networks," *IEEE Network*, vol. 18, no. 4, pp. 36–44, 2004.

[623] M. Kim and B. Noble, "Mobile network estimation," in *Proceedings of 7th Annual International Conference on Mobile Computing and Networking (MobiCom)*. ACM, 2001, pp. 298–309.

[624] A. Balachandran, G. M. Voelker, and P. Bahl, "Wireless hotspots: current challenges and future directions," in *Proceedings of 1st ACM International Workshop on Wireless Mobile Applications and Services on WLAN Hotspots (WMASH)*. ACM, 2003, pp. 1–9.

[625] Y. Bejerano, S.-J. Han, and L. E. Li, "Fairness and load balancing in wireless LANs using association control," in *Proceedings of 10th Annual International Conference on Mobile Computing and Networking (MobiCom)*. ACM, 2004, pp. 315–329.

[626] A. Mishra, V. Brik, S. Banerjee, A. Srinivasan, and W. Arbaugh, "A client-driven approach for channel management in wireless LANs," in *Proceedings of IEEE International Conference on Computer Communications (INFOCOM)*, April 2006, pp. 1–12.

[627] A. Mishra, V. Shrivastava, S. Banerjee, and W. Arbaugh, "Partially overlapped channels not considered harmful," *ACM SIGMETRICS Performance Evaluation Review*, vol. 34, no. 1, pp. 63–74, 2006.

[628] TCI 8067, "Spectrum processor data specification," 2000. [Online]. Available: http://www.tcibr.com/PDFs/8067webs.pdf

[629] B.-J. Kwak, N.-O. Song, and L. E. Miller, "Performance analysis of exponential backoff," *IEEE/ACM Transactions on Networking*, vol. 13, no. 2, pp. 343–355, 2005.

[630] J. So and N. Vaidya, "A routing protocol for utilizing multiple channels in multi-hop wireless networks with a single transceiver," Tech. Rep., October 2004.

[631] M. X. Gong and S. F. Midkiff, "Distributed channel assignment protocols: a cross-layer approach to wireless ad hoc networks," in *Proceedings of IEEE Wireless Communications and Networking Conference (WCNC)*, vol. 4, March 2005, pp. 2195–2200.

[632] Y.-C. Liang, Y. Zeng, E. Peh, and A. T. Hoang, "Sensing-throughput tradeoff for cognitive radio networks," in *Proceedings of IEEE International Conference on Communications (ICC)*, June 2007, pp. 5330–5335.

[633] Y. Pei, A. T. Hoang, and Y.-C. Liang, "Sensing-throughput tradeoff in cognitive radio networks: how frequently should spectrum sensing be carried out?" in *Proceedings of IEEE International Symposium on Personal, Indoor and Mobile Radio Communications (PIMRC)*, September 2007.

[634] Y.-C. Liang, Y. Zeng, E. C. Y. Peh, and A. T. Hoang, "Sensing–throughput tradeoff for cognitive radio networks," *IEEE Transactions on Wireless Communications*, vol. 7, no. 4, pp. 1326–1337, 2008.

[635] R. M. Corless, G. H. Gonnet, D. E. G. Hare, and D. J. Jeffrey, "Lambert's w function in Maple," *Maple Technical Newsletter*, no. 9, pp. 12–22, 1993.

[636] C. Evci and B. Fino, "Spectrum management, pricing, and efficiency control in broadband wireless communications," in *Proceedings of the IEEE*, vol. 89, no. 1, pp. 105–115, 2001.

[637] D. Niyato and E. Hossain, "Spectrum trading: an economics of radio resource sharing in cognitive radio," *IEEE Wireless Communications*, vol. 15, no. 6, pp. 71–80, 2008.

[638] C. E. Caicedo and M. B. H. Weiss, "Spectrum trading: an analysis of implementation issues," in *Proceedings of IEEE International Symposium on New Frontiers in Dynamic Spectrum Access Networks (DySPAN)*, April 2007, pp. 579–584.

[639] S. Zekavat and X. Li, "User-central wireless system: ultimate dynamic channel allocation," in *Proceedings of IEEE International Symposium on New Frontiers in Dynamic Spectrum Access Networks (DySPAN)*, November 2005, pp. 82–87.

[640] A. Attara, S. A. Ghorashib, M. Sooriyabandarac, and A. H. Aghvamia, "Challenges of real-time secondary usage of spectrum," *Computer Networks*, vol. 52, no. 4, pp. 816–830, 2008.

[641] J. Hou, J. Yang, and S. Papavassiliou, "Integration of pricing with call admission control to meet QoS requirements in cellular networks," *IEEE Transactions on Parallel and Distributed Systems*, vol. 13, no. 9, pp. 898–910, 2002.

[642] T. C.-Y. Ng and W. Yu, "Joint optimization of relay strategies and resource allocations in cooperative cellular networks," *IEEE Journal on Selected Areas in Communications*, vol. 25, no. 2, pp. 328–339, 2007.

[643] S.-L. Kim and H.-M. Baek, "On the economic aspects of downlink scheduling in DS-CDMA systems: base-station density perspective," *IEEE Transactions on Vehicular Technology*, vol. 55, no. 5, pp. 1594–1602, 2006.

[644] S. Menon and A. Amiri, "Multiperiod cellular network design via price-influenced simulated annealing (PISA)," *IEEE Transactions on Systems, Man and Cybernetics, Part B*, vol. 36, no. 3, pp. 600–610, 2006.

[645] C. U. Saraydar, N. B. Mandayam, and D. J. Goodman, "Pricing and power control in a multicell wireless data network," *IEEE Journal on Selected Areas in Communications*, vol. 19, no. 10, pp. 1886–1892, 2001.

[646] R. K. Lam, J. C. S. Lui, and D.-M. Chiu, "On the access pricing issues of wireless mesh networks," in *Proceedings of IEEE International Conference on Distributed Computing Systems (ICDCS)*, July 2006.

[647] L. Tan, X. Zhang, L. Andrew, and M. Zukerman, "Price-based max-min fair rate allocation in wireless multi-hop networks," in *Proceedings of IEEE Region 10 TENCON*, November 2005.

[648] M. H. Lin and C. C. Lo, "A location-based incentive pricing scheme for tree-based relaying in multi-hop cellular networks," in *Proceedings of IFIP/IEEE International Symposium on Integrated Network Management (IM)*, May 2005, pp. 339–352.

[649] C. Yi, G. Ge, and H. Ruimin, "On the resource allocation for wireless ad hoc networks," in *Proceedings of IEEE International Conference on Automation and Logistics*, August 2007, pp. 2004–2007.

[650] P. Mani and D. W. Petr, "Investment function: enhanced fairness and performance in multi-hop wireless networks," in *Proceedings of IEEE International Conference on Mobile Adhoc and Sensor Systems (MASS)*, October 2006, pp. 721–728.

[651] Y. Xue, B. Li, and K. Nahrstedt, "Optimal resource allocation in wireless ad hoc networks: a price-based approach," *IEEE Transactions on Mobile Computing*, vol. 5, no. 4, pp. 347–364, 2006.

[652] M. X. Goemans, L. Li, V. S. Mirrokni, and M. Thottan, "Market sharing games applied to content distribution in ad hoc networks," *IEEE Journal on Selected Areas in Communications*, vol. 24, no. 5, pp. 1020–1033, 2006.

[653] O. Ileri, S.-C. Mau, and N. B. Mandayam, "Pricing for enabling forwarding in self-configuring ad hoc networks," *IEEE Journal on Selected Areas in Communications*, vol. 23, no. 1, pp. 151–162, 2005.

[654] P. Marbach and Y. Qiu, "Cooperation in wireless ad hoc networks: a market-based approach," *IEEE/ACM Transactions on Networking*, vol. 13, no. 6, pp. 1325–1338, 2005.

[655] Y. Zhou, D. Wu, and S. M. Nettles, "Authentication, authorization, and accounting real-time secondary market services," in *Proceedings of IEEE International Conference on Communications (ICC)*, vol. 2, May 2005, pp. 1005–1009.

[656] TSB51, "Cellular radio telecommunications intersystem operations: authentication, signaling message encryption and voice privacy," May 1993.

[657] C.-C. Lee, M.-S. Hwang, and W.-P. Yang, "Extension of authentication protocol for GSM," *IEE Proceedings Communications*, vol. 150, no. 2, pp. 91–95, 2003.

[658] R. P. C. Kaufman and M. Speciner, *Network Security: Private Communication in a Public World*. Prentice-Hall, 2002.

[659] Y. Zhou, D. Wu, and S. M. Nettles, "On architecture of authentication, authorization, and accounting for real-time secondary market services," *International Journal of Wireless and Mobile Computing*, to appear.

[660] E. Walter, *Cambridge Advanced Learner's Dictionary*. Cambridge University Press, 2005.

[661] S. C. Maurice, O. R. Phillips, and C. E. Ferguson, *Economic Analysis: Theory and Application*. Richard D Irwin, 1992.

[662] F. P. Kelly, A. Maulloo, and D. Tan, "Rate control for communication networks: shadow prices, proportional fairness and stability," *Journal of the Operational Research Society*, vol. 49, no. 3, pp. 237–252, 1998.

[663] J.-W. Lee, M. Chiang, and A. R. Calderbank, "Price-based distributed algorithms for rate–reliability tradeoff in network utility maximization," *IEEE Journal on Selected Areas in Communications*, vol. 24, no. 5, pp. 962–976, 2006.

[664] Z. Jiang, Y. Ge, and Y. Li, "Max-utility wireless resource management for best-effort traffic," *IEEE Transactions on Wireless Communications*, vol. 4, no. 1, p. 2005, 2005.

[665] P. Liu, R. A. Berry, and M. L. Honig, "A fluid analysis of a utility-based wireless scheduling policy," *IEEE Transactions on Information Theory*, vol. 52, no. 7, pp. 2872–2889, 2006.

[666] F. Meshkati, M. Chiang, H. V. Poor, and S. C. Schwartz, "A game-theoretic approach to energy-efficient power control in multicarrier CDMA systems," *IEEE Journal on Selected Areas in Communications*, vol. 24, no. 6, pp. 1115–1129, 2006.

[667] M. K. H. Yeung and Y.-K. Kwok, "A game theoretic approach to power aware wireless data access," *IEEE Transactions on Mobile Computing*, vol. 5, no. 8, pp. 1057–1073, 2006.

[668] B. Radunovic and J. Y. L. Boudec, "Rate performance objectives of multihop wireless networks," *IEEE Transactions on Mobile Computing*, vol. 3, no. 4, pp. 334–349, 2004.

[669] J. Musacchio and J. Walrand, "WiFi access point pricing as a dynamic game," *IEEE/ACM Transactions on Networking*, vol. 14, no. 2, pp. 289–301, 2006.

[670] J. Liu, B. Li, Y. T. Hou, and I. Chlamtac, "On optimal layering and bandwidth allocation for multisession video broadcasting," *IEEE Transactions on Wireless Communications*, vol. 3, no. 2, pp. 656–667, 2004.

[671] O. Ileri, D. Samardzija, and N. B. Mandayam, "Demand responsive pricing and competitive spectrum allocation via a spectrum server," in *Proceedings of IEEE International Symposium on New Frontiers in Dynamic Spectrum Access Networks (DySPAN)*, November 2005, pp. 194–202.

[672] Y. Xing, R. Chandramouli, and C. Cordeiro, "Price dynamics in competitive agile spectrum access markets," *IEEE Journal on Selected Areas in Communications*, vol. 25, no. 3, pp. 613–621, 2007.

[673] M. J. Osborne, *An Introduction to Game Theory*. Oxford University Press, 2004.

[674] D. Niyato and E. Hossain, "Integration of WiMAX and WiFi: optimal pricing for bandwidth sharing," *IEEE Communications Magazine*, vol. 45, no. 5, pp. 140–146, May 2007.

[675] P. Cotae and M. Aguire, "A game theoretical approach of the total weighted squared correlation in S-CDMA systems," in *Proceedings of IEEE International Conference on Communications (ICC)*, vol. 1, June 2006, pp. 263–268.

[676] L. Lai and H. E. Gamal, "Fading multiple access channels: a game theoretic perspective," in *Proceedings of IEEE International Symposium on Information Theory (ISIT)*, July 2006, pp. 1334–1338.

[677] Q. Jun, Y. Yu, K. Wang, and G. Zhu, "Handoff strategy analysis by Stackelberg model over HSDPA," in *Proceedings of International Conference on Wireless Communications, Networking and Mobile Computing (WiCom)*, September 2007, pp. 1766–1770.

[678] P. Cotae, C. Comsa, and I. Bogdan, "On the Stackelberg equilibrium of total weighted squared correlation in synchronous DS-CDMA systems: algorithm and numerical results," in *Proceedings of International Symposium on Signals, Circuits and Systems (ISSCS)*, vol. 2, July 2005, pp. 669–672.

[679] B. Wang, Z. Han, and K. J. R. Liu, "Distributed relay selection and power control for multiuser cooperative communication networks using buyer/seller game," in *Proceedings of IEEE International Conference on Computer Communications (INFOCOM)*, May 2007, pp. 544–552.

[680] C. Kloeck, H. Jaekel, and F. K. Jondral, "Dynamic and local combined pricing, allocation and billing system with cognitive radios," in *Proceedings of IEEE International Symposium on New Frontiers in Dynamic Spectrum Access Networks (DySPAN)*, November 2005, pp. 73–81.

[681] D. Friedman, D. P. Friedman, and J. Rust, *The Double Auction Market: Institutions, Theories, and Evidence*. Westview Press, 1993.

[682] S. Sengupta and M. Chatterjee, "Sequential and concurrent auction mechanisms for dynamic spectrum access," in *Proceedings of International Conference on Cognitive Radio Oriented Wireless Networks and Communications (CrownCom) 2007*, July–August 2007.

[683] D. Grandblaise, C. Kloeck, K. Moessner *et al.*, "Techno – economic of collaborative based secondary spectrum usage – E^2R research project outcomes overview," in *Proceedings of IEEE International Symposium on New Frontiers in Dynamic Spectrum Access Networks (DySPAN)*, November 2005, pp. 318–327.

[684] FP6, "End-to-end reconfigurability (e2r) integrated project (ip)." [Online]. Available: www.e2r.motlabs.com

[685] M. K. Pereirasamy, J. Luo, M. Dillinger, and C. Hartmann, "An approach for interoperator spectrum sharing for 3G systems and beyond," in *Proceedings of IEEE International Symposium on Personal, Indoor and Mobile Radio Communications (PIMRC)*, vol. 3, September 2004, pp. 1952–1956.

[686] O. Simeone, I. Stanojev, S. Savazzi, Y. Bar-Ness, U. Spagnolini, and R. Pickholtz, "Spectrum leasing to cooperating secondary ad hoc networks," *IEEE Journal on Selected Areas in Communications*, vol. 26, no. 1, pp. 203–213, 2008.

[687] J. N. Laneman and G. W. Wornell, "Distributed space-time-coded protocols for exploiting cooperative diversity in wireless networks," *IEEE Transactions on Information Theory*, vol. 49, no. 10, pp. 2415–2425, 2003.

[688] J. Huang, R. A. Berry, and M. L. Honig, "Distributed interference compensation for wireless networks," *IEEE Journal on Selected Areas in Communications*, vol. 24, no. 5, pp. 1074–1084, 2006.

[689] R. Cendrillon, W. Yu, M. Moonen, J. Verlinden, and T. Bostoen, "Optimal multiuser spectrum balancing for digital subscriber lines," *IEEE Transactions on Communications*, vol. 54, no. 5, pp. 922–933, 2006.

[690] F. Wang, O. Younis, and M. Krunz, "GMAC: a game-theoretic MAC protocol for mobile ad hoc networks," in *Proceedings of IEEE International Symposium on Modeling and Optimization in Mobile, Ad Hoc and Wireless Networks*, April 2006, pp. 1–9.

[691] T. Basar, "Relaxation techniques and asynchronous algorithms for on-line computation of non-cooperative equilibria," *Journal of Economic Dynamics and Control*, vol. 11, no. 4, pp. 531–549, 1987.

[692] W. Wang, Y. Cui, T. Peng, and W. Wang, "Noncooperative power control game with exponential pricing for cognitive radio network," in *Proceedings of IEEE Vehicular Technology Conference (VTC) Spring*, April 2007, pp. 3125–3129.

[693] M. J. Osborne and C. Pitchik, "Equilibrium in hotelling's model of spatial competition," *Econometrica*, vol. 55, no. 4, pp. 911–922, 1987.

[694] G. Isiklar and A. Bener, "Brokering and pricing architecture over cognitive radio wireless networks," in *Proceedings of IEEE Consumer Communications and Networking Conference (CCNC)*, January 2008, pp. 1004–1008.

[695] H. Mutlu, M. Alanyali, and D. Starobinski, "Spot pricing of secondary spectrum usage in wireless cellular networks," in *Proceedings of IEEE International Conference on Computer Communications (INFOCOM)*, April 2008.

[696] P. C. Fishburn and A. M. Odlyzko, "Dynamic behavior of differential pricing and quality of service options for the Internet," *Decision Support Systems*, vol. 28, no. 1–2, pp. 123–136, 2000.

[697] S. M. Ross, *Introduction to Stochastic Dynamic Programming*. Academic Press, 1995.

[698] D. Niyato and E. Hossain, "Competitive spectrum sharing in cognitive radio networks: a dynamic game approach," *IEEE Transactions on Wireless Communications*, vol. 7, no. 7, pp. 2651–2660, 2008.

[699] H. N. Agiza, G.-I. Bischi, and M. Kopel, "Multistability in a dynamic cournot game with three oligopolists," *Mathematics and Computers in Simulation*, vol. 51, no. 1–2, pp. 63–90, 1999.

[700] N. Singh and X. Vives, "Price and quantity competition in a differentiated duopoly," *RAND Journal of Economics*, vol. 15, no. 4, pp. 546–554, 1984.

[701] T. L. Vincent and J. S. Brown, *Evolutionary Game Theory, Natural Selection, and Darwinian Dynamics*. Cambridge University Press, 2005.

[702] L. A. Cox, *Risk Analysis: Foundations, Models and Methods*. Springer, 2001.

[703] D. M. Holthausen, "A risk–return model with risk and return measured as deviations from a target return," *The American Economic Review*, vol. 71, no. 1, pp. 182–188, 1981.

[704] M. R. Yilmaz, "The use of risk and return models for multiattribute decisions with decomposable utilities," *The Journal of Financial and Quantitative Analysis*, vol. 18, no. 3, pp. 279–285, 1983.

[705] B. K. Stone, "Risk, return, and equilibrium: a general single-period theory of asset selection and capital-market equilibrium," *The Accounting Review*, vol. 46, no. 3, pp. 646–647, 1971.

Index

Printed in the United States
by Baker & Taylor Publisher Services